Electric Machinery and Transformers

Second Edition

Bhag S. Guru

Hüseyin R. Hiziroğlu

Department of Electrical and Computer Engineering

GMI Engineering & Management Institute

Flint, Michigan

SAUNDERS COLLEGE PUBLISHING

Harcourt Brace College Publishers

Fort Worth Philadelphia San Diego New York

Orlando San Antonio Toronto Montreal

London Sydney Tokyo

Text Typeface: Palatino
Compositor: CRWaldman
Acquisitions Editor: Emily Barrosse
Assistant Editor: Michelle Slavin
Managing Editor: Carol Field
Project Editor: Linda Boyle
Copy Editor: Donna Walker
Manager of Art and Design: Carol Bleistine
Art Director: Robin Milicevic, Anne Muldrow
Text Designer: Merce Wilczek
Cover Designer: Lawrence Didona
Text Artwork: Vantage Art, Inc.
Director of EDP: Tim Frelick
Production Manager: Joanne Cassetti

Printed in the United States of America

Electric Machinery and Transformers, 2/e

ISBN 0-03-010592-7

Library of Congress Catalog Card Number: 94-067127

4567890123 016 10 987654321

This book was printed on acid-free recycled content paper, containing
MORE THAN
10% POSTCONSUMER WASTE

To

Our Parents and Families

Bhag S. Guru and Hüseyin R. Hiziroğlu

Preface

The first edition of this textbook appeared almost seven years ago and its immediate adoption by many institutions of highest standing, both here and abroad, has been very gratifying. Since its publication, we have received many constructive suggestions from our own students and from fellow professors at other universities. As a result of these suggestions, the textbook has been completely rewritten. Although we have kept the elementary and the fundamental developments of the subject intact, we have added considerably more advanced material in each chapter. In order to promote discussion and appreciation for the elegance of the point under consideration, we have added a large number of **Review Questions** near the end of each chapter. A **Summary** is also included in each chapter to emphasize the salient features of that chapter. We have sprinkled the text with ample examples for immediate reinforcement, as well as further clarification of the topics. We have also added a new set of problems (**Exercises**) at the end of each section. These exercises are expected to nurture confidence and enhance the basic understanding of the material presented in each section. We have also revised and increased the number of problems at the end of each chapter in such a way that they now offer a wide range of challenges for the student. These problems are an important part of the text and should be an integral part of the study of electric machines. We suggest that the students use intuitive reasoning to solve the problems.

Some of the problems in the text require the solution of nonlinear equations and we suggest that the student should be encouraged to solve these problems using software packages such as Mathcad. We have used Mathcad to solve almost all the examples in this textbook. We encourage the use of a software package because it reduces the drudgery of mundane calculations, enables the student to probe further into the intricacies of a machine, and helps the student to focus on "what if" types of queries.

Our basic philosophy in writing the second edition has not changed. We felt then and we strongly feel now that a superficial treatment of the subject based upon mere statements of facts leads only to the memorization of equations which are soon forgotten. In order to motivate the student, we must present the material in a systematic order. In other words, when teaching electric machines our objectives are to

- Explain the physical construction of a machine,
- Shed some light on its windings and their placement,
- Describe the fundamental laws that govern its operation,
- Justify the assumptions imperative to develop the relevant theory, and
- Emphasize its limitations.

Once the student acquires a clear understanding of a machine, we must then develop the necessary equations using as few fundamental laws as possible. Each equation should be put in proper perspective by associating it with the machine's performance. When the operating principles of a machine are explained properly and the corresponding equations are developed from basic laws, the student will learn to

- Appreciate the theoretical development,
- Reduce intimidation, and
- Grasp the powers of reasoning.

The end result of such a teaching process is that, in the future, the student will not hesitate to tackle even more difficult problems with confidence. We have tried our best to incorporate this philosophy in the development of the text.

Our experience dictates that students tend to view the theoretical development as an abstraction and place emphasis on some of the equations, which then become "formulas" to them. In order to make the students appreciate the theory, it is, therefore, the teacher's responsibility to show that the theory can be applied to solve practical problems under various conditions. To attain this goal, a teacher must dwell on his/her expertise in the subject and highlight other areas of application from time to time. This also necessitates the need to stress any new advancements in the area when the fundamentals are under discussion. For example, while explaining the magnetic force between two current-carrying conductors, a teacher must draw attention to magnetically levitated vehicles.

In writing this text we have assumed that a student has a strong background in the areas of linear differential equations, analysis of dc and ac (single- and three-phase) electric circuits, the Laplace transforms and their applications, and the electromagnetic field theory. Since the length of a term, a quarter, and/or a semester varies from institution to institution, the material in this book can be presented in a one-semester course or two-quarter courses in the junior and/or senior years. For institutions that are on a quarter system and offer only one course in electrical machines, we suggest that they concentrate on the fundamentals and drop the more advanced topics such as those discussed in the last two chapters.

Chapters 1 and 2 are written to review the important concepts of electric circuits and electromagnetic fields. Chapter 1 has been expanded to include the discussion of single-phase and three-phase power measurements. The presentation of magnetic circuits in Chapter 2 has been further elaborated to ensure that the student understands the effects of saturation of magnetic materials on the performance of a machine. The nonlinear behavior of magnetic materials due to saturation is, in fact, a blessing in disguise for the stable operation of dc generators.

Chapter 3 is the heart of this text, as it presents the conversion of energy from one form to another. The complete revision of this chapter now includes information on the

- Analysis of magnetically coupled coils (necessary for the operation of transformers),
- Induced emf in a coil rotating in a uniform magnetic field (vital for the operation of dc machines),
- Induced emf in a coil rotating in a time-varying magnetic field (essential for the operation of ac machines), and
- The concept of the revolving field (crucial for the operation of synchronous and induction machines).

Operating principles of single- and three-phase transformers and autotransformers are described in Chapter 4. The types of losses that occur in transformers and the measures to be taken to keep these losses to a minimum for optimal operating conditions are explained. The revision of this chapter includes the winding diagrams of various connections of three-phase transformers, the equivalent circuit representation of each connection of an autotransformer, and the exact analysis of autotransformers.

Our experience in teaching electrical machines points to the fact that the operation of motors and generators must be explained separately in order to avoid confusion. For this reason, Chapters 5 and 6 are devoted to the study of dc generators and dc motors, respectively. Chapters 7 and 8 present information on synchronous generators and synchronous motors. In addition to explaining the construction, the operation, the external characteristics, and the determination of parameters of these machines, we have now included information on how these machines are wound. This information will enable the student to visualize the placement of coils and their interconnections, and the number of parallel paths for the currents in the machine.

Chapter 9 explains the construction and the operation of three-phase induction motors. The criteria for maximum efficiency, maximum power, and maximum torque are developed and the significance of each criterion is emphasized. Also explained in this chapter is the effect of rotor resistance on a motor's performance. Various classifications of induction motors are explained.

Single-phase motors are examined in Chapter 10. Methods to determine the performance of single-phase induction motors with both windings are also included. Some insights into the workings of a shaded-pole motor and a universal motor are also included in this chapter.

Chapter 11 prepares the student to investigate the dynamic behavior of machines. A computer program based upon fourth-order Runga-Kutta algorithm is included to analyze the dynamics of machines numerically. Since we have used

Laplace transforms in our analysis, a table of Laplace transforms of common functions is included in Appendix B for a quick reference.

Different types of permanent-magnet motors, including brushless dc motors and switched reluctance motors, are the topics of discussion in Chapter 12. The pertinent theory to analyze linear induction motors is also included in this chapter.

Except for expressing the speed in revolutions per minute (rpm), and the power output of a motor in terms of horsepower (hp), we have used the International System of Units (SI) in our mathematical development. Since the English System of Units is commonly used in the United States, the conversion from one system to another becomes necessary. To this end we have included Appendix A.

Since the boldface characters are commonly used for vectors and phasor quantities in most of the books, the teacher has to adopt his/her own notation while teaching the course. The problem is compounded when the symbols are not used consistently. This causes confusion among students when they try to correlate the class notes with the symbols used in the text. In order to eliminate such a confusion, we have adopted a consistent notation system in this text. Capital letters are used for dc quantities and for the rms values of the ac variables. The lowercase letters are employed for the instantaneous values of time-dependent variables. For representations of a vector, a phasor, and a complex quantity we have incorporated an arrow (→), a tilde (˜), and a caret (ˆ) on top of the letters, respectively.

We have endeavored to write and publish an accurate book. We have used Mathcad to solve most of the examples in this book and have rounded the results of calculations only for the purpose of displaying them. There may still be some numerical errors in the book. It is also possible that some sections need further clarification while others may be considered to be too verbose. We welcome any comments and suggestions in this regard and will consider them for future revisions.

Acknowledgments

We would like to express our appreciation to the students in the various classes on electric machines that we have taught while using the first edition of the book. They were encouraged to freely criticize and openly express their opinions. We are especially indebted to Dr. Albert Simeon, now retired, for his general advice on various sections while teaching from our book. The suggestions/comments from the students, as well as our colleagues at GMI and other universities, led to the complete revision of the first edition. We are quite confident that the second edition of the book is almost "error-free" and we are grateful to Dr. S. Hossein Mousavinezhad, Western Michigan University, who read each line and checked each calculation. Any errors that may still appear in the text are the responsibility of the authors. However, we will appreciate any errors being brought to our attention.

We would also like to thank the following reviewers whose suggestions were invaluable to us in refining this text:

John Bentley, Mesa Community College
Joseph M. Farren, The University of Dayton
Henslay William Kabisama, California State Polytechnic University-Pomona
Edward J. Milano, University of Hartford
Kwa-Sur Tam, Virginia Polytechnic Institute and State University

We appreciate the following persons and companies that cooperated in providing us with photographs:

Robert Colwell	MagneTek	Owosso, Michigan
Mark Luehmann	Consumers Power Co.	Jackson, Michigan
Timothy Kildea	Toledo Commutator Co.	Owosso, Michigan
John Horvath	Bodine Electric Co.	Chicago, Illinois
Paul Lindhorst	MagneTek	St. Louis, Missouri

And, of course, we cannot forget the efforts put forth by the staff at Saunders College Publishing. They kept us on our toes throughout this immensely complex project. We appreciate their professionalism and sincere concern in not compromising either in the rendering of the art or in the editing of the text. For these reasons, we are specifically grateful to Emily Barrosse, Michelle Slavin, and Linda Boyle.

Finally, we must thank our respective families for their honest sacrifices, active encouragement, unconditional support, absolute understanding, and complete cooperation as we labored to revise this text.

Bhag S. Guru
Hüseyin R. Hiziroğlu

Contents

5 Direct-Current Generators 265

6 Direct-Current Motors 331

7 Synchronous Generators 382

8 Synchronous Motors 450

9 Polyphase Induction Motors 488

CHAPTER 1

Review of Electric Circuit Theory

Three-phase, high voltage, power transformer. (*Courtesy of Consumers Power*)

1.1 Introduction

Without any reservation or exaggeration, we can say that the availability of energy in the electric form has made our lives much more comfortable than ever before in the history of mankind. In fact, we may find it very difficult if not impossible to function if some of the devices that operate on electricity are suddenly taken away from us. Ironically, most of the energy available in the electric form is converted directly or indirectly from some other form of energy. An example of the direct energy conversion process is the conversion of light energy into electric energy by solar cells.

In the indirect energy-conversion process, we may use such resources as oil, natural gas, and coal. By burning these fuels we generate heat, which is then utilized to produce steam in a boiler. The steam propels the blades of a turbine, which in turn rotates the rotor of an **electric generator** that produces electric power. On the other hand, the potential energy of water is converted into mechanical energy by a turbine in a hydroelectric plant. Therefore, what matters the most in the indirect energy conversion process is the mechanical energy that must be supplied to the rotor of an electric generator. This process may also be referred to as an **electromechanical energy conversion** process owing to the conversion of mechanical energy into electric energy.

The main objective of this book is to discuss the basic principles of electromechanical energy conversion. Another goal of this text is to highlight the fact that the electromechanical energy conversion process is a reversible process. That is, we can also transform electric energy into mechanical energy by using a device known as an **electric motor**.

Economically it has been found advantageous to generate electric power at a central location at relatively low-voltage levels, transmit it over a distance via transmission lines at comparatively high-voltage levels, and then distribute it to consumers, again at low-voltage levels. Raising the voltage level at the generating end and lowering it at the consumer's end are done effectively by means of **step-up** and **step-down transformers**, respectively. This voltage change is made to reduce the electrical losses along the length of the transmission line. Even though a transformer does not convert mechanical energy into electric energy or vice versa, its study is essential because it plays a major role in the transmission and distribution of electric energy.

The method of treatment adopted for each type of electric machine discussed in this book is as follows:

(a) Discuss the construction of the machine
(b) Explain how the electromagnetic fields interact as a medium in the energy transfer process
(c) Represent the machine by its electric equivalent circuit
(d) Determine its performance using basic laws of electric circuit theory

It is apparent that in order to comprehend electric machines, we must be familiar with electromagnetic fields and be able to analyze electric circuits. The purpose of this chapter is to review some of the basic laws of electric circuits. In Chapter 2 we review the basic laws of electromagnetic field theory and apply them to the analysis of magnetic circuits. The discussion in this and the next chapter is intended merely as a review, and the reader is expected to have some prior knowledge of the material.

1.2 Direct-Current Circuit Analysis

The fundamental laws of electric circuit theory are Ohm's law, Kirchhoff's current law, and Kirchhoff's voltage law. Although all electric circuits can be analyzed by applying these laws, the resulting equations become cumbersome as the complexity of an electric circuit increases. As you are aware, other circuit analysis techniques, such as the node-voltage method and the mesh-current method, enable us to solve complex but practical problems easily. As you will see in the following chapters, quite often we wish to determine the current, voltage, and/or power that is delivered to some part of a network, which we refer to as the load. In this case, we can simplify our analysis by using Thevenin's theorem.

We begin our discussion by stating Ohm's law, Kirchhoff's laws, Thevenin's theorem, and the maximum power transfer theorem and then show how we can apply them to solve for currents, voltages, and power in a direct-current (dc) electric circuit.

Ohm's Law

Ohm's law states that the voltage drop across a resistor is equal to the product of the current through it and its resistance. That is,

$$V = IR \tag{1.1}$$

where V is the voltage drop across the resistance R and I is the current through it. We have used the **capital letters for the current and the voltage to indicate that I and V are time-invariant quantities** in a dc circuit. These conventions are in accordance with those you may have already used in analyzing dc electric circuits.

Kirchhoff's Current Law

The algebraic sum of all the currents at any node in an electric circuit is equal to zero. That is,

$$\sum_{m=1}^{n} I_m = 0 \tag{1.2}$$

where n is the number of branches forming a node and I_m is the current in the mth branch.

Kirchhoff's Voltage Law

Kirchhoff's voltage law states that the algebraic sum of all the voltages around a closed path in an electric circuit is zero. That is,

$$\sum_{m=1}^{n} V_m = 0 \tag{1.3}$$

where V_m is the voltage across the mth branch in a closed path containing n branches.

The term "algebraic" in the above statements of Kirchhoff's laws alerts us that we must pay due attention to the directions of the currents at a node and the polarities of the voltages in a closed loop. To do so, we follow the standard conventions and we summarize them below for brevity.

1. If the currents entering a node are considered positive, then the currents leaving that node are negative.
2. If we consider the drop in potential as positive while traversing a closed loop, then the rise in potential is negative.
3. A source delivers power to the circuit if the current through it flows from its negative terminal toward the positive terminal. Likewise, a source absorbs power (behaves like a sink) if the direction of the current through it is from its positive terminal toward the negative terminal.

Thevenin's Theorem

A linear circuit containing any number of sources and elements, when viewed from two nodes (terminals), can be replaced by an equivalent voltage source (also known as the **Thevenin voltage**), V_T, in series with an equivalent resistance, R_T (also called **Thevenin resistance**), where V_T is the open-circuit voltage between the two nodes and R_T is the ratio of the open-circuit voltage to the short-circuit current. If the electric circuit contains only independent sources, R_T can be obtained by looking at the terminals with the voltage sources replaced by short circuits and the current sources by open circuits. The open-circuit voltage V_T is obtained by removing the load and leaving the terminals open.

Maximum Power Transfer Theorem

The maximum power transfer theorem states that in a dc electric circuit, maximum power transfer takes place when the load resistance is equal to Thevenin's equivalent resistance.

EXAMPLE 1.1

Determine the value of the load resistance R_L in Figure 1.1**a** for maximum power transfer. What is the maximum power delivered to R_L?

● SOLUTION

Let us first disconnect the load resistance R_L to determine the open-circuit voltage V_{ab}, as indicated in Figure 1.1**b**.

Applying Kirchhoff's voltage law (KVL), we obtain

$$(20 + 30)\, I + 25 - 100 = 0$$

or

$$I = 1.5 \text{ A}$$

The open-circuit voltage is

$$V_T = V_{ab} = 100 - 1.5 \times 20 = 70 \text{ V}$$

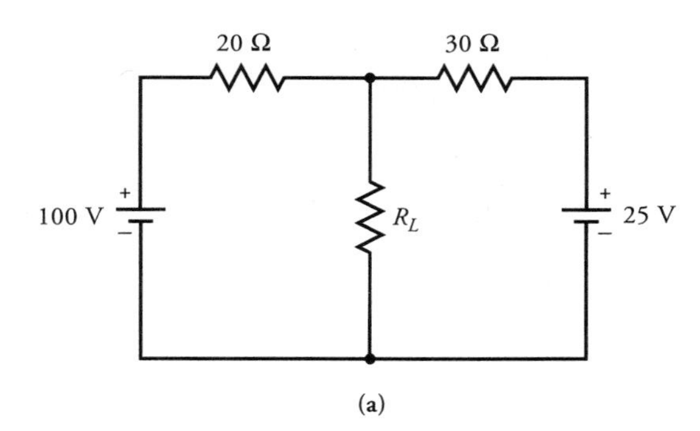

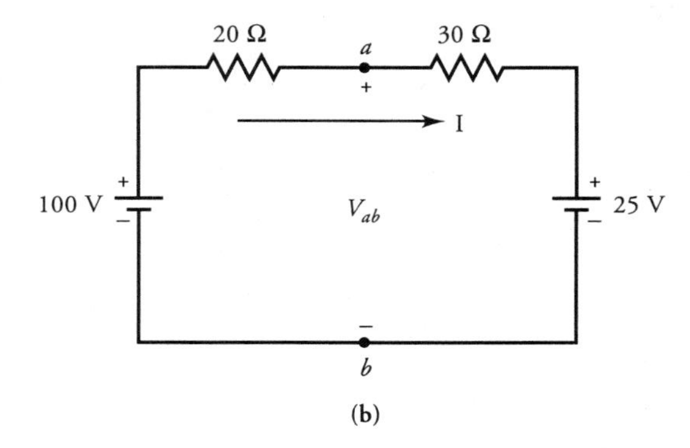

Figure 1.1 **(a)** Circuit for Example 1.1. **(b)** R_L removed to obtain V_{ab}.

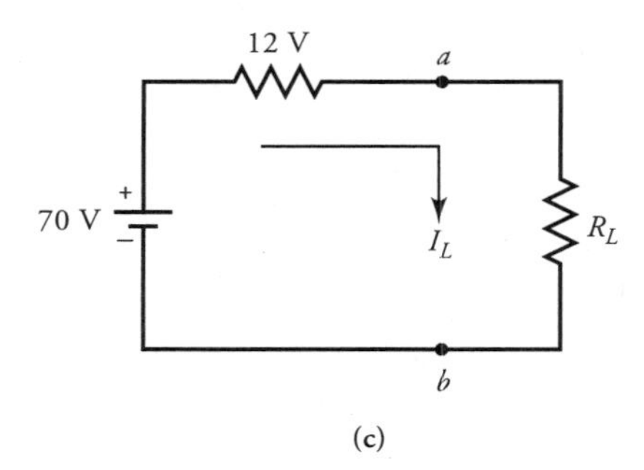

Figure 1.1 **(c)** Thevenin's equivalent circuit.

By shorting the independent voltage sources, we obtain Thevenin's equivalent resistance as

$$R_T = \frac{20 \times 30}{(20 + 30)} = 12\ \Omega$$

Figure 1.1**c** shows the load resistance R_L connected to Thevenin's equivalent circuit. For the maximum power transfer, $R_L = R_T = 12\ \Omega$.

The current through the load resistance is

$$I_L = \frac{70}{24} = 2.917\ \text{A}$$

Finally, the maximum power transferred to the load is

$$P_L = (2.917)^2 \times 12 = 102.08\ \text{W}$$

■

Exercises

1.1. For the dc circuit shown in Figure E1.1, determine (a) the current through each element using the mesh-current method, (b) the voltage at each node with reference to node *a* using the node-voltage method, (c) the power supplied by each source, and (d) the power dissipated by each resistor.

1.2. Use Thevenin's theorem to determine the current through and the power dissipated by the 15-Ω resistance in the circuit shown in Figure E1.1. What resistance must be placed either in series or in parallel with the 15-Ω resistance so that the combination receives the maximum power?

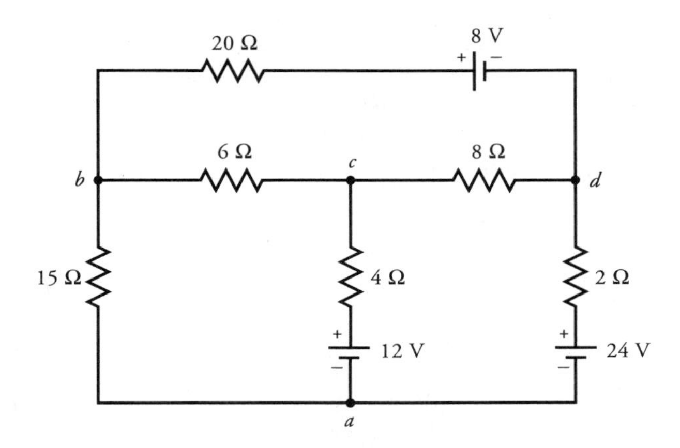

Figure E1.1 Circuit for Exercise 1.1.

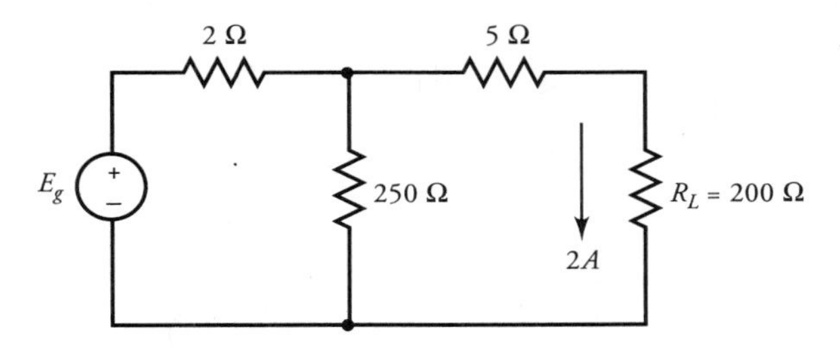

Figure E1.3 Circuit for Exercise 1.3.

1.3. The equivalent circuit of a dc generator connected to a load of 200 Ω is shown in Figure E1.3. Determine E_g if the load current is 2 A. E_g is called the induced electromotive force (emf, or generated voltage) in a dc generator.

1.3 Alternating-Current Circuit Analysis

Of all the various types of electric motors, alternating-current (ac) motors are by far the most popular and most widely used. An ac motor is designed to operate as either a **single-phase motor** or a **three-phase motor**. In integral-horsepower sizes, most motors are designed to operate on a three-phase supply. However, in the fractional-horsepower sizes, the use of the single-phase motor exceeds that of the three-phase motor. For this reason, we review single-phase ac circuits in this section. Three-phase ac circuits are reviewed in the next section.

The general expression for the single-phase current waveform as shown in Figure 1.2**a** is

$$i(t) = I_m \sin(\omega t + \theta) \tag{1.4}$$

where I_m is the maximum value or the amplitude of the current, ω is the angular frequency (in radians/second [rad/s]), and θ is the initial phase shift. The angular frequency ω can be expressed as

$$\omega = 2\pi f \tag{1.5}$$

where f is the frequency of the current waveform in hertz (Hz). The time required to complete one cycle of the waveform is called the **time period** and is given as

$$T = \frac{1}{f} \tag{1.6}$$

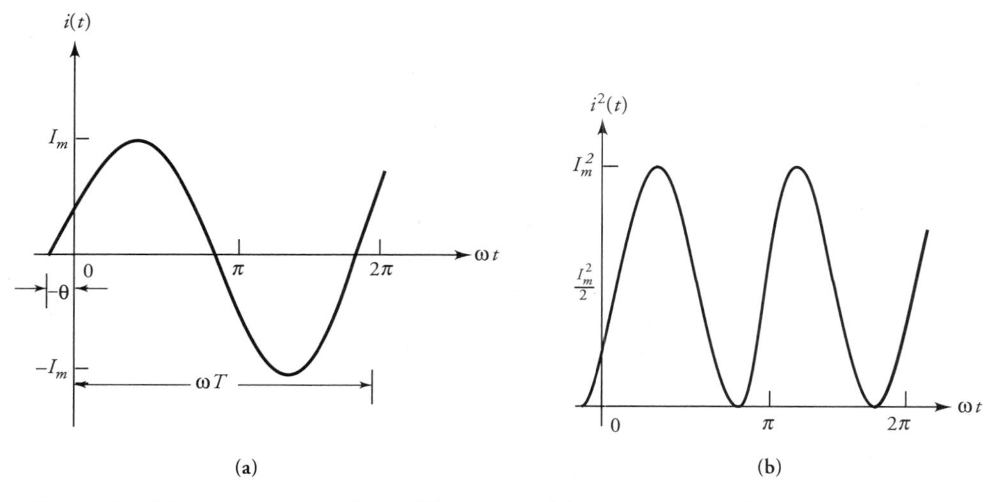

Figure 1.2 **(a)** A current waveform. **(b)** Square of the current waveform.

The average value of the periodic current waveform (Figure 1.2**a**) is

$$I_{avg} = \frac{1}{T}\int_0^T i(t)\,dt = 0 \tag{1.7}$$

The average value of a sinusoidally varying function of the form given in Eq. (4) is always zero.

The root-mean-square (rms) or effective value of the current waveform (Figure 1.2**a**) is

$$I = \sqrt{\frac{1}{T}\int_0^T i^2(t)\,dt} = \frac{I_m}{\sqrt{2}} \tag{1.8}$$

The function $i^2(t)$ is sketched in Figure 1.2**b**, which clearly shows that the squared wave lies entirely above the zero axis. We say that an alternating current has an effective value of 1 A when it produces heat in a certain resistance at the same rate that heat is produced in the same resistance by 1 A of direct current. Note that the rms value of a sinusoidal function is always 70.7% of its maximum value.

Instantaneous Power

The power at any given instant is equal to the product of the voltage and the current at that instant. That is,

$$p(t) = v(t)\,i(t) \tag{1.9}$$

Let $v(t) = V_m \cos(\omega t + \alpha)$ be the voltage across an element in a circuit and $i(t) = I_m \cos(\omega t + \phi)$ be the current through the element. Then the instantaneous power is

$$p(t) = V_m I_m \cos(\omega t + \alpha)\cos(\omega t + \phi) \tag{1.10}$$

and its average value is

$$P = \frac{1}{2} V_m I_m \cos(\alpha - \phi) = VI \cos\theta \tag{1.11}$$

where $V = V_m/\sqrt{2}$ and $I = I_m/\sqrt{2}$ are the rms values of the voltage and the current, and $\theta = \alpha - \phi$ is the phase angle between them, as shown in Figure 1.3. Equation (1.11) states that the average power is equal to the product of the rms

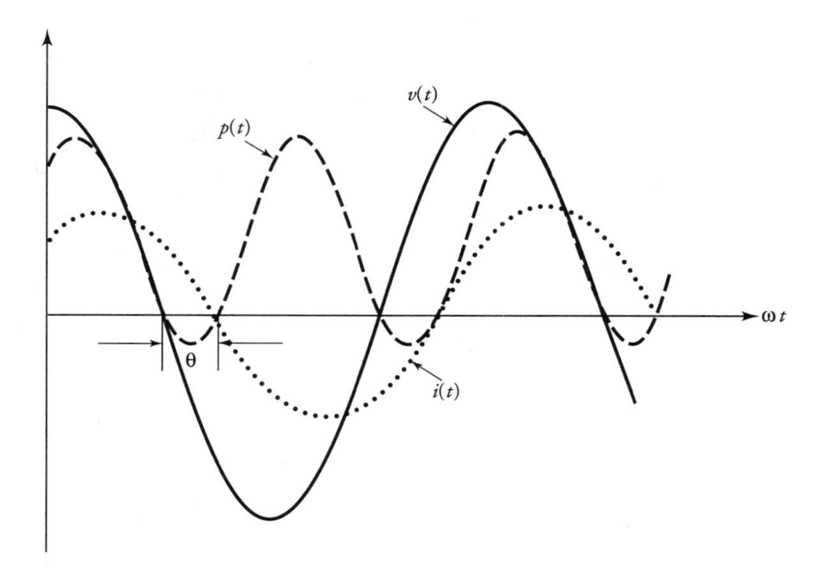

Figure 1.3 Voltage, current, and instantaneous power waveforms.

value of the voltage, the rms value of the current, and the cosine of the phase angle between them. The phase angle θ between the voltage and the current is called the **power factor angle**, and $\cos\theta$ is referred to as the **power factor**.

The power factor is said to be **lagging** when the current in the circuit lags the voltage drop across it, as shown in Figure 1.3. In this case, the circuit is **inductive** in nature. On the other hand, the current **leads** the applied voltage for a **capacitive** circuit and has a **leading** power factor.

Phasor Analysis

In the analysis of electric circuits you were introduced to the concept that a sine function can be represented by a **phasor.** Let us first answer the question: What is a phasor?

From Euler's identity,

$$V_m e^{j(\omega t + \theta)} = V_m \cos(\omega t + \theta) + jV_m \sin(\omega t + \theta)$$

where $j = \sqrt{-1}$.

If a source voltage is given as

$$v(t) = V_m \cos(\omega t + \theta)$$

then we can express it as

$$v(t) = \text{Re}\left[V_m e^{j(\omega t + \theta)}\right]$$

where Re stands for **the real part of** the expression within the brackets. We can also write $v(t)$ as

$$v(t) = \text{Re}\left[\sqrt{2}Ve^{j\theta}e^{j\omega t}\right]$$

where V is the rms value of $v(t)$. Because $\sqrt{2}$ is simply a constant multiplier and ω depends upon the frequency of the applied source, we can temporarily drop these parameters from consideration. If we assume that Re is also implied, we can define a quantity such that

$$\tilde{V} = Ve^{j\theta} = V\underline{/\theta}$$

then **$\tilde{V}$ is said to be a phasor representation of $v(t)$** in terms of its rms value. That is,

$$v(t) = \sqrt{2}V\cos(\omega t + \theta) \quad <=> \quad \tilde{V} = V\underline{/\theta} \tag{1.12}$$

If $\tilde{V}$ and $\tilde{I}$ are the phasor representations of $v(t)$ and $i(t)$, then the phasor equivalences of Ohm's law for resistance R, inductance L, and capacitance C are given in Table 1.1. Note that $j\omega L$ and $1/j\omega C$ are the inductive impedance ($\hat{Z}_L$) and

Table 1.1: Ohm's Law for R, L, and C

	Circuit Symbol	Time-Domain	Phasor-Domain
Resistance, R	R, i, +, −, v_R	$v_R = iR$	$\tilde{V} = \tilde{I}R$
Inductance, L	L, i, +, −, v_L	$v_L = L\dfrac{di}{dt}$	$\tilde{V}_L = j\omega L\tilde{I}$ $= \tilde{I}\hat{Z}_L$
Capacitance, C	C, i, +, −, v_C	$i = C\dfrac{dv_C}{dt}$	$\tilde{I} = j\omega C\tilde{V}_C$ $= \dfrac{\tilde{V}_C}{\hat{Z}_C}$

the capacitive impedance ($\hat{Z}_C$) corresponding to the inductance L and the capacitance C. The impedances can also be given as

$$\hat{Z}_L = jX_L = j\omega L$$

and

$$\hat{Z}_C = -jX_C = -\frac{j}{\omega C}$$

where $X_L = \omega L$ is the inductive reactance and $X_C = 1/\omega C$ is the capacitive reactance.

Kirchhoff's voltage and current laws in the phasor form are

$$\sum_{m=1}^{n} \tilde{V}_m = 0 \tag{1.13}$$

and

$$\sum_{m=1}^{n} \tilde{I}_m = 0 \tag{1.14}$$

A given circuit with sinusoidally varying sources can be transformed into its equivalent circuit in phasor domain and then solved using algebraic manipulations of complex numbers. **Once we have obtained the required circuit variable in the phasor form, we can reconvert it into its proper time-domain representation by multiplying the phasor with $e^{j\omega t}$ and considering only the real part of it.**

In terms of phasors, we define the **complex power** as

$$\hat{S} = \tilde{V}\tilde{I}^* = P + jQ \tag{1.15}$$

where **$\tilde{I}^*$ is the conjugate of $\tilde{I}$, P is the real or the average power, and Q is the reactive power.** For a purely resistive circuit, Q is zero. Q is less than zero for a capacitive circuit and greater than zero for an inductive circuit.

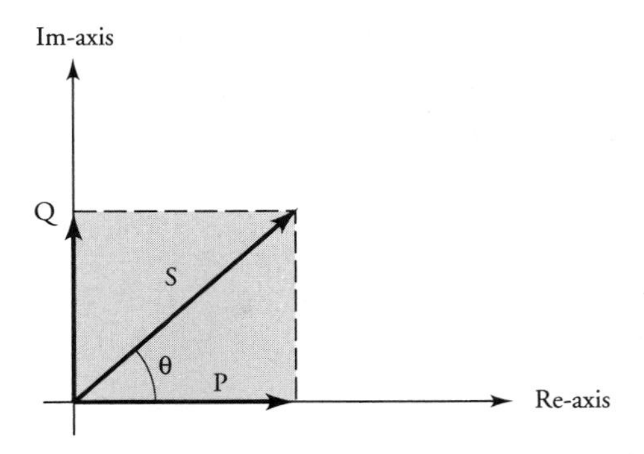

Figure 1.4 Complex power diagram.

The Power Diagram (Triangle)

We can draw a power diagram by plotting P along the real axis and Q along the imaginary axis in the complex plane, as indicated in Figure 1.4, where θ is the power factor angle.

The Phasor Diagram

The phasor diagram is a name given to a sketch in a complex plane of the phasor voltages and the phasor currents in a given circuit. We exploit the phasor diagrams to simplify the analytic work throughout this text. While plotting the phasor diagram, the currents and voltages in a circuit are always understood to have their own amplitude scales but a common angle scale.

EXAMPLE 1.2

Find the current in the inductive circuit shown in Figure 1.5**a**. Draw the power diagram and the phasor diagram, and sketch the input voltage and the current in time domain.

● SOLUTION

The angular frequency is $\omega = 1000$ rad/s. The inductive and capacitive impedances are

$$\hat{Z}_L = j1000 \times 1 \times 10^{-3} = j1\ \Omega$$

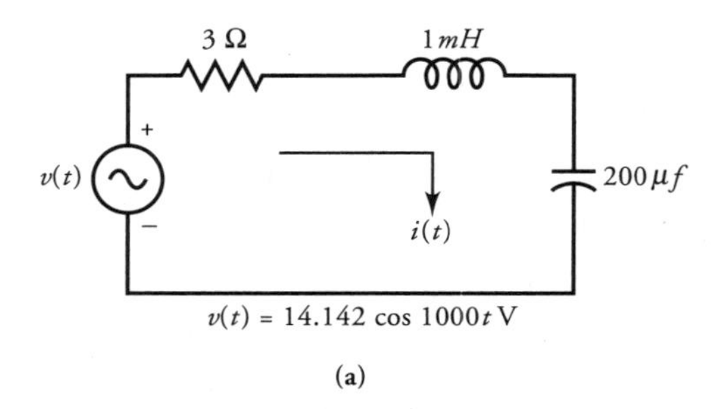

(a)

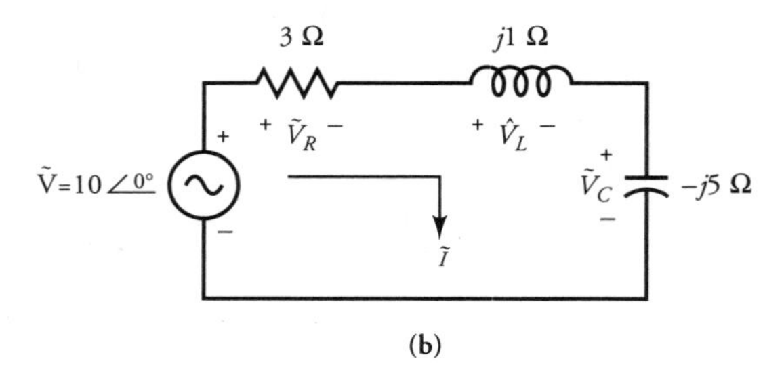

(b)

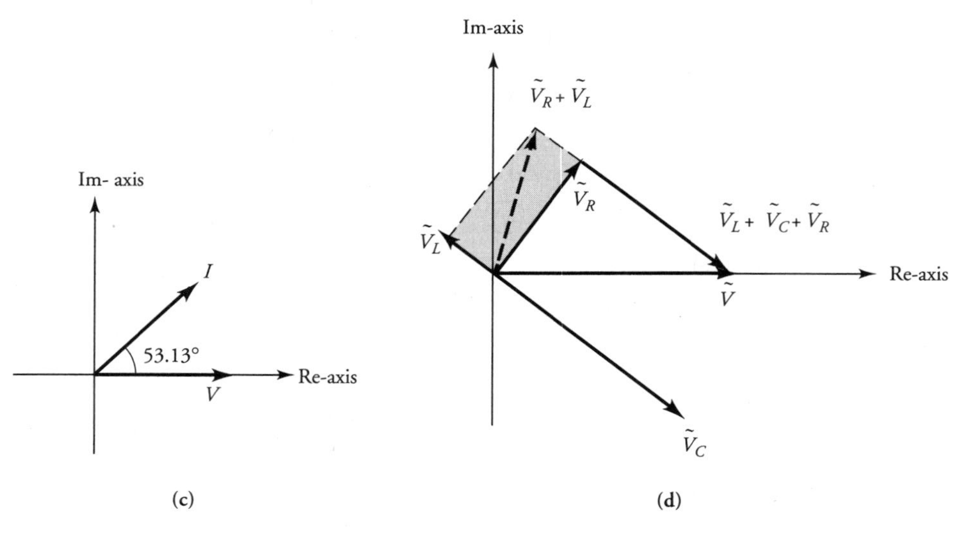

(c) (d)

Figure 1.5 **(a)** Circuit for Example 1.2. **(b)** Phasor equivalent circuit. **(c)** Voltage and current relationship. **(d)** Phasor diagram.

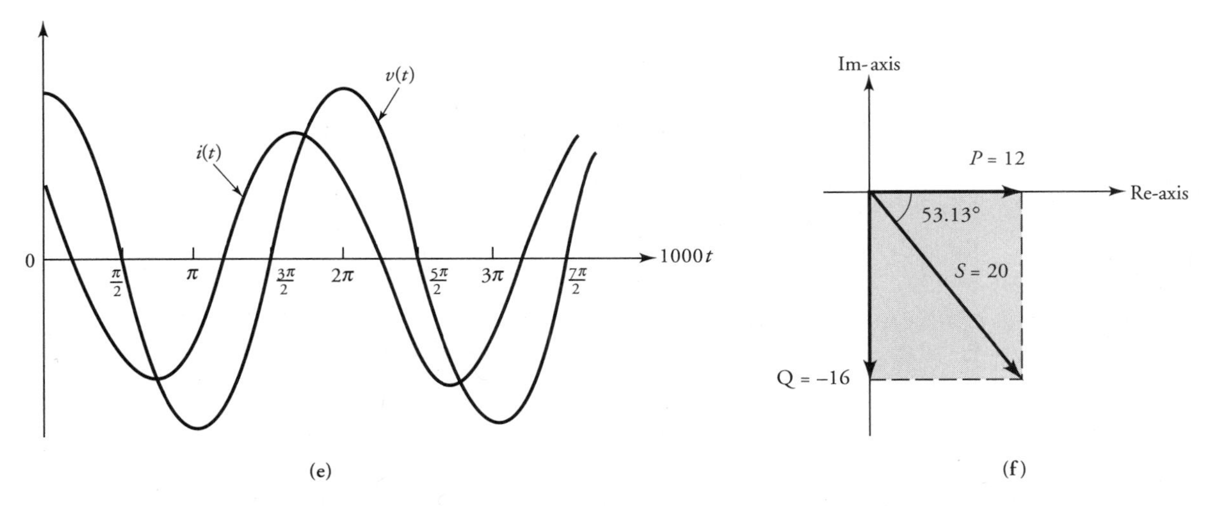

Figure 1.5 **(e)** Voltage and current waveforms. **(f)** Power diagram.

and

$$\hat{Z}_C = -\frac{j}{1000 \times 200 \times 10^{-6}} = -j5\ \Omega$$

We can now redraw the phasor equivalent circuit as shown in Figure 1.5**b**. The impedance of the circuit is

$$\hat{Z} = 3 + j1 - j5 = 3 - j4 = 5\underline{/-53.13^\circ}\ \Omega$$

We can now calculate the current in the circuit as

$$\tilde{I} = \frac{\tilde{V}}{\hat{Z}} = \frac{10\underline{/0^\circ}}{5\underline{/-53.13^\circ}} = 2\underline{/53.13^\circ}\ \text{A}$$

The phasor relation between the applied voltage and the current in the circuit is illustrated in Figure 1.5**c**. Note that the current leads the applied voltage by an angle of 53.13°.

The phasor voltage drops in the circuit are

$$\tilde{V}_R = \tilde{I}R = 6\angle 53.13^\circ \text{ V}$$

$$\tilde{V}_L = \tilde{I}\hat{Z}_L = j2\angle 53.13^\circ = 2\angle 143.13^\circ \text{ V}$$

and

$$\tilde{V}_C = \tilde{I}\hat{Z}_C = -j10\angle 53.13^\circ = 10\angle -36.87^\circ \text{ V}$$

The corresponding phasor diagram is shown in Figure 1.5**d**.

The expression for the current in the time-domain is

$$i(t) = \text{Re}\left[\sqrt{2}e^{j53.13^\circ}e^{j1000t}\right] = 2.828\cos(1000t + 53.13^\circ) \text{ A}$$

The voltage and current waveforms are sketched in Figure 1.5**e**. As you can see, the current leads the voltage by an angle of 53.13°.

The complex power supplied by the source is

$$\hat{S} = \tilde{V}\tilde{I}^* = [10\angle 0^\circ][2\angle -53.13^\circ] = 20\angle -53.13^\circ$$

$$= 12 - j16 \text{ VA.}$$

Hence, the apparent power S is 20 VA, the real power P is 12 W, and the reactive power Q is -16 VAR. The power triangle is sketched in Figure 1.5**f**.

■

Power Factor Correction

Most of the active loads such as induction motors operating on ac power supply have lagging power factors. Any decrease in the power factor results in an increase in the current for the same power output. The increase in the current translates into an increase in the power loss on the transmission line. As the power factor falls below a certain level, the power supply company assesses a power factor penalty to offset the additional power loss on the transmission line.

Because the current in a capacitor leads the voltage drop across it, we can connect capacitors in parallel with the inductive load to improve the overall power factor. The power requirements of the load remain the same because a capacitor does not dissipate power. Power factor improvement with capacitors is called the passive power factor control. Later in this book, we discuss an active power factor control that uses synchronous machines.

EXAMPLE 1.3

A certain load draws a current of 10 A at a lagging power factor of 0.5 from a 120-V, 60-Hz source. The electric supply company imposes a penalty if the power factor drops below 0.8. What size capacitor must be used just to avoid the penalty?

● SOLUTION

If the supply voltage is $\tilde{V} = 120\underline{/0^\circ}$, the current through the load is $\tilde{I}_L = 10\underline{/-60^\circ}$. The phasor diagram for the voltage and the current is shown in Figure 1.6**a**. The complex power absorbed by the load is

$$\hat{S} = \tilde{V}\tilde{I}_L^* = 1200\underline{/60^\circ} = 600 + j1039.23 \text{ VA}$$

The average power required by the load is 600 W, and the associated reactive component of power at 0.5 power factor (pf) lagging is 1039.23 VAR.

Let us now connect a capacitor across the load, as shown in Figure 1.6**b**. Let

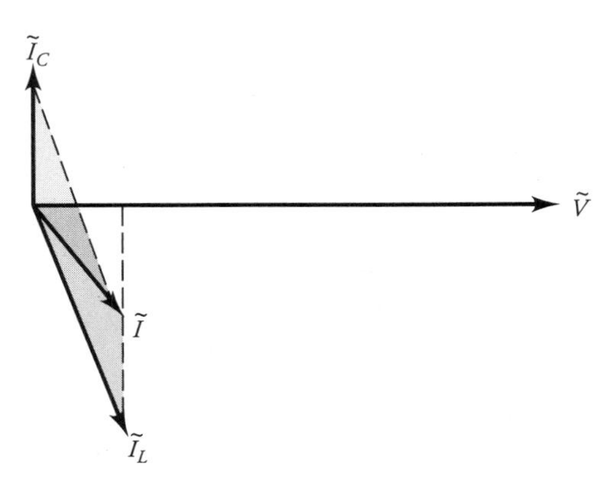

(a) Phasor diagram for the voltage and the current for Example 1.3

Figure 1.6a Power factor correction.

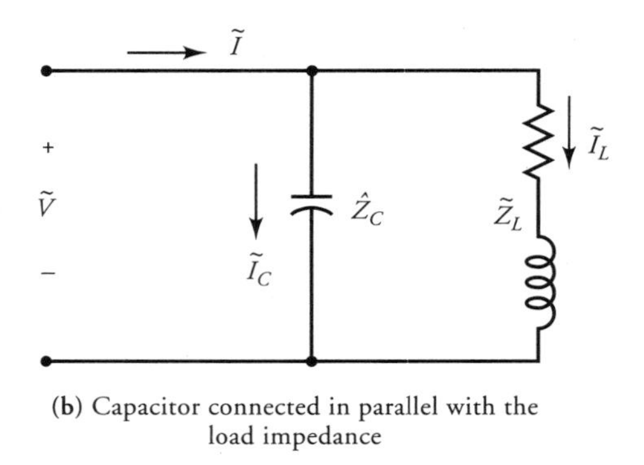

(b) Capacitor connected in parallel with the load impedance

Figure 1.6b Power factor correction.

$\tilde{I}_C$ be the current through the capacitor; then the total current intake from the source $\tilde{I}$ must have a power factor of 0.8 lagging just to avoid the penalty. That is,

$$\tilde{I} = \tilde{I}_L + \tilde{I}_C = I\angle{-36.87^\circ}$$

Because the average power required from the source should still be 600 W, the total current I is

$$I = \frac{600}{120 \times 0.8} = 6.25 \text{ A}$$

The current through the capacitor is

$$\tilde{I}_C = \tilde{I} - \tilde{I}_L = 6.25\angle{-36.87^\circ} - 10\angle{-60^\circ} = j4.91 \text{ A}$$

Thus, the capacitive impedance is

$$\hat{Z}_C = \frac{120}{j4.91} = -j24.44\ \Omega \qquad \text{or} \qquad X_C = 24.44\ \Omega$$

Hence, the required value of the capacitor is

$$C = \frac{1}{2\pi \times 60 \times 24.44} = 108.53\ \mu\text{F}$$

■

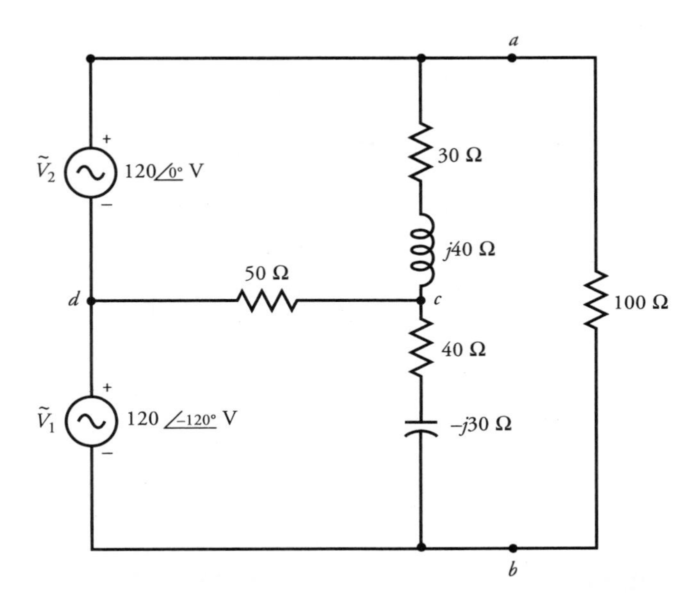

Figure E1.6 Circuit for Exercise 1.6.

Exercises

1.4. A voltage source of $v(t) = 120 \sin(200t + 30°)$ V is impressed across (a) a resistance of 10 Ω, (b) an inductance of 10 mH, and (c) a capacitance of 500 μF. In each case draw the phasor equivalent circuit and determine (i) the current in the circuit, (ii) the apparent power, the real power, and the reactive power supplied by the source, and (iii) sketch the voltage and current waveforms.

1.5. A voltage waveform of $v(t) = 230 \cos(100t)$ is applied across a series circuit consisting of (1) a resistance of 50 Ω and a capacitance of 100 μF and (2) a resistance of 50 Ω and an inductance of 200 mH. For each circuit, draw the phasor equivalent circuit, the phasor diagram, the power triangle, and the voltage and current waveforms in time-domain. What is the parallel equivalent circuit in each case?

1.6. A phasor equivalent circuit is given in Figure E1.6. Determine the voltage at node c with respect to node b using the node-voltage method. Calculate the current in each branch using the mesh-current method. What is the power supplied by each source? What is the total power dissipated in the circuit? Is the total power supplied equal to the total power dissipated?

1.7. In Example 1.3, what size capacitor must be used in order to make the overall power factor to be unity?

1.4 Three-Phase Circuits

This section provides a brief discussion of a balanced three-phase source supplying power to a balanced three-phase load. In practice, the power is distributed to a load via a three-wire transmission line from a remote three-phase generating station. A three-phase generator is always designed to act as a balanced three-phase source. There is no reason to believe that the three-wire transmission line from the generator to the load is not balanced. For unbalanced loads (loads with different phase impedances) the mode of analysis is the same as that discussed in the preceding section. Only when the load is balanced can we simplify our analysis of a three-phase system by representing the three-phase circuit by a per-phase equivalent circuit, as explained below.

Three-Phase Source

A balanced three-phase source can be visualized as if it were composed of three single-phase sources such that (a) the amplitude of each voltage source is the same, and (b) each voltage source is 120° out of phase with each of the other two. Each source is then said to represent one of the three phases of the three-phase source. The three sources can be connected to form either a wye (Y)-connection (Figure 1.7) or a delta (Δ)-connection (Figure 1.8). Three-phase voltages for the Y-connection, in time-domain in and phasor form in terms of rms values, are

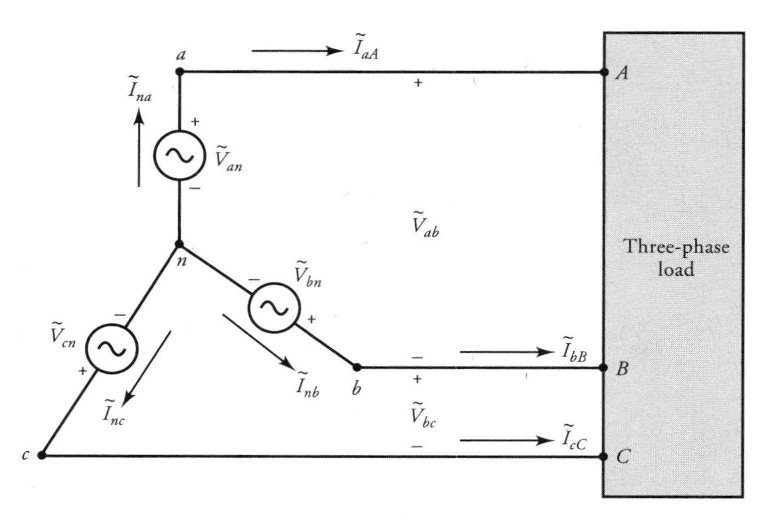

Figure 1.7 A Y-connected three-phase source.

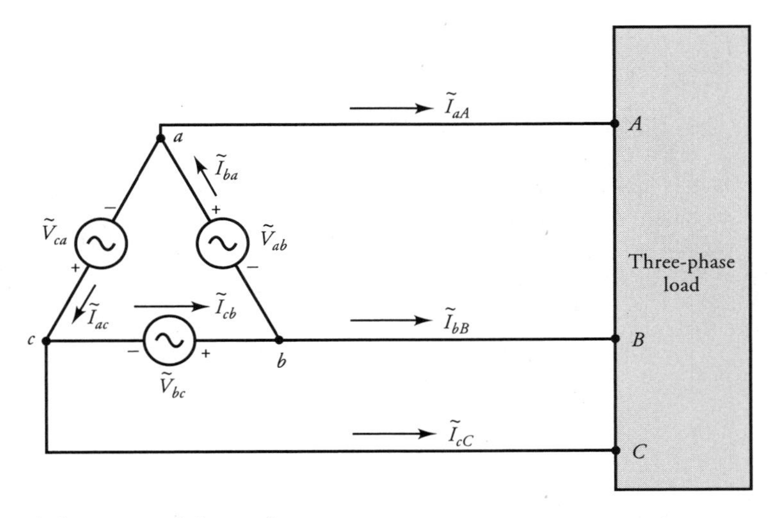

Figure 1.8 A Δ-connected three-phase source.

	Time-Domain		Phasor Form	
	$v_{an} = \sqrt{2}V \cos \omega t$	<=>	$\tilde{V}_{an} = V\angle 0°$	(1.16a)
	$v_{bn} = \sqrt{2}V \cos(\omega t - 120°)$	<=>	$\tilde{V}_{bn} = V\angle -120°$	(1.16b)
	$v_{cn} = \sqrt{2}V \cos(\omega t + 120°)$	<=>	$\tilde{V}_{cn} = V\angle 120°$	(1.16c)

where V is the rms value of each **phase voltage** and ω is the angular frequency. The subscript n indicates that the phase voltages are with respect to the common (neutral) terminal n. The voltage waveforms are sketched in Figure 1.9. Note that the voltage of phase b lags while the voltage of phase c leads with respect to the voltage of phase a. This is termed a **positive or clockwise phase sequence**. The phase sequence is said to be **negative or counterclockwise** if the voltage of phase b leads while the voltage of phase c lags with respect to the voltage of phase a.

The **line voltage** $\tilde{V}_{ab}$, from line a to line b, for the positive phase sequence is

$$\tilde{V}_{ab} = \tilde{V}_{an} - \tilde{V}_{bn} = \sqrt{3}V\angle 30° = V_\ell\angle 30° \quad (1.17a)$$

$$\tilde{V}_{bc} = \tilde{V}_{bn} - \tilde{V}_{cn} = \sqrt{3}V\angle -90° = V_\ell\angle -90° \quad (1.17b)$$

$$\tilde{V}_{ca} = \tilde{V}_{cn} - \tilde{V}_{an} = \sqrt{3}V\angle 150° = V_\ell\angle 150° \quad (1.17c)$$

where $V_\ell = \sqrt{3}V$ is the magnitude of the line voltage (or line-to-line voltage). Thus, **in a Y-connected balanced three-phase system, the magnitude of the line**

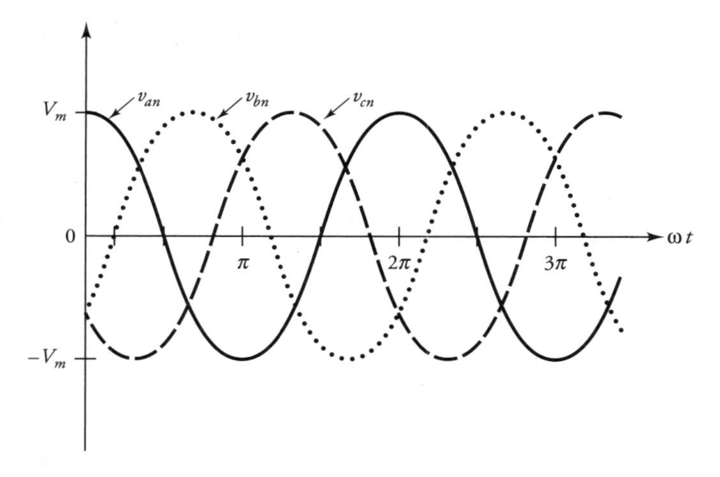

Figure 1.9 Voltage waveforms for the positive phase sequence.

voltage is $\sqrt{3}$ times the phase voltage. For the positive phase sequence, the line voltage leads the phase voltage by 30°, as portrayed by the phasor diagram of Figure 1.10.

Similarly, we can show that for a Y-connected source, the line voltage lags the phase voltage by 30° if the phase sequence is negative.

As you can see from Figure 1.7, the line current in a Y-connected source is the same as the phase current. At node n, the algebraic sum of the currents must be zero. Thus,

$$\tilde{I}_{na} + \tilde{I}_{nb} + \tilde{I}_{nc} = 0 \tag{1.17d}$$

For a Δ-connected source, the phase voltage is the same as the line voltage. For the positive phase sequence, the line or the phase voltages are

$$\tilde{V}_{ab} = V\angle 0^\circ \tag{1.18a}$$

$$\tilde{V}_{bc} = V\angle -120^\circ \tag{1.18b}$$

and

$$\tilde{V}_{ca} = V\angle 120^\circ \tag{1.18c}$$

Because the sources form a closed loop, the algebraic sum of the voltages is zero, in accordance with Kirchhoff's voltage law. That is,

$$\tilde{V}_{ab} + \tilde{V}_{bc} + \tilde{V}_{ca} = 0 \tag{1.18d}$$

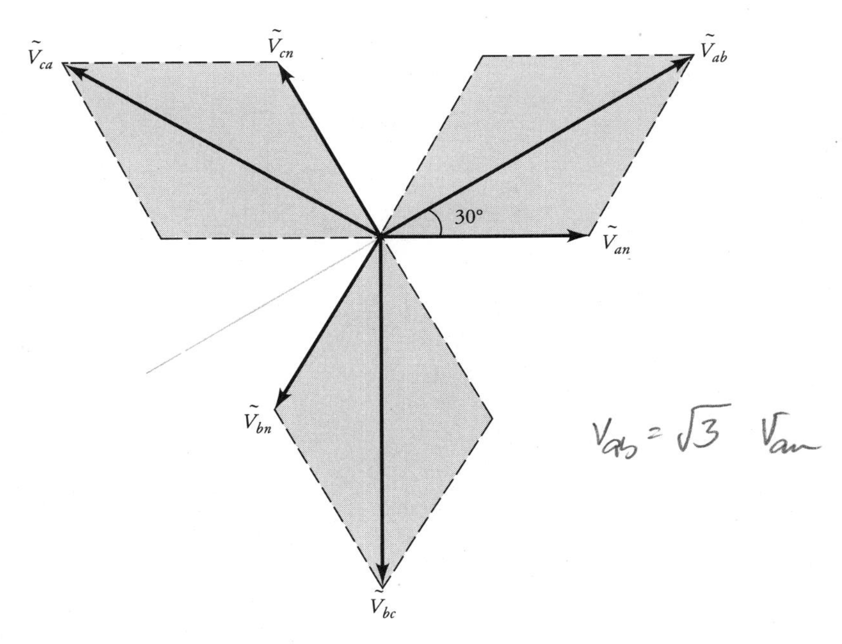

Figure 1.10 Phase and line voltages for the positive phase sequence, Y-connected source.

From Figure 1.8, the line currents are

$$\tilde{I}_{aA} = \tilde{I}_{ba} - \tilde{I}_{ac} = \sqrt{3}\tilde{I}_{ba}\underline{/-30^\circ} = \sqrt{3}I\underline{/\phi - 30^\circ} = I_\ell\underline{/\phi - 30^\circ} \quad (1.19a)$$

where we have assumed that $\tilde{I}_{ba} = I\underline{/\phi}$, I is the rms value of the phase current, and $I_\ell = \sqrt{3}I$ is the magnitude of the line current. Thus, **for a Δ-connected source, the line current is $\sqrt{3}$ times the phase current and lags the phase voltage by 30° for the positive phase sequence**, as illustrated in Figure 1.11. The other two line currents are

$$\tilde{I}_{bB} = I_\ell\underline{/\phi - 150^\circ} \quad (1.19b)$$

$$\tilde{I}_{cC} = I_\ell\underline{/\phi + 90^\circ} \quad (1.19c)$$

In the same way, we can show that the line current leads the phase current by 30° if the phase sequence is negative.

Three-Phase Load

A three-phase load is said to be balanced if the load impedance of each phase is the same. Just like the source voltages, the load impedances can either be Y- or

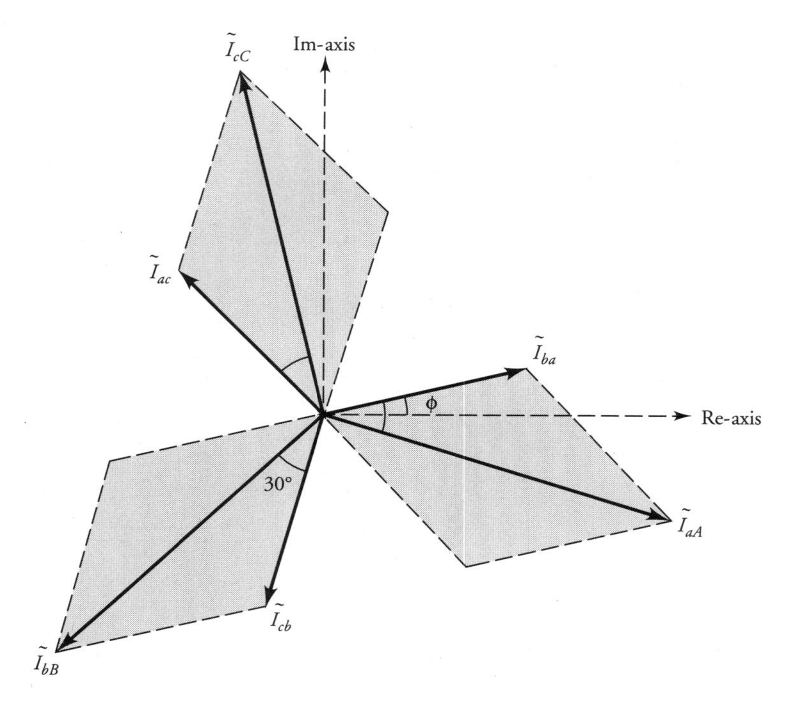

Figure 1.11 Relationships between phase and line currents for a Δ-connected source with positive phase sequence.

Δ-connected, as shown in Figure 1.12. This gives rise to four possible source-to-load connections: Y-Y, Y-Δ, Δ-Y, and Δ-Δ. If we master how to analyze a Y-Y connection, we can analyze them all by making Δ-Y transformations whenever necessary.

Δ-to-Y Transformation

A balanced Δ-connected load with a phase impedance $\hat{Z}_{\Delta}$ can be transformed into a balanced Y-connected load with a phase impedance $\hat{Z}_{Y}$ by using the following equation.

$$\hat{Z}_Y = \frac{\hat{Z}_\Delta}{3}$$

In a Y-connected load, the line current is the same as the phase current. However, if the load is Δ-connected, we can show that the magnitude of the line current is $\sqrt{3}$ times the magnitude of the phase current. The line current leads or lags the

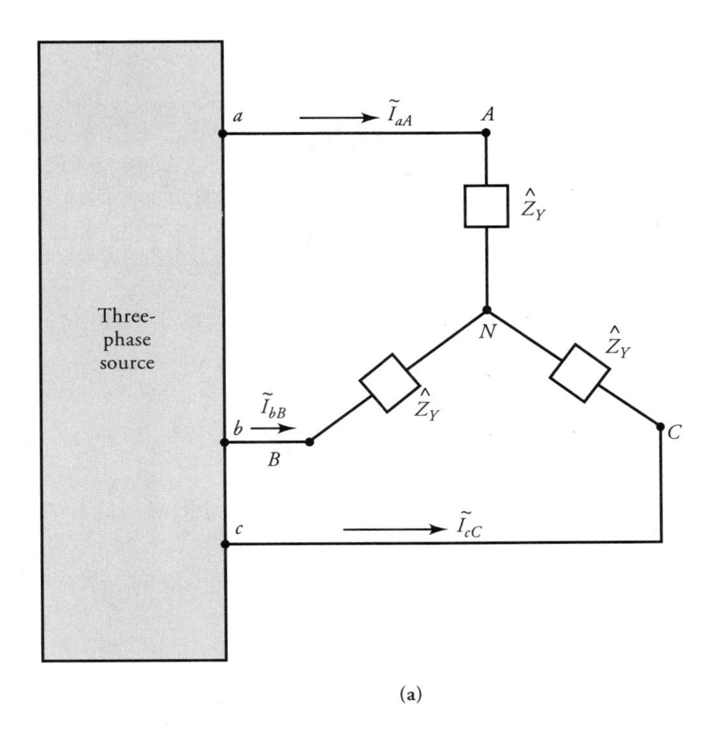

(a)

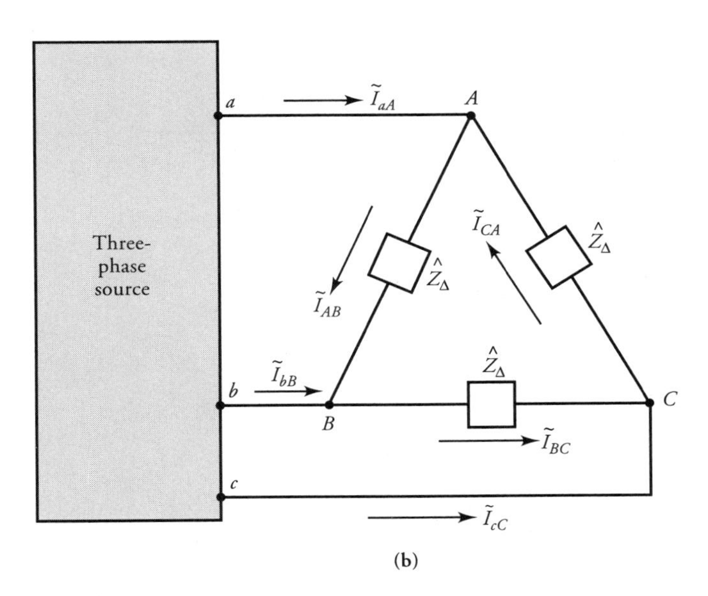

(b)

Figure 1.12 (a) Y-connected, and **(b)** Δ-connected three-phase load.

phase current by an angle of 30°, depending upon whether the phase sequence is negative or positive.

If we have a balanced three-phase source, balanced loads, and balanced line impedances, we can visualize a short circuit between the neutral nodes of the source and the load in a Y-Y connected system. Such a visualization permits us to reduce the three-phase problem into three single-phase problems, all identical except for the consistent difference in phase angle. In other words, once we have solved one of the three single-phase problems, we know the solutions to the other two. Thus, we work the problem on a **per-phase** basis, as outlined by the following example.

EXAMPLE 1.4

A balanced three-phase, 866-V, 60-Hz, Y-connected source feeds a balanced, Δ-connected load via a 100-km long three-wire transmission line. The impedance of each wire of the transmission line is $1 + j2\ \Omega$. The per-phase impedance of the load is $177 - j246\ \Omega$. If the phase sequence is positive, determine the line and the phase currents, the power absorbed by the load, and the power dissipated by the transmission line.

● SOLUTION

By making Δ-to-Y transformation, $\hat{Z}_Y = 59 - j82\ \Omega$, we can represent the balanced three-phase circuit on a per-phase basis, as shown in Figure 1.13**a**. The per-phase voltage, assuming phase-a as the reference, is

$$\tilde{V} = \frac{866}{\sqrt{3}} \approx 500\underline{/0^\circ}\ \text{V}$$

The total impedance on a per-phase basis is

$$\hat{Z} = 1 + j2 + 59 - j82 = 60 - j80 = 100\underline{/-53.13^\circ}\ \Omega$$

The current in the circuit is

$$\tilde{I} = \frac{500}{60 - j80} = 5\underline{/53.13^\circ}\ \text{A}$$

Because the current leads the applied voltage, the power factor (pf) is leading and has a magnitude of

$$\text{pf} = \cos(53.13^\circ) = 0.6\ \text{(lead)}$$

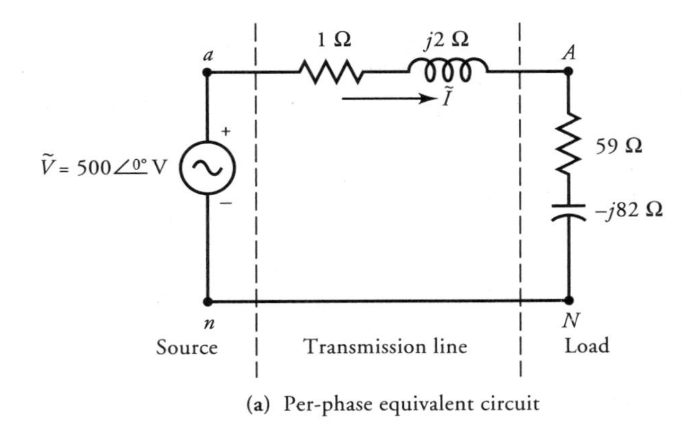

(a) Per-phase equivalent circuit

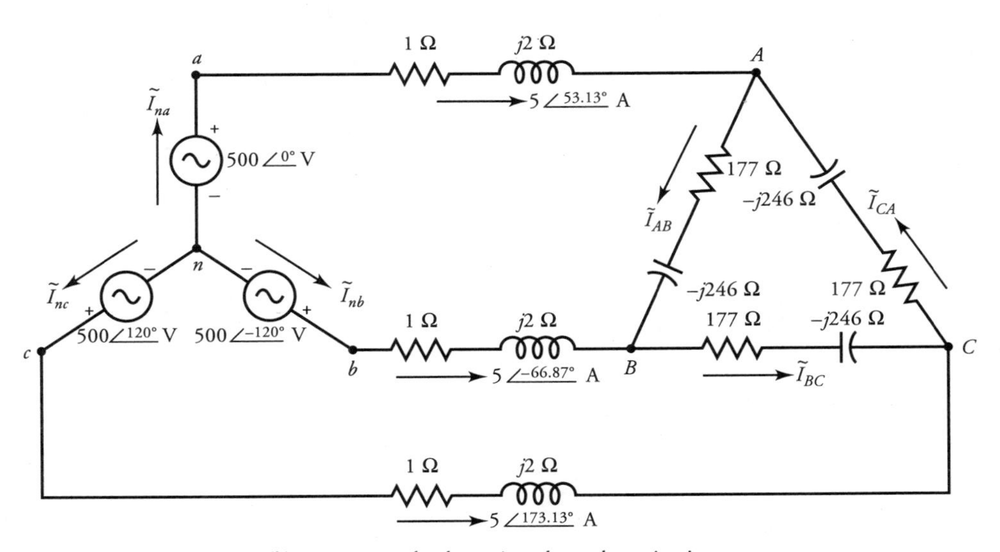

(b) Currents and voltages in a three-phase circuit

Figure 1.13

Because the source is Y-connected, the line current and the phase current for the source are the same. Thus, for a positive phase sequence, source currents are

$$\tilde{I}_{na} = 5\underline{/53.13^\circ}\text{ A}$$

$$\tilde{I}_{nb} = 5\underline{/-66.87^\circ}\text{ A}$$

and

$$\tilde{I}_{nc} = 5\underline{/173.13^\circ}\text{ A}$$

as shown in Figure 1.13**b**. For a Δ-connected load, the load current $\tilde{I}_{AB}$ is

$$\tilde{I}_{AB} = \frac{\tilde{I}}{\sqrt{3}\angle -30^\circ} = 2.887\angle 83.13^\circ \text{ A}$$

Similarly, the other phase currents through the load are

$$\tilde{I}_{BC} = 2.887\angle -36.87^\circ \text{ A}$$

and

$$\tilde{I}_{CA} = 2.887\angle -156.87^\circ \text{ A}$$

The line or phase voltages at the load end are

$$\begin{aligned}\tilde{V}_{AB} &= \tilde{I}_{AB}\hat{Z}_{AB} = 2.887\angle 83.13^\circ \times [177 - j246] \\ &= 874.93\angle 28.87^\circ \text{ V}\end{aligned}$$

Likewise, the other line or phase voltages are

$$\tilde{V}_{BC} = 874.93\angle -91.13^\circ \text{ V}$$

and

$$\tilde{V}_{CA} = 874.93\angle 148.87 \text{ V}$$

The average power dissipated in phase AB of the load is

$$P_{AB} = I_{AB}^2 \times 177 = 1475.25 \text{ W}$$

The power dissipated by each of the other two phases is also 1475.25 W. Thus, the total power dissipated by the load is

$$P_{\text{Load}} = 3 \times 1475.25 = 4425.75 \text{ W}$$

The total power dissipated by the three-wire transmission line is

$$P_{\text{Line}} = 3 \times 5^2 \times 1 = 75 \text{ W}$$

Hence, the total power supplied by the three-phase source is

$$\begin{aligned}P_{\text{Source}} &= P_{\text{Load}} + P_{\text{Line}} \\ &= 4425.75 + 75 = 4500.75 \text{ W}\end{aligned}$$

■

Exercises

1.8. A balanced three-phase Y-connected source has a line voltage of 208 V and feeds two balanced Y-connected loads. The per-phase impedances of the two loads are $20 + j70\ \Omega$ and $50 + j30\ \Omega$. Calculate the power supplied to each load by the source. What is the power factor of each load? What is the overall power factor?

1.9. In Exercise 1.8 three Δ-connected capacitors are connected in parallel with the two loads in order to improve the power factor to 0.8 lagging. What must be the impedance of each capacitor? What is the size of each capacitor if the line frequency is 60 Hz?

1.10. A balanced three-phase Δ-connected source has a line voltage of 120 V. It is connected to a load via a three-wire transmission line. The per-phase impedance of the Δ-connected load is $30 + j120\ \Omega$. The impedance of each wire of the transmission line is $2 + j4\ \Omega$. How much power is dissipated by the load? By the transmission line? What is the power factor of the load? What is the overall power factor?

1.5 Measurement of Power

The last item on our agenda is to review the measurement of power in dc circuits, single-phase circuits, and three-phase circuits. Let us examine each case separately.

Measurement of Power in a Direct-Current Circuit

Since both the voltage and the current in a dc circuit are constant, the power supplied by the source or dissipated by the load is also constant. If a source of V volts supplies a current of I amperes to a load, the power supplied by the source is equal to the product VI. Therefore, we can determine the power in a dc circuit by measuring the voltage and the current by means of a voltmeter and an ammeter, respectively. The following example demonstrates how to compute power in a dc circuit.

EXAMPLE 1.5

Consider the circuit shown in Figure 1.14**a**. Calculate the power dissipated by the 25-Ω resistor. Draw a schematic that enables you to measure the power. What are the readings on the voltmeter and the ammeter?

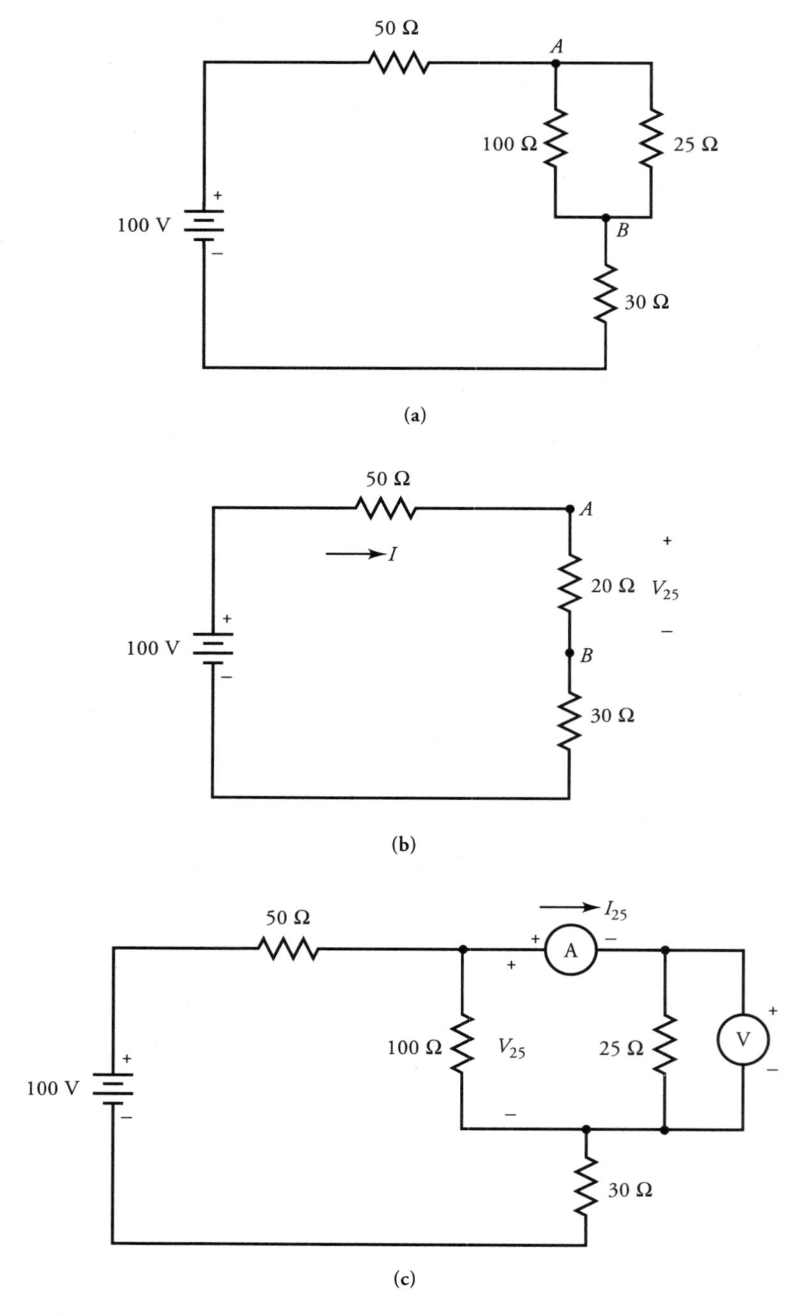

Figure 1.14 **(a)** Circuit for Example 1.5. **(b)** An equivalent circuit. **(c)** Measurements of current through and voltage drop across the 25-Ω resistor.

● SOLUTION

Since the 25-Ω resistor is in parallel with a 100-Ω resistor, we combine them to obtain an equivalent resistance of 20 Ω, as shown in Figure 1.14**b**. The total resistance in the circuit is 100 Ω. Thus, the current supplied by the source is 1 A.

The schematic that enables us to measure the power dissipated by the 25-Ω resistance is given in Figure 1.14**c**. The voltage drop across the 25-Ω resistance is the same as that across the equivalent resistance of 20 Ω. That is, $V_{25} = 1 \times 20 = 20$ V. Therefore, the voltmeter reading is 20 V.

The current through the 25-Ω resistor is $I_{25} = (1 \times 20)/25 = 0.8$ A. Hence, the ammeter reads 0.8 A.

The power, being the product of measured values of the voltage and the current, is $P_{25} = 20 \times 0.8 = 16$ W. Thus, the power dissipation by the 25-Ω resistor is 16 W. ■

The Wattmeter

A wattmeter is a single instrument that performs the combined functions of an ammeter and a voltmeter. It is calibrated to read the average power directly. It is used to measure the average power in ac circuits. The coil that measures the current is called the **current coil** (CC), and the coil that measures the voltage is known as the **potential coil** or the **voltage coil** (VC). The like polarity terminals of the current and the voltage coils are marked either with dots (•) or ± signs. The significance of these markings is that the current must either enter or leave the like polarity terminals at any time.

If the average value of the ac power is equal to the constant value of a dc power, the wattmeter readings are the same. This permits us to calibrate a wattmeter on direct current and then use it on alternating current.

Figure 1.15 shows two possible ways to connect the wattmeter in a circuit properly. In practice, the current coil has a very small resistance and the potential coil carries a very small current. Therefore, each coil dissipates some power, however small it may be. If the wattmeter is connected as shown in Figure 1.15**a**, the wattmeter measurement also includes the power dissipated by the potential coil. Therefore, even when the load is disconnected completely, the wattmeter still shows the power dissipated by its potential coil. In order to measure the small amount of power supplied to the load accurately, we must subtract the power loss in its potential coil. If we use the connection as shown in Figure 1.15**b**, our power measurement now includes the power loss in the current coil. In our calculations, however, we always assume an ideal wattmeter; that is, the resistance of the current coil is zero and the current through the potential coil is vanishingly small.

Measurement of Power in a Single-Phase Circuit

In an ac circuit, both the voltage $v(t)$ and the current $i(t)$ pulsate with time, as depicted in Figure 1.3, where $i(t)$ is shown lagging $v(t)$ by an angle θ. Because the

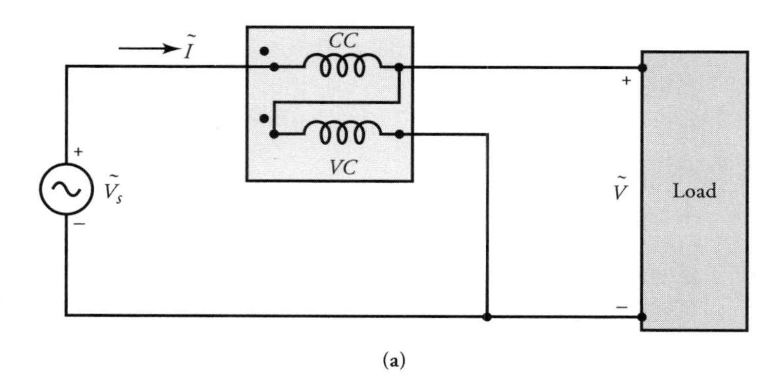

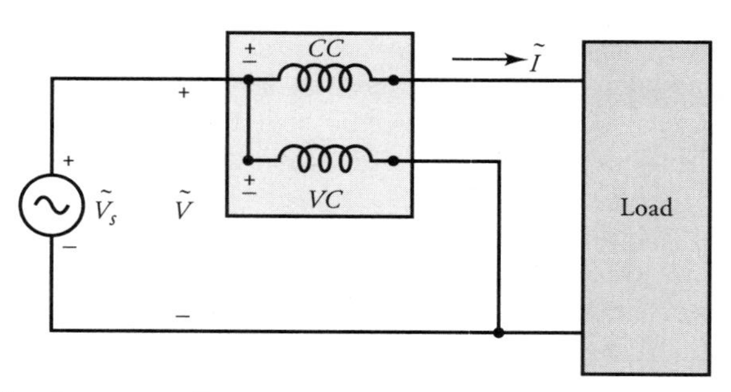

Figure 1.15 Wattmeter connections. (b)

instantaneous power is $p(t) = v(t)\ i(t)$, it pulsates twice as fast. Note that the instantaneous power is positive when $v(t)$ and $i(t)$ are both positive or both negative. The instantaneous power is negative only when one of them (voltage or current) is positive and the other is negative. Therefore, the power becomes negative twice per cycle as shown in Figure 1.3.

The wattmeter is calibrated to read the average value of the power. If V is the rms value of the voltage across the potential coil of the wattmeter, I is the rms value of the current through the current coil, and θ is the phase shift between the two, the wattmeter reading is

$$P = VI \cos \theta \tag{1.20a}$$

In terms of phasors we can express Eq. (1.20a) as

$$P = \text{Re}[\tilde{V}\tilde{I}^*] \tag{1.20b}$$

Equation (1.20b) allows us to determine the average power supplied by the source or dissipated by the load in terms of phasor quantities. Equation (1.20a), however, permits us to measure the power factor (pf = cos θ) of the circuit by means of a wattmeter, voltmeter, and ammeter. The following example demonstrates how to do so.

EXAMPLE 1.6

Two loads are connected in parallel via a transmission line to a 117-V, single-phase ac generator as shown in Figure 1.16. A wattmeter, voltmeter, and ammeter are connected at the load site to obtain the power consumed by the load and its power factor. Determine (a) the readings on the three meters and (b) the power factor of the load.

● SOLUTION

The two impedances in parallel can be replaced by an equivalent impedance

$$\hat{Z}_L = \frac{40 \times j30}{40 + j30} = 24\underline{/53.13^\circ} = 14.4 + j19.2\ \Omega$$

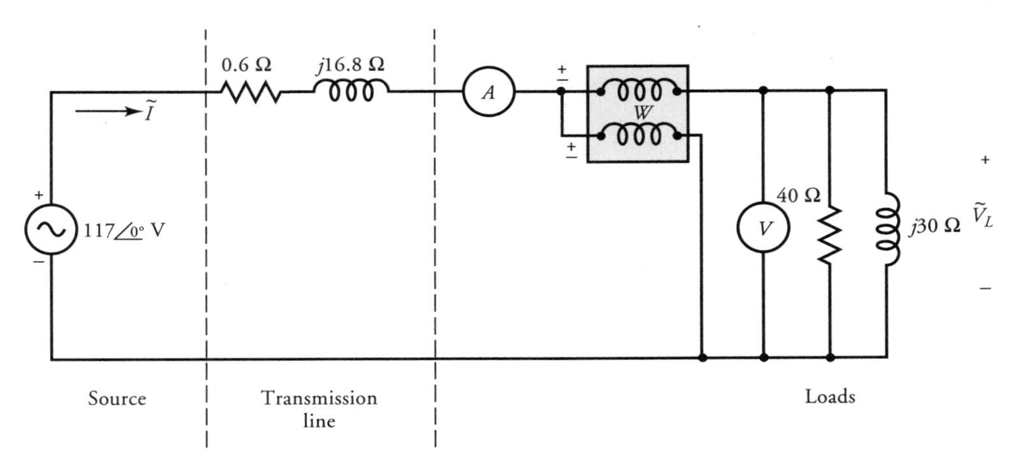

Figure 1.16 Two parallel loads connected to a source via a transmission line.

The total impedance in the circuit is

$$\tilde{Z} = 14.4 + j19.2 + 0.6 + j16.8 = 15 + j36\ \Omega$$

Thus, the current supplied by the source is

$$\tilde{I} = \frac{117}{15 + j36} = 3\underline{/-67.38^\circ}\ \text{A}$$

The ammeter reading is 3 A and the current through the current coil in phasor form is $3\underline{/-67.38^\circ}$ A. The voltage drop across the load is

$$\tilde{V}_L = \tilde{I}\hat{Z}_L = [3\underline{/-67.38^\circ}][14.4 + j19.2] = 72\underline{/-14.25^\circ}\ \text{V}$$

The voltmeter reads 72 V, and the phasor voltage across the potential coil of the wattmeter is $72\underline{/-14.25^\circ}$ V.

The reading on the wattmeter is

$$\begin{aligned} P &= \text{Re}[(72\underline{/-14.25^\circ})(3\underline{/67.38^\circ})] \\ &= \text{Re}[216\underline{/53.13^\circ}] = 129.6\ \text{W} \end{aligned}$$

Note that P does not account for the power loss on the transmission line. The power factor of the load is

$$\text{pf} = \cos\theta = \frac{P}{V_L I} = \frac{129.6}{72 \times 3} = 0.6\ \text{(lag)}$$

The corresponding phase angle is -53.13°. The negative sign accounts for the fact that the load current $\tilde{I}$ lags the load voltage $\tilde{V}_L$.

■

Measurement of Power in a Three-Phase Circuit

When the neutral point of the source is connected to the neutral point of the load by a wire known as the **neutral wire**, the circuit is usually referred to as a **three-phase four-wire** system. In this case, the power in each phase can be measured by connecting a wattmeter in that phase as shown in Figure 1.17.

The readings on the three wattmeters—W_1, W_2, and W_3—are

$$P_1 = \text{Re}[\tilde{V}_{an}\tilde{I}^*_{aA}] \tag{1.21a}$$

$$P_2 = \text{Re}[\tilde{V}_{bn}\tilde{I}^*_{bB}] \tag{1.21b}$$

and

$$P_3 = \text{Re}[\tilde{V}_{cn}\tilde{I}^*_{cC}] \tag{1.21c}$$

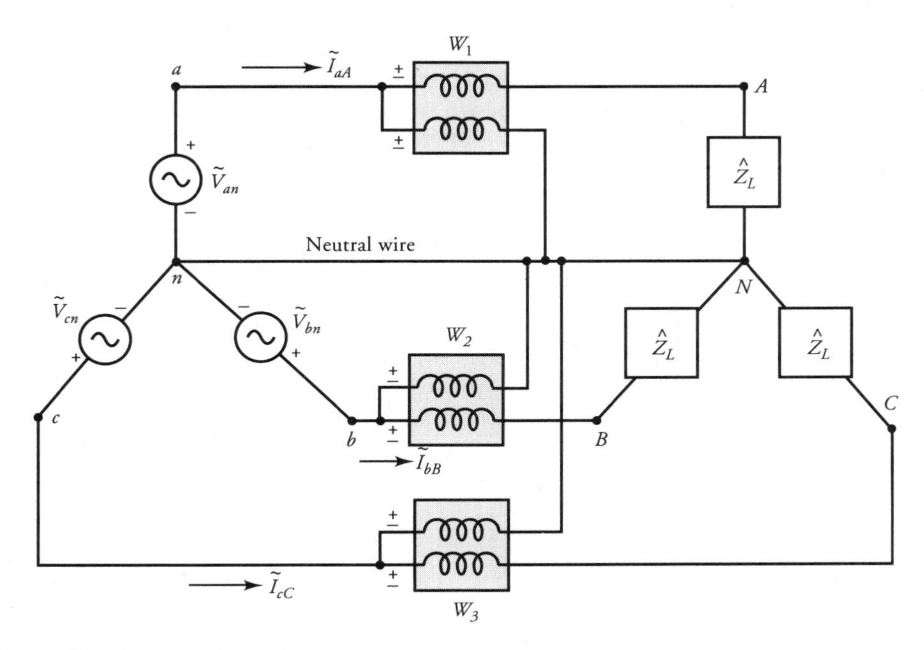

Figure 1.17 A three-phase four-wire system.

respectively. The total power is the sum of the three wattmeter readings. In a balanced three-phase system, all wattmeters record the same reading. This method is suitable only when the load and the source are both Y-connected.

Two-Wattmeter Method

When the source and/or the load is Δ-connected or we do not have access to the neutral wire, the circuit is said to represent a **three-phase three-wire** system and the power is usually measured by means of two wattmeters as shown in Figure 1.18**a**. In this case, the total power is equal to the algebraic sum of the two wattmeter readings whether the three-phase system is balanced or not. For the connections shown in Figure 1.18**a**, the readings on wattmeters W_1 and W_2 are

$$P_1 = \text{Re}[\tilde{V}_{ac}\tilde{I}^*_{aA}] \tag{1.22a}$$

and

$$P_2 = \text{Re}[\tilde{V}_{bc}\tilde{I}^*_{bB}] \tag{1.22b}$$

respectively.

Let us now consider a balanced three-phase system with positive phase sequence. The phasor diagram for the voltages and the currents in the system is

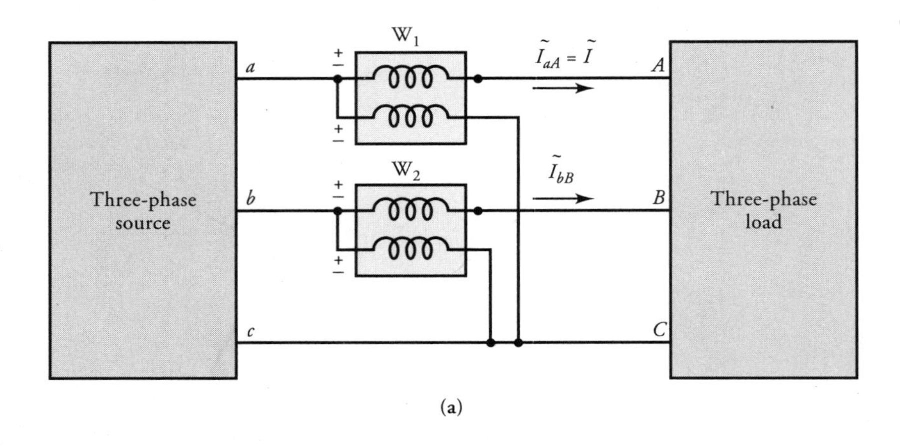

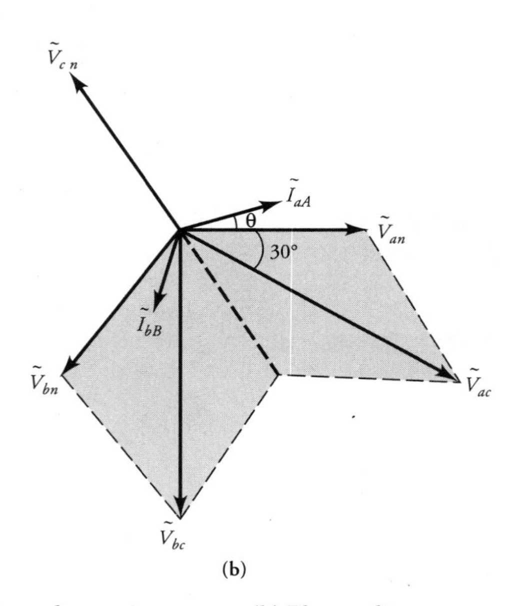

Figure 1.18 **(a)** Three-phase three-wire system. **(b)** Phasor diagram.

shown in Figure 1.18**b**, where the phase voltage $\tilde{V}_{an} = V\angle 0^\circ$ has been assumed as the reference voltage. The two wattmeter readings become

$$P_1 = \mathrm{Re}[(\sqrt{3}V\angle -30^\circ)(I\angle -\theta)] = \sqrt{3}VI\cos(30^\circ + \theta) \tag{1.23a}$$

$$P_2 = \mathrm{Re}[(\sqrt{3}V\angle -90^\circ)(I\angle -\theta + 120^\circ)] = \sqrt{3}VI\cos(30^\circ - \theta) \tag{1.23b}$$

Thus, the total power is

$$P = P_1 + P_2 = \sqrt{3}VI[\cos(30^\circ + \theta) + \cos(30^\circ - \theta)]$$

$$= 3VI\cos\theta \tag{1.24a}$$

$$= \sqrt{3}V_\ell I_\ell \cos\theta \tag{1.24b}$$

where V_ℓ is the magnitude of the line voltage between any two lines and I_ℓ is the line current. Note that $VI\cos\theta$ is the power in each phase of a balanced three-phase system. Equation (1.24b) is given in terms of line currents and line voltages and is valid for both Δ- and Y-connected systems. In a Y-connected system, $I_\ell = I$ and $V_\ell = \sqrt{3}V$. On the other hand, $I_\ell = \sqrt{3}I$ and $V_\ell = V$ in a Δ-connected system.

From Eqs. (1.23a) and (1.23b) we can make the following observations:

(a) The two wattmeter readings are equal when $\theta = 0$ or the power factor is 100%. In this case, the load is purely resistive.
(b) Wattmeter W_1 reads zero when $\theta = 60^\circ$ or the power factor is 50% leading. For all leading power factors below 50%, wattmeter W_1 reads negative.
(c) Wattmeter W_2 reads zero if $\theta = -60^\circ$ or the power factor is 50% lagging. For all lagging power factors below 50%, wattmeter W_2 reads negative.
(d) For both leading and lagging power factors above 50% but below 100% both wattmeter readings are positive but not equal.

In order to read the negative power using deflection-type wattmeters, we usually reverse the connections to the current coil. Electronic wattmeters are designed to show the minus sign to indicate negative power.

From Eqs. (1.23a) and (1.23b) we can compute the reactive (or quadrature) power Q, and the power factor angle θ as

$$Q = \sqrt{3}(P_2 - P_1) \tag{1.25a}$$

and

$$\theta = \tan^{-1}\left[\frac{\sqrt{3}(P_2 - P_1)}{P_2 + P_1}\right] \tag{1.25b}$$

Since the two wattmeters can be connected between any two lines of a three-phase system, we can express Eq. (1.25b) in a general form as

$$\theta = \tan^{-1}\left[\frac{\sqrt{3}P_d}{P_s}\right] \tag{1.25c}$$

where P_d is the algebraic difference of the two wattmeter readings and P_s is the algebraic sum. The sign of the power factor angle θ can be easily verified from the type of load. The power factor angle θ must be negative for an inductive load $(R + jX)$ and positive for a capacitive load $(R - jX)$.

EXAMPLE 1.7

A balanced three-phase, 1351-V, 60-Hz, Δ-connected source with a negative phase sequence feeds a balanced Y-connected load with a per-phase impedance of $360 + j150\ \Omega$ as shown in Figure 1.19**a**. What are the readings on the two wattmeters? Compute the total power and the power factor of the load.

● SOLUTION

By transforming a Δ-connected source into an equivalent Y-connected source, we obtain the magnitude of the phase voltage as

$$V = \frac{1351}{\sqrt{3}} = 780 \text{ V}$$

If we assume phase-*a* voltage as the reference, we can draw an equivalent circuit on a per-phase basis, as depicted in Figure 1.19**b**. The current in the circuit is

$$\tilde{I} = \frac{780}{360 + j150} = 2\underline{/-22.62^\circ} \text{ A}$$

Since the current lags the applied voltage by $-22.62°$, the power factor angle is $-22.62°$.

The phasor diagram for the phase and line voltages and currents is sketched in Figure 1.19**c** for the negative phase-sequence. Note that $\tilde{I}$ is equal to the line current $\tilde{I}_{aA}$.

The reading on wattmeter W_1 is

$$P_1 = \text{Re}[\tilde{V}_{ab}\tilde{I}^*_{aA}] = \text{Re}[(1351\underline{/-30^\circ})(2\underline{/22.62^\circ})] = 2679.62 \text{ W}$$

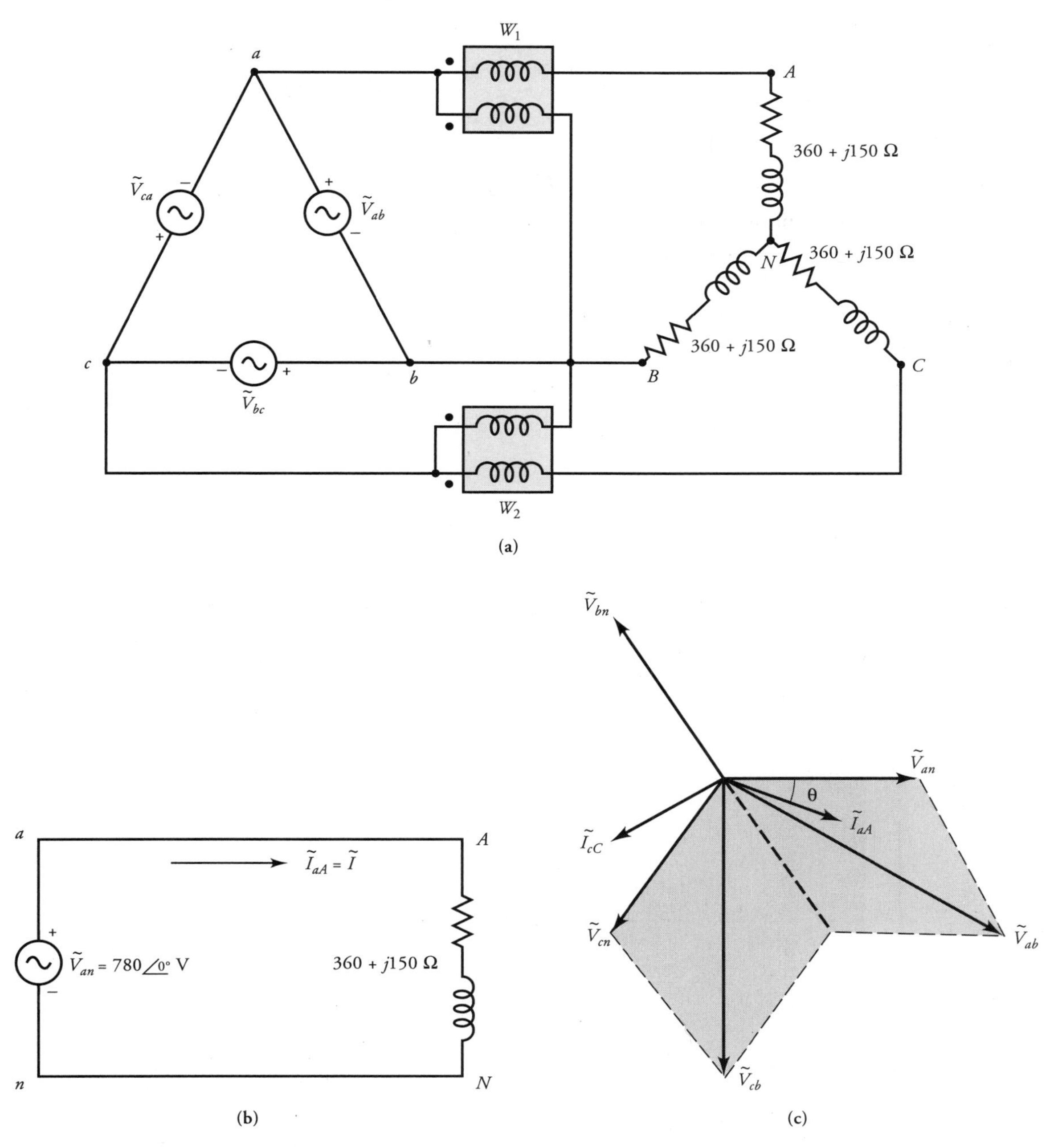

Figure 1.19 **(a)** Circuit for Example 1.7, **(b)** Per-phase equivalent circuit and **(c)** Phasor diagram.

The reading on wattmeter W_2 is

$$P_2 = \text{Re}[\tilde{V}_{cb}\tilde{I}^*_{cC}] = \text{Re}[(1351\underline{/-90°})(2\underline{/142.62°})] = 1640.38 \text{ W}$$

The total power delivered to the load is

$$P = 2679.62 + 1640.38 = 4320 \text{ W}$$

Because the readings on both the wattmeters are positive, the power factor must be greater than 50%. In addition, the power factor must be lagging because the load is inductive.

Even though we already know the power factor angle, we can verify it by using Eq. (1.25b):

$$\theta = \tan^{-1}\left[\frac{\sqrt{3}(-1039.24)}{4320}\right] = -22.62°$$

Thus, the power factor of the load is

$$\text{pf} = \cos\theta = \cos(22.62°) = 0.923, \text{ or } 92.3\% \text{ (lag)}$$

We can also verify the total power consumed by the load as

$$P = 3I^2R = 3 \times 2^2 \times 360 = 4320 \text{ W}$$

■

EXERCISES

1.11. Consider the circuit shown in Figure E1.11. Calculate the power dissipated by the 10-Ω resistor. Draw a schematic that enables you to measure the power. What are the readings on the voltmeter and the ammeter?

1.12. An equivalent circuit of a 120-V, 60-Hz, single-phase induction motor is given in Figure E1.12. Sketch a circuit that assists you in measuring the power input to the motor, the current intake by the motor, and the power factor of the motor. Find the reading on each meter.

1.13. The current, voltage, and power to a capacitive circuit are measured and found to be 3 A, 450 V, and 810 W, respectively. Determine (a) the series equivalent circuit and (b) the parallel equivalent circuit.

1.14. Three identical impedances of $2373 + j1500\ \Omega$ are Δ-connected to a 1732-V, 60-Hz, Δ-connected balanced source via a three-wire transmission line. Each line can be represented by an equivalent resistance of 75 Ω as shown in Figure E1.14. What must be the readings on the two wattmeters, the

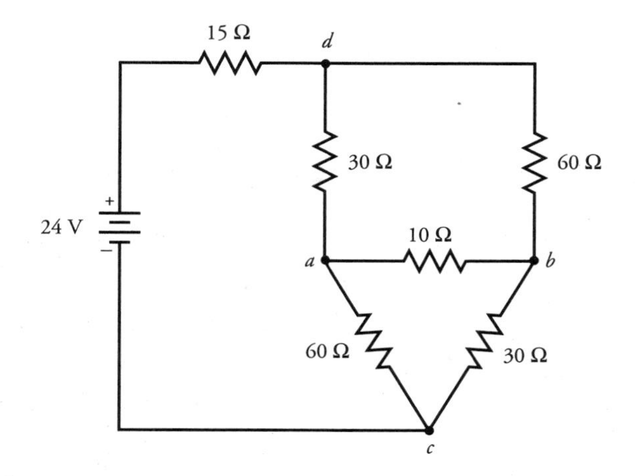

Figure E1.11

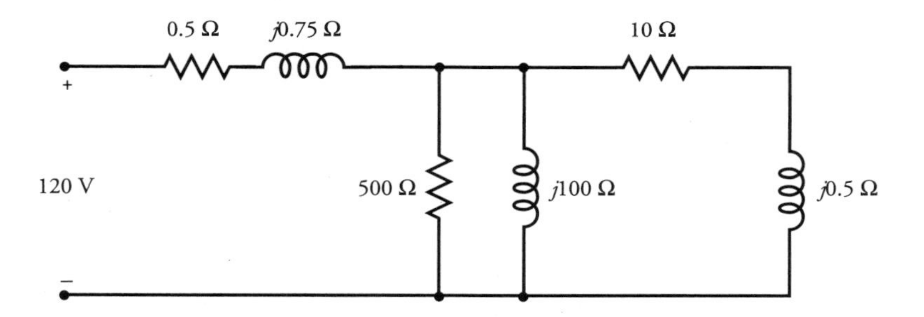

Figure E1.12

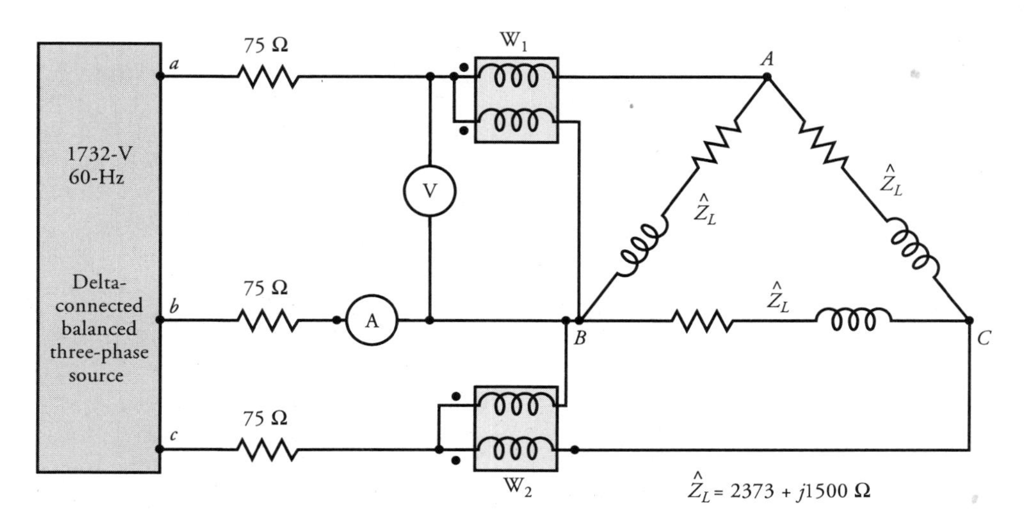

Figure E1.14

voltmeter, and the ammeter? What is the power factor of the load? Assume that the phase sequence is positive.

1.15. The current, voltage, and power factor of a balanced three-phase load are measured and found to be 8.66 A, 208 V, and 0.8 lagging, respectively. If the load is Y-connected, find (a) the series equivalent circuit and (b) the parallel equivalent circuit.

1.16. Repeat Exercise 1.15 for a Δ-connected load.

SUMMARY

In this chapter we reviewed some of the techniques used in analyzing electric circuits. We first stated experimental laws of electric circuit theory and then explained how to apply them in order to solve dc circuits, single-phase circuits, and three-phase circuits.

The two important theorems, Thevenin's theorem and the maximum power transfer theorem, were stated and applied.

We reviewed the concept of phasors and explained how a voltage and current can be represented in phasor form. Choosing one phasor as a reference, we demonstrated how a phasor diagram can be drawn and used. The concepts of complex power, real power, and reactive power were reviewed. The passive technique of power factor correction using capacitors was highlighted.

Students are usually overwhelmed by three-phase circuits. There seems to be something about the three-phase circuits that baffles them. We have attempted to take the mystery out of three-phase circuits by explaining the positive and negative phase-sequences, the difference between Y- and Δ-connected circuits, and Δ-to-Y transformations.

Our experience over the years has indicated that students tend to get perplexed when it comes to connecting a wattmeter in the circuit. By explaining how a wattmeter works, we aimed to overcome apprehension about this measuring device. We also shed some light on the two-wattmeter method to measure total power in a three-phase circuit.

Some of the important equations presented in this chapter are summarized below for easy reference.

Ohm's law: $\tilde{V} = \tilde{I}\hat{Z}$

Kirchhoff's current law: $\sum_{i=1}^{n} \tilde{I}_i = 0$

Kirchhoff's voltage law: $\sum_{i=1}^{n} \tilde{V}_i = 0$

Complex power: $\hat{S} = \tilde{V}\tilde{I}^* = P + jQ$

Maximum power transfer theorem: $\hat{Z}_L = \hat{Z}_T^*$

Balanced three-phase system

Positive phase-sequence: $\tilde{V}_{an} = V\angle\theta$, $\tilde{V}_{bn} = V\angle\theta - 120°$, $\tilde{V}_{cn} = V\angle\theta + 120°$

Negative phase-sequence: $\tilde{V}_{an} = V\angle\theta$, $\tilde{V}_{bn} = V\angle\theta + 120°$, $\tilde{V}_{cn} = V\angle\theta - 120°$

Y-connected source: $\tilde{V}_\ell = \sqrt{3}\tilde{V}\angle\pm 30°$ (+ for positive phase-sequence)

$\tilde{I}_\ell = \tilde{I}$

Δ-connected source: $\tilde{V}_\ell = \tilde{V}$

$\tilde{I}_\ell = \sqrt{3}\tilde{I}\angle\pm 30°$ (− for positive phase-sequence)

Three-phase power: $P = 3VI \cos\theta = \sqrt{3}V_\ell I_\ell \cos\theta$

Review Questions

1.1. Electrons are passing a reference point on a conductor at the rate of 3×10^{23} per hour. What is the current in amperes?

1.2. A 60-W bulb carries a current of 0.5 A when connected to a dc source. What is the source voltage? What is the total charge that passes through its filament in 2 hours?

1.3. A 100-Ω resistor is rated at 0.5 W. What is the maximum voltage it can safely withstand? What is the maximum current that can safely flow through it? Can it safely sustain the maximum current when the maximum voltage is applied across it?

1.4. If we expend 1 joule of energy in moving 1 coulomb of charge from one point to the other in a circuit, the potential difference between the two points is ______.

1.5. State the differences between a linear and a nonlinear circuit.

1.6. Can we apply Ohm's law to nonlinear circuits? Justify your answer.

1.7. State Kirchhoff's current and voltage laws. What does the term "algebraic sum" mean?

1.8. The current through a resistance when connected to a 24-V battery is 1.2 A. Determine the resistance that must be inserted in series to reduce the current to 0.8 A.

1.9. A dc generator can be represented by a voltage source in series with a resistance. The open-circuit (no-load) voltage of the generator is 124 V. A load resistance of 24 Ω draws a current of 5 A. What is the internal resistance of the generator?

1.10. A dc generator delivers power to a load over a 1.25-Ω transmission line. The load voltage is 220 V. If the voltage drop in the line is not to exceed 10% of the load voltage, what is the maximum power that can be delivered to the load?

1.11. Two parallel resistors carry currents of 2 A and 5 A. If the resistance of one is 100 Ω, what is the resistance of the other?

1.12. How many different values of resistances can be obtained with three resistors of 1 kΩ, 5 kΩ, and 200 Ω?

1.13. A 1200-W electric heater operates on a 120-V, 60-Hz ac supply. What is the rms current through the heater? What is the maximum instantaneous current? What is the resistance of the heater? Sketch $v(t)$ and $i(t)$ curves as a function of time.

1.14. A series circuit consists of a resistance of 10 Ω, an inductance of 0.5 mH, and a capacitance of 200 μF. What is the impedance of the circuit at a frequency of 60 Hz? Is the impedance inductive or capacitive?

1.15. A $120\underline{/-30^\circ}$ V voltage source supplies a current of $2\underline{/-20^\circ}$ A to a load. Is the load inductive or capacitive? What is the power factor of the load? Compute the average power, reactive power, and apparent power delivered to the load.

1.16. An ac circuit can be represented by a voltage source of $25\underline{/45^\circ}$ V in series with an impedance of $15 + j20$ Ω. In order to transfer maximum power to the load, what must be the load impedance? What is the current in the circuit? How much power is transferred to the load?

1.17. The phase-b voltage of a three-phase balanced source is $\tilde{V}_{bn} = 120\underline{/-60^\circ}$ V. What are the other phase voltages for (a) a positive phase-sequence and (b) a negative phase-sequence?

1.18. The line voltage of a balanced three-phase source is $\hat{V}_{cb} = 17.32\underline{/-30^\circ}$ kV. What are the phase and line voltages if the phase sequence is positive?

1.19. What are the differences between apparent power, real power, and reactive power?

1.20. Why do we express apparent power in VA, real power in W, and reactive power in VAR?

1.21. A student suggested that she can improve the power factor by connecting a resistor instead of a capacitor in parallel with an inductive load. What do you think? What are its ramifications?

1.22. In the power factor correction techniques, we always use capacitors in parallel with the load. Why can't we use capacitors in series with the load?

1.23. Why is the power factor correction necessary?

1.24. Is it more economical to use a Δ-connected capacitor bank instead of a Y-connected capacitor bank for power factor correction in three-phase circuits?

1.25. A student suggests that the power factor of a load can be determined by using the two wattmeter readings as

$$\cos\theta = \left[\frac{P_1 + P_2}{\sqrt{3}V_\ell I_\ell}\right]$$

Under what conditions does the above equation yield the exact power factor?

Problems

1.1. An electric heater takes 576 W when connected to a dc source of a certain voltage. If the voltage is increased by 10%, the current through the heater is 5.28 A. Compute (a) the original voltage and (b) the resistance of the heater.

1.2. A dc motor draws 20 A from a 120-V dc supply. The motor operates 10 hours each day. If the energy cost is 10 cents per kilowatt-hour (kWh), what is the monthly bill for operating this motor? Assume 30 days in a month.

1.3. A remote dc generator (sending end) transmits power over a transmission line having a resistance of 0.2 Ω to a 10-kW load (receiving end). The load voltage is 125 V. What is the voltage at the sending end? How much power is lost by the transmission line?

1.4. Obtain the current in each branch of the circuit shown in Figure P1.4 using (a) node-voltage method and (b) mesh-current method.

1.5. In the circuit of Figure P1.5, what value of R_L dissipates a maximum power from the circuit? What is the maximum power dissipated by it?

1.6. What is the power dissipated by each resistor in Figure P1.6? How much power is supplied by each source?

1.7. Use the superposition theorem to find the current in the 10-Ω resistance of the circuit given in Figure P1.7.

1.8. Find the current through and the voltage drop across each element in the circuit shown in Figure P1.8, using (a) node-voltage method and (b) mesh-current method. What must be the power rating of each resistor?

1.9. Use the superposition theorem to find the current through the 200-Ω resistance in the circuit given in Figure P1.8. How much power is dissipated by it?

1.10. An equivalent circuit of a transformer is given in Figure P1.10. If the load impedance is $200 + j300$ Ω and the load voltage is 120 V, find the applied voltage $\tilde{V}_S$. Calculate the power output to the load and power supplied by

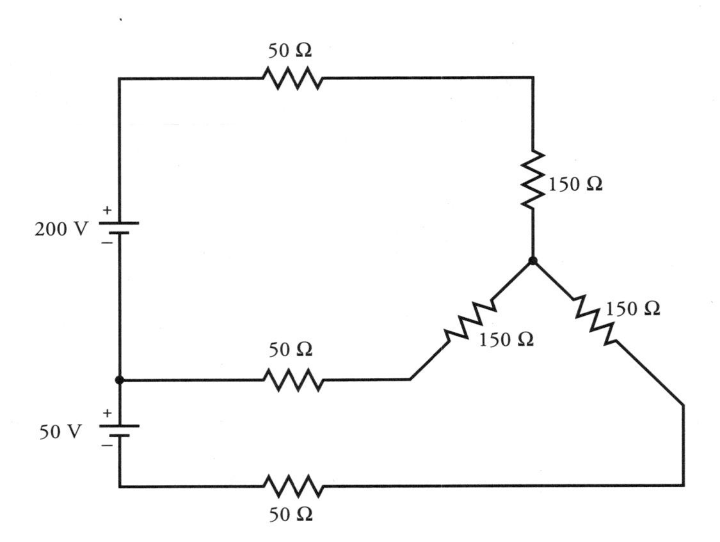

Figure P1.4

the source. What is the efficiency of the transformer? [*Hint:* Efficiency is defined as the ratio of the output power to the input power.]

1.11. A 120-V, 60-Hz source is applied across an inductive load of $100 + j700\ \Omega$. What is the power factor of the load? What must be the size of the capacitor that can be placed in parallel with the load in order to improve the power factor to 0.95 lagging? Draw the phasor diagram to illustrate the improvement in the power factor.

1.12. A certain load takes 10 kW at 0.5 pf lagging when connected to a 230-V, 50-Hz source. The supply company charges a penalty when the power factor falls below 0.8. What size of capacitor must be used in order to avoid the penalty?

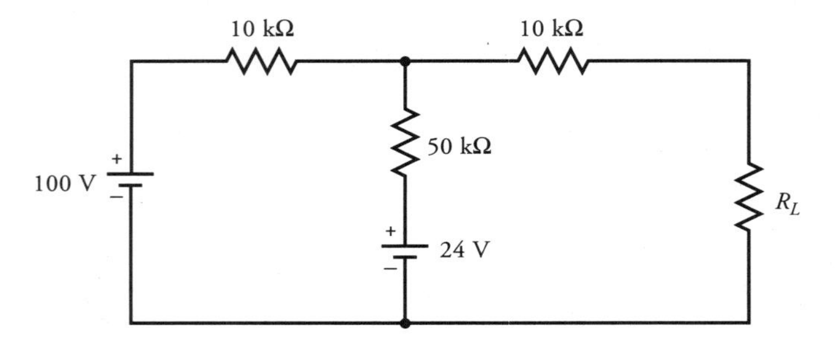

Figure P1.5

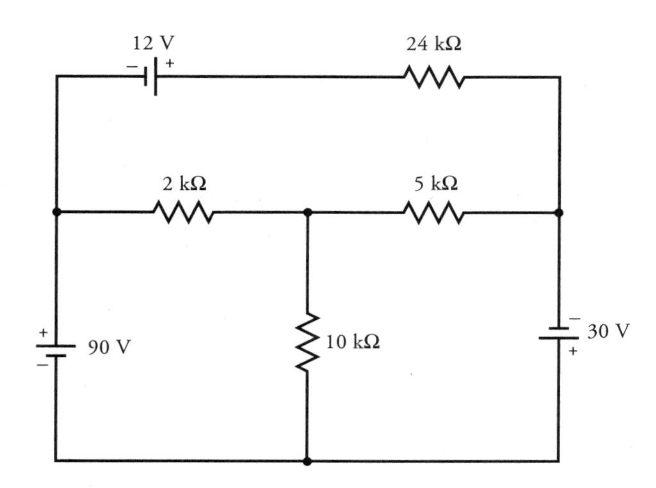

Figure P1.6

1.13. An electric heater takes 1000 W at unity power factor from a 120-V, 60-Hz supply. When a single-phase induction motor is connected across the supply, the power factor of the total load becomes 0.8 lagging. If the power requirement of the motor is 1750 W, what is its power factor? What size of capacitor must be connected to improve the power factor to 0.9 lagging? Draw the phasor diagram in each case.

1.14. A single-phase generator is at a distance of 100 km from the load. The impedance of the transmission line is $2 + j4$ mΩ per kilometer. The load requires 10 kW at 860 V. Find the voltage of the generator and the power loss in the line when the power factor is (a) unity, (b) 0.8 lagging, and (c) 0.8 leading. Draw the phasor diagram in each case.

1.15. The current, voltage, and power to an inductive load are measured and

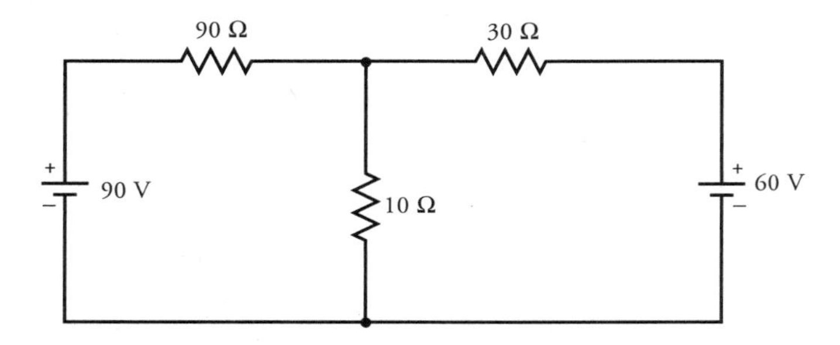

Figure P1.7

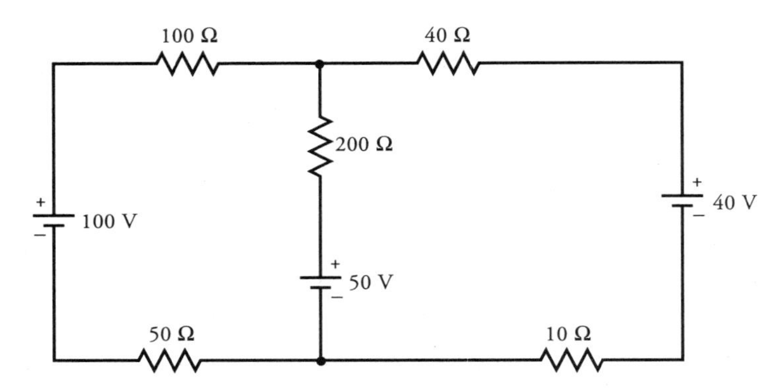

Figure P1.8

found to be 10 A, 120 V, and 800 W, respectively. Represent the load by (a) a series equivalent circuit and (b) a parallel equivalent circuit.

1.16. The current, voltage, and power factor of a load are measured and found to be 5 A, 230 V, and 0.8 leading. Represent the load by (a) a series equivalent circuit, and (b) a parallel equivalent circuit.

1.17. A wattmeter is connected in a circuit as shown in Figure P1.17. What must be the reading on the wattmeter?

1.18. A 220-V, 60-Hz, Y-connected, balanced three-phase source with a positive phase-sequence supplies an unbalanced load as shown in Figure P1.18. What is the current in the neutral wire? What is the total power dissipated by the load? Draw a schematic that helps measure the power requirement of each phase.

1.19. If the neutral wire in Figure P1.18 is disconnected, what must be the current in each phase? What are the three-phase voltages? Draw a schematic that

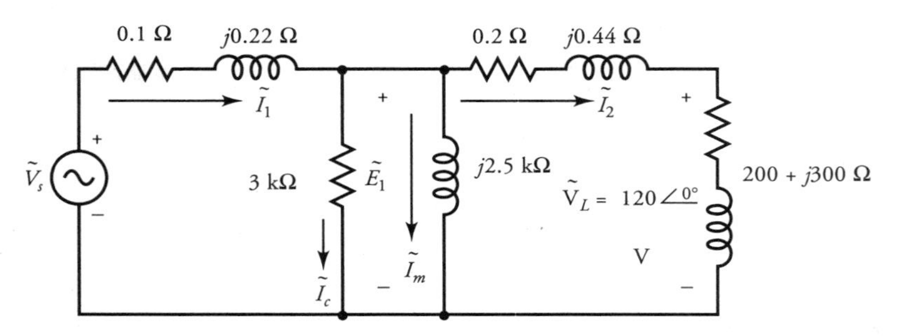

Figure P1.10

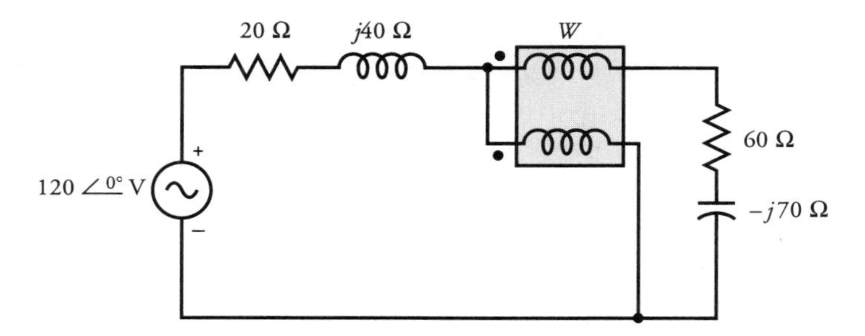

Figure P1.17

enables you to measure the total power requirement. Is any change in the total power required by the load?

1.20. A balanced three-phase load dissipates 12 kW at a power factor of 0.8 lagging from a balanced 460-V, 60-Hz, Y-connected three-phase supply. For a Y-connected load, obtain an equivalent (a) series circuit and (b) parallel circuit.

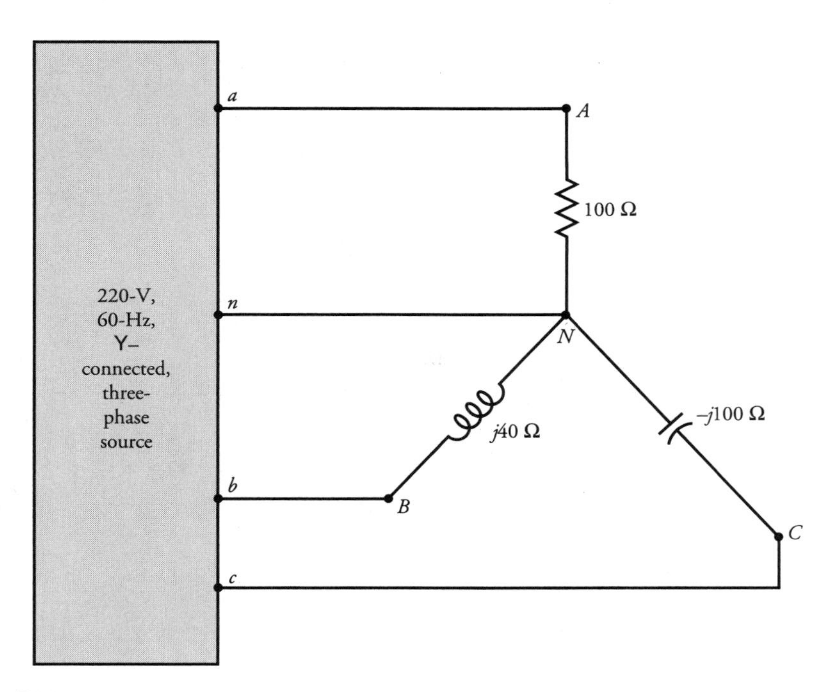

Figure P1.18

1.21. A balanced three-phase load dissipates 15 kW at a power factor of 0.8 leading from a balanced 208-V, 50-Hz, Δ-connected source. For a Δ-connected load, obtain an equivalent (a) series circuit and (b) parallel circuit.

1.22. A balanced three-phase load is connected to a balanced three-phase supply via a transmission line as shown in Figure P1.22. What must be the readings on the two wattmeters, the ammeter, and the voltmeter? What is the total power supplied to the load? Using the wattmeter readings, verify the power factor of the load.

1.23. A clever student played a trick on his partner by connecting an electronic wattmeter as shown in Figure P1.23. What must be the reading on the wattmeter?

1.24. The load on a 460-V, 60-Hz, Y-connected, balanced three-phase source consists of three equal Y-connected impedances of $100 + j100\ \Omega$ and three equal Δ-connected impedances of $300 - j300\ \Omega$. Compute the line current, power, and the power factor for the total load.

1.25. A balanced, Y-connected, three-phase, 230-V, 60-Hz source supplies 600 W to a Y-connected load at 0.7 pf lagging. Three equal Y-connected capacitors are placed in parallel with the load to improve the power factor to 0.85 lagging. What must be the size of each capacitor? What must be the size of each capacitor if they are Δ-connected?

1.26. A Y-connected load with an impedance of $12 - j15\ \Omega$ per phase is connected to a balanced Δ-connected 208-V, 60-Hz, three-phase, positive phase-sequence supply. Determine (a) the line and phase voltages, (b) the line and phase currents, and (c) the apparent, active, and reactive powers in the load. Draw the phasor diagram and the power diagram.

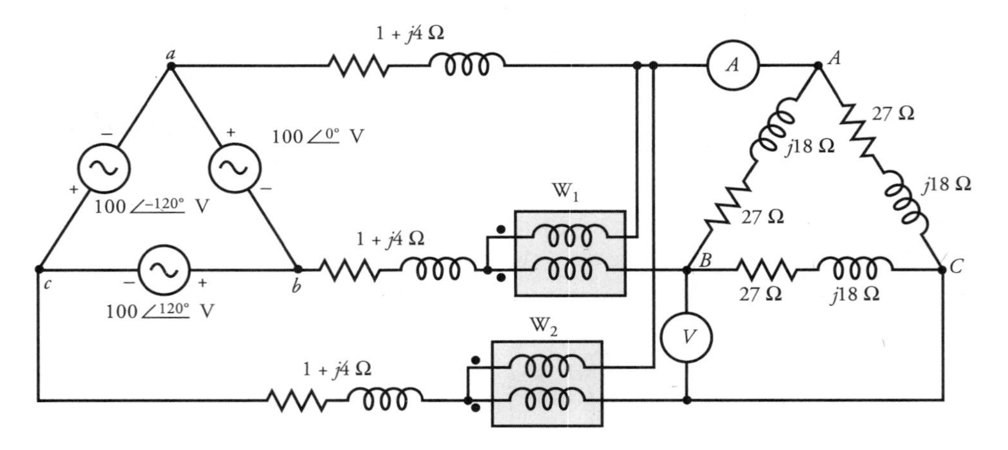

Figure P1.22

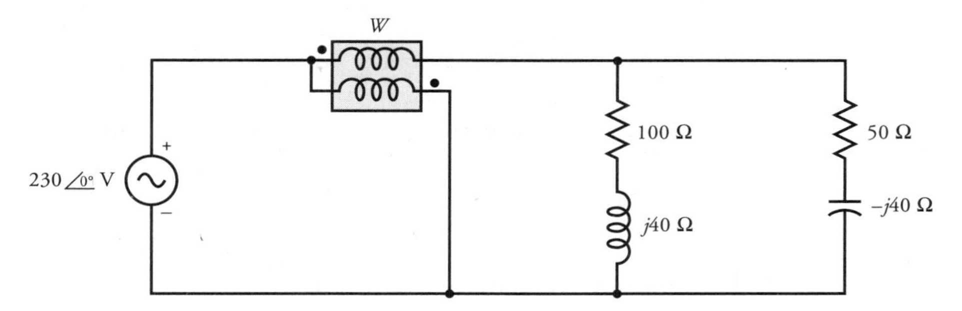

Figure P1.23

1.27. The total power delivered to a balanced, Y-connected, three-phase load is 720 kW at a lagging power factor of 0.8. The line voltage at the load is 3464 V. The impedance of the transmission line is $2.5 + j10.2\ \Omega$. What is the line voltage at the sending end of the transmission line? How much power is dissipated by the transmission line? What is the total power supplied by the source? Determine the load impedance per phase.

1.28. A balanced Y-connected three-phase source supplies power to an unbalanced Δ-connected load. The load impedances per phase are $\hat{Z}_{AB} = 40 + j30\ \Omega$, $\hat{Z}_{BC} = 20\ \Omega$, and $\hat{Z}_{CA} = 10 + j10\ \Omega$. The line voltage is 173.21 V and the phase sequence is positive. Calculate (a) the phase currents, (b) the line currents, (c) the power dissipated in each phase, and (d) the total power supplied by the source. If the current coils of two wattmeters are connected in lines a and b and the potential coils are connected to line c, what must be the reading on each wattmeter? Is the power measured by the two wattmeters the same as that consumed by the unbalanced load?

1.29. The per-phase impedance of a Δ-connected load is $150 + j90\ \Omega$. The power to the load is supplied from a Y-connected source via a transmission line. The impedance per line is $5 + j10\ \Omega$. The line voltage at the load is 600 V. What is the line voltage of the source? What is the total power supplied by the source? If the current coils of two wattmeters are connected in lines a and c and the potential coils are connected to line b, what must be the reading on each wattmeter? Assume positive phase-sequence.

1.30. A balanced three-phase load dissipates 48 kW at a leading power factor of 0.8. The line voltage at the load is 240 V. Represent the load impedance by four equivalent circuits.

1.31. A 1732-V three-phase source delivers power to two three-phase machines connected in parallel. One machine absorbs 240 kVA at a leading power factor of 0.8, and the other machine requires 150 kW at a lagging power factor of 0.707. What is the current supplied by the source? What is the power supplied by the source? What is the overall power factor? Represent each machine by a Y-connected equivalent circuit.

1.32. A 1732-V, 60-Hz, three-phase source delivers power via a transmission line to two machines connected in parallel. The impedance per line is $0.5 + j5\ \Omega$. One machine can be represented by an equivalent Δ-connected per-phase impedance of $150 + j450\ \Omega$. The other machine is Y-connected and has a per-phase impedance of $40 - j320\ \Omega$. Compute the power supplied by the source. What is the power requirement of each machine? How much power is dissipated by the transmission line?

1.33. Repeat Problem 1.32 for a Δ-connected load with a per-phase impedance of 300 Ω connected in parallel with the machines.

1.34. Repeat Problem 1.32 for a Δ-connected capacitor bank with an impedance of $-j1023\ \Omega$ per phase connected in parallel with the machines. What is the size of each capacitance?

CHAPTER 2

Review of Basic Laws of Electromagnetism

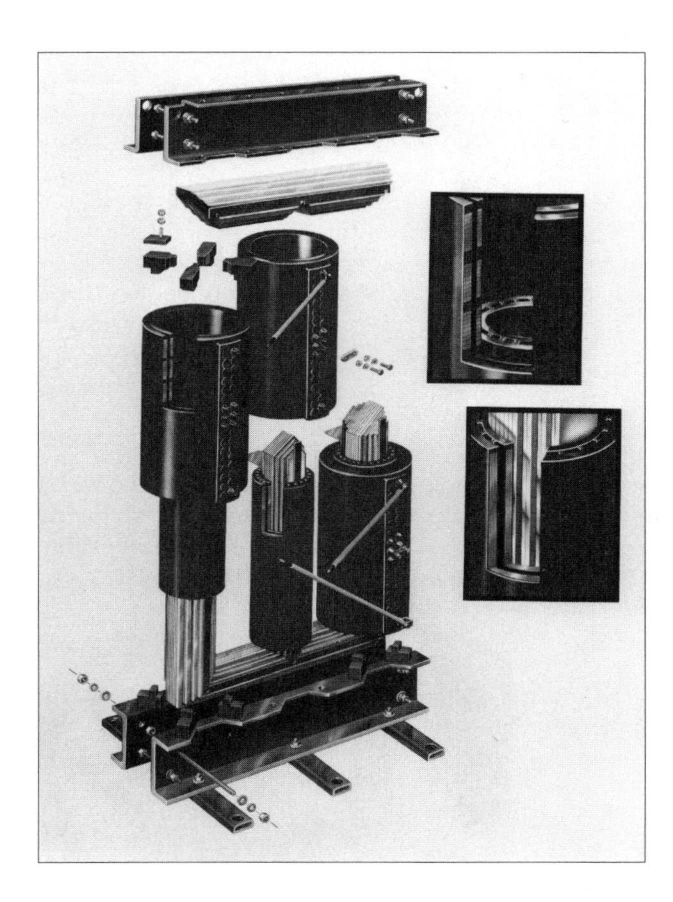

An exploded view of a power transformer, highlighting its magnetic core and cast windings. (*Courtesy of Square D Company*)

2.1 Introduction

We can best understand and predict the behavior and characteristics of an electric machine if we understand not only its physical construction but also the role of the magnetic field in that machine because almost all practical energy conversion devices use magnetic field as a medium. The magnetic field may be set up by a winding (coil) or a permanent magnet. If the magnetic field is produced by a winding, it can be either of constant magnitude (dc) or a function of time (ac).

In transformers, the ac magnetic field helps to transfer energy from the primary (input) side to the secondary (output) side. No electrical connection is needed between the two sides of the transformer. The energy transfer process is based upon the principle of induction.

However, in dc machines and synchronous machines it is the uniform (constant) magnetic field that facilitates the conversion of electric energy to mechanical energy (motor action) or mechanical energy to electrical energy (generator action). In fact, as we will discover, if a motor action exists in a machine, there exists simultaneously a generator action. In other words, one action cannot exist without the other.

It is therefore evident that the study of electric machines requires a basic understanding of electromagnetic fields. The purpose of this chapter is to review the basic laws of electromagnetism, with which the reader is expected to have some familiarity. A detailed discussion of these laws can be found in any book on electromagnetic field theory.

2.2 Maxwell's Equations

The fundamental theory of electromagnetic fields is based on the four Maxwell's equations. These equations are, in fact, generalizations of laws based upon experiments. Our aim in this chapter is not to trace the history of these experiments but to present them in the form that is most useful from the application point of view.

The four Maxwell's equations are:

$$\nabla \times \boldsymbol{E} = -\frac{\partial \boldsymbol{B}}{\partial t} \quad \text{or} \quad \oint_c \boldsymbol{E} \cdot d\boldsymbol{\ell} = -\int_s \frac{\partial \boldsymbol{B}}{\partial t} \cdot d\boldsymbol{s} \tag{2.1a}$$

$$\nabla \times \boldsymbol{H} = \boldsymbol{J} + \frac{\partial \boldsymbol{D}}{\partial t} \quad \text{or} \quad \oint_c \boldsymbol{H} \cdot d\boldsymbol{\ell} = \oint_s \boldsymbol{J} \cdot d\boldsymbol{s} + \int_s \frac{\partial \boldsymbol{D}}{\partial t} \cdot d\boldsymbol{s} \tag{2.1b}$$

$$\nabla \cdot \boldsymbol{B} = 0 \quad \text{or} \quad \oint_s \boldsymbol{B} \cdot d\boldsymbol{s} = 0 \tag{2.1c}$$

$$\nabla \cdot \boldsymbol{D} = \rho \quad \text{or} \quad \oint_s \boldsymbol{D} \cdot d\boldsymbol{s} = \int_v \rho \, dv \tag{2.1d}$$

where $\boldsymbol{E}$ is the electric field intensity in volts/meter (V/m)
$\boldsymbol{H}$ is the magnetic field intensity in amperes/meter (A/m)
$\boldsymbol{B}$ is the magnetic flux density in teslas (T) or webers/meter^2 (Wb/m^2)
$\boldsymbol{D}$ is the electric flux density in coulombs/meter^2 (C/m^2)
$\boldsymbol{J}$ is the volume current density in amperes/meter^2 (A/m^2)
ρ is the volume charge density in coulombs/meter^3 (C/m^3).

In Eqs.(2.1a) and (2.1b) s is the open surface bounded by a closed contour c. However, in Eqs. (2.1c) and (2.1d) v is the volume bounded by a closed surface s.

The above equations are tied together by the law of conservation of charge, which is also known as the **equation of continuity**. That is,

$$\nabla \cdot \boldsymbol{J} = -\frac{\partial \rho}{\partial t} \tag{2.2}$$

In addition to Maxwell's equations and the equation of continuity, we must include the **Lorentz force equation**

$$\boldsymbol{F} = q[\boldsymbol{E} + \boldsymbol{v} \times \boldsymbol{B}] \tag{2.3}$$

which defines the force experienced by a charge q moving with a velocity $\boldsymbol{v}$ through an electric field $\boldsymbol{E}$ and a magnetic field $\boldsymbol{B}$.

We must mention here that $\boldsymbol{E}$ and $\boldsymbol{H}$ are the fundamental fields and $\boldsymbol{D}$ and $\boldsymbol{B}$ are the derived fields. The derived fields are related to the fundamental fields through the following constitutive relations:

$$\boldsymbol{D} = \epsilon \boldsymbol{E} = \epsilon_r \epsilon_0 \boldsymbol{E} \tag{2.4a}$$

$$\boldsymbol{B} = \mu \boldsymbol{H} = \mu_r \mu_0 \boldsymbol{H} \tag{2.4b}$$

where ϵ is the permittivity
μ is the permeability
ϵ_r is the dielectric constant
μ_r is the relative permeability of the medium. ϵ_0 and μ_0 are the permittivity and the permeability of free space. In the International System (SI) of units,

$$\epsilon_0 = 8.854 \times 10^{-12} \approx \frac{10^{-9}}{36\pi} \text{ farad/meter (F/m)} \tag{2.5a}$$

$$\mu_0 = 4\pi \times 10^{-7} \text{ henry/meter (H/m)} \tag{2.5b}$$

Induced Electromotive Force

Equation (2.1a) is a special case of Faraday's law of induction. It represents the induced electromotive force (emf) in a **stationary** closed loop due to a time rate of change of magnetic flux density. That is,

$$e_t = -\int_s \frac{\partial \boldsymbol{B}}{\partial t} \cdot d\boldsymbol{s} \tag{2.6}$$

Since the closed path is stationary, the induced emf is also known as the **transformer emf**. For this reason, Eq. (2.6) is called the **transformer equation** in the integral form.

If a conductor is also moving with a velocity $\boldsymbol{v}$ in a magnetic field $\boldsymbol{B}$, an additional emf will be induced in it, which is given by

$$e_m = \oint_c (\boldsymbol{v} \times \boldsymbol{B}) \cdot d\boldsymbol{\ell} \tag{2.7}$$

and is known as the **motional emf**, or the **speed voltage**. Since this emf is induced by the motion of a conductor in a magnetic field, it is also said to be the emf induced by **flux-cutting action**.

For a closed loop moving in a magnetic field, the total induced emf must be equal to the sum of the transformer emf and the motional emf. That is,

$$e = -\int_s \frac{\partial \boldsymbol{B}}{\partial t} \cdot d\boldsymbol{s} + \int_c (\boldsymbol{v} \times \boldsymbol{B}) \cdot d\boldsymbol{\ell} \tag{2.8a}$$

The above equation is a mathematical definition of **Faraday's law of induction**. It can be written in concise form as

$$e = -\frac{d\Phi}{dt} \tag{2.8b}$$

where

$$\Phi = \int_s \boldsymbol{B} \cdot d\boldsymbol{s} \tag{2.9}$$

is the total flux passing through the loop. Equation (2.8a) or (2.8b) yields the induced emf in a closed loop having only one turn. If there are N turns in the loop, the induced emf is N times as much.

EXAMPLE 2.1

A 1000-turn coil is placed in a magnetic field that varies uniformly from 100 milliwebers (mWb) to 20 mWb in 5 seconds (s). Determine the induced emf in the coil.

● SOLUTION

Change in flux: $\Delta\Phi = 20 - 100 = -80$ mWb
Change in time: $\Delta t = 5$ s

$$\text{Induced emf: } e = -N\frac{d\Phi}{dt} \approx -N\frac{\Delta\Phi}{\Delta t} = -1000 \times \frac{-80 \times 10^{-3}}{5} = 16 \text{ V}$$

■

EXAMPLE 2.2

A square loop with each side 10 cm in length is immersed in a magnetic field intensity of 100 A/m (peak) varying sinusoidally at a frequency of 50 megahertz (MHz). The plane of the loop is perpendicular to the direction of the magnetic field. A voltmeter is connected in series with the loop. What is the reading on the voltmeter?

● SOLUTION

Since the loop is stationary, we can use the transformer equation, Eq. (2.6), to determine the induced emf in the loop. Let us assume that the magnetic field intensity is along the z-axis and the loop is in the xy plane, as shown in Figure 2.1. The magnetic field intensity is

$$\boldsymbol{H} = 100 \sin \omega t \, \boldsymbol{a}_z \text{ A/m}$$

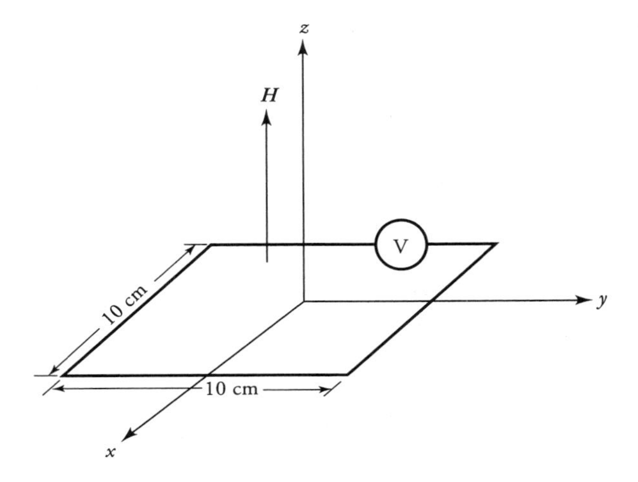

Figure 2.1 Figure for Example 2.1.

where $\omega = 2\pi f$ and $f = 50$ MHz and $\boldsymbol{a}_z$ is the unit vector in the z-direction. Assuming that the permeability of the medium is the same as that of the free space, the magnetic flux density is

$$\boldsymbol{B} = 100\ \mu_0 \sin\ \omega t\ \boldsymbol{a}_z\ \text{T}$$

Since

$$\frac{\partial \boldsymbol{B}}{\partial t} = 100\ \mu_0 \omega \cos\ \omega t\ \boldsymbol{a}_z = 100 \times 4\pi \times 10^{-7} \times 2\pi \times 50 \times 10^6 \cos\ \omega t\ \boldsymbol{a}_z$$

$$= 39{,}478.4 \cos\ \omega t\ \boldsymbol{a}_z$$

and $\quad d\boldsymbol{s} = dx\ dy\ \boldsymbol{a}_z$, the induced emf is

$$e = -39{,}478.4 \cos\ \omega t \int_{-.05}^{.05} dx \int_{-.05}^{.05} dy = -394.784 \cos\ \omega t\ \text{V}$$

Since an ac voltmeter reads only the rms value of a time-varying voltage, the reading on the voltmeter is

$$E = \frac{394.784}{\sqrt{2}} = 279.15\ \text{V}$$

■

EXAMPLE 2.3

A square loop with each side equal to $2a$ meters (m) is rotating with an angular velocity of ω radians/second (rad/s) in a magnetic field that varies as $\boldsymbol{B} = B_m \sin\ \omega t\ \boldsymbol{a}_z$ (T). The axis of the loop is at a right angle to the magnetic field. Determine the induced emf in the loop.

● SOLUTION

The position of the loop at any time t is shown in Figure 2.2. The total flux passing through the loop is

$$\Phi = \int_s \boldsymbol{B} \cdot d\boldsymbol{s} = \int_{-a}^{a} \int_{-a}^{a} B_m \sin\ \omega t\ \cos\ \omega t\ dx\ dy = 2a^2 B_m \sin\ 2\omega t\ \text{Wb}$$

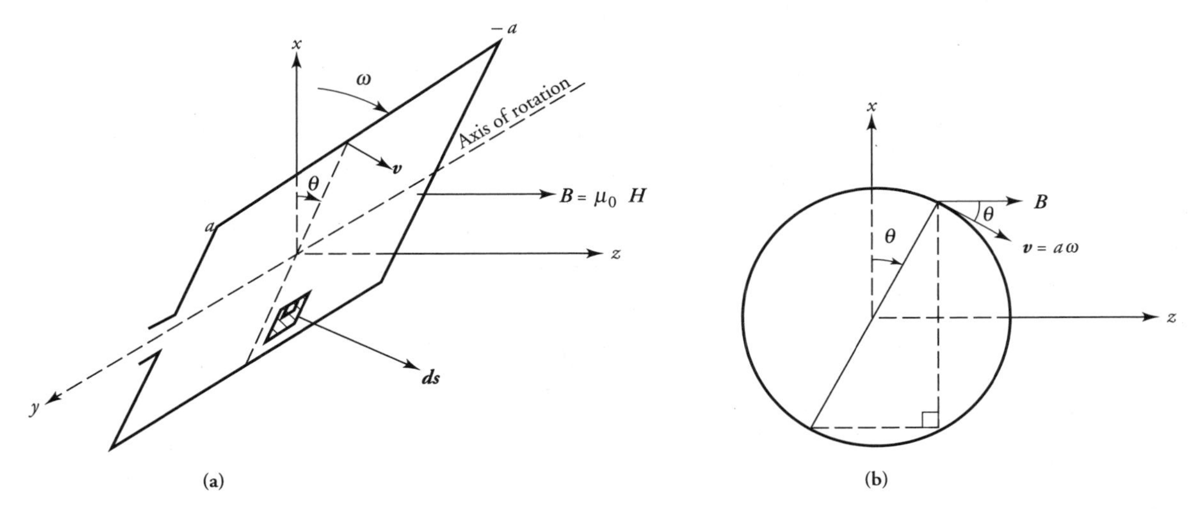

Figure 2.2 **(a)** A square loop rotating on its axis in a magnetic field. **(b)** Cross-section of the loop.

The induced emf, from Eq. (2.8b), is

$$e = -\frac{d\Phi}{dt} = -4\omega a^2 B_m \cos 2\omega t \text{ V}$$

Note that the induced emf in the loop pulsates with twice the angular frequency. ■

Ampere's Law

Equation (2.1b) is a mathematical definition of Ampere's law. It states that the line integral of a magnetic field around a closed loop is equal to the total current enclosed. The total current is the sum of the conduction current $\int \boldsymbol{J} \cdot \boldsymbol{ds}$ and the displacement current $\int (\partial \boldsymbol{D}/\partial t) \cdot \boldsymbol{ds}$. If we limit our discussion to the current flow within a conductor at low frequencies, the displacement current becomes negligible and can be dropped. As a consequence, Eq. (2.1b) becomes

$$\oint_c \boldsymbol{H} \cdot \boldsymbol{d\ell} = \int_s \boldsymbol{J} \cdot \boldsymbol{ds} = I_{\text{enc}} \tag{2.10}$$

where I_{enc} is the total current enclosed by the contour c. The following example illustrates the application of Ampere's law.

EXAMPLE 2.4

A very long cylindrical conductor of radius b carries a uniformly distributed current I. Determine the magnetic field intensity (a) within the conductor and (b) outside the conductor.

SOLUTION

A cylindrical conductor carrying current in the z-direction is shown in Figure 2.3**a**. For a uniform current distribution, the current density is

$$J = \frac{I}{\pi b^2} \boldsymbol{a}_z \ \text{A/m}^2$$

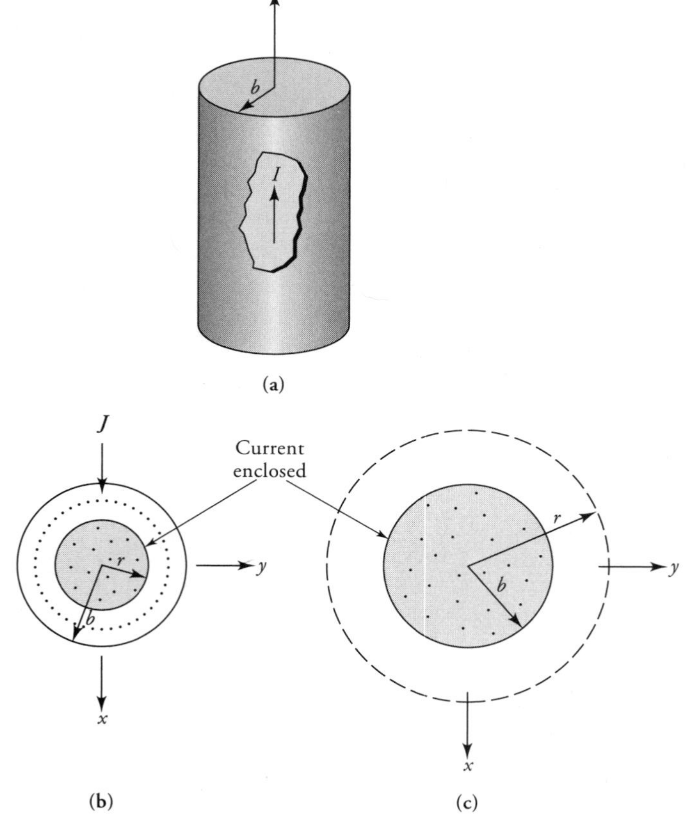

Figure 2.3 **(a)** A very long conductor carrying uniformly distributed current. **(b)** Closed path within the conductor. **(c)** Closed path outside the conductor.

(a) Magnetic field intensity within the cylinder: For any radius r such that $r \leq b$, as shown in Figure 2.3**b**, the current enclosed is

$$I_{\text{enc}} = \int_s \boldsymbol{J} \cdot d\boldsymbol{s} = I\left[\frac{r}{b}\right]^2 \text{ A}$$

The symmetry of the problem dictates that for a z-directed current, the magnetic field intensity must be in the ϕ direction. In addition, the magnitude of the magnetic field intensity must be the same at any point on a cylinder of radius r. Thus,

$$\oint_c \boldsymbol{H} \cdot d\boldsymbol{\ell} = \int_0^{2\pi} H_\phi r \, d\phi = 2\pi r H_\phi$$

Hence, from Eq. (2.10), we have

$$H_\phi = \frac{Ir}{2\pi b^2} \text{ A/m} \qquad \text{for} \qquad r \leq b$$

(b) Magnetic field intensity outside the cylinder: Since the point of observation is outside at a radius r, as shown in Figure 2.3**c**, such that $r \geq b$, the total current enclosed is I. Thus, the magnetic field intensity is

$$H_\phi = \frac{I}{2\pi r} \text{ A/m} \qquad \text{for} \qquad r \geq b$$

■

Ampere's Force Law

When a charge q moves with a velocity $\boldsymbol{v}$ in a magnetic field $\boldsymbol{B}$, the force exerted by the magnetic field on the charge, from Eq. (2.3), is

$$\boldsymbol{F} = q\boldsymbol{v} \times \boldsymbol{B}$$

Because flow of a charge in a conductor constitutes the current in that conductor, the above equation is also expressed in terms of the current in a conductor as

$$\boldsymbol{F} = \int_c I \, d\boldsymbol{\ell} \times \boldsymbol{B} \tag{2.11}$$

where I is the current in the conductor, $d\boldsymbol{\ell}$ is the elemental length of the conductor, and $\boldsymbol{B}$ is the external magnetic field in the region. Equation (2.11) is customarily referred to as **Ampere's force law** and is used to compute the developed torque in all electric motors.

It is evident from Eq. (2.11) that the force experienced by a current-carrying conductor depends upon (a) the strength of $\boldsymbol{B}$ field, (b) the magnitude of the current in the conductor, and (c) the length of the conductor. Since the force acting on a conductor is maximum when the magnetic field is perpendicular to the current-carrying conductor, all rotating machines are designed to house current-carrying conductors perpendicular to the magnetic field.

EXAMPLE 2.5

A straight conductor carrying current I is placed in a uniform magnetic field, as shown in Figure 2.4. Determine the force acting on the conductor.

● SOLUTION

Note that only a part of the conductor that is exposed to the magnetic field contributes to the total force experienced by it. That is,

$$\boldsymbol{F} = \int_{-b}^{b} IB\, dz\, (\boldsymbol{a}_z \times \boldsymbol{a}_y) = -2BIb\, \boldsymbol{a}_x \text{ newtons (N)}$$

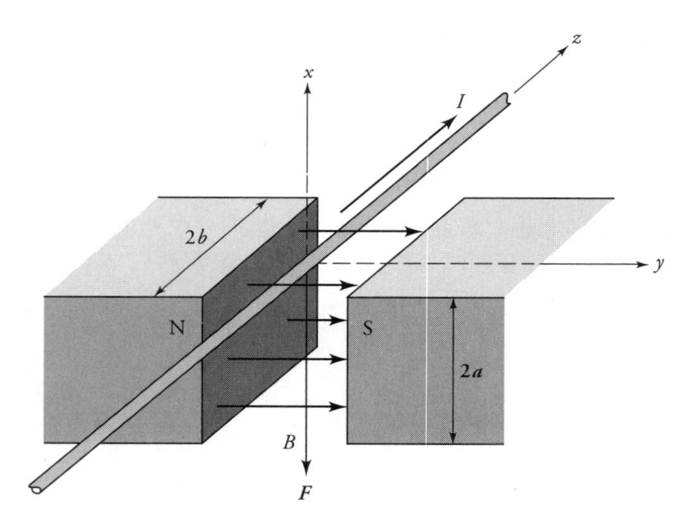

Figure 2.4 Force on a current-carrying conductor in a uniform B field.

where $\boldsymbol{a}_x$, $\boldsymbol{a}_y$, and $\boldsymbol{a}_z$ are the unit vectors in the x, y, and z directions. The negative sign in the above result indicates that the force on the conductor is acting downward as indicated in the figure. The magnitude of the force is $2b\,BI$, where $2b$ is the length of the conductor exposed to the magnetic field.

■

Torque on a Current Loop

In order to develop torque in an electric machine, conductors are joined together to form loops. One such loop is shown in Figure 2.5**a**. The loop is oriented perpendicular to the magnetic field. Using Eq. (2.11) we can determine the force acting on each side of the loop. We find that the forces on the opposite sides of the loop are equal in magnitude but opposite in direction, as shown. Since the lines of action of these forces are the same, the net force acting on the loop is zero.

Let us now orient the loop in such a way that the unit normal to the plane (surface) of the loop makes an angle θ with the $\boldsymbol{B}$ field as shown in Figure 2.5**b**. The force acting on each side is still the same. The lines of action of forces acting on each side of length W still coincide. Thus, the net force in the x direction is still zero. However, the lines of action of the forces acting on each side of length L do not coincide. If the loop is free to rotate along the axis as indicated in the figure, these forces tend to rotate the loop in the counterclockwise direction as outlined below.

For a constant magnetic field, the force acting on the top conductor is

$$\boldsymbol{F}_a = -BIL\,\boldsymbol{a}_y$$

and that on the bottom conductor is

$$\boldsymbol{F}_b = BIL\,\boldsymbol{a}_y$$

Since the torque is $\boldsymbol{r} \times \boldsymbol{F}$, the torques on the top and the bottom conductors are

$$\boldsymbol{T}_a = BIL(W/2)\sin\theta\,\boldsymbol{a}_x$$

and

$$\boldsymbol{T}_b = BIL(W/2)\sin\theta\,\boldsymbol{a}_x$$

Thus, the total torque on the loop is

$$\boldsymbol{T} = BIA\sin\theta\,\boldsymbol{a}_x$$

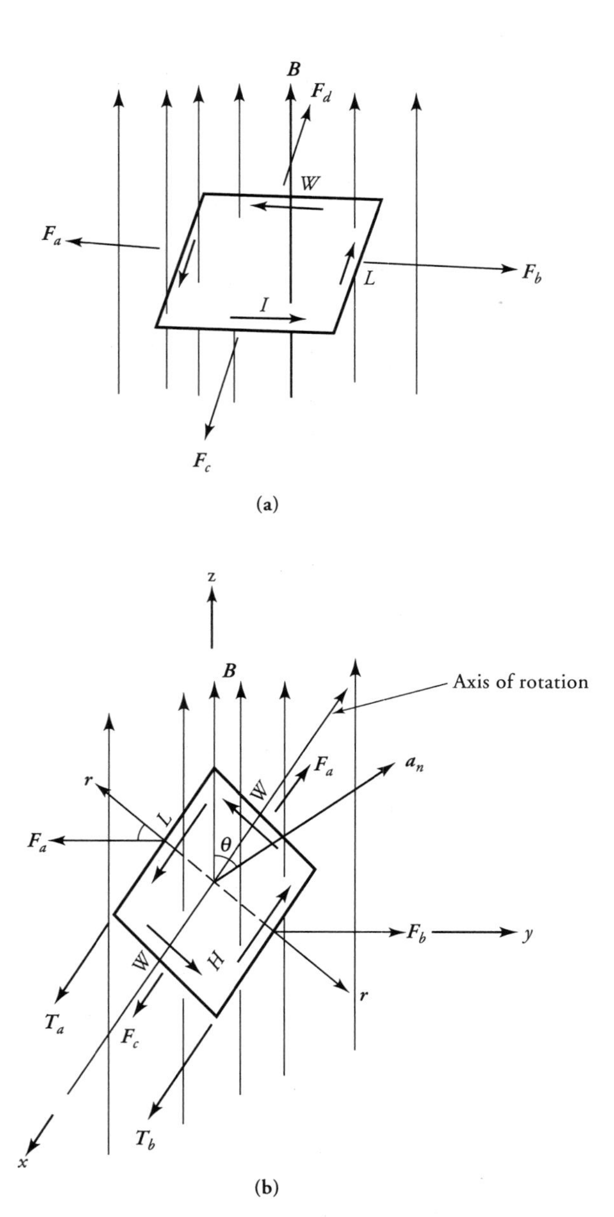

Figure 2.5 **(a)** A current loop perpendicular to a uniform magnetic field. **(b)** A current loop at an angle with a uniform magnetic field.

where $A = LW$ is the area of the loop. This torque tends to rotate the loop so as to make the plane of the loop perpendicular to the $\boldsymbol{B}$ field. In other words, the normal to a current-carrying, free to rotate, loop always seeks to align itself with the $\boldsymbol{B}$ field.

If instead of a single-turn loop we had an N-turn coil, the torque acting on the coil would have been

$$\boldsymbol{T} = BIAN \sin\theta \, \boldsymbol{a}_x \tag{2.12}$$

This is the fundamental equation that governs the development of torque in all rotating machines.

Exercises

2.1. Use Eq. (2.12) to show that $\boldsymbol{T}$ can be expressed as $\boldsymbol{T} = \boldsymbol{m} \times \boldsymbol{B}$ where $\boldsymbol{m} = NIA\,\boldsymbol{a}_n$.

2.2. Two parallel wires are separated by a distance d and carry equal currents I in opposite directions. Compute the force per unit length experienced by either wire.

2.3. Obtain expressions for the motional emf and the transformer emf induced in the square loop of Example 2.3. If the effective resistance of the loop is R, determine the torque that must be applied to keep it rotating at a uniform angular velocity of ω rad/s.

2.4. If the magnetic field in Example 2.3 is given as $\boldsymbol{B} = B_0\,\boldsymbol{a}_z$ (T), calculate the (a) motional emf, (b) transformer emf, and (c) total induced emf. Verify your results using Faraday's law of induction. The total resistance in the loop is R. Determine the average power dissipated by R and the average torque developed if the loop rotates with an angular velocity of ω rad/s. Show that $P = T_{avg}\,\omega$.

2.3 Magnetic Materials and Their Properties

Let us perform an experiment with a cylindrical coil of length L, usually called a solenoid, carrying current I as shown in Figure 2.6**a**. When you solve Problem 2.2, you find that the magnetic flux density at the center of the solenoid is twice as much as at either end as illustrated in Figure 2.6**b**. If we place small samples of various substances into this field, we will discover that the magnetic force experienced by these samples is maximum near the ends of the solenoid where the gradient, dB_z/dz, is large. In order to continue the experiment, let us assume that we always place the sample at the upper end of the solenoid and observe the force experienced by it. Observation reveals that the force on a particular substance is

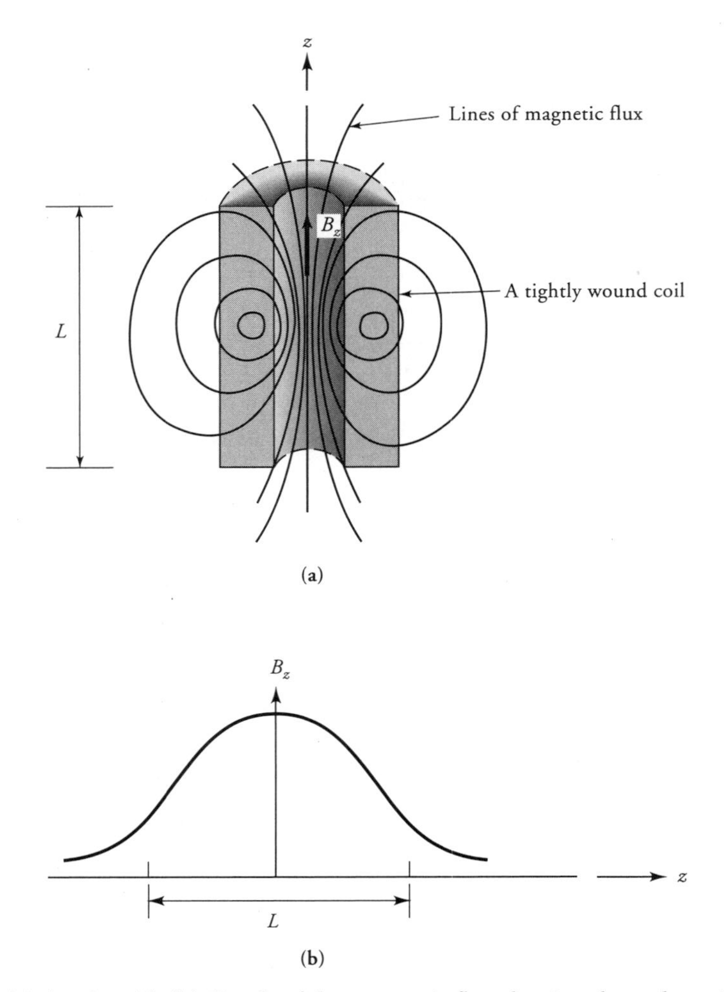

Figure 2.6 **(a)** A solenoid. **(b)** Graph of the magnetic flux density along the axis of the solenoid.

proportional to the mass of the sample and is independent of its shape as long as the sample size is not too large. We also observe that some samples are attracted toward the region of stronger field and other samples are repelled.

Diamagnetic Materials

Those substances that experience a feeble force of repulsion are called **diamagnetic**. Our experiment shows that bismuth, silver, and copper are diamagnetic.

The permeability of a diamagnetic material is slightly less than the permeability of free space. The permeabilities of some diamagnetic materials are given in Table 2.1.

Table 2.1: Relative Permeabilities of Some Diamagnetic Materials

Material	Relative Permeability
Bismuth	0.999 981
Beryllium	0.999 987
Copper	0.999 991
Methane	0.999 969
Silver	0.999 980
Water	0.999 991

Paramagnetic Materials

Two types of substances experience a force of attraction. Substances that are pulled toward the center of the solenoid with a feeble force are called **paramagnetic**. These substances exhibit slightly greater permeabilities than that of free space. A list of some paramagnetic materials and their relative permeabilities is given in Table 2.2.

Table 2.2: Relative Permeabilities of Some Paramagnetic Materials

Material	Relative Permeability
Air	1.000 304
Aluminum	1.000 023
Oxygen	1.001 330
Manganese	1.000 124
Palladium	1.000 800
Platinum	1.000 014

Since the force experienced by a paramagnetic and a diamagnetic substance is quite feeble, for all practical purposes we can group them together and refer to them as nonmagnetic materials. It is a common practice to assume that **the permeability of all nonmagnetic materials is the same as that of free space**. These materials are of no practical use in the construction of magnetic circuits.

Ferromagnetic Materials

Substances like iron are literally sucked in by the magnetic force of attraction in our above-mentioned experiment. These substances are called **ferromagnetic**. The magnetic force of attraction experienced by a ferromagnetic material may be 5000 times that experienced by a paramagnetic material.

To describe fully the magnetic properties of materials, we need the concept of quantum mechanics, which is considered to be beyond the scope of this book. However, we can use the theory of magnetic domains containing magnetic dipoles to explain ferromagnetism.

Magnetic Dipoles

We know that electrons orbit the nucleus at constant speed. Since the current is the amount of charge that passes through a given point per second, an orbiting electron gives rise to a ring of current. When we multiply the current by the surface area enclosed by it, we obtain what is known as an **orbital magnetic moment**.

An electron is also continually spinning around its own axis at a fixed rate. The spinning motion involves circulating charge and bestows an electron with a **spin magnetic moment**.

The net magnetic moment of the atom is obtained by combining the orbital and spin moments of all the electrons by taking into account the directions of those moments. The net magnetic moment produces a far field which is similar to that produced by a current loop (a magnetic dipole).

Ferromagnetism

The behavior of ferromagnetic materials such as iron, cobalt, and nickel can be easily explained in terms of **magnetic domains**. A magnetic domain is a very small region in which all the magnetic dipoles are perfectly aligned as depicted in Figure 2.7. The direction of alignment of the magnetic dipoles varies from one domain to the next. Owing to these random alignments, a virgin material is in a nonmagnetized state.

When the magnetic material is placed in an external magnetic field, all the dipoles tend to align along that magnetic field. One way to place the magnetic material in a magnetic field is to wind a current-carrying wire around it as indicated in Figure 2.8. We can expect that some domains in the magnetic material were already more or less aligned in the direction of the field. Those domains have the tendency to grow in size at the expense of the neighboring domains. The growth of a domain merely changes its boundaries. The movement of domain boundaries, however, depends upon the grain structure of the material.

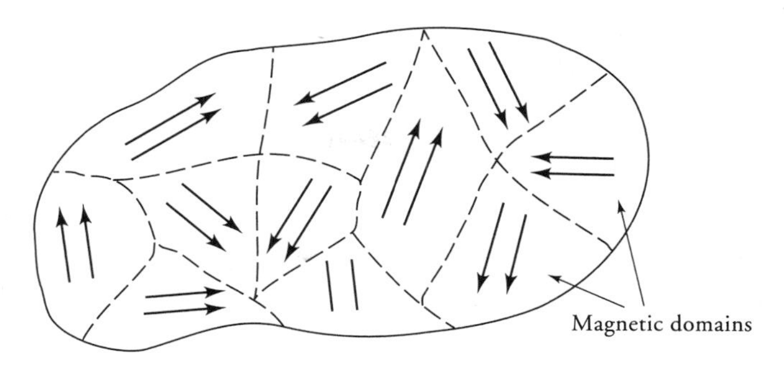

Figure 2.7 Random orientation of magnetic dipoles in a nonmagnetized ferromagnetic material.

We can also expect that some domains will rotate their magnetic dipoles in the direction of the applied field. As a result, the magnetic flux density within the material increases.

The current in the coil in Figure 2.8 establishes the ***H*** field within the material, which can be considered as an independent variable. The applied ***H*** field creates the ***B*** field. As long as the ***B*** field in the material is weak, the movement of the domain walls is reversible. As we continue increasing ***H*** by increasing the current in the coil, the ***B*** field in the material becomes stronger as more magnetic dipoles align themselves with the ***B*** field. If we measure the ***B*** field within the material, we find that it increases rather slowly at first, then more rapidly, then very slowly again, and finally flattens off as depicted by the solid line in Figure 2.9. The curve thus generated is known as the **magnetization characteristic** or simply the ***B-H* curve** of a magnetic material. Each magnetic material has a different magnetization

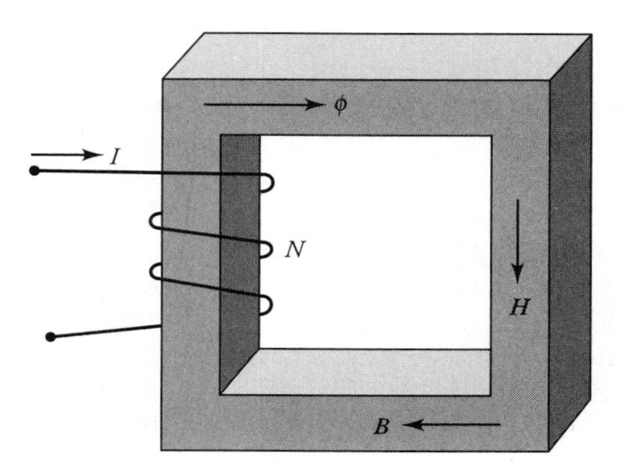

Figure 2.8 A wrapped coil establishes flux in a magnetic material.

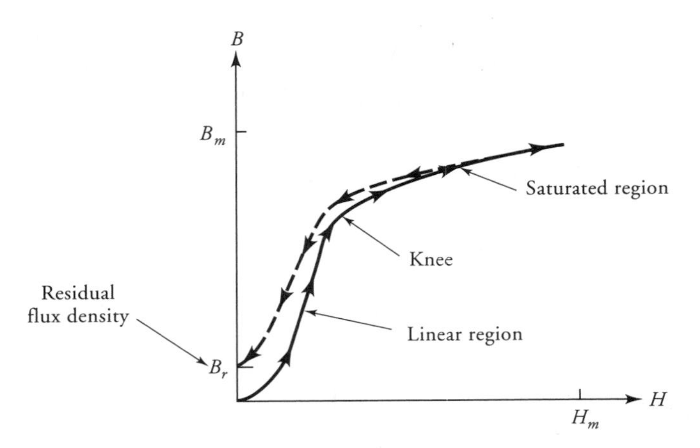

Figure 2.9 Magnetization characteristic of a ferromagnetic material.

characteristic. The place where the curve departs from the straight line is known as the **knee** of the curve. The magnetic saturation sets in as we go above the knee. The permeability of the magnetic material at any point on the curve is the ratio of $\boldsymbol{B}$ to $\boldsymbol{H}$ at that point. The general behavior of the permeability as a function of magnetic flux density is given in Figure 2.10.

The Hysteresis Loop

Figure 2.11 shows a technique to determine the magnetization characteristic of a magnetic material. Shown in the figure is a ferromagnetic material wound with two coils. When a time-varying current source is connected to one coil, it establishes a time-varying flux in the ring, which in turn induces an emf in the other coil in accordance with Faraday's law of induction. The induced emf in the other coil helps determine the changes in the flux and thereby in the magnetic flux density inside the ring. In other words, the applied current is a measure of the magnetic field intensity ($\boldsymbol{H}$), and the induced emf is a measure of the magnetic flux density ($\boldsymbol{B}$) in the ring.

When the B-H curve begins leveling off, we assume that almost all of the magnetic dipoles in the magnetic material have aligned themselves in the direction of the $\boldsymbol{B}$ field. At that time, the flux density in the magnetic material is maximum, B_m, and the material is said to be saturated. The corresponding value of the magnetic field intensity is H_m. A magnetic material is said to be fully saturated when its permeability becomes almost the same as that of free space. At that flux density, the magnetic material behaves no differently from any other nonmagnetic material. Therefore, operating a magnetic material in a fully saturated condition is of no interest in the design of electric machines.

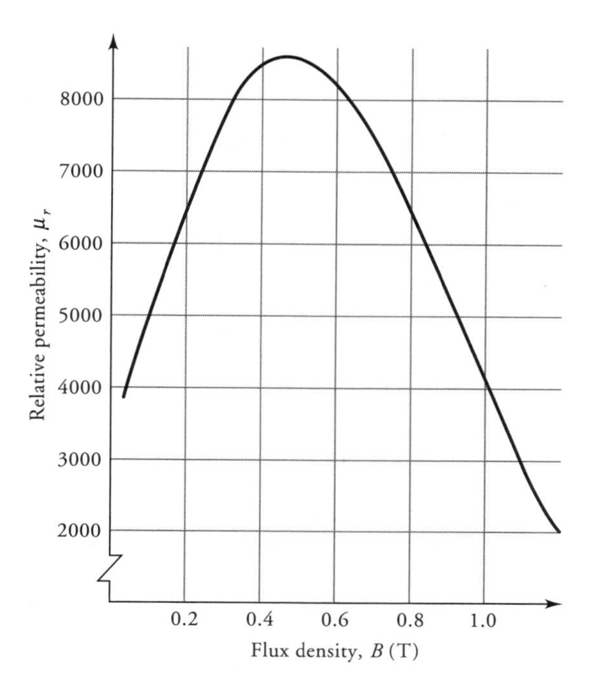

Figure 2.10 Permeability of silicon-steel as a function of flux density.

If we now start lowering the ***H*** field by decreasing the current in the coil, we find that the curve does not retrace itself but follows a different path as shown in Figure 2.12. In other words, we find that the ***B*** field does not decrease as rapidly as it increased. This irreversibility is called **hysteresis**, which simply means that ***B*** lags ***H***. The curve shows that even when the ***H*** field is reduced to zero, some

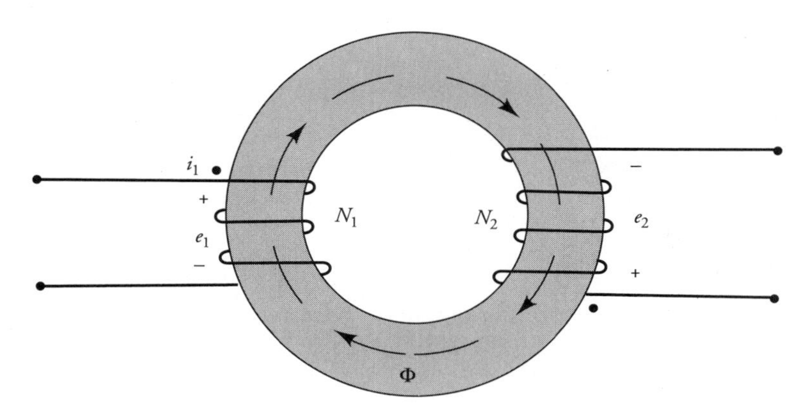

Figure 2.11 A magnetic circuit with two coils.

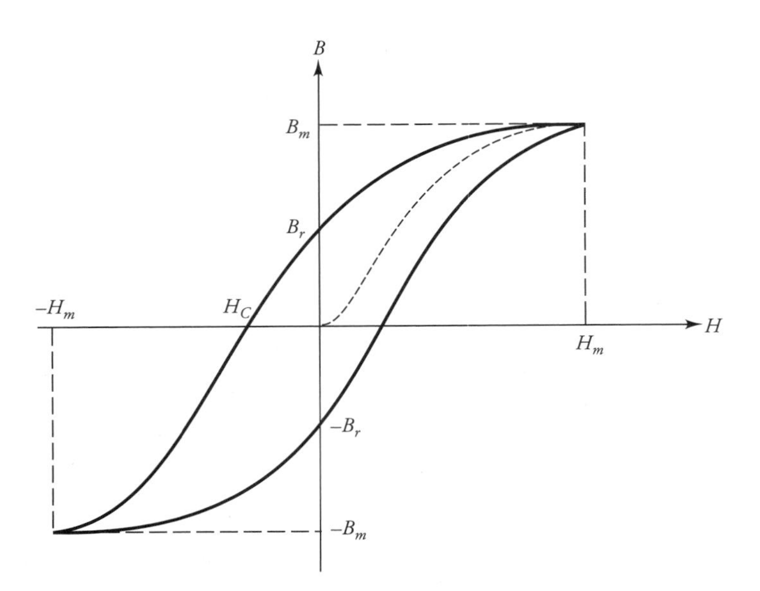

Figure 2.12 Hysteresis loop.

magnetic flux density still exists in the material. This is called the **remanence** or the **residual flux density**, B_r. In other words, the magnetic material has been magnetized and acts like a magnet. This is due to the fact that once the magnetic domains are aligned in a certain direction in response to an externally applied magnetic field, some of them tend to stay that way. The higher the residual flux density, the better is the magnetic material for applications requiring permanent magnets. A material that retains high residual flux density is said to be a **hard** magnetic material. Hard magnetic materials are used in the design of permanent-magnet motors.

To reduce the flux density in the magnetic material to zero, we must reverse the direction of the current in the coil. The value of $\boldsymbol{H}$ that brings $\boldsymbol{B}$ to zero is known as **coercive force**, H_C. Any further increase in $\boldsymbol{H}$ in the reverse direction causes the magnetic material to be magnetized with the opposite polarity. If we now continue to increase $\boldsymbol{H}$ in that direction, the $\boldsymbol{B}$ field increases rapidly at first and then flattens off as saturation approaches. The magnitude of the maximum flux density in either direction is the same.

As the $\boldsymbol{H}$ field is brought to zero by decreasing the current in the coil, the magnetic flux density in the magnetic material is again equal to its residual magnetism but in the opposite direction. Again, we must reverse the direction of the current in the coil in order to bring the magnetic flux density in the magnetic material back to zero. The current in the coil is now in the same direction as it was at the beginning of the experiment. Any further increase in the current starts magnetizing the specimen with the original polarity.

We have gone through a complete process of magnetization. The curve so traced is known as the **hysteresis loop**. The shape of the hysteresis loop depends upon the type of magnetic material. On the basis of the hysteresis loop, magnetic materials can be classified as hard or soft. As mentioned earlier, hard magnetic materials exhibit high remanence and large coercive force. Magnetically **soft materials** possess very low remanence and low coercive force and are used in the construction of ac machines such as transformers and induction motors, in order to minimize **hysteresis loss**. We discuss hysteresis loss in detail later in this chapter.

2.4 Magnetic Circuits

Since magnetic flux lines form a closed path and the magnetic flux entering a boundary is the same as the magnetic flux leaving a boundary, we can draw an analogy between the magnetic flux and the current in a closed conducting circuit. In a conducting circuit the current flows exclusively through the conductor without any leakage through the region surrounding the conductor. On the other hand, the magnetic flux cannot be completely confined to follow a given path in a magnetic material. However, if the permeability of the magnetic material is very high compared with that of the material surrounding it, such as free space, most of the flux is confined to the highly permeable material. This leads to the concentration of magnetic flux within a magnetic material with almost negligible flux in the region surrounding it. Magnetic shielding is based upon such a behavior of the magnetic flux. The channelling of the flux through a highly permeable material is very similar to the current flow through a conductor. For this reason, we call the path followed by the flux in a magnetic material a **magnetic circuit**. Magnetic circuits form an integral part of such devices as rotating machines, transformers, electromagnets, and relays, to name just a few.

Let us consider a simple magnetic circuit in the form of a toroid wound with closely spaced helical winding having N turns as shown in Figure 2.13. The inner and the outer radii of the core of the toroid are a and b, respectively. The toroid is assumed to have a rectangular cross-section with height h. From the geometry of the problem we expect the magnetic field intensity to be in the ϕ direction inside the core of the toroid.

Therefore, at any radius r such that $a \leq r \leq b$, we have

$$\oint_c \boldsymbol{H} \cdot d\boldsymbol{\ell} = 2\pi r H_\phi$$

The total current enclosed by the closed path is NI. In the analysis of magnetic circuits, the total current enclosed is commonly referred to as the applied magnetomotive force (mmf). Even though in the SI system of units the turn is a dimensionless quantity, we use ampere-turn (A·t) as the unit of mmf to differentiate it from the basic unit of current.

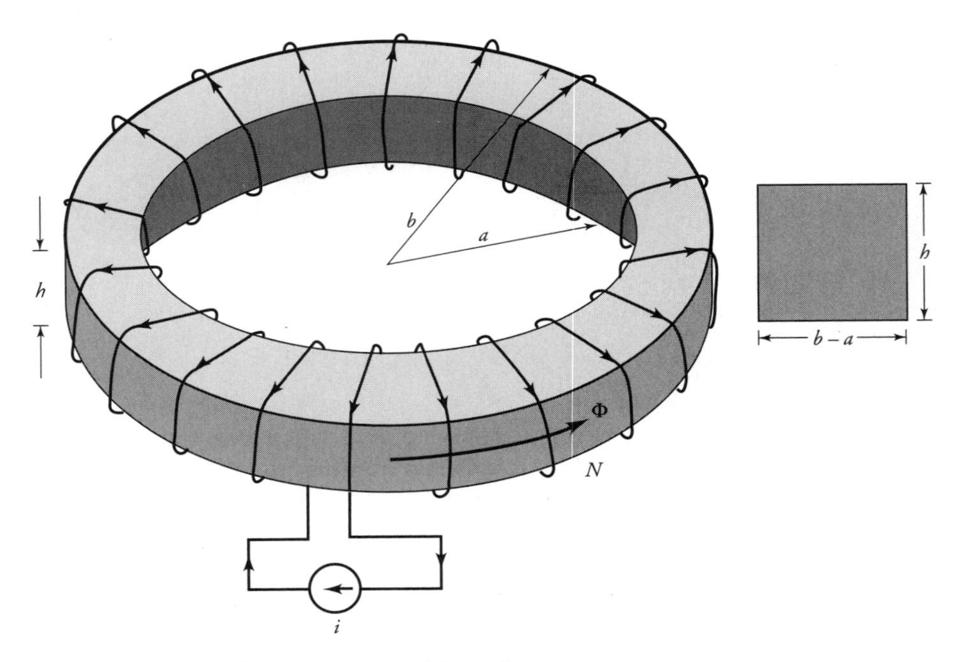

Figure 2.13 A uniformly wound toroidal winding.

Thus, from Ampere's law, we have

$$H_\phi = \frac{NI}{2\pi r} \tag{2.13}$$

The magnetic flux density within the core is

$$B_\phi = \mu H_\phi = \frac{\mu NI}{2\pi r} \tag{2.14}$$

The total flux in the core is

$$\Phi = \int_s \mathbf{B} \cdot d\mathbf{s} = \frac{\mu NI}{2\pi} \int_a^b \frac{1}{r}\, dr \int_0^h dz = \frac{\mu NI}{2\pi} h \ln (b/a) \tag{2.15}$$

When the toroidal core is made of a very highly permeable magnetic material and the winding is concentrated over only a small portion of the toroid, a large portion of the magnetic flux still circulates through the core. A fraction of the total flux produced by the coil does complete its path through the medium surrounding the magnetic circuit and is referred to as the **leakage flux**. In the design of magnetic circuits, an attempt is always made to keep the leakage flux as small as economically possible. Consequently, in the analysis of magnetic circuits, we disregard the leakage flux.

In the case of a toroid, we find that the magnetic field intensity and thereby the magnetic flux density are inversely proportional to the radius of the circular path. In other words, the magnetic flux density is maximum at the inner radius of the toroid and minimum at the outer radius. In the analysis of magnetic circuits, we usually assume that **the magnetic flux density is uniform within the magnetic material, and its magnitude is equal to the magnetic flux density at the mean radius**.

The toroid forms a continuous closed path for the magnetic circuit. However, in applications such as rotating machines, the closed path is often broken by an air-gap. The magnetic circuit now consists of a highly permeable magnetic material in series with an air-gap as depicted in Figure 2.14. Since it is a series circuit, the magnetic flux in the magnetic material is equal to the magnetic flux in the air-gap. The spreading of the magnetic flux in the air-gap, known as **fringing**, is inevitable as shown in the figure. However, if the length of the air-gap is very

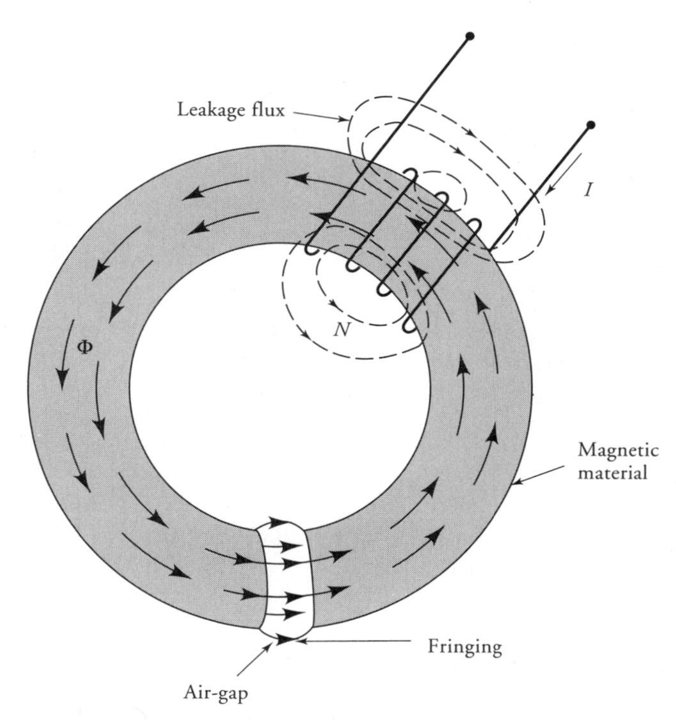

Figure 2.14 A magnetic circuit with an air gap.

small compared with its other dimensions, most of the flux lines are well confined between the opposite surfaces of the magnetic core at the air-gap and the fringing effect is negligible.

In the analysis of magnetic circuits, we assume the following:

(a) The magnetic flux is restricted to flow through the magnetic material with no leakage.
(b) There is no spreading or fringing of the magnetic flux in the air-gap regions.
(c) The magnetic flux density is uniform within the magnetic material.

Let us now consider a magnetic circuit as shown in Figure 2.15. If the coil has N turns and carries a current I, the applied mmf is NI. Thus,

$$\mathcal{F} = NI = \oint_c \mathbf{H} \cdot d\boldsymbol{\ell}$$

If the magnetic field intensity is considered to be uniform within the magnetic material, then the above equation becomes

$$H\ell = NI \tag{2.16}$$

where ℓ is the mean length of the magnetic path as indicated in Figure 2.15**a**.

The magnetic flux density in the magnetic material is

$$B = \mu H = \frac{\mu NI}{\ell}$$

where μ is the permeability of the magnetic material.

The flux in the magnetic material is

$$\Phi = \int_s \mathbf{B} \cdot d\mathbf{s} = BA = \frac{\mu NIA}{\ell}$$

where A is the cross-sectional area of the magnetic material. The above equation can also be written as

$$\Phi = \frac{NI}{\dfrac{\ell}{\mu A}} = \frac{\mathcal{F}}{\dfrac{\ell}{\mu A}} \tag{2.17}$$

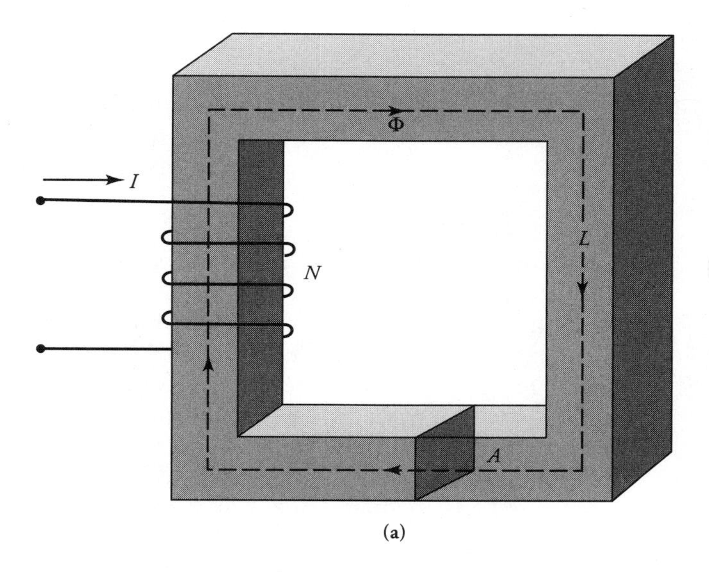

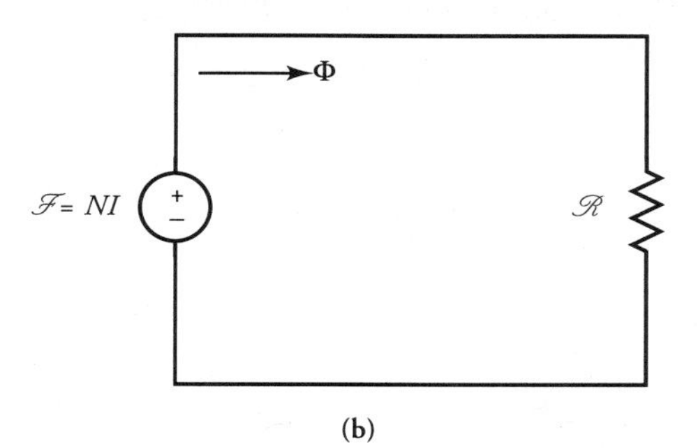

Figure 2.15 (a) Magnetic circuit with mean length L and cross-sectional area A. **(b)** Its equivalent circuit.

By considering the magnetic flux and the applied mmf in the magnetic circuit analogous to the current and the applied emf in an electric circuit, the quantity in the denominator of Eq. (2.17) must be like the resistance in the electric circuit. This quantity is defined as the **reluctance** of the magnetic circuit and is denoted by $\mathcal{R}$ and has the units of ampere-turns per weber (A·t/Wb). Thus,

$$\mathcal{R} = \frac{\ell}{\mu A} \tag{2.18}$$

In terms of reluctance $\mathcal{R}$, we can rearrange Eq. (2.17) as

$$\Phi\mathcal{R} = NI \tag{2.19}$$

Equation (2.19) is known as **Ohm's law** for the magnetic circuit. Comparing the expression for the resistance of a conductor ($R = \ell/\sigma A$) with that for the reluctance of a magnetic circuit ($\mathcal{R} = \ell/\mu A$), we find that the permeability of a magnetic material is analogous to the conductivity of a conductor. The higher the permeability of the magnetic material, the lower is its reluctance. For the same applied mmf, the flux in a highly permeable material is higher than that in a low-permeability material. This result should not surprise us because it is in accordance with our assumptions. The correlations between the reluctance and the resistance, the flux and the current, and the applied mmf and the emf enable us to represent the magnetic circuit in terms of an equivalent reluctance circuit as shown in Figure 2.15**b**.

When the magnetic circuit consists of two or more sections of magnetic material, as portrayed in Figure 2.16**a**, it can be represented in terms of reluctances as shown in Figure 2.16**b**. The total reluctance can be obtained from series and parallel combinations of reluctances of individual sections because the reluctances obey the same rules as the resistances.

If H_i is the magnetic field intensity in the ith section of a magnetic circuit and ℓ_i is the mean length, then the total mmf drop in the magnetic circuit must be equal to the applied mmf. That is,

$$\sum_{i=1}^{n} H_i\ell_i = NI \tag{2.20}$$

This equation is a statement of Kirchhoff's mmf law for a magnetic circuit.

It appears from Eq. (2.20) that a magnetic circuit can always be analyzed using an equivalent reluctance circuit. However, this is true only for a linear magnetic circuit. **A magnetic circuit is linear if the permeability of each magnetic section is constant**. For a ferromagnetic material, the permeability is a function of magnetic flux density as explained earlier. When the permeability of a magnetic ma-

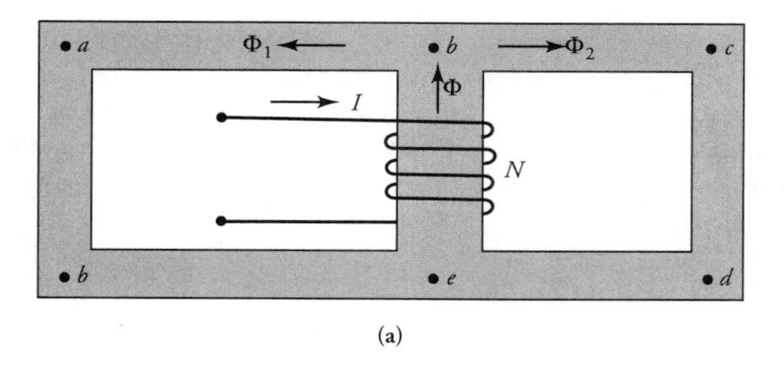

(a)

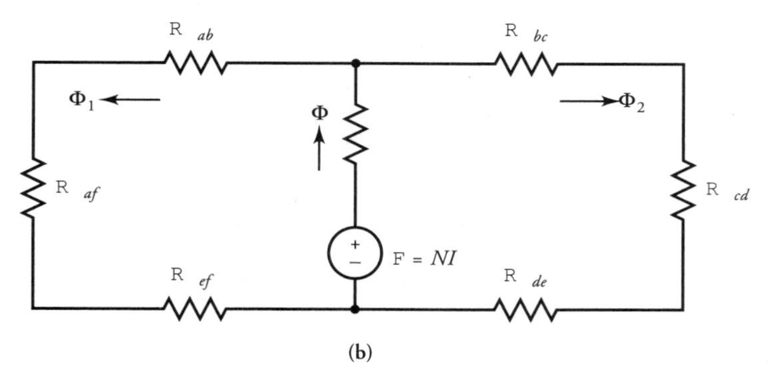

(b)

Figure 2.16 **(a)** A series-parallel magnetic circuit of uniform thickness. **(b)** Its equivalent circuit

terial varies with the flux density, the magnetic circuit is said to be **nonlinear**. All devices using ferromagnetic materials such as iron form nonlinear magnetic circuits.

Magnetic Circuit Problems

Basically two types of problems pertain to the analysis of magnetic circuits. The first type requires the determination of the applied mmf to establish a given flux density in a magnetic circuit. The other problem deals with the calculation of magnetic flux density and thereby the flux in a magnetic circuit when the applied mmf is given.

For a linear magnetic circuit, the solution to either problem can be obtained using the equivalent reluctance circuit because the permeability of the magnetic material is constant.

In a nonlinear magnetic circuit, it is relatively simple and straightforward to determine the required mmf to maintain a certain flux density in the magnetic circuit. In this case, we calculate the flux density in each magnetic section and then obtain H from the B-H curve. Knowing H, we can determine the mmf drops across each magnetic section. The required mmf is simply the sum of the individual mmf drops in accordance with Eq. (2.20).

The second type of problem in a nonlinear circuit may be solved using iterative technique. In this case, we make an educated guess for the mmf drop in one of the magnetic regions and then obtain the total mmf requirements. Then we compare the computed mmf with the applied mmf and make another educated guess if we are far off. By iterating in this way, we soon arrive at a situation in which the error between the calculated mmf and the applied mmf is within permissible limits. What constitutes a permissible limit is another debatable point. In our discussion, we use $\pm 2\%$ as the permissible limit for the error if it is not specified. A computer program can be written to reduce the error even further. The following are examples of linear and nonlinear magnetic circuits.

EXAMPLE 2.6

An electromagnet of square cross-section similar to the one shown in Figure 2.14 has a tightly wound coil with 1500 turns. The inner and the outer radii of the magnetic core are 10 cm and 12 cm, respectively. The length of the air-gap is 1 cm. If the current in the coil is 4 A and the relative permeability of the magnetic material is 1200, determine the flux density in the magnetic circuit.

● SOLUTION

Since the permeability of the magnetic material is constant, we can use the reluctance method to determine the flux density in the core.

The mean radius is 11 cm, and the mean length of the magnetic path is

$$\ell_m = 2\pi \times 11 - 1 = 68.12 \text{ cm}$$

Neglecting the effect of fringing, the cross-sectional area of the magnetic path is the same as that of the air-gap. That is

$$A_m = A_g = 2 \times 2 = 4 \text{ cm}^2$$

The reluctance of each region is

$$\mathcal{R}_m = \frac{68.12 \times 10^{-2}}{1200 \times 4\pi \times 10^{-7} \times 4 \times 10^{-4}} = 1.129 \times 10^6 \text{ A·t/Wb}$$

$$\mathcal{R}_g = \frac{1 \times 10^{-2}}{4\pi \times 10^{-7} \times 4 \times 10^{-4}} = 19.894 \times 10^6 \text{ A·t/Wb}$$

Total reluctance in the series circuit is

$$\mathcal{R} = \mathcal{R}_m + \mathcal{R}_g = 21.023 \times 10^6 \text{ A·t/Wb}$$

Thus, the flux in the magnetic circuit is

$$\Phi = \frac{1500 \times 4}{21.023 \times 10^6} = 285.402 \times 10^{-6} \text{ Wb}$$

Finally, the flux density in either the air-gap or the magnetic region is

$$B_m = B_g = \frac{285.402 \times 10^{-6}}{4 \times 10^{-4}} = 0.714 \text{ T}$$

■

EXAMPLE 2.7

A series-parallel magnetic circuit with its pertinent dimensions in centimeters is given in Figure 2.17. If the flux density in the air-gap is 0.05 T and the relative permeability of the magnetic region is 500, calculate the current in the 1000-turn coil using the fields approach.

● SOLUTION

Since the flux density in the air-gap is given, we can calculate the flux in the air-gap. Since the magnetic sections *def* and *chg* are in series with the air-gap, they carry the same flux. Thus, we can compute the mmf drop for each of these sections as tabulated below.

Section	Flux (mWb)	Area (cm²)	B (T)	H (A·t/m)	ℓ (cm)	mmf (A·t)
fg	0.12	24	0.05	39,788.74	0.5	198.94
def	0.12	24	0.05	79.58	28.0	22.28
chg	0.12	24	0.05	79.58	31.5	25.07
Total mmf drop for magnetic sections *fg*, *def*, and *chg*						246.29

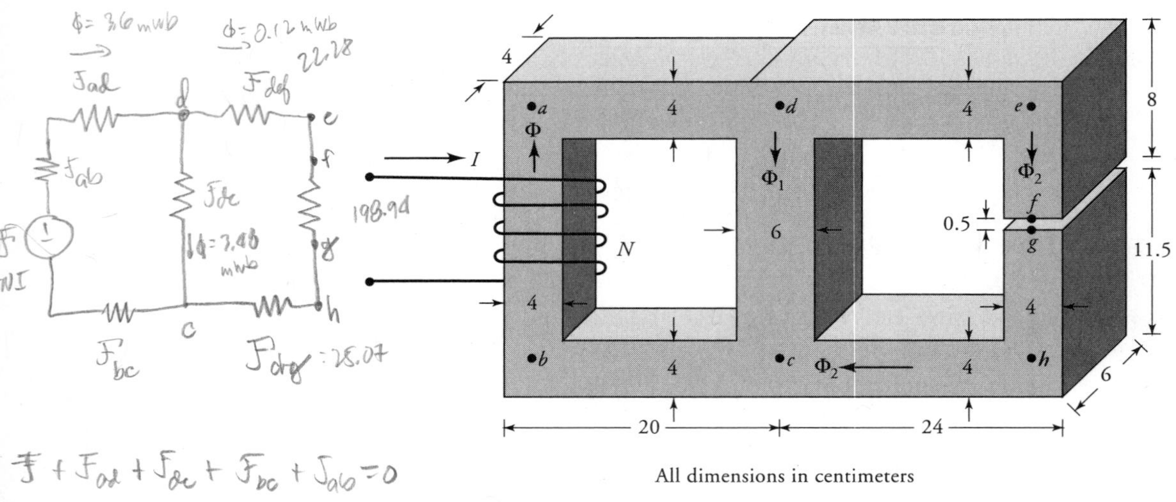

Figure 2.17 A magnetic circuit for Example 2.7.

Since the mmf drop for the region *dc* is the same as that for the combined regions *fg*, *def*, and *chg*, we can determine the flux in region *dc* by working backward. The flux in the region *dabc* is the sum of the fluxes in regions *dc* and *fg*. The mmf drop for each of these regions is tabulated below.

Section	Flux (mWb)	Area (cm²)	B (T)	H (A·t/m)	ℓ (cm)	mmf (A·t)
dc	3.48	36	0.967	1539.31	16	246.29
ad	3.60	16	2.25	3580.99	18	644.58
ab	3.60	16	2.25	3580.99	16	572.96
bc	3.60	16	2.25	3580.99	18	644.58
Total mmf drop for the magnetic circuit						2108.41

The current in the coil: $I = 2108.41/1000 \approx 2.11$ A ■

EXAMPLE 2.8

A magnetic circuit with its pertinent dimensions in millimeters is given in Figure 2.18. The magnetization characteristic of the magnetic material is shown in

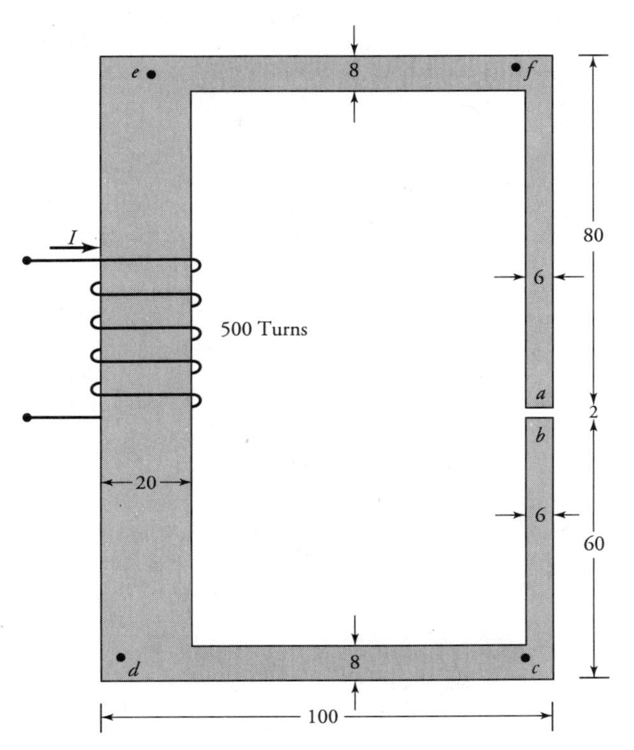

Figure 2.18 Magnetic circuit for Example 2.8.

Figure 2.19. If the magnetic circuit has a uniform thickness of 20 mm and the flux density in the air-gap is 1.0 T, find the current in the 500-turn coil.

● SOLUTION

Since the permeability of the magnetic material depends upon the flux density, we cannot compute the reluctance unless the flux density is known. Problems of this type can be easily solved using the fields approach.

Since the flux density in the air-gap is known, we compute the flux in the air-gap as

$$\Phi_{ab} = 1.0 \times 6 \times 20 \times 10^{-6} = 0.12 \times 10^{-3} \text{ Wb}$$

Since the given magnetic structure forms a series magnetic circuit, the flux in each section of the circuit is the same. We can now compute the mmf drop for each section as tabulated below:

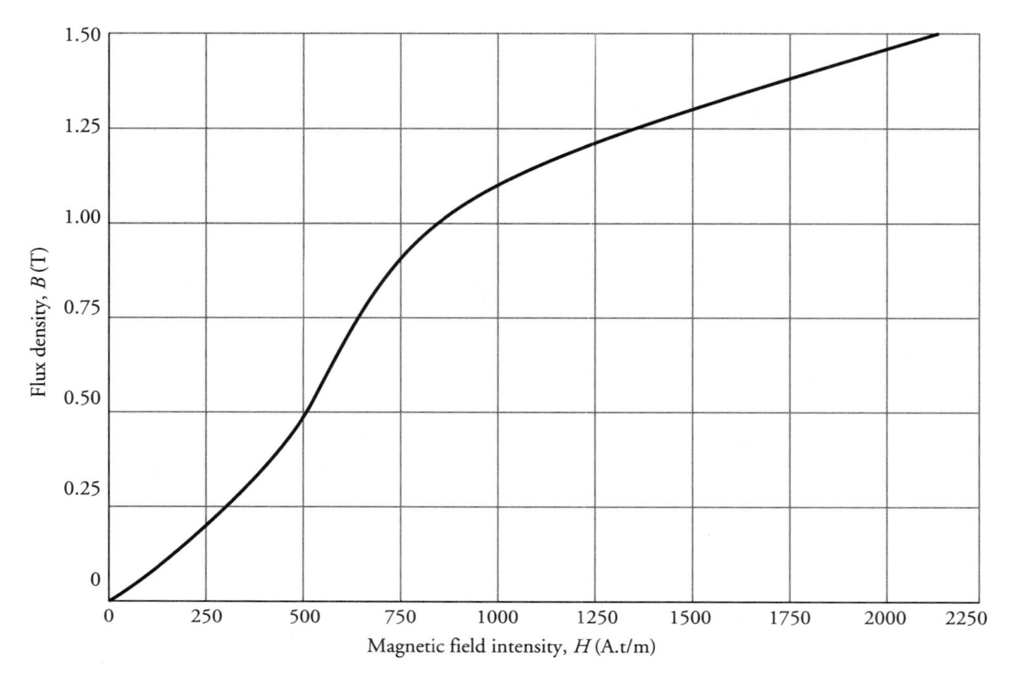

Figure 2.19 Magnetization characteristic of a magnetic material.

Section	Flux (mWb)	Area (mm^2)	B (T)	H (A·t/m)	ℓ (mm)	mmf (A·t)
ab	0.12	120	1.00	795,774.72	2	1591.55
bc	0.12	120	1.00	850.00	56	47.60
cd	0.12	160	0.75	650.00	87	56.55
de	0.12	400	0.30	350.00	134	46.90
ef	0.12	160	0.75	650.00	87	56.55
fa	0.12	120	1.00	850.00	76	64.60
Total mmf drop in the magnetic circuit						1863.75

Therefore, the current in the 500-turn coil is

$$I = \frac{1863.75}{500} = 3.73 \text{ A}$$

■

EXAMPLE 2.9

A magnetic circuit with its mean lengths and cross-sectional areas is given in Figure 2.20. If a 600-turn coil carries a current of 10 A, what is the flux in the series magnetic circuit? Use the magnetization curve given in Figure 2.19 for the magnetic material.

● SOLUTION

The applied mmf $= 600 \times 10 = 6000$ A·t. Since the magnetic circuit is nonlinear, we have to use the iterative method to determine the flux. In the absence of any other information, let us assume that 50% of the total mmf drop takes place in the air-gap. We can then calculate the total mmf drop as follows:

First Iteration

Section	Flux (mWb)	Area (cm^2)	B (T)	H (A·t/m)	ℓ (cm)	mmf (A·t)
ab	0.942	10	0.942	750,000	0.4	3000
bc	0.942	10	0.942	780	30.0	234
cd	0.942	15	0.628	570	20.0	114
da	0.942	10	0.942	780	30.0	234
Total mmf drop in the series magnetic circuit						3582

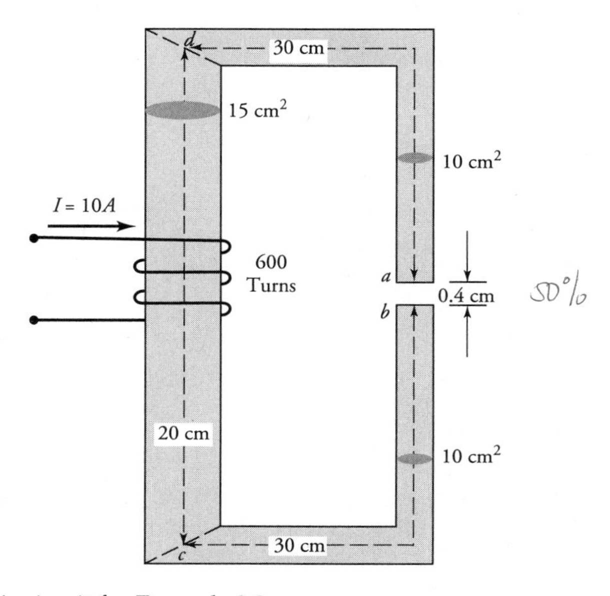

Figure 2.20 Magnetic circuit for Example 2.9.

It is now evident that most of the applied mmf appears as a drop across the air-gap. The ratio of the mmf drop across the air-gap to the total mmf drop is 0.838 (3000/3582). In other words, the mmf drop across the air-gap seems to be 83.8% of the applied mmf. However, any increase in the mmf drop in the air-gap increases the flux density in each magnetic region. Owing to nonlinear magnetic behavior, the increase in mmf drop in each magnetic section may also be considerable. Thus, instead of 83.8% of the total mmf drop across the air-gap, let us assume that the mmf drop is only 80%. Hence, we begin our second iteration with an mmf drop of 4800 A·t (0.8 × 6000) across the air-gap.

Second Iteration

Section	Flux (mWb)	Area (cm^2)	B (T)	H (A·t/m)	ℓ (cm)	mmf (A·t)
ab	1.508	10	1.508	1,200,000	0.4	4800.0
bc	1.508	10	1.508	2,175	30.0	652.5
cd	1.508	15	1.005	850	20.0	170.0
da	1.508	10	1.508	2,175	30.0	652.5
Total mmf drop in the series magnetic circuit						6275.0

The error is still 4.58% and is not within the desirable limit. From the above table, we can conclude that most of the extra mmf drop of 275 A·t is across the air-gap. If we reduce the mmf drop across the air-gap to 4600 A·t or so, it is possible to bring the error well within ±2%. Let us perform one more iteration to do so.

Third Iteration

Section	Flux (mWb)	Area (cm^2)	B (T)	H (A·t/m)	ℓ (cm)	mmf (A·t)
ab	1.445	10	1.445	1,150,000	0.4	4600
bc	1.445	10	1.445	1,950	30.0	585
cd	1.445	15	0.963	820	20.0	164
da	1.445	10	1.445	1,950	30.0	585
Total mmf drop in the series magnetic circuit						5934

The error is now −1.1% and is well within the desirable limit. Therefore, no further iteration is necessary. The flux in the magnetic structure is 1.445 mWb.

■

Exercises

2.5. Find the current in the 1000-turn coil for the magnetic circuit given in Figure 2.17 using the reluctance method. Also draw its analogous electric circuit.

2.6. Find the mmf necessary to establish a flux of 10 mWb in a magnetic ring of circular cross-section. The inner diameter of the ring is 20 cm and the outer diameter is 30 cm. The relative permeability of the magnetic material is 1200. What is the reluctance of the magnetic path?

2.7. The magnetic core with its pertinent dimensions is shown in Figure E2.7. The thickness of the magnetic material is 10 cm. What must be the current in a 100-turn coil to establish a flux of 6.5 mWb in leg *c*? Use the magnetization curve given in Figure 2.19.

2.8. If the 100-turn coil in Figure E2.7 carries a current of 10 A, determine the flux in each leg of the magnetic circuit.

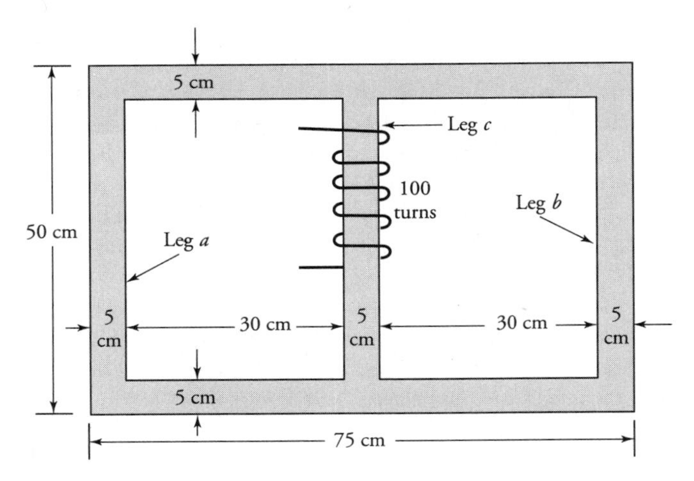

Figure E2.7 Magnetic core with parallel branches.

2.5 Self and Mutual Inductances

In Chapter 1 we reviewed circuit analysis techniques that aid us in analyzing an energy conversion device because each device can be represented by an equivalent electric circuit.

In this chapter, we found that a magnetic circuit is the heart of each energy conversion device. The magnetic circuit consists of a coil or coils wrapped around a highly magnetic material. Since the coil is made of a conducting material such as copper, it has a finite resistance. In the design of an electric machine, an attempt is always made to keep the resistance of the coil as small as possible in order to minimize the power loss in the coil. The resistance of the coil, however small it may be, forms a part of the equivalent electric circuit.

The other part of the equivalent electric circuit stems from the flux in the magnetic core of the energy conversion device and its interaction with the coils wrapped around it. In this section our aim is to develop mathematical relations that enable us to represent a magnetic circuit by self and mutual inductances.

Self Inductance

Let us consider a magnetic circuit as shown in Figure 2.21. A changing current in the coil establishes a changing magnetic flux in the magnetic material. The changing magnetic flux induces an emf in the coil in accordance with Faraday's law of induction. That is,

$$e = N\frac{d\Phi}{dt} \tag{2.21}$$

where Φ is the flux in the magnetic circuit, N is the number of turns in the coil, and $N\Phi$ are the **flux linkages**. Note that **we have dropped the negative sign in the above equation because we have marked the polarity of the induced emf in the coil**. Since the induced emf opposes the applied voltage as indicated, it is also known as the **back emf**.

The **self inductance** or simply the **inductance** of a coil is defined as the ratio of a differential change in the flux linkages to the differential change in the current. That is,

$$L = N\frac{d\Phi}{di} \tag{2.22}$$

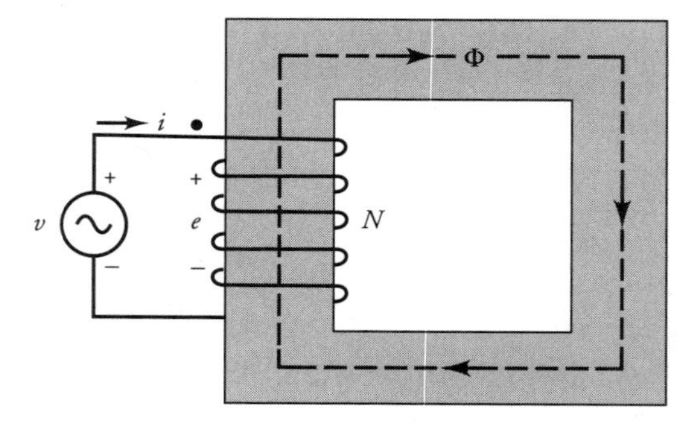

Figure 2.21 Induced emf in a coil.

where L is the inductance of the coil in henry (H).

We can express the induced emf, Eq. (2.21), in terms of the inductance as

$$e = N \frac{d\Phi}{di} \frac{di}{dt} = L \frac{di}{dt} \tag{2.23}$$

This is the form of Faraday's law of induction that is commonly used in the analysis of electric circuits. Equation (2.23) states that **the inductance of a coil is 1 H if the current changing at the rate of 1 A/s induces an emf of 1 V in the coil.**

The inductance concept is extremely useful because it provides a short-cut to express the induced emf in the coil directly in terms of the current producing it. Without this concept, we must first determine the flux in the magnetic circuit caused by the current and then apply Faraday's law of induction to obtain the induced emf.

Since $\mathcal{F} = Ni$ is the applied mmf and $\mathcal{R} = \ell/\mu A$ is the reluctance of the magnetic circuit, the flux in the magnetic core is

$$\Phi = \frac{Ni}{\mathcal{R}}$$

Thus, the inductance of the coil, from Eq. (2.22), is

$$L = \frac{N^2}{\mathcal{R}} = \mathcal{P} N^2 \tag{2.24}$$

where $\mathcal{P} = 1/\mathcal{R} = \mu A/\ell$ is called the **permeance** of the magnetic circuit. Since the turn is a dimensionless quantity, the unit of permeance from the above equation is the henry (H).

From Eq. (2.24) it is clear that the inductance depends upon the physical dimensions of the magnetic circuit and the permeability of the magnetic material. If the coil is wound over a nonmagnetic material, the inductance is constant. On the other hand, if the core is made of a ferromagnetic material, the inductance is not actually constant but a function of magnetic flux density in the core. However, it is usually assumed constant in the analysis of electric machines under the assumption that we are operating in the linear region of the B-H curve, where the permeability is nearly constant.

During our discussion of magnetic circuits we obtained the exact expression for the total flux in the toroid, Eq. (2.15), as

$$\Phi = \frac{\mu Ni}{2\pi} h \ln (b/a)$$

Thus, the exact expression for the inductance of the toroidal winding is

$$L = N\frac{d\Phi}{di} = \frac{\mu}{2\pi}N^2h\ln(b/a)$$

The equivalent electric circuit of the toroid is shown in Figure 2.22, where R is the resistance of the coil. Even though the coil resistance is a distributed parameter, it is a common practice to represent it as a lumped parameter.

Mutual Inductance

Let us now consider the magnetic circuit as shown in Figure 2.23**a**. A time-varying current i_1 in coil-1 establishes a magnetic flux Φ_1 when coil-2 is left open. This flux induces an emf in coil-1. A part of the flux, Φ_{21}, produced by coil-1 is shown linking coil-2. This flux induces an emf in coil-2. Thus,

$$e_1 = N_1\frac{d\Phi_1}{dt} = L_1\frac{di_1}{dt} \tag{2.25}$$

$$e_2 = N_2\frac{d\Phi_{21}}{dt} = N_2\frac{d\Phi_{21}}{di_1}\frac{di_1}{dt} = M_{21}\frac{di_1}{dt} \tag{2.26}$$

where L_1 is the self inductance of coil-1,

and
$$M_{21} = N_2\frac{d\Phi_{21}}{di_1} \tag{2.27}$$

is called the **mutual inductance from coil-1 to coil-2**.

In a similar manner, if coil-2 carries a current i_2 while coil-1 is left open and creates a flux Φ_2, as shown in Figure 2.23**b**, then Φ_2 induces emf in coil-2. A part

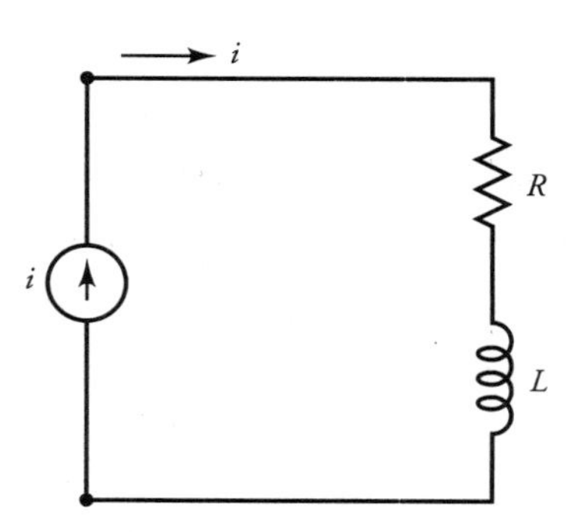

Figure 2.22 A series equivalent circuit of a toroid shown in Figure 2.13.

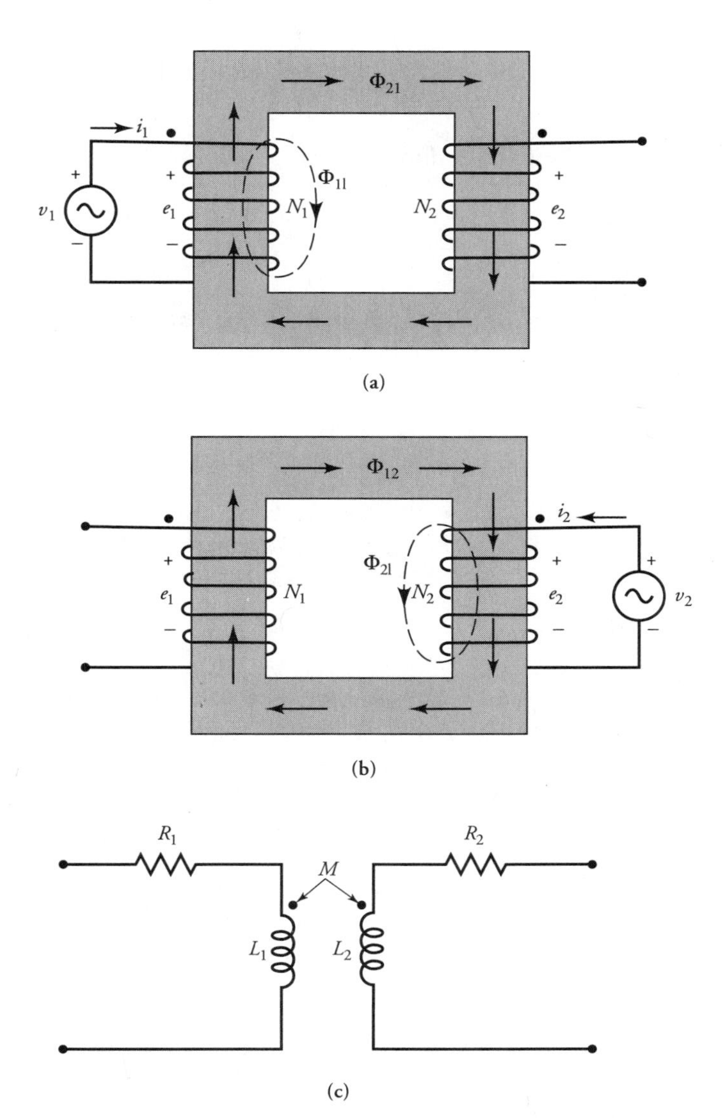

Figure 2.23 Self and mutual inductance of two coils and their equivalent circuit.

of the flux Φ_{12} shown in the figure links coil-1 and induces an emf in coil-1. That is,

$$e_2 = N_2 \frac{d\Phi_2}{dt} = L_2 \frac{di_2}{dt} \tag{2.28}$$

$$e_1 = N_1 \frac{d\Phi_{12}}{dt} = N_1 \frac{d\Phi_{12}}{di_2}\frac{di_2}{dt} = M_{12} \frac{di_2}{dt} \tag{2.29}$$

where L_2 is the self inductance of coil-2,

and
$$M_{12} = N_1 \frac{d\Phi_{12}}{di_2} \tag{2.30}$$

is called the **mutual inductance from coil-2 to coil-1**.

From Eqs. (2.27) and (2.30), we obtain

$$M_{12}M_{21} = N_1 \frac{d\Phi_{21}}{di_1} N_2 \frac{d\Phi_{12}}{di_2}$$

If we write $\Phi_{21} = k_1\Phi_1$, where k_1 defines the fraction of the flux of coil-1 linking coil-2, and $\Phi_{12} = k_2\,\Phi_2$, where k_2 determines the portion of the flux of coil-2 linking coil-1, then the above equation can be expressed in terms of the self inductances as

$$M_{12}M_{21} = k_1 k_2 L_1 L_2 \tag{2.31}$$

In a linear system, we expect $M_{12} = M_{21} = M$, where M is the **mutual inductance** of coils 1 and 2. Equation (2.31) then reduces to

$$M = k\sqrt{L_1 L_2} \tag{2.32}$$

where $k = \sqrt{k_1 k_2}$ is known as the **coefficient of coupling** or the **coupling factor** between the two coils. From our discussion, we realize that k can have values only between 0 (**magnetically isolated coils**) and 1 (**tightly coupled coils**). For the magnetic circuit of Figure 2.23 we see that $k \to 1$ as the leakage flux $\to 0$. In order to minimize the leakage flux, the two coils are usually wound over each other. When more than two coils are coupled together by a common magnetic circuit, each pair of coils has a separate coupling coefficient.

The equivalent circuit of the two coupled coils is given in Figure 2.23**c**. In order to conform to the polarities of the induced emfs in Figures 2.23**a** and **b**, we have placed a dot on one end (terminal) of each coil. The ramification of this convention is that when an increasing current enters the dotted end of one coil, it makes the dotted end of each coil positive with respect to its unmarked end. In other words, the dotted ends are like polarity terminals.

EXAMPLE 2.10

Two identical 500-turn coils are wound on the same magnetic core. A current changing at the rate of 2000 A/s in coil-1 induces a voltage of 20 V in coil-2. What is the mutual inductance of this arrangement? If the self inductance of coil-1 is 25 mH, what percentage of the flux set up by coil-1 links coil-2?

● SOLUTION

From Eq. (2.26), the mutual inductance between the coils is

$$M = \frac{20}{2000} = 0.01 \text{ H}$$

Since the two coils are identical, their self inductances must be the same. Then, from Eq. (2.32), we have

$$M = k\sqrt{L_1 L_2} = kL$$

where $L_1 = L_2 = 25$ mH. Thus, the coefficient of coupling is

$$k = \frac{0.01}{0.025} = 0.4$$

Hence, with this arrangement of coils, only 40% of the flux produced by coil-1 links coil-2.

■

Exercises

2.9. A coil sets up 100 mWb of flux. Of this total flux, 30 mWb are linked with another coil. What is the coefficient of coupling between the two coils?

2.10. The mutual inductance between two coils is 0.5 H. A current waveform as shown in Figure E2.10 is applied to one coil. Sketch the induced emf in the other coil. If the coefficient of coupling is 0.8 and the two coils are identical, determine the self inductance of each coil.

2.11. The coefficient of coupling between two coils is 0.9. The inductance of one coil is 2-mH and that of the other is 10 mH. If the open-circuit voltage

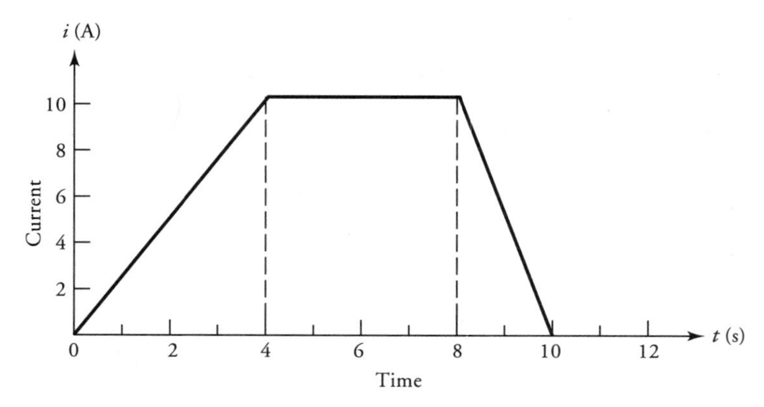

Figure E2.10 Current waveform for Exercise 2.10.

induced in a 10-mH coil is 100 cos 200t V, determine the induced voltage in the other coil.

2.12. Show that Eq. (2.32) can also be expressed as $M = kN_1N_2\mathcal{P}$.

2.6 Magnetically Coupled Coils

Two or more magnetically coupled coils can be connected in series and/or in parallel. The effective inductance of the magnetically coupled coils depends not only upon the orientation of the coil but also on the direction of the flux produced by each coil. A two-winding transformer, an autotransformer, and an induction motor are examples of magnetically coupled coils (circuits). We discuss each of these topics in greater length later in this book.

Series Connection

In a series connection, the magnetically coupled coils can be connected in **series aiding** or **series opposing**. The coils are said to be connected in **series aiding** when they produce the flux in the same direction. In other words, the total flux linking either coil is more than its own flux.

An equivalent circuit of two coils connected in series aiding is given in Figure 2.24**a**. If the current in the circuit at any time t is $i(t)$, then the voltage drop across each coil is

$$v_1 = L_1 \frac{di}{dt} + M \frac{di}{dt} \tag{2.33}$$

$$v_2 = L_2 \frac{di}{dt} + M \frac{di}{dt} \tag{2.34}$$

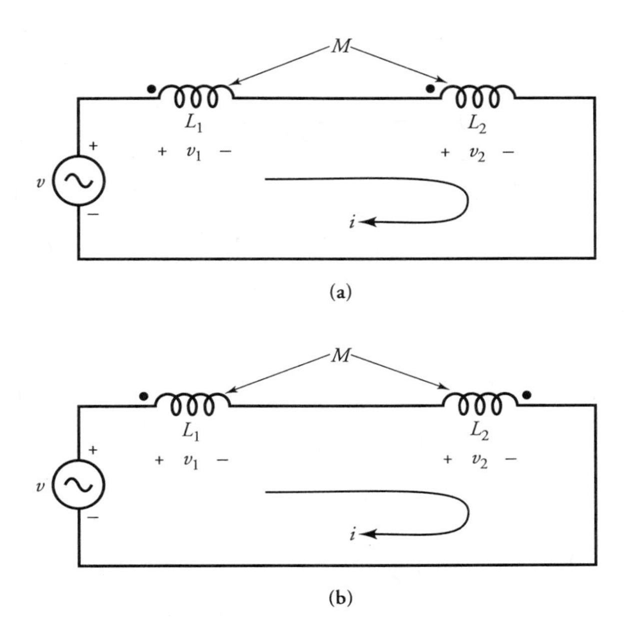

Figure 2.24 Magnetically coupled coils, **(a)** series aiding and **(b)** series opposing

where L_1 and L_2 are the self inductances of coils 1 and 2 and M is the mutual inductance between them. To simplify the mathematical development, we have ignored the resistance of each coil.

From Kirchhoff's voltage law, we have

$$v = v_1 + v_2 = [L_1 + L_2 + 2M]\frac{di}{dt} \tag{2.35}$$

Thus, the effective inductance of two coils connected in series aiding is

$$\text{Series aiding: } L = L_1 + L_2 + 2M \tag{2.36}$$

On the other hand, when the flux produced by one coil opposes the flux produced by the other coil, the coils are said to be connected in **series opposing**. By following a similar development we can show that the effective inductance of a series opposing connection, Figure 2.24**b**, is

$$\text{Series opposing: } L = L_1 + L_2 - 2M \tag{2.37}$$

EXAMPLE 2.11

The effective inductances when two coils are connected in series aiding and series opposing are 2.38 H and 1.02 H, respectively. If the inductance of one coil is 16 times the inductance of the other, determine (a) the inductance of each coil, (b) the mutual inductance, and (c) the coefficient of coupling.

● SOLUTION

From the given information, we have

$$L_1 + L_2 + 2M = 2.38 \text{ H}$$

and

$$L_1 + L_2 - 2M = 1.02 \text{ H}$$

By adding and subtracting the above equations, we obtain

$$L_1 + L_2 = 1.7 \text{ H} \qquad \text{and} \qquad M = 0.34 \text{ H}$$

Let $L_1 = 16\,L_2$, then from $L_1 + L_2 = 1.7$ H, we get

$$L_1 = 1.6 \text{ H} \qquad \text{and} \qquad L_2 = 0.1 \text{ H}$$

The coefficient of coupling is

$$k = \frac{0.34}{\sqrt{1.6 \times 0.1}} = 0.85$$

■

Parallel Connection

Two magnetically coupled coils connected in parallel are shown in Figure 2.25. Let us assume that the two coils produce the flux in the same direction when the currents enter at the dotted terminals. The coils are then said to be connected in **parallel aiding.** If i_1 and i_2 are the currents through coils 1 and 2 with inductances L_1 and L_2, then the current supplied by the source and their time-rate of change are

$$i = i_1 + i_2 \tag{2.38}$$

$$\frac{di}{dt} = \frac{di_1}{dt} + \frac{di_2}{dt} \tag{2.39}$$

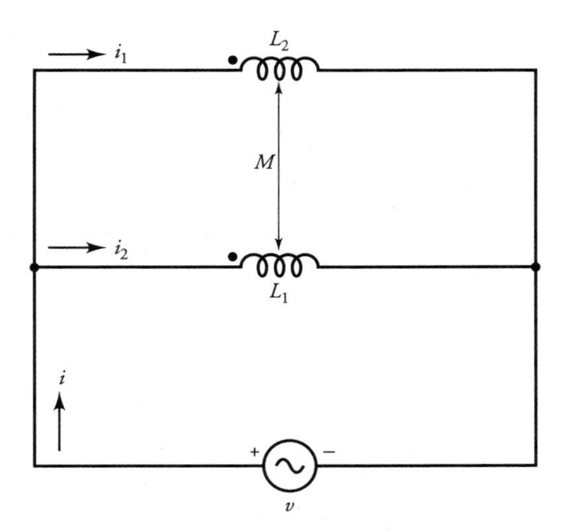

Figure 2.25 Magnetically coupled coils connected in parallel-aiding.

The voltage drop across coil-1 is

$$v = L_1 \frac{di_1}{dt} + M \frac{di_2}{dt}$$

Substituting for di_1/dt from Eq. (2.39) in the above equation, we obtain

$$v = L_1 \frac{di}{dt} - [L_1 - M] \frac{di_2}{dt} \tag{2.40}$$

Similarly, the voltage drop across coil-2 is

$$v = M \frac{di}{dt} + [L_2 - M] \frac{di_2}{dt} \tag{2.41}$$

Equating Eqs. (2.40) and (2.41) and solving for di/dt, we have

$$\frac{di}{dt} = \left[\frac{L_1 + L_2 - 2M}{L_1 - M}\right] \frac{di_2}{dt} \tag{2.42}$$

If the equivalent inductance of this arrangement is L, then

$$v = L \frac{di}{dt} \tag{2.43}$$

From Eqs. (2.40) and (2.43), we get

$$\frac{di}{dt} = -\left[\frac{L_1 - M}{L - L_1}\right]\frac{di_2}{dt} \tag{2.44}$$

We now have two expressions, Eqs. (2.42) and (2.44), for the rate of change of source current di/dt. Since each expression must yield the same rate of change of the source current, by equating them and after some rearranging, we obtain

$$L = \left[\frac{L_1L_2 - M^2}{L_1 + L_2 - 2M}\right] \tag{2.45}$$

as the equivalent inductance of two parallel aiding magnetically coupled coils.

It is left to the reader to verify that the equivalent inductance of two magnetically coupled coils connected in **parallel opposing** is

$$L = \left[\frac{L_1L_2 - M^2}{L_1 + L_2 + 2M}\right] \tag{2.46}$$

EXAMPLE 2.12

Two magnetically coupled coils with self inductances of 1.6 H and 0.1 H are connected in parallel. The mutual inductance is 0.34 H. Calculate the effective inductance when (a) the coils are connected in parallel aiding and (b) parallel opposing.

● SOLUTION

(a) From Eq. (2.45), the effective inductance for the parallel aiding connection is

$$L = \frac{1.6 \times 0.1 - 0.34^2}{1.6 + 0.1 - 2 \times 0.34} = 0.0435 \text{ H, or } 43.5 \text{ mH}$$

(b) The equivalent inductance for the parallel opposing connection, from Eq. (2.46), is

$$L = \frac{1.6 \times 0.1 - 0.34^2}{1.6 + 0.1 + 2 \times 0.34} = 0.0187 \text{ H, or } 18.7 \text{ mH}$$

■

Exercises

2.13. Verify Eqs. (2.37) and (2.46).

2.14. Two coils with inductances of 2 H and 0.5 H are connected in series aiding. If the coefficient of coupling between the coils is 0.4, determine the effective inductance of the coils. What is the effective inductance if the coils are connected in series opposing?

2.15. The inductances of two coils are 20 mH and 80 mH. The mutual inductance of the coils is 4 mH. Determine (a) the coefficient of coupling, (b) the equivalent inductance when they are connected in series aiding, (c) the equivalent inductance when they are connected in series opposing, (d) the equivalent inductance when they are connected in parallel aiding, and (e) the equivalent inductance when they are connected in parallel opposing.

2.16. Two coils with inductances of 2 H and 6 H are connected in parallel. The equivalent inductance is 1 H. What is the mutual inductance? Are the coils connected in parallel aiding or parallel opposing if the coefficient of coupling is above 50%? What is the coefficient of coupling?

2.7 Magnetic Losses

In the analysis of an electric machine we usually divide the losses in the machine into three major categories—the copper losses, the mechanical losses, and the magnetic losses. The copper loss takes place in the winding of each machine. The friction loss arises from the rotation of the rotating part (rotor) of a machine. It can be divided into three parts: bearing friction, brush friction, and windage (wind friction). The two sources of magnetic loss (**iron loss** or **core loss**) are the eddy-current loss and the hysteresis loss.

Eddy-Current Loss

When a time-varying magnetic flux links a coil, it induces an emf in the coil in accordance with Faraday's law of induction. As the coil is wound over a magnetic material, an emf is also induced in the magnetic material by the same time-varying flux. Since a magnetic material is also a good conductor of electricity, the induced emf along a closed path inside the magnetic material sets up a current along that path. The location and the path of this current are such that it encloses the magnetic flux that produces it. In fact, there are as many closed paths surrounding the magnetic field inside the magnetic material as one can imagine. Figure 2.26**a** shows some of the paths for the **induced currents** in a solid magnetic material when the flux density is increasing with time. Because the swirling pattern of these circulating currents in a magnetic material resembles the eddy currents of air or water, they are called **eddy currents**.

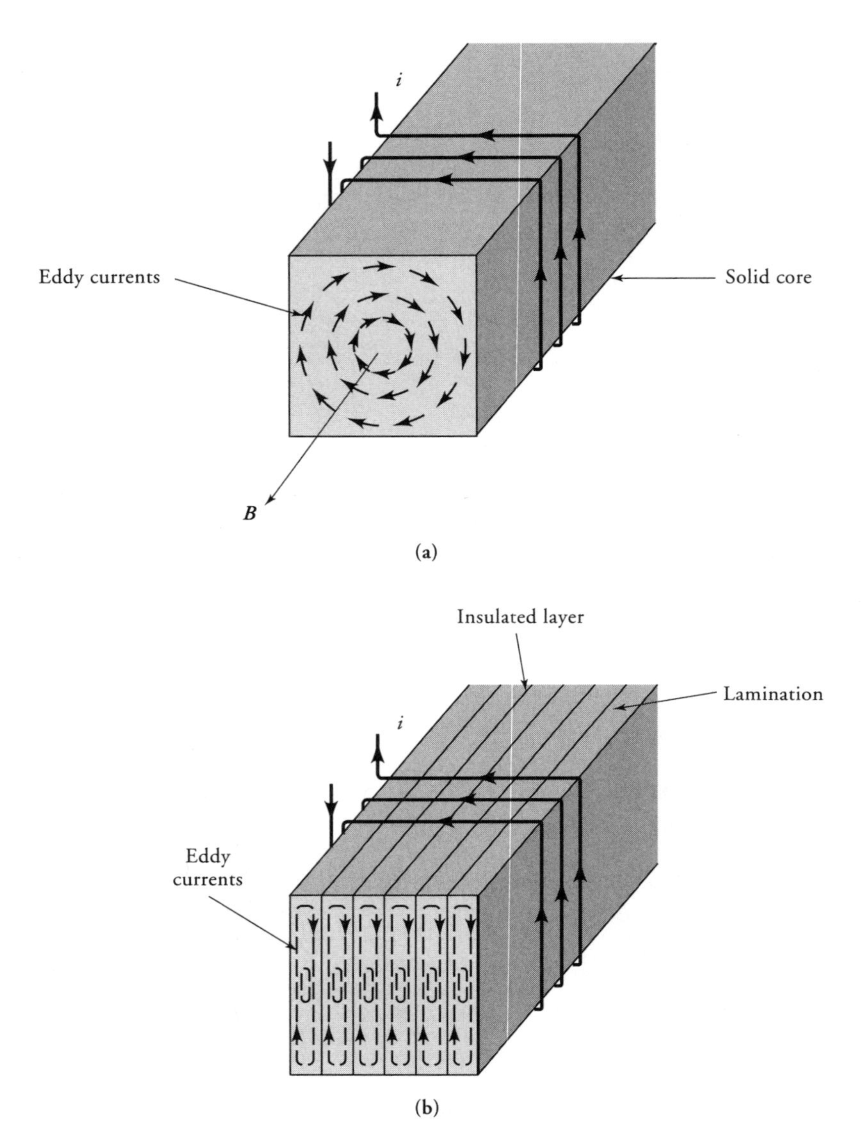

Figure 2.26 Eddy currents in **(a)** solid and **(b)** laminated magnetic core when the current in the coil is increasing with time.

As the flux in the magnetic material increases, so does the induced emf in each circular path. The increase in the induced emf results in an increase in the eddy current in that path. As a consequence of this current, energy is converted into heat in the resistance of the path. By summing up the power loss in each loop within the magnetic material, we obtain the total power loss in the magnetic material caused by the eddy currents. This power loss is called **eddy-current loss**.

The eddy currents establish their own magnetic flux, which tends to oppose the original magnetic flux. Therefore, the eddy currents not only result in the eddy-current loss within the magnetic material but also exert a demagnetization effect on the core. Consequently, we have to apply more mmf to produce the same flux in the core. The demagnetization increases as we approach the axis of the magnetic core. The overall effect of demagnetization is the crowding of the magnetic flux toward the outer surface of the magnetic material. This coerces the inner part of the core to be magnetically useless.

The adverse effects of eddy currents can be minimized if the magnetic core can be made highly resistive in the direction in which the eddy currents tend to flow. We can accomplish this in practice by building up the magnetic core by stacking thin pieces of the magnetic materials. The thin piece, known as a **lamination**, is coated with varnish or shellac and is commercially available in thicknesses varying from 0.36 mm to 0.70 mm. The thin coating makes one lamination fairly well insulated electrically from the other. The magnetic core built with laminations (Figure 2.26**b**) forces the eddy currents to follow long and narrow paths within each lamination. The net result is the reduced eddy currents in the magnetic material.

If we assume that the flux density varies sinusoidally with time but is uniform at any instant over the cross-section of the magnetic core, we can show that the average eddy-current loss is

$$P_e = k_e f^2 \delta^2 B_m^2 V \tag{2.47}$$

where P_e is the eddy-current loss in watts (W), k_e is a constant that depends upon the conductivity of the magnetic material, f is the frequency in hertz (Hz), δ is the lamination thickness in meters, B_m is the maximum flux density in teslas (T), and V is the volume of the magnetic material in cubic meters (m^3).

Hysteresis Loss

Each time the magnetic material is made to traverse its hysteresis loop, it produces a power loss that is commonly referred to as the **hysteresis loss**. The hysteresis loss stems from the molecular friction as the magnetic domains are forced to reverse their directions by the applied mmf.

After several cycles of magnetization, the hysteresis loop becomes symmetric as shown in Figure 2.27. The energy per unit volume (energy density) supplied to the magnetic field when the flux density is changed from $-B_r$ (point a) to B_m (point c) along the path abc is

$$w_1 = \int_{-B_r}^{B_m} \mathbf{H} \cdot d\mathbf{B}$$

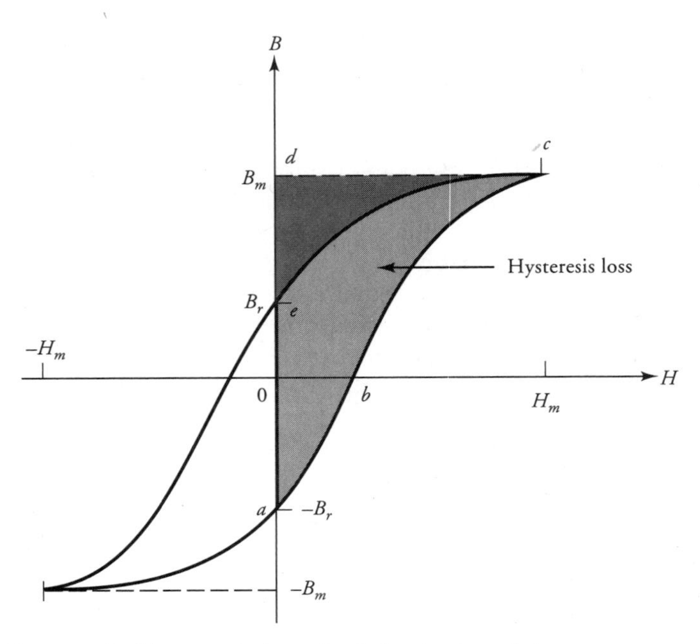

Figure 2.27 Symmetrical hysteresis loop depicting hysteresis loss.

which is simply the hatched area *abcdea*. As the flux density is now decreased from B_m (point *c*) to B_r (point *e*) along the path *ce*, the energy density released by the magnetic field is

$$w_2 = \int_{B_m}^{B_r} \mathbf{H} \cdot d\mathbf{B}$$

which is simply the shaded area *cdec*. We have traced half the hysteresis loop. The loss in the energy density during this half of the loop is $w_1 - w_2$. The other half, being identical, results in the same loss in the energy density. Thus, the total loss in the energy density is

$$w = 2[w_1 - w_2] \tag{2.48}$$

which represents the total area of the hysteresis loop. Equation (2.48) yields the loss in energy per unit volume per cycle.

In general, Eq. (2.48) cannot be evaluated analytically because it is not possible to express H as a simple function of B. However, we can plot the hysteresis to some scale and measure its area.

By testing various ferromagnetic materials, Charles Steinmetz proposed that the hysteresis loss can be expressed as

$$P_h = k_h f B_m^n V \tag{2.49}$$

where P_h is the hysteresis loss in watts, k_h is a constant that depends upon the magnetic material, and n is the Steinmetz exponent. From experiments, it has been found that n varies from 1.5 to 2.5. Other quantities are as previously defined.

The magnetic loss is the sum of the eddy-current loss and the hysteresis loss. Note that the hysteresis loss does not depend upon the lamination thickness. For a given magnetic material, we can express the magnetic loss as

$$P_m = K_e f^2 B_m^2 + K_h f B_m^n \tag{2.50}$$

where $K_e = k_e \delta^2 V$, $K_h = k_h V$, and P_m is the total magnetic loss.

If the magnetic losses have to be separated into eddy-current loss and hysteresis loss, we need three independent equations to determine three unknown quantities K_e, K_h, and n. To do so, the magnetic losses are measured at two frequencies and two flux densities as demonstrated by the following example.

EXAMPLE 2.13

The following data were obtained on a thin sheet of silicon steel. Compute the hysteresis loss and the eddy-current loss.

Frequency (Hz)	Flux Density (T)	Magnetic Loss (W/kg)
25	1.1	0.4
25	1.5	0.8
60	1.1	1.1

● SOLUTION

Substituting the values in Eq. (2.50), we have

$$30.25K_e + 1.1^n K_h = 0.016 \tag{2.51}$$

$$56.25K_e + 1.5^n K_h = 0.032 \tag{2.52}$$

$$72.6K_e + 1.1^n K_h = 0.02 \tag{2.53}$$

Subtracting Eq. (2.51) from Eq. (2.53), we get

$$K_e = 94.451 \times 10^{-6}$$

Substituting for K_e in Eqs. (2.51) and (2.52) and solving for n, we obtain

$$n = 2.284$$

Finally, from Eq. (2.53), $K_h = 0.0106$.

The computed eddy-current and hysteresis losses are

Frequency (Hz)	Flux Density (T)	Hysteresis Loss (W/kg)	Eddy-Current Loss (W/kg)
25	1.1	0.329	0.071
25	1.5	0.669	0.133
60	1.1	0.791	0.411

These calculations show that the eddy-current loss is smaller than the hysteresis loss. In a laminated core, the hysteresis loss is usually greater than the eddy-current loss. ■

Exercises

2.17. The eddy-current loss in an induction machine is 200 W and the hysteresis loss is 400 W when operating from a 50-Hz supply. Find the total magnetic loss when the machine is operated from a 60-Hz source with a decrease of 10% in the flux density. Assume $n = 1.6$.

2.18. The following data were obtained on a ring of cast iron. Determine the hysteresis and the eddy-current losses at both frequencies and at both flux densities.

Frequency (Hz)	Flux Density (T)	Magnetic Loss (W/kg)
50	0.8	2.4
50	1.2	5.2
60	0.8	3.0

2.19. A sample of a magnetic material is known to have a hysteresis energy loss of 300 joules per cycle per cubic meter when the maximum flux density is 1.2 T. Find the hysteresis loss in the magnetic material if the flux density is 0.8 T, the volume of the magnetic material is 40 cm^3, and the frequency of operation is 60 Hz. The Steinmetz constant is 1.6.

2.20. In a given specimen of a magnetic material, the magnetic loss was 21.6 W at a frequency of 25 Hz and a flux density of 0.8 T. When the same specimen was tested at a frequency of 60 Hz, the magnetic loss increased to 58.4 W at the same flux density. What are the hysteresis and eddy-current losses at each frequency? The Steinmetz constant is 1.5.

2.8 Permanent Magnets

The rapid development of new permanent magnetic materials and their commercial availability have increased their use in the design of dc and synchronous machines. In all machines using permanent magnets to set up the required magnetic flux, it is desirable that a magnetic material for permanent magnets have the following characteristics:

(a) A large retentivity (residual flux density) so that the magnet is "strong" and provides the needed flux

(b) A large coercivity so that it cannot be easily demagnetized by stray magnetic fields

An ideal permanent magnetic material exhibits a flat-topped, wide hysteresis curve so that the residual magnetism stays at a high level when the applied field is removed. In other words, the area enclosed by the hysteresis curve is very large. In practice, very few magnetic materials satisfy these requirements.

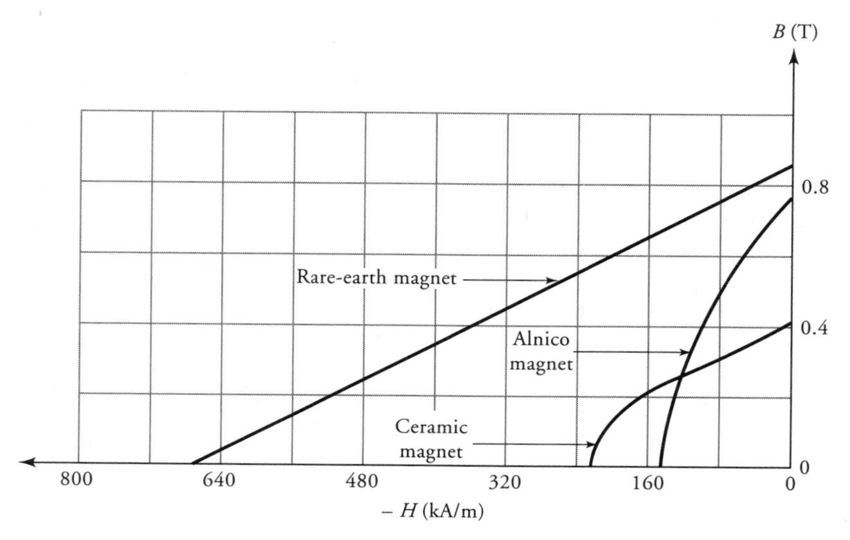

Figure 2.28 Demagnetization characteristics of Alnio, ceramic, and rare-earth magnets.

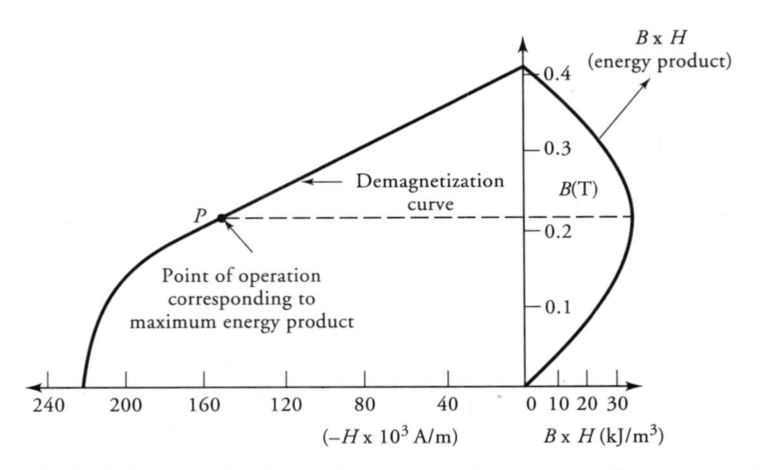

Figure 2.29 Typical demagnetization and energy-product curves of a permanent magnet.

For proper use of a permanent magnet, we concentrate our attention on its demagnetization characteristics, that is, the magnetic behavior of a material in the second quadrant of the hysteresis loop. Figure 2.28 shows the demagnetization characteristics of Alnico, ceramic, and samarium-based rare-earth cobalt magnets. As you can see, a dramatic difference exists between the demagnetization curves of magnetic materials belonging to different groups of alloys.

In the design of magnetic circuits using permanent magnets, we may also like to operate the magnet where it can supply the maximum energy. As mentioned earlier, energy density is simply the area of the hysteresis loop ($\boldsymbol{B} \cdot \boldsymbol{H}$). This area,

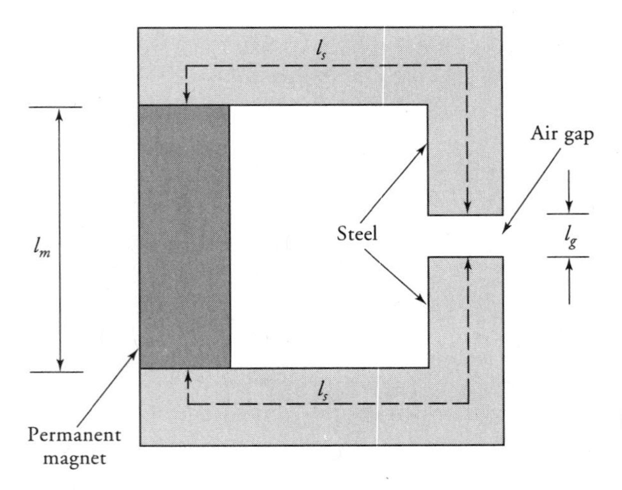

Figure 2.30 A series magnetic circuit using a permanent magnet.

usually called the **energy product**, along with the demagnetization curve, is shown in Figure 2.29. The operating point at its maximum energy product level is indicated by the dashed line in the figure.

Figure 2.30 shows a series magnetic circuit that employs a permanent magnet to set up the necessary flux in the air-gap region. The L-shaped sections are usually made of a highly permeable magnetic material and are needed to channel the flux toward the air-gap. The application of Ampere's law to the closed magnetic circuit yields

$$H_m \ell_m + H_s \ell_s + H_g \ell_g + H_s \ell_s = 0 \tag{2.54}$$

where the subscripts m, s, and g are used to identify the quantities in the permanent magnet, yoke, and air-gap regions, respectively. In a series magnetic circuit, the magnetic flux is essentially the same. Therefore,

$$B_m A_m = B_g A_g = B_s A_s$$

Because $B = \mu H$, we can write

$$H_g = \frac{B_m A_m}{\mu_0 A_g} \quad \text{and} \quad H_s = \frac{B_m A_m}{\mu_s A_s}$$

Equation (2.54) can now be expressed as

$$H_m = -\left[\frac{\ell_g A_m}{\mu_0 \ell_m A_g} + \frac{2\ell_s A_m}{\mu_s \ell_m A_s}\right] B_m \tag{2.55}$$

The above equation is called the **operating line**, and its intersection with the demagnetization curve yields the **operating point**. If the permeability of the L-sections is very high, Eq. (2.55) may then be approximated as

$$H_m = -\frac{\ell_g A_m}{\mu_0 \ell_m A_g} B_m \tag{2.56}$$

Note that H_m, in the second quadrant of the hysteresis loop, is a negative quantity. Therefore, we can drop the negative sign in Eq. (2.56) and take H_m to be the magnitude of the coercive force corresponding to the flux density B_m in the magnet.

If the area of the air-gap is the same as that of the magnet, that is, $A_g = A_m$, then the flux density in the air-gap is equal to the flux density in the magnet ($B_g = B_m$). Thus, from Eq. (2.56), we have

$$B_g = B_m = \frac{\mu_0 \ell_m}{\ell_g} H_m$$

or

$$B_g^2 = \mu_0 \frac{\ell_m A_m}{\ell_g A_g} H_m B_m$$

or

$$B_g H_g V_g = B_m H_m V_m \tag{2.57}$$

where V_g and V_m are the volumes of the air-gap and the magnet, respectively. The above equation emphasizes the fact that the **energy available in the air gap is maximum when the point of operation corresponds to the maximum energy product of the magnet.**

EXAMPLE 2.14

The physical dimensions of the magnetic circuit of Figure 2.30 are as follows: $\ell_g = 1$ cm, $A_g = A_s = A_m = 10$ cm^2, $\ell_s = 50$ cm, and $\mu_r = 500$ for steel. What is the minimum length of the magnet required to maintain the maximum energy in the air-gap?

SOLUTION

Substituting the given values in Eq. (2.55), we have

$$H_m = -\frac{9549.3}{\ell_m} B_m \tag{2.58}$$

To maintain the maximum energy in the air-gap, the point of operation must correspond to the maximum energy product of the magnet. From the demagnetization and the energy product curves shown in Figure 2.31, we obtain $B_m = 0.23$ T and $H_m = -144$ kA/m. Substituting these values in Eq. (2.58), the minimum length of the magnet is 1.53 cm. ■

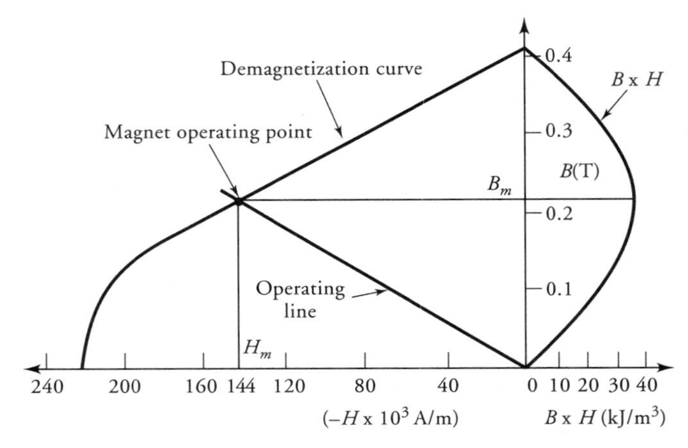

Figure 2.31 Demagnetization and energy-product curves of a magnet for Example 2.14.

EXAMPLE 2.15

Calculate the induced emf in a 10-turn coil rotating at 100 rad/s in a permanent magnet system shown in Figure 2.32**a**. The rotor and the yoke are made of an alloy with a relative permeability of 3000. The demagnetization curve of the magnet is given in Figure 2.32**b**.

● SOLUTION

Figure 2.32**c** shows that there are two magnetic paths for the flux. From symmetry, the flux in each path would be half of the total flux. That is, if Φ is the flux per path, the total flux supplied by each magnet is 2Φ. Focusing our attention on one magnetic path, we have

$$\Phi = B_m A_m = B_s A_s = B_g A_g = B_a A_a \tag{2.59}$$

where the subscripts m, s, g, and a are used to identify the magnet, yoke, air-gap, and rotor sections of the closed magnetic path. The mean lengths of the magnetic sections are:

Magnet: $\ell_m = 52 - 42 = 10$ mm

Yoke: $\ell_s = 2.5 + 2.5 + (2\pi \times 54.5/4) = 90.61$ mm

Air-gap: $\ell_g = 42 - 40 = 2$ mm

Rotor: $\ell_a = 17.5 + 17.5 + (2\pi \times 22.5/4) = 70.34$ mm

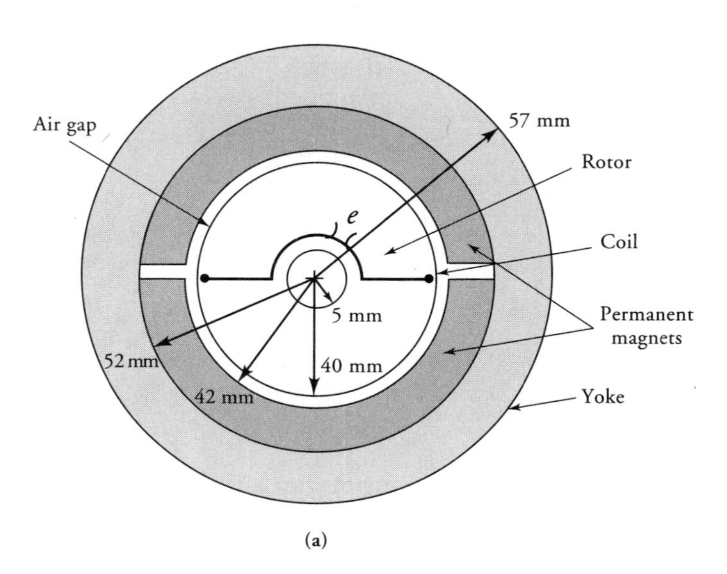

Figure 2.32 **(a)** Magnetic circuit for Example 2.15.

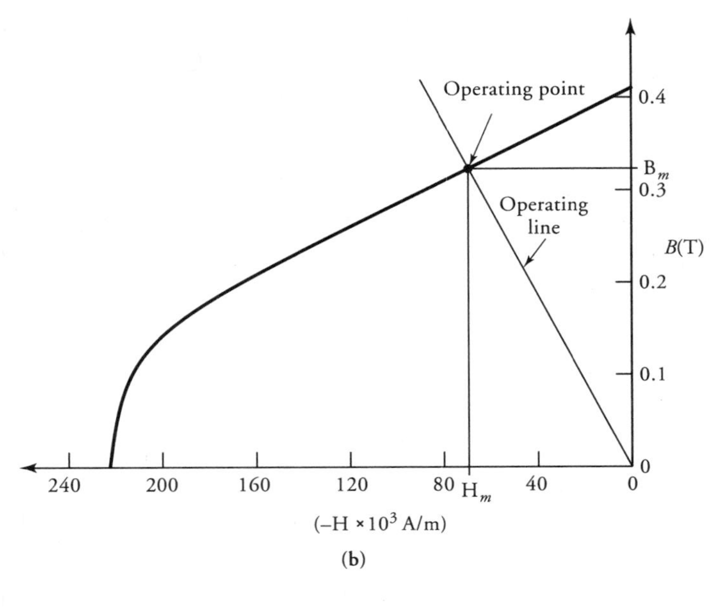

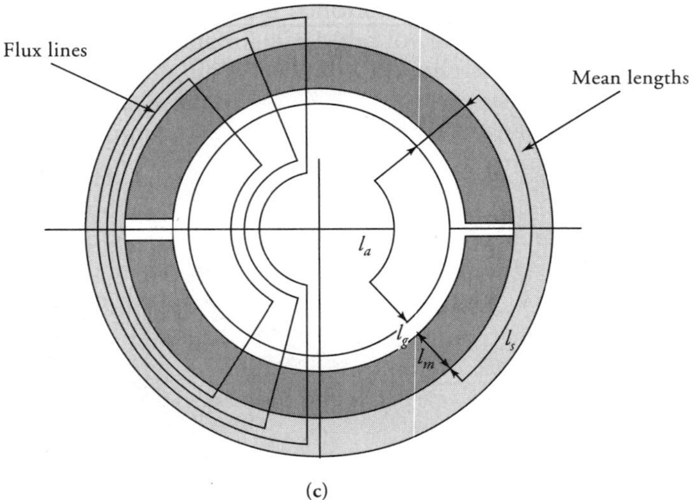

Figure 2.32 (b) Demagnetization curve of the magnet. **(c)** Flux distribution and mean lengths of the magnetic paths.

The cross-sectional areas are:

Magnet: $A_m = 50 \times (52 + 42)\pi/4 = 3691.37 \text{ mm}^2$

Yoke: $A_s = 5 \times 50 = 250 \text{ mm}^2$

Air-gap: $A_g = 50 \times (42 + 40)\pi/4 = 3220.13 \text{ mm}^2$

Rotor: $A_a = 35 \times 50 = 1750 \text{ mm}^2$

The application of Ampere's law yields

$$H_m\ell_m + H_g\ell_g + H_a\ell_a + H_g\ell_g + H_m\ell_m + H_s\ell_s = 0$$

or

$$2H_m\ell_m = -[H_s\ell_s + H_a\ell_a + 2H_g\ell_g]$$

The above equation can be expressed in terms of B_m, using Eq. (2.59), as

$$2H_m\ell_m = -\left[\frac{\ell_s A_m}{\mu_s A_s} + \frac{\ell_a A_m}{\mu_a A_a} + 2\frac{\ell_g A_m}{\mu_0 A_g}\right] B_m$$

Substituting the values, we obtain operating line as

$$H_m = -202{,}158.28\ B_m$$

The intersection of the operating line with the demagnetization curve of the magnet yields $B_m = 0.337$ T.

Hence, the flux per path set up by the magnet is

$$\Phi = B_m A_m = 0.337 \times 3691.37 \times 10^{-6} = 1.244 \times 10^{-3}\ \text{Wb}$$

Thus, the total flux supplied by each magnet is

$$\Phi_t = 2\Phi = 2.488\ \text{mWb}$$

As the coil is rotating with a uniform angular velocity of 100 rad/s, the flux linking the coil is maximum (Φ_t) when the plane of the coil is perpendicular to the magnetic flux. On the other hand, the flux linking the coil is zero when the plane of the coil is parallel to the magnetic field. Therefore, we can write a general expression for the flux linking the coil as

$$\Phi_C = 2.488 \cos 100t\ \text{mWb}$$

From Faraday's law of induction, the induced emf in the 10-turn coil is

$$e = -N\frac{d\Phi_C}{dt} = 2.488 \sin 100t\ \text{V}$$

■

Exercises

2.21. Plot the energy product curve for the rare-earth magnet with its demagnetization curve given in Figure 2.28.

2.22. Repeat Example 2.15 if the magnets are replaced by rare-earth magnets.

2.23. A rare-earth magnet is molded in the form of a toroid with a diameter of 20 cm and the cross-sectional area of 2 cm^2. What must be the length of the air-gap for which the flux density is 0.8 T?

2.24. What must be the induced emf in the coil of Exercise 2.22 if the mmf drops in the rotor and the yoke are ignored? Compute the percent error in the induced emf as a result of this approximation.

SUMMARY

In this chapter we reviewed the fundamentals of electromagnetism. We stated the four Maxwell's equations, which are, in fact, generalizations of laws based upon experimental evidence. We introduced the concepts of motional emf and transformer emf. We stated that the total emf induced in a loop that changes its position as a function of time when immersed in a time-varying magnetic field is the sum of the motional emf and the transformer emf. The induced emf can also be determined by means of Faraday's law of induction.

When the current is confined to flow in a conductor and the frequency of interest is low, the conduction current density is many magnitudes higher than the displacement current density. Therefore, we dropped the displacement current from our considerations when applying Ampere's circuital law.

We used the Lorentz force equation to determine the force experienced by a current-carrying conductor. When the current flows through a loop placed in a magnetic field, a torque tends to rotate it so as to make the plane of the loop perpendicular to the magnetic field.

We highlighted the differences between diamagnetic, paramagnetic, and ferromagnetic materials. For all practical purposes, diamagnetic and paramagnetic materials can be treated as nonmagnetic materials because their permeabilities differ very slightly from that of free space. In the study of electric machines it is the ferromagnetic materials that are of interest to us. Virtually all ferromagnetic materials have nonlinear properties. When subjected to time-varying magnetic fields, the ferromagnetic materials contribute to the magnetic losses, the eddy-current loss, and the hysteresis loss. The eddy-current loss can be reduced by using laminated magnetic core, whereas the hysteresis loss depends upon the chemistry of the magnetic material. The hysteresis loss is smaller for soft magnetic materials than for hard magnetic materials. For this reason, we use soft magnetic materials for transformers, induction machines, and synchronous machines. Hard magnetic materials, on the other hand, make good permanent magnets and are used as the stators for the dc machines.

We explained in great length how to analyze a magnetic circuit because the magnetic circuit is the heart of each electric machine. Each magnetic circuit can be represented by an equivalent reluctance circuit.

Another important role of a magnetic circuit is to enable us to transport energy from one part to another without any direct electric connection. We showed how to represent coils by their self and mutual inductances when they share a common flux.

Some of the important equations presented in this chapter are summarized below for easy reference.

Transformer emf: $e_t = -\int_s \frac{\partial \mathbf{B}}{\partial t} \cdot d\mathbf{s}$

Motional emf: $e_m = \oint_c (\mathbf{v} \times \mathbf{B}) \cdot d\boldsymbol{\ell}$

Total induced emf: $e = e_t + e_m = -\frac{d\Phi}{dt}$

Ampere's law: $\oint_c \mathbf{H} \cdot d\boldsymbol{\ell} = I$

Torque on a loop: $\mathbf{T} = \mathbf{m} \times \mathbf{B}$

Self inductance: $L = \frac{N^2}{\mathcal{R}}$ where $\mathcal{R} = \frac{\ell}{\mu A}$

Mutual inductance: $M = k\sqrt{L_1 L_2}$

Ohm's law for a magnetic circuit: $\mathcal{F} = \Phi\mathcal{R}$ where $\mathcal{F} = NI$

Kirchhoff's laws for a magnetic circuit:

$$\sum_{i=1}^{n} H_i \ell_i = \mathcal{F}$$

and

$$\sum_{i=1}^{n} \Phi_i = 0$$

Inductance of two coupled coils

Series aiding: $L = L_1 + L_2 + 2M$

Series opposing: $L = L_1 + L_2 - 2M$

Parallel aiding: $L = \dfrac{L_1L_2 - M^2}{L_1 + L_2 - 2M}$

Parallel opposing: $L = \dfrac{L_1L_2 - M^2}{L_1 + L_2 + 2M}$

Review Questions

2.1. State Maxwell's equations. Give a brief account of each equation and mention the law on which each equation is based.

2.2. Show that the conduction current in a conductor is very large compared with the displacement current. Use copper as the conductor and 60 Hz as the power frequency.

2.3. At what frequency is the conduction-current density in copper equal in magnitude to the displacement-current density?

2.4. Does Maxwell's equation $\nabla \times \boldsymbol{E} = -\partial \boldsymbol{B}/\partial t$ completely describe Faraday's law of induction? If not, why not? If yes, explain.

2.5. What is the significance of the transformer equation?

2.6. What is the significance of the motional emf?

2.7. Can the motional emf and the transformer emf take place in a loop at the same time? If yes, what must be the conditions? If not, explain why not.

2.8. A very long, hollow conductor carries a current along its axis. What is the magnetic field intensity inside the conductor?

2.9. A very long, hollow conductor carries a current in its circumferential direction. What is the magnetic field inside it?

2.10. A 10-turn coil is in the xy-plane. The $\boldsymbol{B}$ field is in the z direction and is increasing at a rate of 10 T/hr. If the area of the loop is 20 cm^2, what is the induced emf in the coil? What is the flux linking the loop?

2.11. The plane of a 10-turn coil makes an angle of 60° with the magnetic field. If the magnetic flux density is 1.2 T, what is the torque acting on the loop? Assume that the area of the loop is 10 cm^2.

2.12. A single loop of wire lying in the plane of the paper carries a current in the clockwise direction. What effect is noticed if a compass is placed within the loop? Has this loop any properties in common with those of a bar magnet?

2.13. What are the magnetic losses? How can the eddy-current loss be made small? How can the hysteresis loss be reduced?

2.14. Why is it important to make the coefficient of coupling between any two coils as close to unity as possible?

2.15. If the flux in a machine does not vary with time, should the core of that machine be laminated?
2.16. Explain ferromagnetism in your own words.
2.17. Why is alnico a better permanent magnet material than ceramic? Which has the greater retentivity?
2.18. Why do we have to worry about the leakage flux in magnetic calculations whereas in almost all electrical problems leakage current can be ignored?
2.19. What happens to the reluctance of a magnetic circuit when an air-gap is cut in it? How does an air-gap affect the inductance of a magnetic circuit?
2.20. Why can't we always use the reluctance approach to find the flux established by a certain mmf in a magnetic circuit?
2.21. State the condition when the mmf is numerically equal to the current producing it.
2.22. Is it possible to have a flux of 0.5 Wb in a magnetic circuit without having a flux density of 0.5 T?
2.23. Is energy required to maintain a steady magnetic field after it has been established?
2.24. When a steady current through a coil can establish a certain magnetic field, why do we need permanent magnets?
2.25. Why do we need ferromagnetic materials in the design of machines when the flux can be established in free space?
2.26. Consider two identical magnetic circuits. One is made of a ferromagnetic material with a relative permeability of 1000 and the other uses a nonmagnetic material. For the same applied emf, compute the ratio of (a) reluctances, (b) inductances, (c) magnetic field intensities, (d) magnetic flux densities, and (e) energy densities.
2.27. What is a magnetic dipole? How can we explain the permeability of magnetic materials in terms of the magnetic dipole moments?
2.28. Explain the terms "retentivity" and "coercivity." Does high retentivity imply high coercivity and vice versa?
2.29. After magnetic saturation is reached, is it possible to increase further the flux density by increasing the applied mmf? Explain.
2.30. Why do hard magnetic materials tend to retain high residual flux density?
2.31. Can you think of any useful application of hysteresis?
2.32. If two parallel conductors are carrying currents in the same direction, do they experience a force of attraction or repulsion?
2.33. State Faraday's law of induction. What is the significance of the negative sign?
2.34. Is it true that the induced emf in a coil is maximum when the flux linking it is maximum? If yes, explain. If not, why not?
2.35. A conductor is moving in a steady magnetic field with a uniform velocity. What do you conclude if the induced emf in the conductor is zero?
2.36. A commercially available inductor is rated at 115 V, 0.3 A. If the resistance of the coil is neglected, what is its inductance? What happens if the inductor is connected across a 115-V dc supply?

PROBLEMS

2.1. A rectangular loop carrying current I_2 is placed close to a straight conductor carrying current I_1 as shown in Figure P2.1. Obtain an expression for the magnetic force experienced by the loop.

2.2. A cylinder of radius a and length L wound closely and tightly with N turns of very fine wire is said to form a solenoid (inductor). If the wire carries a steady current I, find the magnetic flux density at any point on the axis of the solenoid. What is the magnetic flux density at the center of the cylinder? Also obtain expressions for the magnetic flux density at its ends.

2.3. A rectangular conducting loop has a sliding side moving to the right as shown in Figure P2.3. The loop is placed in a uniform magnetic field $\boldsymbol{B}$, which is normal to its plane. Calculate (a) the motional emf, (b) the transformer emf, and (c) the total induced emf.

2.4. Solve Problem 2.3 using Faraday's law of induction. What must be the external force applied to the sliding side to keep it moving with a uniform velocity?

2.5. If the magnitude of the $\boldsymbol{B}$ field in Problem 2.3 varies as $B_0 \cos \omega t$, what is the total voltage induced in the rectangular loop?

2.6. A dc generator is constructed by having a metal cart travel around a set of perfectly conducting rails forming a large circle. The rails are L m apart and there is a uniform magnetic $\boldsymbol{B}$ field normal to the plane as shown in Figure P2.6. The cart has a mass m. It is driven by a rocket engine having a constant thrust F_0. A resistor R is connected as a load. Obtain an expression for the current as a function of time. What is the current after the generator attains the steady-state condition?

2.7. A conducting wire of length L is pivoted at one end and rotates in the

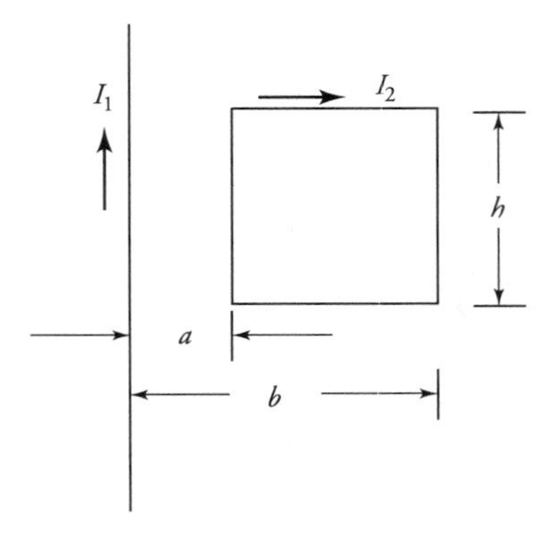

Figure P2.1 A conductor and a loop.

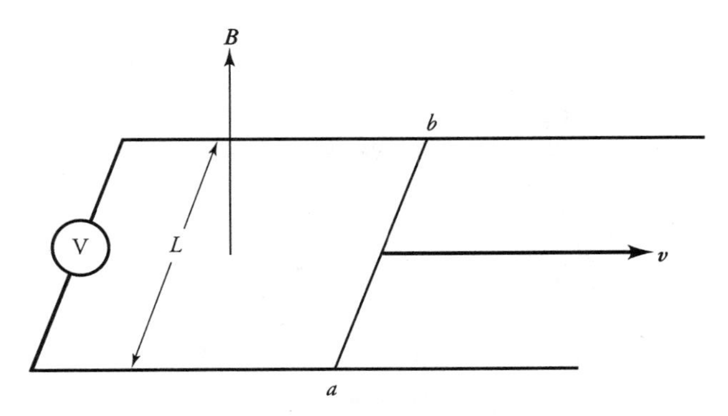

Figure P2.3 Induced emf in a loop with a sliding side.

xy-plane at an angular velocity ω. If a constant $\boldsymbol{B}$ field is directed along the z-direction, determine the induced emf between the two ends of the wire. Which end of the wire is positive with respect to the other?

2.8. A 200-turn circular coil has a mean area of 10 cm^2, and the plane of the coil makes an angle of 30° with the uniform magnetic flux density of 1.2 T. Calculate the torque acting on the coil if it carries a current of 50 A.

2.9. A 10-turn coil, 10 cm × 20 cm, is placed in a magnetic field of 0.8 T. The coil carries a current of 15 A and is free to rotate about its long axis. Plot the torque experienced by the coil against the displacement angle of the coil for one complete rotation.

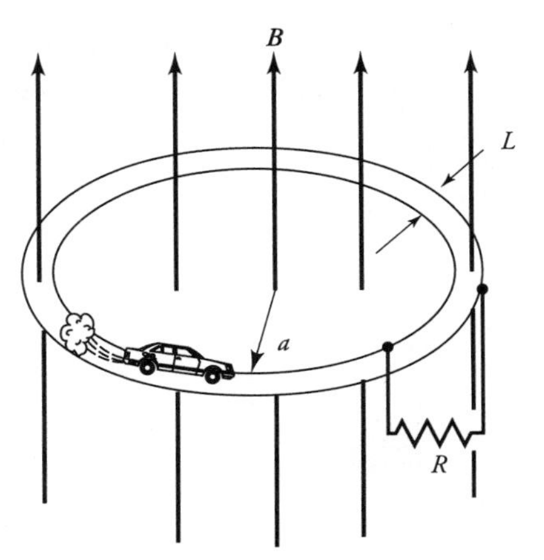

Figure P2.6 A direct-current generator.

2.10. A D'Arsonval meter is designed to have a coil of 25 turns mounted in a magnetic field of 0.2 T. The coil is 4 cm long and 2.5 cm broad. The restoring torque of the meter is accomplished by a spring and is proportional to the deflection angle θ. The spring constant is 50 μNm/deg. The scale covers 50° of arc and is divided into 100 equal parts. Calculate the current through the coil (a) per degree of deflection, (b) per scale division, and (c) for full-scale deflection. The meter's design is such that the magnetic field is always in the radial direction with respect to the axis of the coil.

2.11. A linear conductor with its ends at (-3, -4, 0) m and (5, 12, 0) m carries a current of 250 A. If the magnetic flux density in free space is 0.2 $\boldsymbol{a}_z$ T, determine the magnetic force acting on the conductor.

2.12. A metallic rod 1.2 m in length and having a mass of 500 gm is suspended by a pair of flexible leads in a magnetic field of 0.9 T as shown in Figure P2.12. Determine the current needed to remove the tension in the supporting leads.

2.13. A uniformly distributed toroidal coil of 400 turns is wound over an iron ring of square cross-section and carries a current of 200 A. The inner radius is 10 cm and the outer radius is 12 cm. The relative permeability of the iron ring is 1500. Determine (a) the flux in the ring, (b) the reluctance of the ring, and (c) the equivalent inductance of the toroid.

2.14. We wish to establish an air-gap flux density of 1.0 T in the magnetic circuit shown in Figure 2.14. If $A_g = A_m = 40$ cm^2, $\ell_g = 0.5$ mm, $\ell_m = 1.2$ m, $N = 100$ turns, and $\mu_r = 2500$, determine the current in the coil using (a) the reluctance concept and (b) the field equations.

2.15. Figure P2.15 shows a magnetic circuit made of a magnetic material for

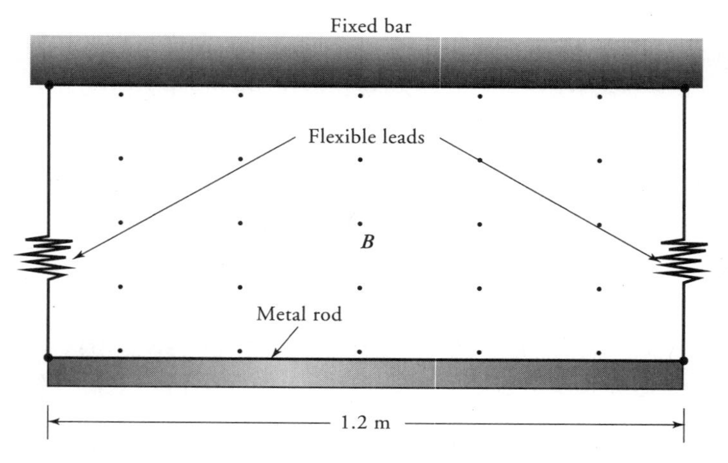

Figure P2.12 A metal bar suspended with flexible leads.

which the magnetization curve given in Figure 2.19 holds. What must be the current in the 1600-turn coil to establish a flux density of 0.8 T in the air-gap? State all the assumptions. What is the reluctance of each magnetic section? Calculate the total reluctance and the inductance of the magnetic circuit.

2.16. A magnetic circuit is given in Figure P2.16. What must be the current in the 1600-turn coil to set up a flux density of 0.1 T in the air-gap? What is the inductance of the magnetic circuit? What is the energy stored in it? All dimensions are in centimeters.

2.17. A magnetic circuit in Figure P2.17 is made of silicon steel, for which the relative permeability is given in Figure 2.19. The outer legs have 500 turns each. It is required to establish a flux of 3.6 mWb in the air-gap by applying equal currents to both windings. What is the current in each winding?

2.18. The cross-section of a 2-pole dc machine with pertinent dimensions in millimeters is given in Figure P2.18. Analyze the magnetic circuit and determine the required mmf per pole to set up a flux of 5.46 mWb in the air-gap. The *B-H* curve for the magnetic material is given in Figure 2.19. The active length (stack length) of the machine is 150 mm. Using the equivalent reluctance circuit, verify the mmf requirements per pole.

2.19. The magnetic circuit of a 4-pole dc machine with an active length of 56 mm is shown in Figure P2.19. Determine the mmf per pole needed to establish a flux density of 1.0 T in the air-gap. Assume that the effective arc of the air-gap is the same as the width of the pole. Use the *B-H* curve as given in Figure 2.19. Using the equivalent reluctance circuit, verify the mmf requirements per pole.

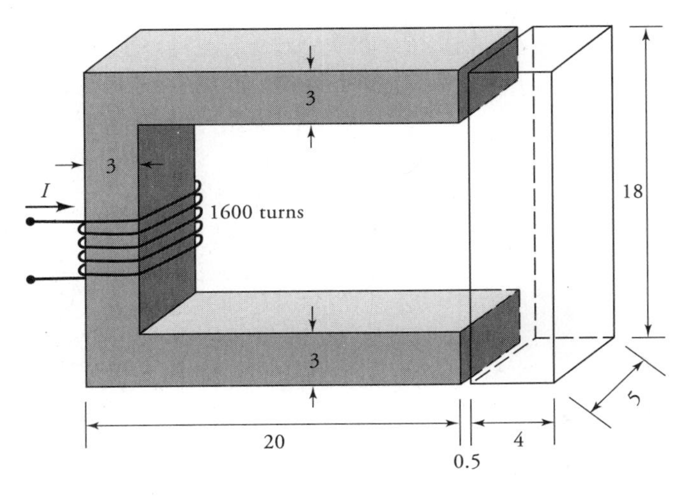

Figure P2.15 Magnetic circuit for Problem 2.15.

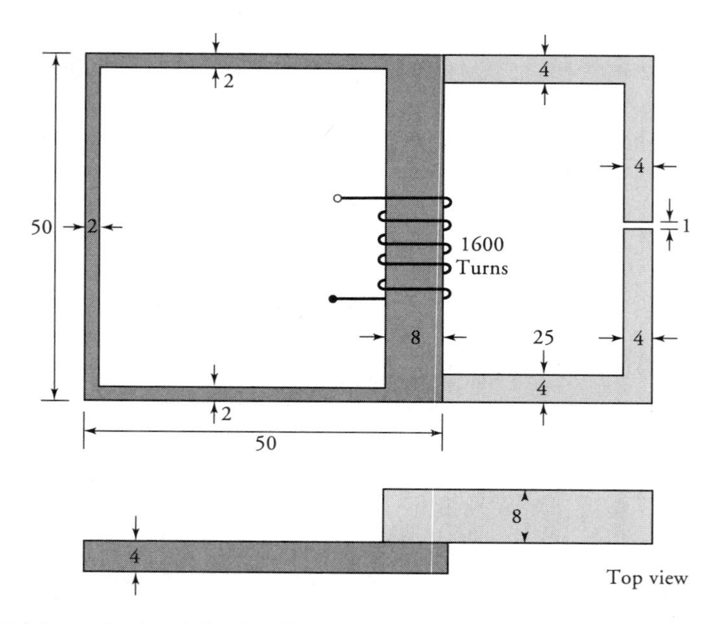

Figure P2.16 Magnetic circuit for Problem 2.16.

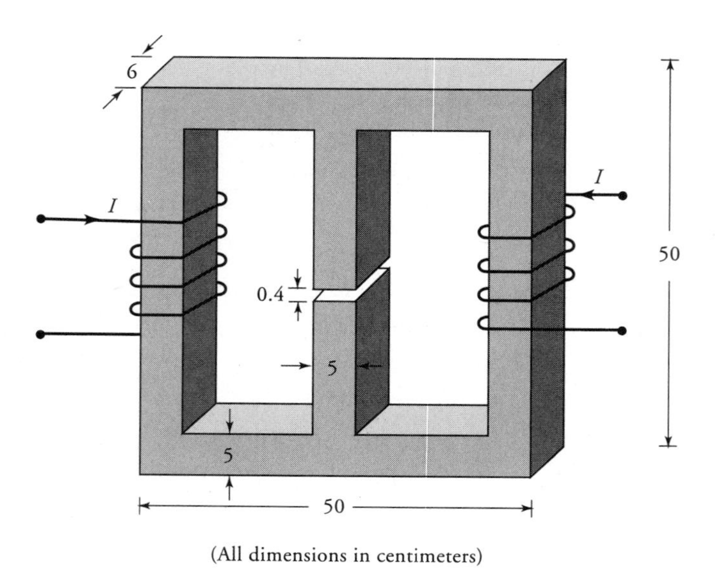

Figure P2.17 Magnetic circuit for Problem 2.17.

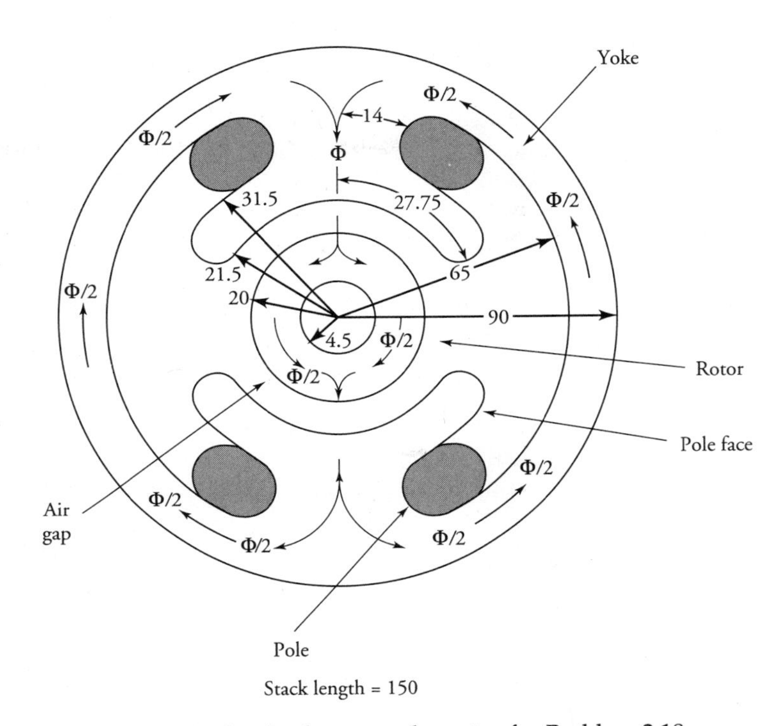

Figure P2.18 The magnetic circuit of a two-pole motor for Problem 2.18.

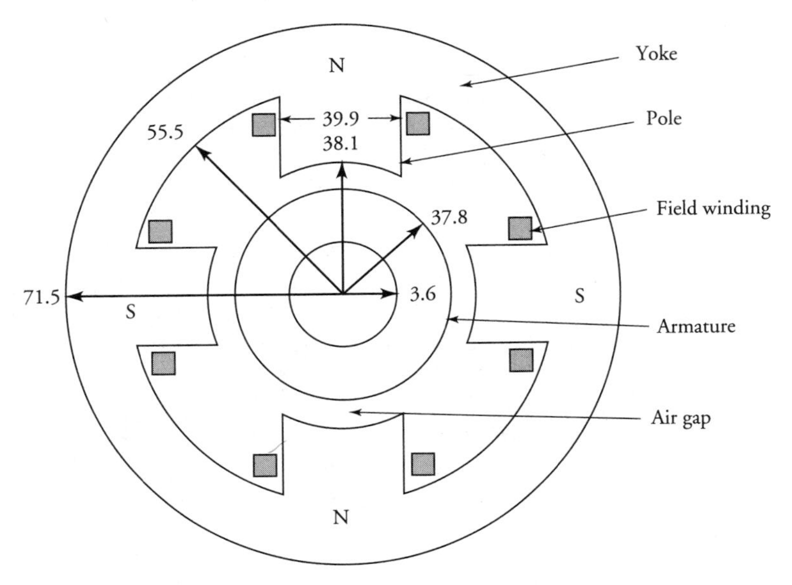

Figure P2.19 Magnetic circuit of a 4-pole motor for Problem 2.19.

2.20. An iron ring with a 50-cm mean diameter has a square cross-section of 4 cm^2. It is wound with two coils having 1000 and 400 turns. Calculate (a) the self inductance of each coil, and (b) the mutual inductance between them. Assume $\mu_r = 500$. State all the assumptions.

2.21. Two coils having 100 and 500 turns are loosely wound on a magnetic material with a 40-cm^2 cross-section and a mean length of 120 cm. Only 60% of the flux produced by one coil links the other coil. If the current changes linearly from 0 to 100 A in 10 ms in the 100-turn coil, determine the induced emf in the 500-turn coil. What is the mutual inductance between them? Assume $\mu_r = 600$.

2.22. Coils A and B are wound on a nonmagnetic core. An emf of 5 V is induced in coil-A when the flux linking it changes at the rate of 10 mWb/s. A current of 5 A in coil-B causes a flux of 2 mWb to link coil-A. Determine the mutual inductance between them. [*Note:* For a linear magnetic circuit, $M = N_1\Phi_{12}/I_2 = N_2\Phi_{21}/I_1$.]

2.23. Two coils are tightly wound on a magnetic core having a mean length of 80 cm and a core diameter of 2.4 cm. The relative permeability of the magnetic material is 1000. If the mutual inductance between the coils is 25 mH and one coil has 200 turns, determine the number of turns on the other coil.

2.24. A 3-mH coil is magnetically coupled to another 12-mH coil. The coefficient of coupling is 0.5. Calculate the effective inductance when the coils are connected in (a) series aiding, (b) series opposing, (c) parallel aiding, and (d) parallel opposing.

2.25. The effective inductances when two coils are connected in series aiding and series opposing are 220 mH and 40 mH. If the inductance of one coil is 2.25 times the inductance of the other, determine the inductance of each coil, the mutual inductance between them, and the coefficient of coupling.

2.26. A coil of inductance 90 mH is magnetically coupled to another coil of inductance 40 mH. The effective inductance when the two coils are connected in parallel aiding is 39.375 mH. When connected in parallel opposing, the effective inductance is 7.159 mH. Determine the mutual inductance between them.

2.27. For the coupled circuit shown in Figure P2.27, find the branch currents and the power supplied by each source.

2.28. For the coupled circuit given in Figure P2.28, determine (a) the current through each coil, (b) the voltage drop across each coil, and (c) the power supplied by the source.

2.29. Determine (a) the current in each branch, (b) the voltage drop across each coil, and (c) the average power supplied by the source if the coefficient of coupling between the two coils in Figure P2.29 is 0.8.

2.30. Coil-1 with a resistance of 0.2 Ω and a self inductance of 20 mH is tightly ($k = 1$) coupled to coil-2. The resistance and the self inductance of coil-2 are 0.8 Ω and 80 mH, respectively. Coil-2 is connected to an inductive load of 30 Ω in series with a 20-mH inductance. If the rms value of the impressed

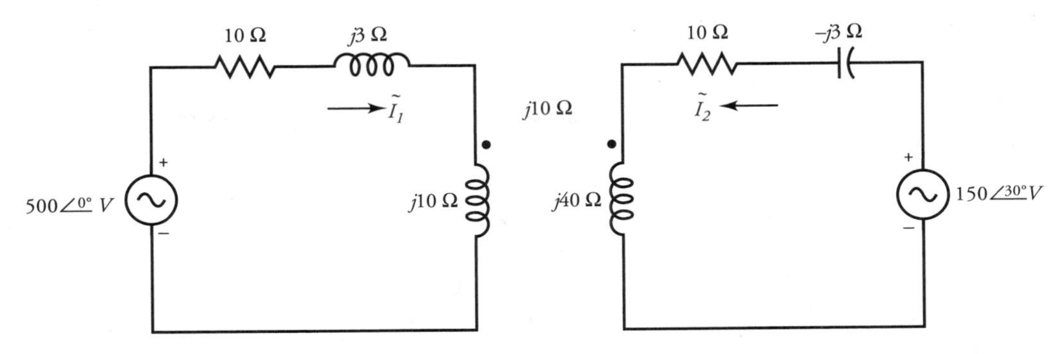

Figure P2.27 Coupled circuit for Problem 2.27.

voltage on coil-1 is 480 V and its frequency is 2000 rad/s, what is the power output (supplied to the load)? What is the power input? Find the efficiency, that is, the ratio of power output to power input.

2.31. In the circuit given in Figure P2.31, the switch is closed at $t = 0$. Determine the voltage drop across each coil as a function of time. What is the steady-state value of each voltage drop? Determine the energy stored in the magnetic system 1.5 ms after the switch is closed.

2.32. In the circuit given in Figure P2.32, the switch is closed at $t = 0$. Calculate the current in the circuit as a function of time. What is the steady-state current in the circuit? Sketch the instantaneous energy in the coupled coils as a function of time.

2.33. The eddy-current loss in a machine is 150 W at a frequency of 60 Hz and a flux density of 1.2 T. (a) What is the eddy-current loss when the frequency is raised to 400 Hz and the flux density remains unchanged? (b) What is

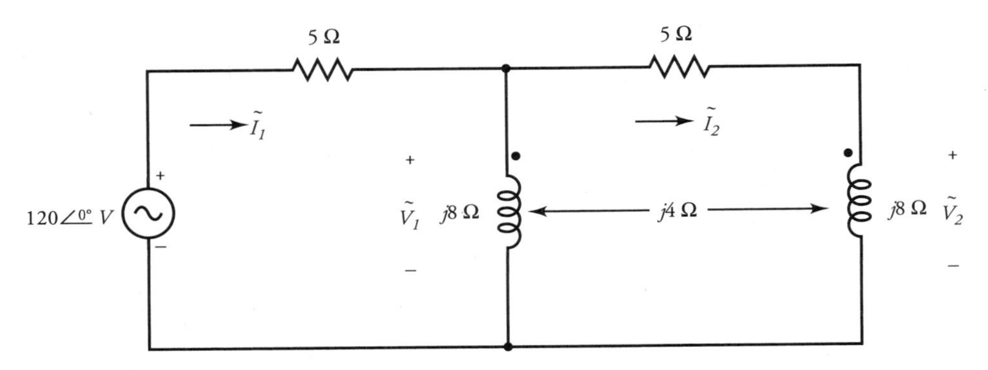

Figure P2.28 Coupled circuit for Problem 2.28.

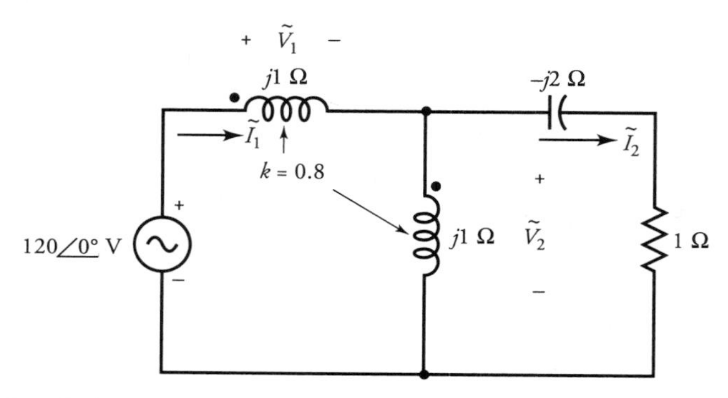

Figure P2.29 Coupled circuit for Problem 2.29.

the loss at a frequency of 60 Hz when the flux density is reduced by 25%? (c) What is the loss when the flux density is reduced by 25% and the frequency is raised to 400 Hz?

2.34. The hysteresis loss in a machine is 100 W at a frequency of 50 Hz and a flux density of 0.8 T. The Steinmetz coefficient is 1.5. (a) What is the loss if the frequency is increased to 400 Hz but the flux density remains the same? (b) What is the loss if the frequency remains the same but the flux density is increased by 25%? (c) What is the loss if the frequency is raised to 400 Hz and the flux density is increased by 25%?

2.35. In order to determine the hysteresis loss in a magnetic material, data points were taken through a complete cycle of hysteresis loop and plotted according to the following scales: H: 1 cm = 50 A·t/m, and B: 1 cm = 0.2 T. The area of the hysteresis loop is 6.25 cm^2. If the test frequency is 60 Hz and the volume of the magnetic material is 450 cm^3, find the hysteresis loss. If the density of the magnetic material is 7.8×10^3 kg/m^3, calculate the hysteresis loss in watts/kg.

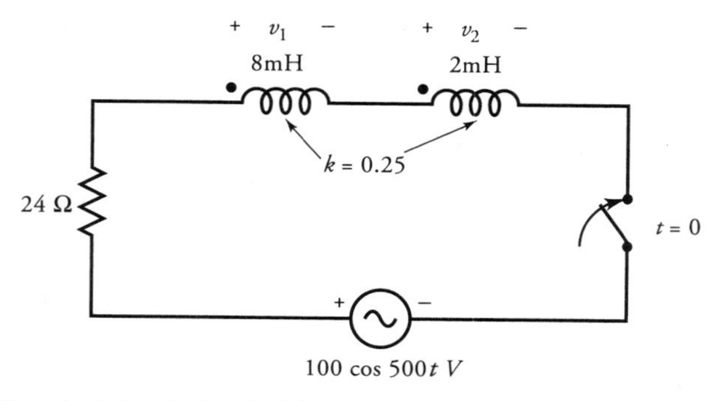

Figure P2.31 Coupled circuit for Problem 2.31.

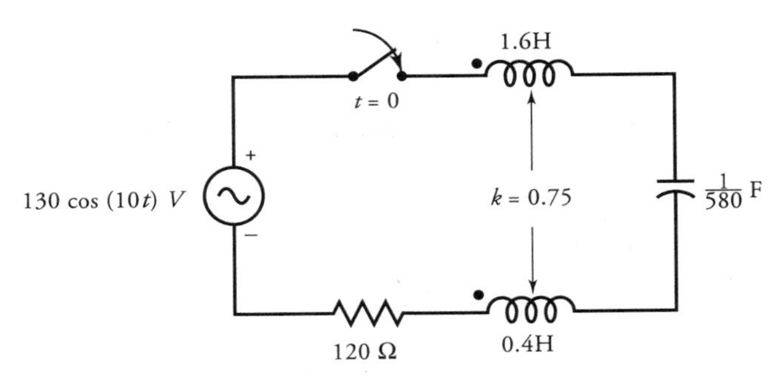

Figure P2.32 Circuit for Problem 2.32.

2.36. A hysteresis loop is plotted according to the scales 1 cm = 100 A·t/m and 1 cm = 0.1 T. The area of the loop is 20 cm^2. Compute the hysteresis loss for the specimen in joules per cubic meter per cycle.

2.37. The following data points were taken on a thin sheet of steel.

Frequency (Hz)	Flux Density (T)	Magnetic Loss (W/kg)
50	1.0	3.60
50	1.2	4.98
60	1.0	4.80

Compute the eddy-current and hysteresis losses in the specimen at (a) 50 Hz, 1 T, (b) 50 Hz, 1.2 T, (c) 60 Hz, 1 T, and (d) 60 Hz, 1.2 T.

2.38. The following data points were taken on a thin sheet of steel.

Frequency (Hz)	Flux Density (T)	Magnetic Loss (W/kg)
60	1.0	1.92
60	1.5	4.22

Compute the eddy-current and hysteresis losses in the specimen at both flux densities. The Steinmetz coefficient is 1.75.

2.39. A magnetic circuit using two rare-earth magnets is given in Figure P2.39 with its dimensions in centimeters. The thickness is 5 cm. Find the flux density and the total flux in the air-gap if the permeability of the magnetic material is infinite.

2.40. Repeat Problem 2.39 if the relative permeability of the magnetic material and the rotor is 500.

2.41. A magnetic circuit using rare-earth magnets is shown in Figure P2.41. The arc of the magnet subtends an angle of 120°, as indicated. The axial length is 150 mm. What must be the length of each magnet so that it operates at its maximum energy-product level? Assume that there is no mmf drop in

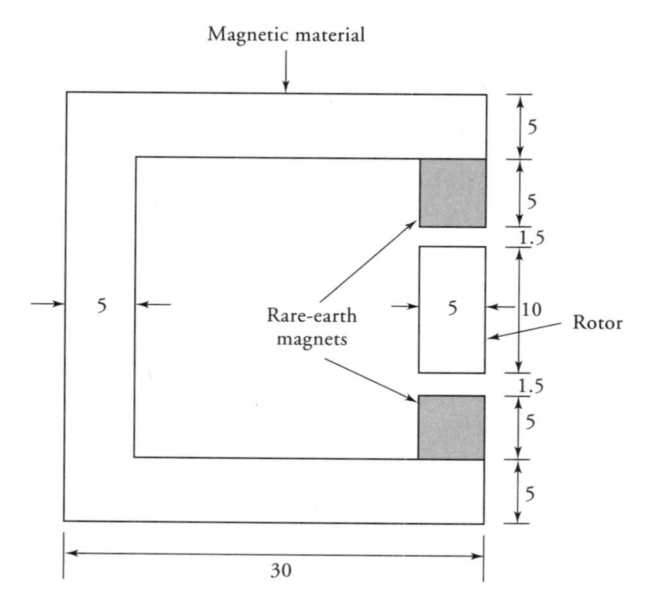

Figure P2.39 Magnetic circuit for Problem 2.39.

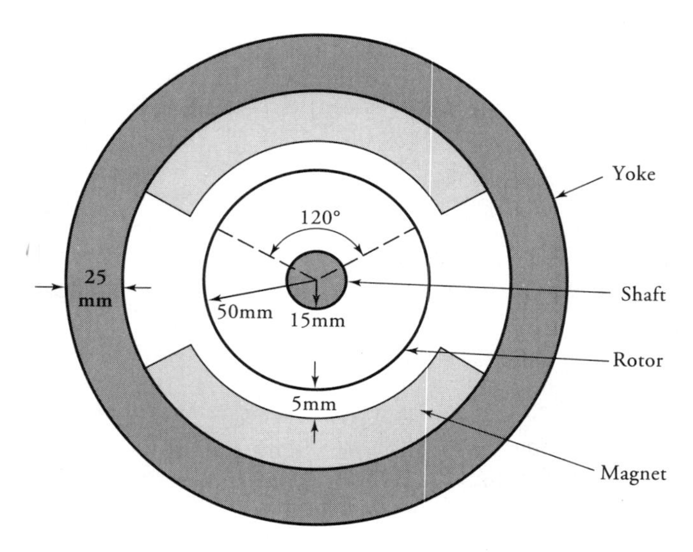

Figure P2.41 Magnetic circuit for Problem 2.41.

the rotor and the yoke. What is the induced emf in a 100-turn coil if it rotates at an angular velocity of 120 rad/s? What is the frequency of the induced voltage?

2.42. If the relative permeability of the magnetic material used for the rotor and the yoke in Problem 2.41 is 500 and the length of the magnet is the same as computed in Problem 2.41, what is the flux density in the air-gap? Also compute the induced emf in the 100-turn coil rotating at 120 rad/s.

2.43. A magnetic circuit using a ceramic magnet is made in the form of a toroid. The mean radius of the toroid is 10 cm, and its cross-sectional area is 3 cm^2. It has an air-gap of 3 cm. Find the flux density and the flux in the air-gap.

2.44. A magnetic circuit using a permanent magnet is made in the form of a toroid. The mean radius of the toroid is 8 cm, and the cross-sectional area is 2 cm^2. The retentivity is 1.6 T, and the coercivity is -80 kA/m. The demagnetization curve is basically a straight line. What must be the length of the air-gap so that the magnet operates at its maximum energy level? What are the flux density and the flux in the air-gap?

2.45. If the air-gap in Problem 2.44 is increased to 4.65 cm, what must be the flux density and the flux in the air-gap?

CHAPTER 3

Principles of Electromechanical Energy Conversion

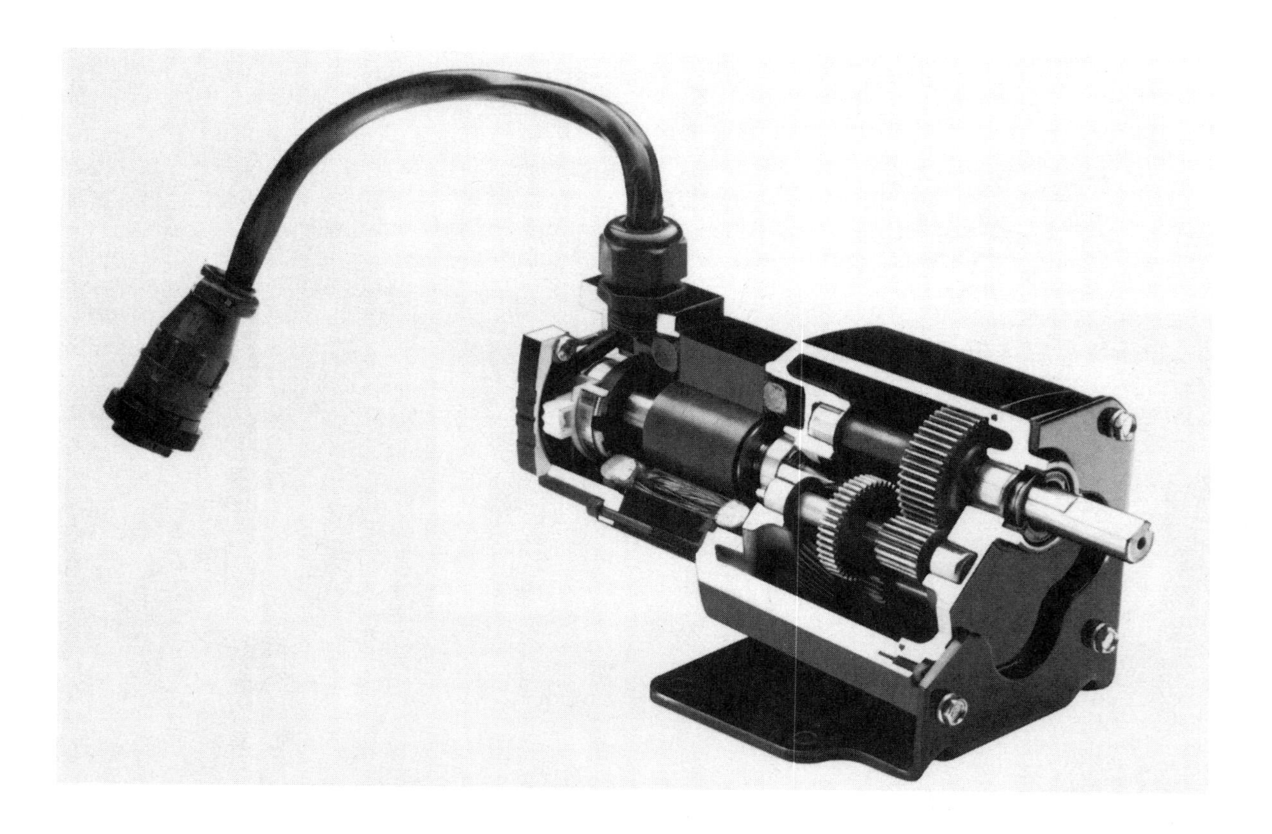

Sectional view of a brushless gearmotor. (*Courtesy of Bodine Electric Company*)

3.1 Introduction

We all know that energy exists in many forms, and we use numerous devices on a daily basis that convert one form of energy into another. When we speak of electromechanical energy conversion, however, we mean either the conversion of electric energy into mechanical energy or vice versa. For example, an electric motor converts electric energy into mechanical energy. On the other hand, an electric generator transforms mechanical energy to electric energy.

Electromechanical energy conversion is a reversible process except for the losses in the system. The term "reversible" implies that the energy can be transferred back and forth between the electrical and the mechanical systems. However, each time we go through an energy conversion process, some of the energy is converted into heat and is lost from the system forever. In this chapter, our aim is to explore the basic principles of electromechanical energy conversion.

When a current-carrying conductor is placed in a magnetic field, it experiences a force that tends to move it. If the conductor is free to move in the direction of the magnetic force, the magnetic field aids in the conversion of electric energy into mechanical energy. This is essentially the principle of operation of all electric motors. On the other hand, if an externally applied force makes the conductor move in a direction opposite to the magnetic force, the mechanical energy is converted into electric energy. Generator action is based upon this principle. In both cases, the magnetic field acts as a medium for the energy conversion.

The energy transfer process also takes place when the electric field is used as the medium. Consider the two oppositely charged plates of a capacitor which are separated by a dielectric medium. A force of attraction exists between the two plates that tends to move them together. If we let one plate move in the direction of the force, we are essentially converting electric energy into mechanical energy. On the other hand, if we apply an external force on one plate and try to increase the separation between them, we are then converting mechanical energy into electric energy. Electrostatic transducers, such as an electrostatic microphone and an electrostatic voltmeter, use electrostatic fields for the conversion of energy.

The conversion of energy from one form into another satisfies the principle of conservation of energy. Therefore, the input energy W_i is equal to the sum of the useful output energy W_o, the loss in energy as heat W_ℓ, and the change in the stored energy in the field W_f. That is,

$$W_i = W_o + W_\ell + W_f \tag{3.1}$$

The energy flow diagram is shown in Figure 3.1. The output energy and the loss in energy are considered positive quantities. The change in the stored energy, on the other hand, may be positive or negative, depending upon whether it is increasing or decreasing. In the above equation, if W_i represents the electrical

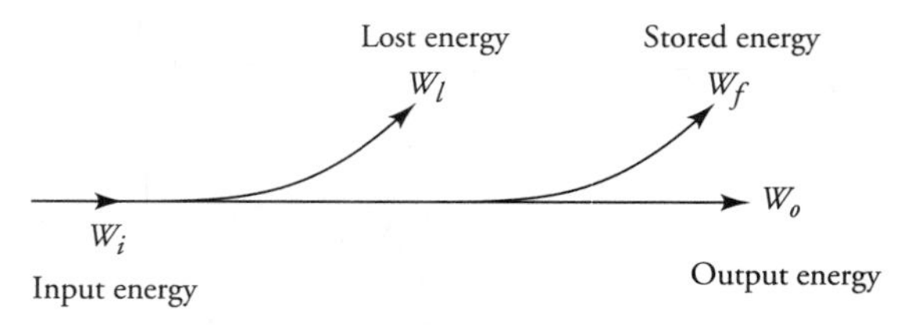

Figure 3.1 Energy flow diagram.

energy input, then W_o may be the electric equivalent of the mechanical energy output.

The system is said to be conservative or lossless if the loss in energy in the system is zero. In that case, Eq. (3.1) becomes

$$W_i = W_o + W_f \tag{3.2}$$

Note that there is no such restriction that the input energy has to be either electric or mechanical. In fact, in some electric machines, such as a synchronous machine, the input energy is both mechanical and electric. The output of a machine is usually either mechanical or electric. If the output energy of a system is zero, then the input energy must either (a) increase the stored energy of the system, (b) be dissipated as heat by the system, or (c) both.

3.2 Electric Field as a Medium

In the study of electrostatic fields, you may have derived many different equations to obtain the electrostatic energy of a charged system. Our aim in this section is to show how to calculate the force on one of the objects in a charged system from knowledge of the electrostatic energy in that system.

To simplify our development, let us consider a parallel plate capacitor as shown in Figure 3.2. If x is the separation between the plates and A is the cross-sectional area of each plate, then the electric field intensity $\boldsymbol{E}$ (V/m) in the region between the plates is

$$\boldsymbol{E} = -\frac{V}{x}\,\boldsymbol{a}_x \tag{3.3}$$

where V is the potential difference between the plates and $\boldsymbol{a}_x$ is the unit vector in the x-direction as shown. If we assume that the charge is uniformly distributed over each plate, then the total charge on the top plate is

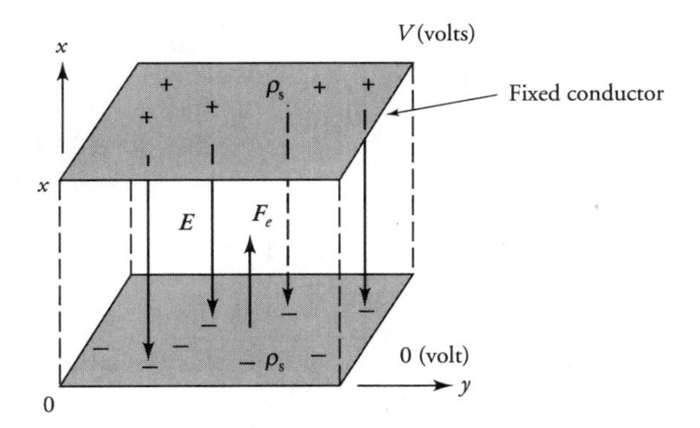

Figure 3.2 Force of attraction between two charged parallel plates.

$$Q = DA = \frac{\epsilon A}{x} V \tag{3.4}$$

where $\boldsymbol{D} = \epsilon \boldsymbol{E}$ is the electric flux density (C/m^2).

Then, the capacitance of the parallel-plate capacitor is

$$C = \frac{\epsilon A}{x} \tag{3.5}$$

The electric energy stored in the capacitor is

$$W_e = \frac{1}{2} C V^2 \tag{3.6}$$

A force of attraction exists between the two plates in accordance with Coulomb's law. Let us assume that the top plate is fixed and the lower plate is free to move. The force of attraction between the two plates tends to move the lower plate in the x-direction. If the lower plate moves a distance dx in time dt, then the change in the electrostatic energy in the system under lossless conditions, from Eq. (3.2), is

$$dW_e = dW_i - dW_o \tag{3.7}$$

where dW_e is the change in the stored energy in the capacitor. The change in the input energy is

$$dW_i = V\,dQ \tag{3.8}$$

because the electric energy input in time dt is $VI\,dt$ and $dQ = I\,dt$. The mechanical energy output in time dt is

$$dW_o = F_e\,dx \tag{3.9}$$

where F_e is the electric force acting on the bottom plate. Thus, we can rearrange and write Eq. (3.7) as

$$F_e\,dx = V\,dQ - dW_e \tag{3.10}$$

In an electrostatic system, both the energy stored in the system and the charge are functions of the applied voltage and the separation between the plates. Therefore, the differential changes in the stored energy and the charge can be written as

$$dW_e = \frac{\partial W_e}{\partial x}\,dx + \frac{\partial W_e}{\partial V}\,dV \tag{3.11}$$

$$dQ = \frac{\partial Q}{\partial x}\,dx + \frac{\partial Q}{\partial V}\,dV \tag{3.12}$$

Substituting Eqs. (3.11) and (3.12) in Eq. (3.10) and dividing both sides by dx, we get

$$F_e = \left[V\frac{\partial Q}{\partial x} - \frac{\partial W_e}{\partial x}\right] + \left[V\frac{\partial Q}{\partial V} - \frac{\partial W_e}{\partial V}\right]\frac{dV}{dx}$$

However, F_e must be independent of the incremental changes in dx and dV because they are quite arbitrary. That is, $dV/dx = 0$. Consequently, the electric field intensity in the region between the two plates is constant. Thus,

$$F_e = V\frac{\partial Q}{\partial x} - \frac{\partial W_e}{\partial x} \tag{3.13}$$

Equation (3.13) is a general equation to determine the force acting on a charged body in a charged system when the electric field in the medium is held constant. However, for a parallel-plate capacitor, $Q = CV$ and

$$\frac{\partial Q}{\partial x} = C\frac{\partial V}{\partial x} + V\frac{\partial C}{\partial x}$$

From Eq. (3.6),

$$\frac{\partial W_e}{\partial x} = \frac{1}{2}V^2\frac{\partial C}{\partial x} + CV\frac{\partial V}{\partial x}$$

Therefore, for a parallel-plate capacitor, Eq. (3.13) becomes

$$F_e = \frac{1}{2} V^2 \frac{\partial C}{\partial x} \tag{3.14}$$

However, from Eq. (3.5), we have

$$\frac{\partial C}{\partial x} = -\frac{\epsilon A}{x^2}$$

Substituting for $\partial C/\partial x$ in Eq. (3.14), we obtain the force acting on the bottom plate as

$$F_e = -\frac{1}{2} \epsilon A \left[\frac{V}{x}\right]^2 = -\frac{1}{2\epsilon A} Q^2 \tag{3.15}$$

The negative sign highlights the fact that F_e is a force of attraction. We can consider two special cases as follows:

Case 1. An Isolated System

In this case we are dealing with an isolated system in which the **charge is constant**. Therefore, the rate of change of charge with displacement is zero; That is, $\partial Q/\partial x = 0$. Hence, the electric force acting on a conductor, from Eq. (3.13), is

$$F_e = -\frac{\partial W_e}{\partial x} \tag{3.16}$$

when the **charge is constant**.

Case 2. System with Fixed Potential

In this case all of the free charges exist on the surfaces of conductors, and each conductor is maintained at a fixed potential by means of external sources of energy. In this case, we can write

$$V \frac{\partial Q}{\partial x} = \frac{\partial QV}{\partial x} = 2 \frac{\partial W_e}{\partial x}$$

because $W_e = \frac{1}{2} QV$ in a system of two conductors.

We can now express Eq. (3.13) as

$$F_e = \frac{\partial W_e}{\partial x} \tag{3.17}$$

when the **potential is held constant.**

EXAMPLE 3.1

The upper plate of a parallel-plate capacitor is held stationary while the lower plate is free to move as shown in Figure 3.3. The surface area of each plate is 20 cm^2, and the separation is 5 mm. Determine the mass of an object suspended from the lower plate that keeps it stationary when the potential difference between the two plates is 10 kV. What is the energy stored in the electric field?

● SOLUTION

To keep the lower plate stationary, the net force acting on the plate must be zero. That is,

$$mg = F_e$$

where F_e is the magnitude of the electric force of attraction experienced by the lower plate.

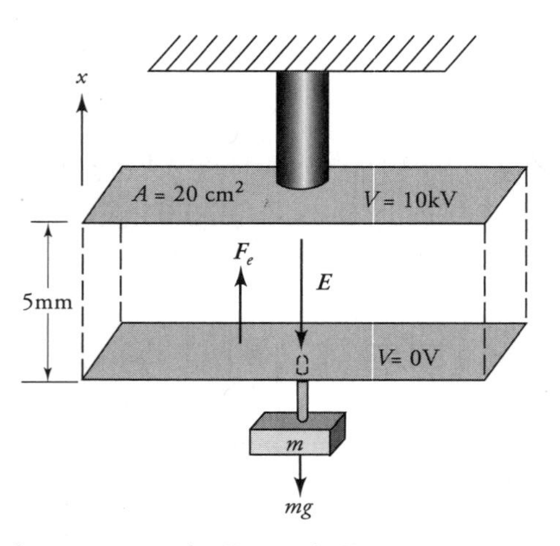

Figure 3.3 A parallel-plate capacitor for Example 3.1.

However, F_e from Eq. (3.15) is

$$F_e = \frac{1}{2} \times \frac{10^{-9}}{36\pi} \times 20 \times 10^{-4} \times \left[\frac{10 \times 10^3}{5 \times 10^{-3}}\right]^2$$

$$= 35.37 \times 10^{-3}\ \text{N} \quad \text{or} \quad 35.37\ \text{mN}$$

Hence,
$$m = \frac{35.37 \times 10^{-3}}{9.81} = 3.61 \times 10^{-3}\ \text{kg} \quad \text{or} \quad 3.61\ \text{g}$$

The energy in the system is

$$W_f = \frac{1}{2} CV^2 = \frac{\epsilon A}{2x} V^2 = \frac{10^{-9} \times 20 \times 10^{-4}}{2 \times 36\pi \times 5 \times 10^{-3}} [10 \times 10^3]^2$$

$$= 177\ \mu\text{J}$$

■

EXAMPLE 3.2

The region between a parallel-plate capacitor is partially filled with a dielectric slab, and the capacitor is charged to a potential of V volts. The width of each plate is w. The dielectric slab is then withdrawn to the position shown in Figure 3.4. Calculate the force tending to pull the slab.

● SOLUTION

The electrostatic energy stored in the parallel-plate capacitor is

$$W_e = \frac{1}{2} \int_v D \cdot E\, dv$$

$$= \frac{1}{2} \epsilon_o w\, d \left[\frac{V}{d}\right]^2 [\epsilon_r x + (b - x)]$$

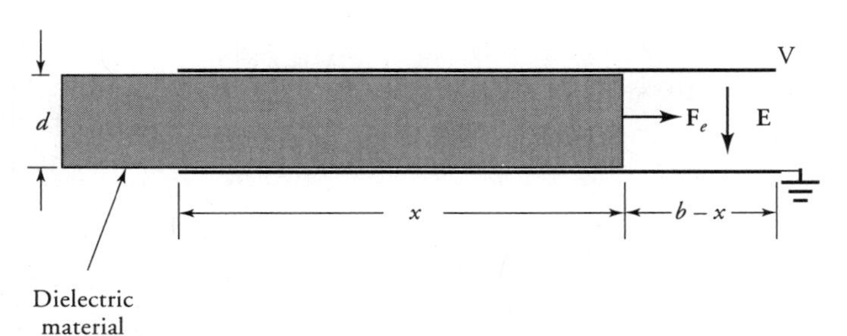

Figure 3.4 Force acting on a partially-withdrawn dielectric slab in a parallel-plate capacitor.

Because the potential is held constant, we can use Eq. (3.17) to obtain the force acting on the dielectric material as

$$F_e = \frac{w}{2d} \epsilon_o [\epsilon_r - 1] V^2$$

Note that the force is in the direction of increasing x.

■

The amount of force developed by an electric system is usually very small even when the applied voltage is high and the physical dimensions of the system are quite large. When the magnetic field is used as a medium, a system with the same physical dimensions develops a force many orders of magnitude higher than a system using an electric field as a medium, as explained in the next section.

Exercises

3.1. Two parallel plates, each having dimensions of 20 cm by 20 cm, are held 2 mm apart in air. If the potential difference between the plates is 2 kV, determine (a) the energy stored in the capacitor and (b) the force acting on each plate.

3.2. What is the magnitude of the force in Example 3.2 if $\epsilon_r = 9$, $x = 10$ cm, $b = 20$ cm, $w = 5$ cm, $d = 2$ mm, and the potential difference is 2 kV? What is the energy in the system?

3.3 Magnetic Field as a Medium

Consider a magnetic circuit with mean length ℓ and cross-sectional area A as shown in Figure 3.5. Let the current through an N-turn coil be $i(t)$ when a voltage source $v(t)$ is impressed across its terminals. The current $i(t)$ establishes a flux $\Phi(t)$ in the magnetic circuit, which induces an electromotive force (emf) $e(t)$ in the coil in accordance with Faraday's law of induction. To sustain the flux in the core of the magnetic circuit, the applied source must supply electrical energy. In a conservative system, the electrical energy input in time dt is

$$dW_i = vi\,dt = -ei\,dt$$

However, the induced emf in the coil is

$$e = -\frac{d\lambda}{dt}$$

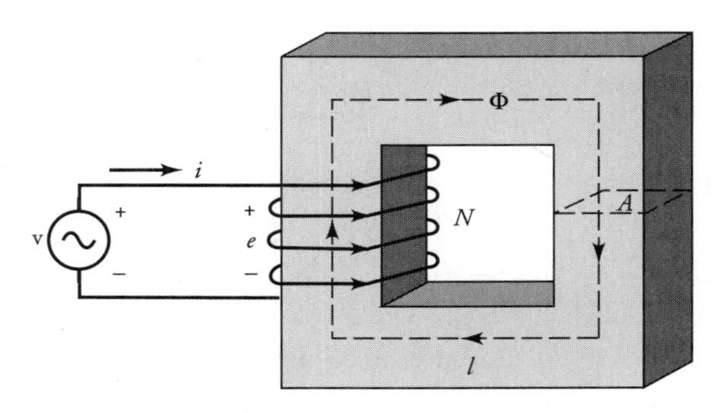

Figure 3.5 A magnetic circuit of mean-length l and cross-sectional area A.

where $\lambda = N\Phi$ represents the total flux linkages. Therefore, the input energy in time interval dt can be written as

$$dW_i = i\,d\lambda = Ni\,d\Phi$$

Hence, the energy input to set up a flux $\Phi(t)$ in an N-turn coil is

$$W_i = \int i\,d\lambda = N \int i\,d\Phi \tag{3.18}$$

In a conservative system with no output energy, the energy supplied by the source must be stored in the system as magnetic energy. In a linear system, the energy stored in the system is

$$W_m = \frac{1}{2} N\Phi i = \frac{1}{2} Li^2 \tag{3.19}$$

where the inductance L is

$$L = \frac{N\Phi}{i}$$

Since $Ni = H\ell$ and $d\Phi = A\,dB$, where H is the magnetic field intensity and B is the magnetic flux density, we can express Eq. (3.18) as

$$W_i = \ell A \int H\,dB \tag{3.20}$$

which yields the energy that must be supplied to set up a flux density B in the magnetic core.

EXAMPLE 3.3

If the relationship between the total flux linkages and the current in the coil for the magnetic circuit shown in Figure 3.5 is given as $\lambda = 6i/(2i + 1)$ weber-turns (Wb·t), determine the energy stored in the magnetic field for $0 \le \lambda \le 2$ Wb·t.

● SOLUTION

From the given relationship between the total flux linkages and the current in the coil, we obtain

$$i = \frac{\lambda}{6 - 2\lambda}$$

Thus, the energy stored in the magnetic field, from Eq. (3.18), is

$$W_m = \int_0^2 \frac{\lambda\, d\lambda}{6 - 2\lambda} = 0.648 \text{ J}$$

■

Let us obtain expressions for the magnetic force in terms of the magnetic stored energy. By analogy with the electrostatic force, we now consider the constant flux and the constant current cases separately.

Case 1. Constant Flux

Let us say a magnetic circuit is moved by a distance dx in time dt in a region where the flux is held constant. From Eq. (3.18), the energy supplied by the source must be zero because the initial and final values of the flux are the same. Then, for a conservative system, we have

$$F_m\, dx + dW_m = 0$$

Consequently, the mechanical force acting on the magnetic circuit is

$$F_m = -\frac{dW_m}{dx} \tag{3.21a}$$

when the **flux is held constant**. This equation clearly shows that the rate of decrease in the stored magnetic energy with respect to the displacement determines the force developed by the magnetic device.

Case 2. Constant Current

For a change in flux $d\Phi$ when I is held constant, the energy input from Eq. (3.18) is

$$dW_i = I\,d\Phi$$

Since

$$dW_m = \frac{1}{2} I\,d\Phi$$

the energy input can be expressed as

$$dW_i = 2\,dW_m$$

Thus, for a conservative system

$$F_m\,dx + dW_m = 2\,dW_m$$

Hence, the magnetic force is

$$F_m = \frac{dW_m}{dx} \tag{3.21b}$$

when the **current is held constant**.

Since the energy W_m can be either a function of the current I and the displacement $x[W_m(I, x)]$ or a function of flux Φ and the displacement $x[W_m(\Phi, x)]$, a convenient way to express Eqs. (3.21a) and (3.21b) is to express them in terms of partial derivatives as

$$F_m = \frac{\partial W_m(\Phi, x)}{\partial x} \tag{3.22a}$$

when the **flux is held constant**, and

$$F_m = \frac{\partial W_m(I, x)}{\partial x} \tag{3.22b}$$

when the **current is held constant**.

Let us now express the magnetic force in terms of the inductance of the magnetic circuit. The rate of change of stored magnetic energy, from Eq. (3.19) when the flux is constant, is

$$\frac{dW_m}{dx} = \frac{1}{2} N\Phi \frac{\partial i}{\partial x} = -\frac{1}{2} i^2 \frac{\partial L}{\partial x}$$

Therefore, the magnetic force, from Eq. (3.21a), is

$$F_m = \frac{1}{2} i^2 \frac{\partial L}{\partial x} \tag{3.23}$$

We can also show that Eq. (3.23) is valid when the current is held constant. This equation is quite simple to use for linear magnetic circuits because we can determine the inductance of the magnetic circuit using the reluctance concept.

Magnetic Circuit with Air-Gap

In the design of electric machines, the magnetic circuit is often broken by the presence of an air-gap. Consider a magnetic circuit shown in Figure 3.6 with two air-gaps. The continuity of the flux in the magnetic circuit dictates that a force of attraction must exist between the two members of the magnetic circuit. As the magnetic force tends to move one magnetic piece closer to the other, the decrease in the field energy in the air-gap is responsible for the development of the force. The energy in each air-gap, from Eq. (3.20), is

$$W_g = \frac{1}{2} \mu_0 H_0^2 A x = \frac{1}{2\mu_0} B_0^2 A x$$

where $B_0 = \mu_0 H_0$ is the magnetic flux density in the air-gap.

The force developed in each air-gap, when the flux Φ ($\Phi = BA$) is held constant, from Eq. (3.22a), is

$$F_g = -\frac{1}{2} \mu_0 H_0^2 A = -\frac{1}{2\mu_0} B_0^2 A \tag{3.24}$$

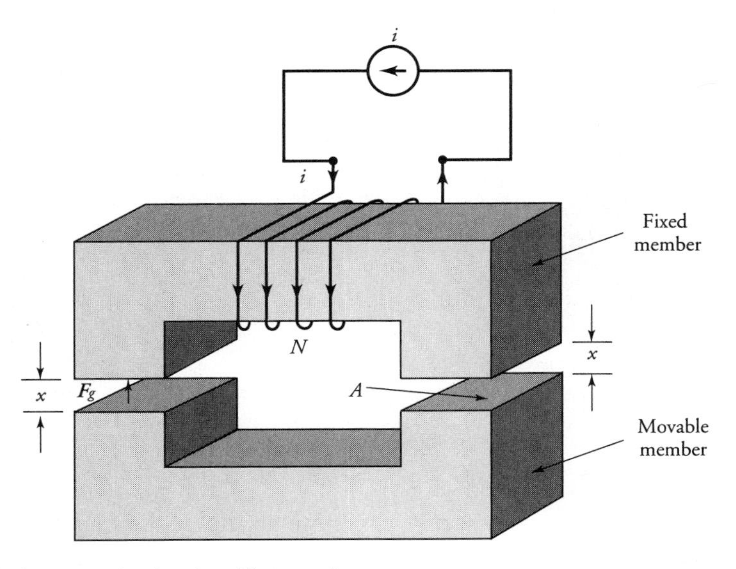

Figure 3.6 A magnetic circuit with two air-gaps.

This expression is very similar to the one obtained for the energy conversion using the electric field as a medium [Eq. (3.15)]. Once again, the presence of the negative sign in the above equation indicates that F_g is a force of attraction per air-gap that tends to decrease x.

EXAMPLE 3.4

The magnetic circuit of Figure 3.7 is excited by a 100-turn coil wound over the central leg. Determine the current in the coil that is necessary to keep the movable part suspended at a distance of 1 cm. What is the energy stored in the system? The relative permeability and the density of the magnetic material are 2000 and 7.85 g/cm^3, respectively.

● SOLUTION

Since the permeability of the magnetic material is constant, we can use the reluctance concept to determine the inductance of the magnetic circuit when the movable part is at a distance x. An equivalent circuit in terms of the reluctances is shown in Figure 3.8.

The mean length for each of the outer legs including a part of the movable

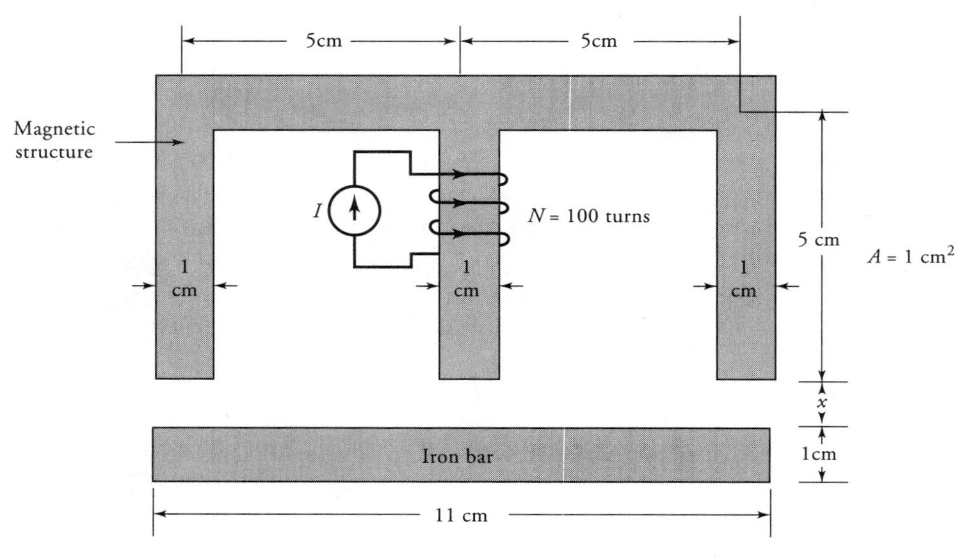

Figure 3.7 A magnetic circuit for Example 3.4.

part is 16 cm. The mean length of the central leg is 5.5 cm. Using $\mathcal{R} = \ell/\mu A$, we can compute the reluctance of each part as

$$\mathcal{R}_o = 6.366 \times 10^5 \text{ H}^{-1}$$

$$\mathcal{R}_g = 7.958 \times 10^9 x \text{ H}^{-1}$$

and

$$\mathcal{R}_c = 2.188 \times 10^5 \text{ H}^{-1}$$

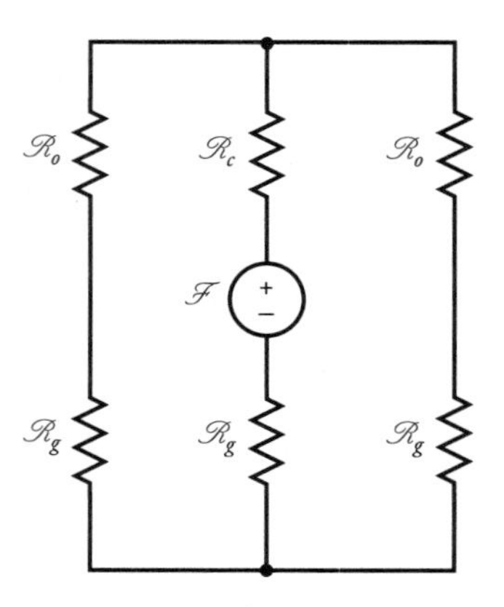

Figure 3.8 An equivalent reluctance circuit for the magnetic circuit given in Figure 3.7.

The applied mmf is $\mathcal{F} = 100\,I$ where I is the required current in the coil. The total reluctance as viewed from the magnetomotive source is

$$\mathcal{R} = \mathcal{R}_c + \mathcal{R}_g + 0.5(\mathcal{R}_o + \mathcal{R}_g) = 5.371 \times 10^5 + 11.937 \times 10^9 x$$

Hence, the inductance is

$$L = \frac{N^2}{\mathcal{R}} = \frac{1}{53.71 + 1{,}193{,}700x}$$

The magnetic force acting on the movable part, from Eq. (3.23), is

$$F_m = -\frac{596{,}850}{[53.71 + 1{,}193{,}700x]^2} I^2$$

The negative sign only highlights the fact that the force is acting in the upward direction. Therefore, the magnitude of the force of attraction for $x = 1$ cm is

$$F_m = 4.15 \times 10^{-3} I^2 \text{ N}$$

On the other hand, the force due to gravity experienced by the movable part having a volume of 11 cm^3 is

$$F_g = mg = 7.85 \times 10^{-3} \times 11 \times 9.81 = 0.847 \text{ N}$$

For the movable part to be stationary, the force of gravity must equal the magnetic force. Equating the two forces, we obtain

$$I = 14.28 \text{ A}$$

The inductance of the magnetic circuit at $x = 1$ cm is 83.4 μH. Thus, the energy stored in the magnetic field is

$$W_f = \frac{1}{2} LI^2 = 0.5 \times 83.4 \times 10^{-6} \times 14.28^2 = 8.5 \text{ mJ}$$

■

EXAMPLE 3.5

Determine the minimum amount of current required to keep the magnetic plate at a distance of 1 mm from the pole faces of an electromagnet having 1000 turns when the torque exerted by the spring at an effective radius of 20 cm is 20 N·m as shown in Figure 3.9. Assume that each pole face is 3 cm square and the magnetomotive force (mmf) requirements for the electromagnet and the magnetic plate are negligible in comparison with the air-gap.

● SOLUTION

The force exerted by the spring on the magnetic plate is

$$F_s = \frac{20}{0.2} = 100 \text{ N}$$

To keep the magnetic plate in equilibrium, the electromagnet should exert an attractive force of 100 N on the magnetic plate. If the length of the air-gap is x, then its reluctance is

$$\mathcal{R} = \frac{2x}{4\pi \times 10^{-7} \times 9 \times 10^{-4}} = 1.768 \times 10^9 x \text{ H}^{-1}$$

The inductance is

$$L = \frac{N^2}{\mathcal{R}} = \frac{1000^2}{1.768x \times 10^9} = \frac{565.49 \times 10^{-6}}{x} \text{ H}$$

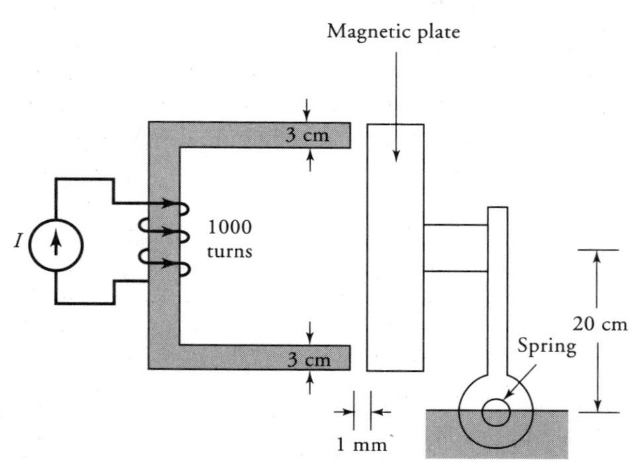

Figure 3.9 A magnetic circuit for Example 3.5.

Hence, the magnetic force exerted by the electromagnet, from Eq. (3.23), is

$$F_m = \frac{1}{2} I^2 \frac{\partial L}{\partial x} = -282.74 \times 10^{-6} \left[\frac{I}{x}\right]^2$$

Thus, the force of attraction at a distance of $x = 1$ mm is $282.74\, I^2$. By setting the force of attraction equal to 100 N, we obtain

$$I = 0.595 \text{ A}$$

■

Exercises

3.3. Repeat Example 3.4 using the field concept. [*Hint:* Assume B_o as the flux density in the outer legs; then $2B_o$ is the flux density in the central leg. Calculate the force acting on the movable part in terms of B_o.]

3.4. A magnetic circuit of a plunger is shown in Figure E3.4. Calculate the force acting on the plunger when the distance x is 2 cm and the current in the

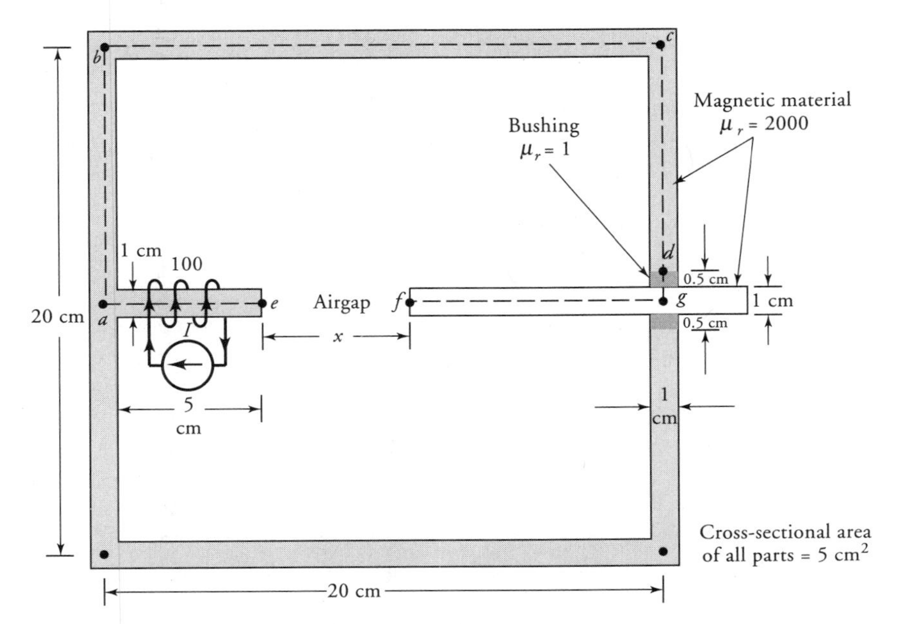

Figure E3.4 A magnetic circuit of a plunger for Exercise 3.4.

100-turn coil is 5 A. The relative permeability of the magnetic material is 2000. Assume that the bushing is nonmagnetic.

3.4 A Coil in a Uniform Magnetic Field

As pointed out in the preceding chapter, an emf is induced in a coil when it rotates in a uniform (constant) magnetic field. In fact, it is the relative motion between conductors and the constant magnetic field that is responsible for the induced emf. Therefore, it does not matter whether the field is stationary and the coil rotates, or the coil is fixed in position and the field is made to rotate. Direct-current (dc) machines and synchronous machines are based upon this principle.

The design of a dc machine calls for the establishment of a stationary magnetic field in which the coils rotate. On the other hand, the coils are held stationary and the magnetic field rotates in all synchronous machines. For that reason, a synchronous machine is referred to as an inside-out machine. The advantages and drawbacks of each design are covered in detail in the following chapters.

A rotating machine has two essential parts: a stationary part (stator) and a rotating part (rotor). The rotor of a dc machine is usually referred to as an **armature**. The outer diameter of the rotor is smaller than the inner diameter of the stator so that it can rotate freely inside the stator. Therefore, a rotating machine is a device with one continuous and uniform air-gap.

Both the stator and the rotor are made from highly permeable magnetic materials so that the reluctance of each is negligible compared with the reluctance of the air-gap. Hence, almost all the mmf in the magnetic circuit of a rotating machine is consumed to establish the required flux in the air-gap.

The constant magnetic field can be set up by either an electromagnet or a permanent magnet. An electromagnet is formed by winding a coil around a magnetic material. A machine with electromagnets is called the **wound machine**. A machine is said to be a **permanent magnet (PM) machine** when the field is set up by permanent magnets. The advantage of a wound machine is that we can control the flux in the machine by controlling the current in the coil. The PM machine has the advantages that (a) its size is smaller and (b) its efficiency is higher than that of a wound machine of the same power rating. For the sake of discussion and without any loss of generality, we will use permanent magnets to establish the magnetic field.

Generator Action

Figure 3.10 shows a cross-section of a single-turn coil being rotated in the clockwise direction in a constant magnetic field set up by two permanent magnets. This is an ideal 2-pole machine in which each magnet spans one-half the circumference. That is, the maximum possible arc that a pole can subtend in a 2-pole machine is

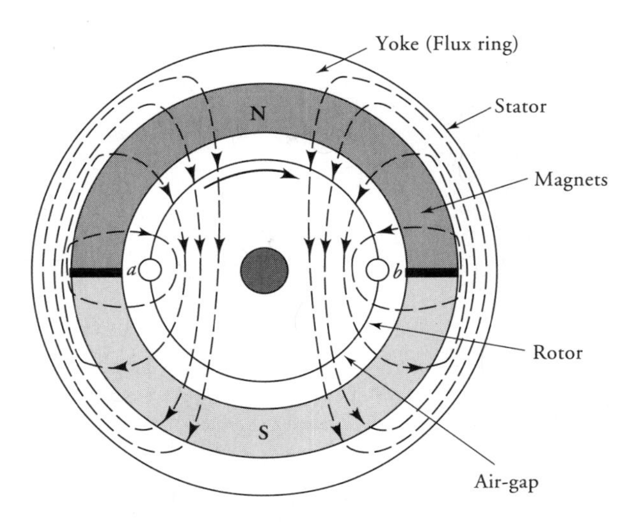

Figure 3.10 A two-pole rotating machine.

180° (mechanical). We will soon show that (a) the mechanical angle θ_m in a 2-pole machine is also equal to the electrical angle θ and (b) the maximum arc a pole can subtend is always 180° electrical.

The two sides of the coil are placed diametrically opposite (180° apart) each other. Therefore, when one side of the coil just enters (leaves) the region under the north pole, the other does the same under the south pole. When the arc subtended by the coil is equal to the 180° electrical, it is referred to as a **full-pitch** coil.

In a "real" machine, the pole does not subtend an angle of 180° electrical, nor is the coil full-pitch. The angle subtended by the pole is usually between 120° and 135° electrical. This is especially true for the wound machines. When the pitch of the coil is less than 180° electrical, it is referred to as a **fractional pitch** coil. Even though it is not shown in Figure 3.10, the periphery of the rotor has a plurality of slots. If there are, for example, ten slots per pole on the rotor, the coil may span as many as nine slots. In this case, the pitch of the coil is 162° electrical ($180 \times 9/10$).

Let us now rotate the coil in the clockwise direction. When the coil is in position as indicated by Figure 3.11**a**, the flux linking the coil is maximum and its rate of change is zero. Therefore, no voltage is induced in the coil.

As the coil moves to position in Figure 3.11**b**, the flux linking the coil is reduced. This change in the flux induces an emf in the coil in accordance with Faraday's law of induction. To determine the direction of the induced emf, visualize a resistance connected between the two ends *a* and *b* of the coil. There should now be a current in the coil. The direction of the current should be such that it establishes a magnetic field that opposes the change in the flux passing through the coil. Because the flux linking the coil is reducing, the current in the coil must

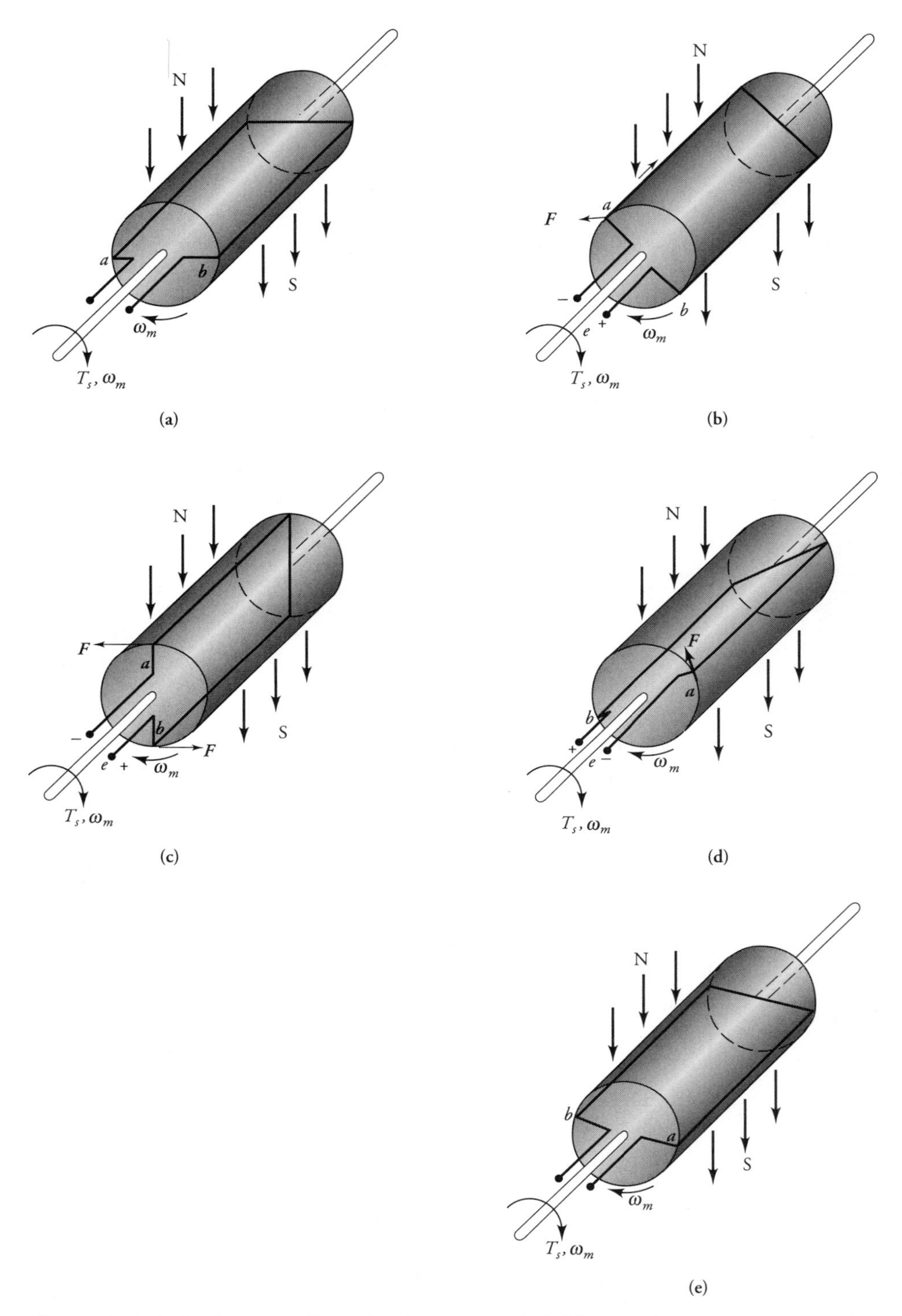

Figure 3.11 A single-turn coil rotating in a magnetic field set up by permanent magnets.

be from a to b, as indicated in the figure, in order to oppose the reduction in the flux passing through the coil. Thus, the induced emf between coil-ends b and a is positive.

Let us now assume that the coil has moved to its vertical position as illustrated in Figure 3.11**c**. The mechanical angle of rotation of the coil is 90°. The flux linking the coil is minimum. However, the rate of change of flux linking the coil is maximum. Therefore, the induced emf and the current through the resistance are maximum. This can be easily understood by considering the position of the coil just before 90°.

Figure 3.11**d** shows a position when the coil has rotated an angle greater than 90°. Since the flux linking the coil is increasing, the direction of the current in the coil due to induced emf should oppose the change. Thus, the current in the coil is still from a to b as shown in the figure. As the coil moves toward the position shown in Figure 3.11**e**, the flux linking the coil is increasing and its rate of change is decreasing. When the coil attains the position in Figure 3.11**e**, the flux linking the coil is maximum and the induced emf is zero.

As the conductor a of the coil has moved under the north pole, the induced emf in the coil has changed from zero to maximum and then back to zero. The rotating coil in a constant magnetic field, therefore, functions as a source of time-varying emf. In simple words, the machine under consideration is a **2-pole alternating-current (ac) generator**.

When a machine is drawn with its poles as shown in Figure 3.12**a**, it is known as the **developed diagram**. A developed diagram enables us to visualize what is happening under each pole.

If Φ_P is the flux per pole (Figure 3.12**b**), the flux linking the coil can be expressed as a cosine function (Figure 3.12**c**). That is,

$$\Phi = \Phi_P \cos \theta$$

where θ is the angular position of the coil in electrical degrees. Thus, the induced emf in the coil is

$$e = -\frac{d\Phi}{dt} = \Phi_P \sin \theta \frac{d\theta}{dt}$$

where $d\theta/dt$ is the angular frequency, ω, of the coil. Thus, we can rewrite the above equation as

$$e = \Phi_P \omega \sin \omega t \tag{3.25}$$

In the 2-pole generator under discussion, a positive half-cycle of the induced emf is generated when the conductor a rotates under the north pole in the clockwise direction. The other half-cycle is generated when the conductor b moves

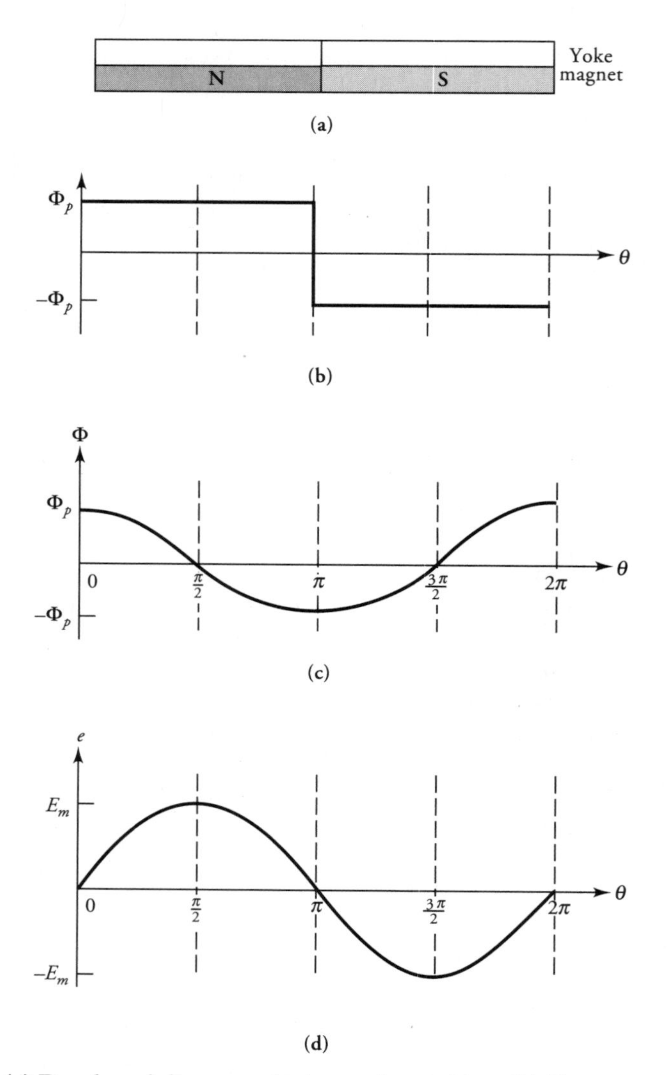

Figure 3.12 **(a)** Developed diagram of a two-pole machine. **(b)** Flux per pole set up by the magnets. **(c)** Flux passing through a coil rotating at a constant speed. **(d)** Induced emf in the coil.

under the north pole. In other words, one cycle (360° electrical) of the waveform is generated when the coil has rotated by one revolution (360° mechanical) as depicted in Figure 3.12**d**. If θ_m represents the mechanical angle of rotation, then $\theta = \theta_m$ in a 2-pole machine.

In order to gain access to the rotating terminals and be able to connect a resistance so that a current can actually flow through it, the rotating ends of the coil are connected to a pair of **slip rings** as indicated in Figure 3.13. The outer part of each slip ring is a conductor to which one end of the coil is connected. The inner part of the slip ring is an insulator that insulates its outer part from the shaft of the rotor. Spring-loaded brushes ride over the slip ring and provide means for external load connections. An alternating current flows through R as the coil is made to rotate in a stationary magnetic field.

When the machine is specifically designed to provide an alternating current to the load (an ac generator), we can eliminate the slip rings and brushes by mounting the magnets on the rotor and placing the coil inside the slots of the stationary member. All ac generators (**synchronous generators,** or **alternators**) are, in fact, designed in this fashion. Since the emf is now being induced in the winding wound on the stationary member, it is quite customary to refer to the stator of a synchronous machine as an armature.

If the two slip rings in Figure 3.13 are replaced by a **split ring**, as shown in Figure 3.14**a**, the coil-end a is permanently connected to one part of the split ring and the coil-end b to the other. As the coil rotates in the stationary magnetic field, the upper brush is always connected to the part of the split ring that is negative. Thus, the polarity of the upper brush is always negative. On the other hand, the lower brush always has a positive polarity. In other words, the current in the load resistance is always from A to B. The load current, therefore, is a full-wave rectified alternating current as shown in Figure 3.14**b**. The split-ring mechanism is com-

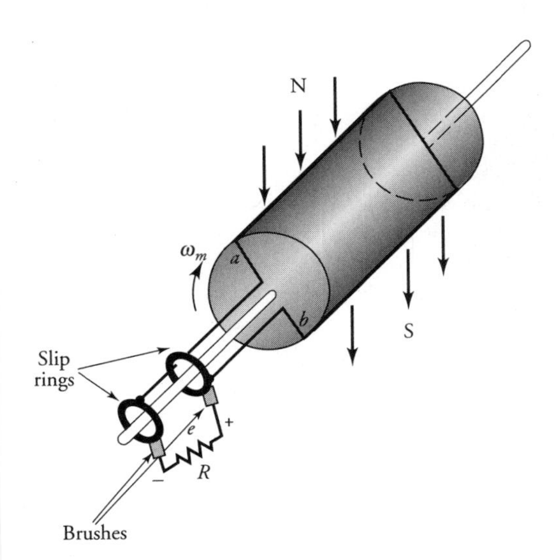

Figure 3.13 An elementary generator with slip rings and brushes.

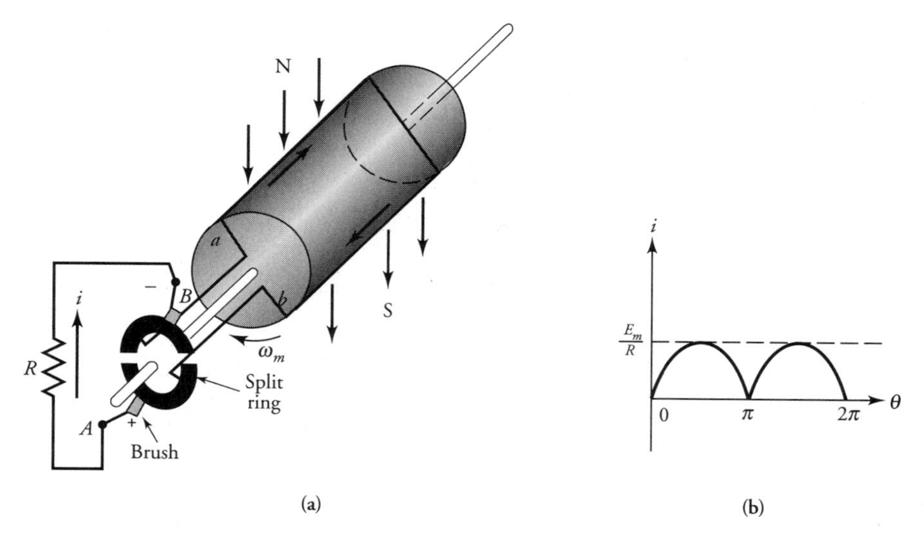

Figure 3.14 (a) A split-ring converts an ac generator into a dc generator. **(b)** The current waveform through the load resistance.

monly referred to as a **commutator**. Its function is to convert alternating current to direct current. This process of conversion is called the **commutation process**. A commutator converts an ac generator into a **dc generator**. In fact, all dc machines have similar construction.

Multipole Machines

Figure 3.15**a** shows a 4-pole machine, where the coil is wound to span a pole. Therefore, the distance between the two sides of the coil is one-fourth the circumference of the rotor. If we now rotate the coil in the clockwise direction, starting at the position indicated, there are two complete cycles of induced emf per revolution. Thus, θ is 720° while θ_m is still 360°. Stated differently, the electrical angle of the induced emf is twice as much as the mechanical angle of rotation. The flux per pole, the flux linking the coil, and the induced emf in the coil are shown in Figures 3.15**b**, 3.15**c**, and 3.15**d**, respectively.

Figure 3.16 shows a 6-pole machine where the coil now spans one-sixth the circumference of the rotor. Once again, the coil spans one pole as shown. Therefore, it is a full-pitch coil. As we rotate the coil we expect three cycles of induced emf per revolution. Therefore, for $\theta_m = 360°$, $\theta = 1080°$. In other words, the electrical angle of the induced emf is three times the mechanical angle of rotation. The flux per pole, the flux linking the coil, and the induced emf are also shown in the figure.

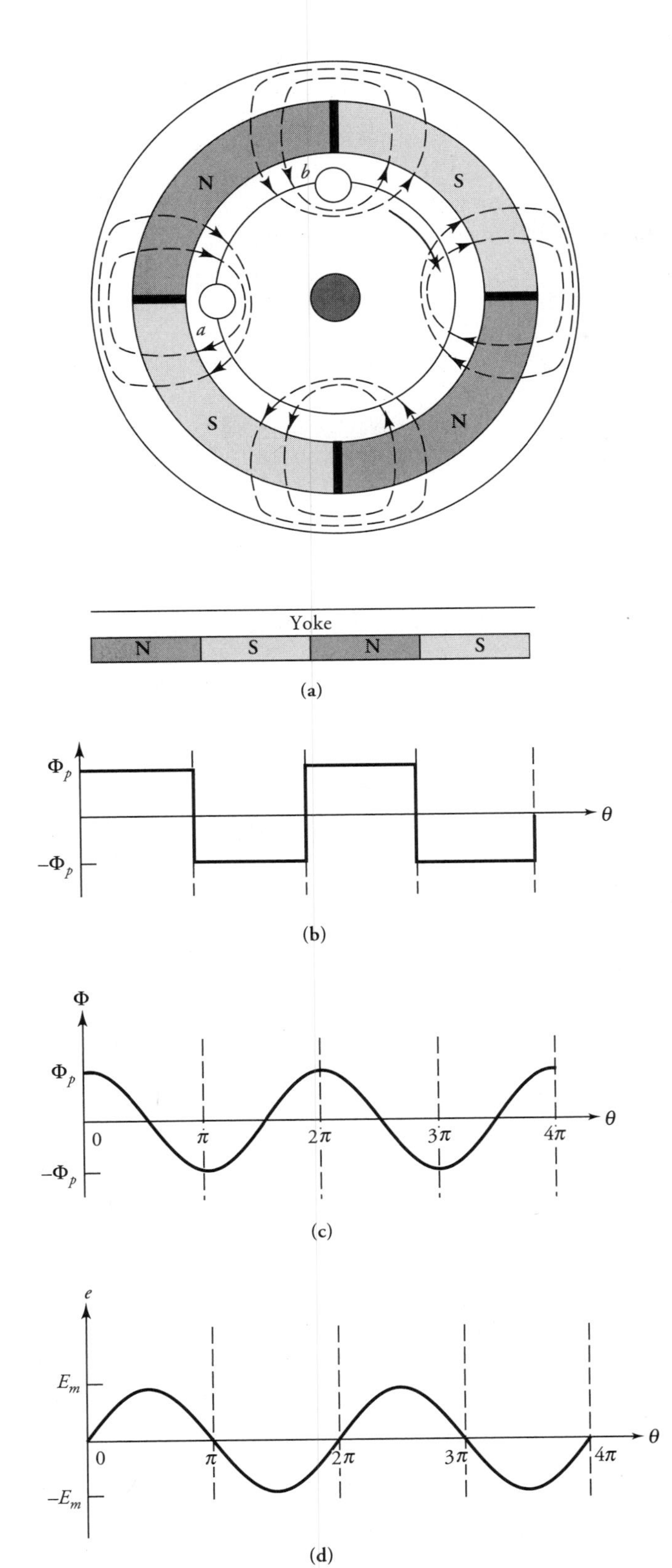

Figure 3.15 **(a)** Developed diagram of a 4-pole alternating-current generator. **(b)** Flux per pole set up by the magnets. **(c)** Flux passing through the coil rotating at a constant speed. **(d)** Induced emf in the coil.

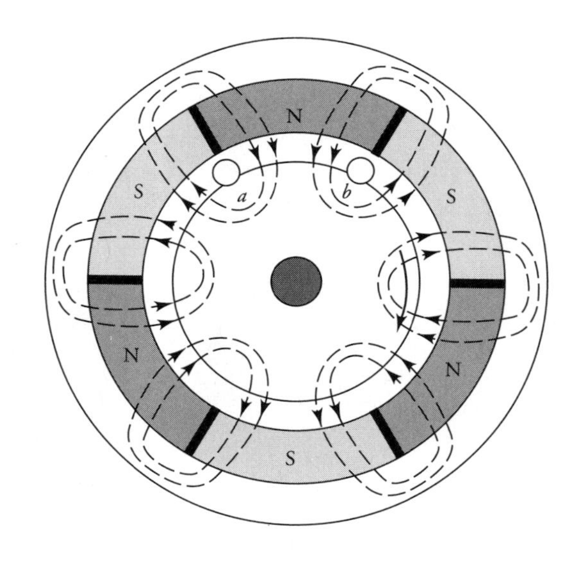

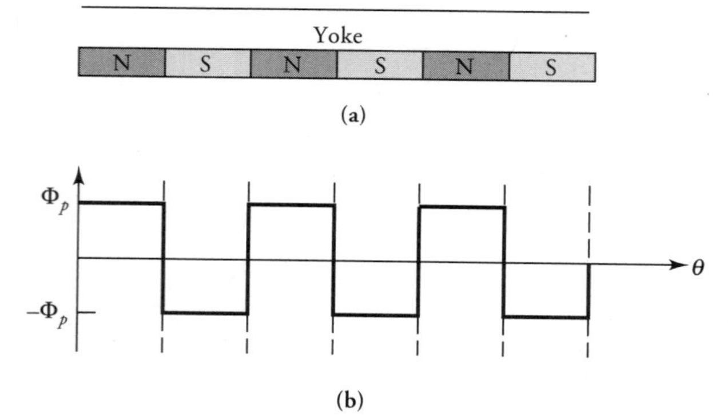

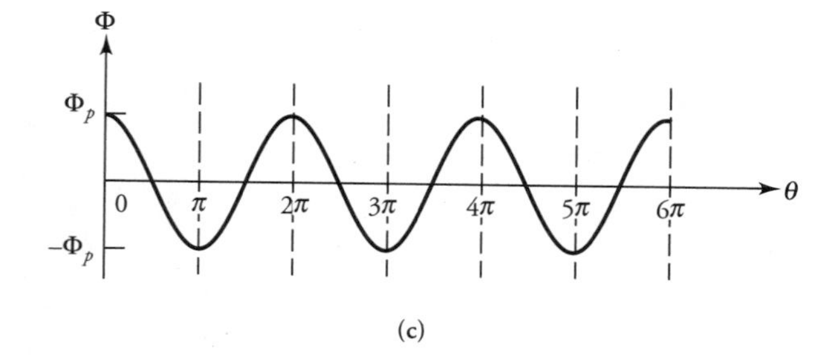

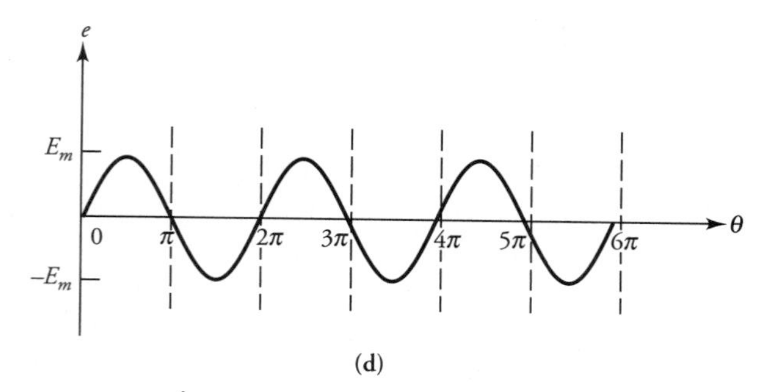

Figure 3.16 A six-pole ac generator. **(a)** The developed diagram. **(b)** Flux per pole set up by the magnets. **(c)** Flux passing through the coil rotating at a constant speed. **(d)** Induced emf in the coil.

From the above discussion it is evident that a pair of poles is responsible for one cycle of the induced emf. If there are P poles in a machine, then

$$\theta = \frac{P}{2}\,\theta_m \tag{3.26}$$

The above equation establishes a link between the mechanical angle of rotation and the angle of the induced emf. This is one of the most important relationships in the study of rotating machines. Differentiating Eq. (3.26), we obtain

$$\omega = \frac{P}{2}\,\omega_m \tag{3.27}$$

where ω is the angular frequency (rad/s) of the induced emf, and ω_m is the angular velocity (rad/s) of the rotor. By setting $\omega = 2\pi f$, where f is the frequency (Hz) of the induced emf, we get

$$f = \frac{P}{4\pi}\,\omega_m \tag{3.28}$$

If the coil is rotating at a speed of N_m revolutions per minute (rpm), then the angular velocity of rotation is

$$\omega_m = \frac{2\pi}{60}\,N_m \tag{3.29}$$

We can, therefore, rewrite Eq. (3.28) as

$$f = \frac{P}{120}\,N_m \tag{3.30a}$$

or

$$N_m = \frac{120f}{P} \tag{3.30b}$$

The maximum value of the induced emf in a single-turn coil, from Eq. (3.25), is

$$\begin{aligned} E_m &= \Phi_P\omega \\ &= \frac{P}{2}\,\Phi_P\omega_m \\ &= \frac{2\pi P}{120}\,\Phi_P N_m \end{aligned} \tag{3.31}$$

In a dc machine, the average value of the induced voltage in a single-turn coil is

$$\begin{aligned} E_c &= \frac{2}{\pi} E_m \\ &= \frac{P}{\pi} \Phi_P \omega_m \\ &= \frac{P}{30} \Phi_P N_m \end{aligned} \tag{3.32}$$

This is another important equation, and our study of dc machines will begin with it.

Force on a Conductor

When a current-carrying conductor is placed in a magnetic field, it experiences a force that tends to impart motion to the conductor in accordance with the Lorentz force equation. That is,

$$\boldsymbol{F} = \int_c i \, d\boldsymbol{\ell} \times \boldsymbol{B}$$

where $d\boldsymbol{\ell}$ is the length of the current-carrying element and $\boldsymbol{B}$ is the magnetic flux density. From the above equation, the force acting on a linear current-carrying conductor placed in a magnetic field, as shown in Figure 3.17, is

$$\boldsymbol{F} = i\boldsymbol{L} \times \boldsymbol{B} \tag{3.33}$$

where L is the length of the conductor exposed to the magnetic field.

Motor Action

Consider a 2-pole machine similar to the one shown in Figure 3.10. Instead of rotating the coil, let us inject a current through the coil by connecting it to a constant voltage source as shown in Figure 3.18. A current-carrying conductor

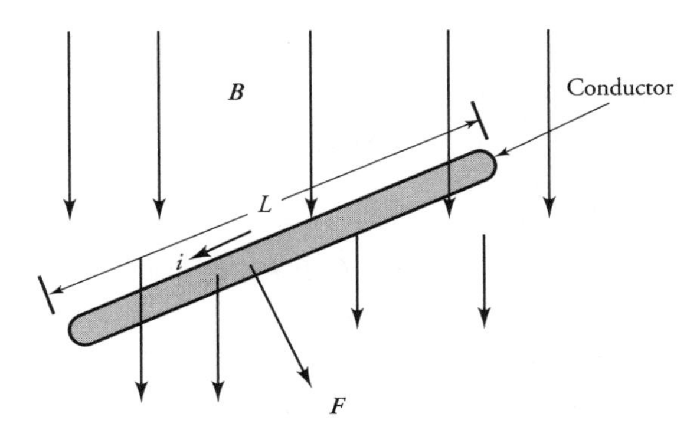

Figure 3.17 A current-carrying conductor immersed in a magnetic field experiences a force as shown.

immersed in a magnetic field experiences a force as outlined above. For the direction of the current shown, the force on the upper conductor a of the coil is

$$\boldsymbol{F} = iLB\,\boldsymbol{a}_{\Phi} \tag{3.34a}$$

where L is the effective length of the conductor. In all machines, the length of the magnet (electromagnet) is equal to the length of the conductor. Note that the

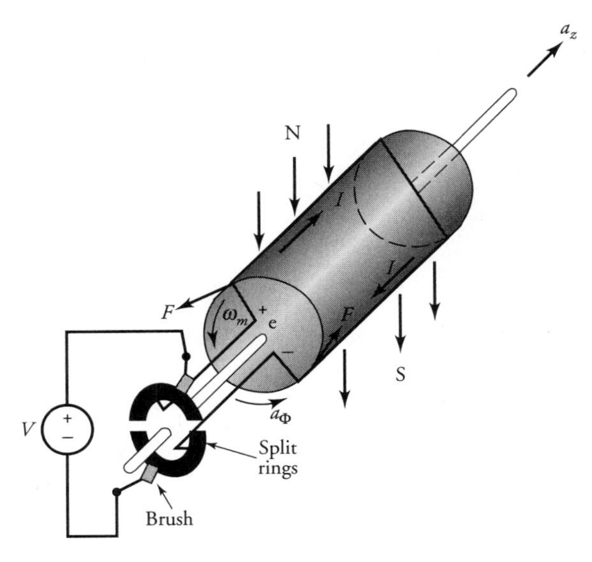

Figure 3.18 A two-pole dc motor.

length of the conductor L is perpendicular to the magnetic flux density $\mathbf{B}$. Similarly, the force exerted on the conductor b of the coil is

$$\mathbf{F} = iLB\,\mathbf{a}_{\phi} \tag{3.34b}$$

Both forces are acting in the same direction. Therefore, the magnitude of the total force experienced by the two conductors is

$$F_e = 2iBL \tag{3.34c}$$

The above force exerts a torque on the single-turn coil. That is,

$$T_e = F_e r = 2BiLr \tag{3.35}$$

where r is the radius at which each conductor is located. The resulting torque has the tendency to rotate the rotor in the counterclockwise direction. This is known as the **motor action**.

In order to keep the coil rotating in the clockwise direction (**generator action**), it must be coupled to a prime mover that provides a torque equal to the torque developed by the machine. In other words, the external torque applied (T_s) must be equal and opposite to the internal torque developed (T_e).

As soon as the coil starts rotating in the counterclockwise direction, an emf is induced in the coil in accordance with Faraday's law of induction. The polarity of the induced emf is also shown in the figure. In a motor, the induced emf is usually referred to as the **counter emf** or the **back emf**. The counter emf opposes the applied voltage.

In summary, a current-carrying conductor immersed in a magnetic field always experiences a force acting on it (motor action), and a conductor moving in a magnetic field has emf induced in it (generator action). **In any rotating machine, both actions are present at the same time.**

EXAMPLE 3.6

The rotor of a 4-pole generator is wound with a 100-turn coil. If the flux per pole is 4.5 mWb and the rotor turns at a speed of 1800 rpm, determine (a) the frequency of the induced emf in the rotor, (b) the maximum value of the induced voltage, (c) the rms value of the induced voltage for an ac generator, and (d) the average value of the induced voltage in a dc generator.

● SOLUTION

$$\omega_m = \frac{2 \times \pi \times 1800}{60} = 188.496 \text{ rad/s}$$

(a) The frequency of the induced emf, from Eq. (3.30a), is

$$f = \frac{4 \times 1800}{120} = 60 \text{ Hz} \qquad \text{or} \qquad \omega = 2\pi f \approx 377 \text{ rad/s}$$

(b) The maximum value of the induced emf per turn, from Eq. (3.31), is

$$E_m = \frac{2\pi \times 4}{120} \times 4.5 \times 10^{-3} \times 1800 = 1.6965 \text{ V}$$

Thus, the induced emf in the 100-turn coil is

$$E_{mc} = 100 \times 1.6965 = 169.65 \text{ V}$$

(c) The effective (rms) value of the induced emf is

$$E_{eff} = \frac{E_{mc}}{\sqrt{2}} = \frac{169.65}{\sqrt{2}} \approx 120 \text{ V}$$

(d) The average value of the induced voltage is

$$E_c = \frac{2}{\pi} E_{mc} = \frac{2}{\pi} \times 169.65 = 108 \text{ V}$$

■

Exercises

3.5. If the frequency of the induced emf in an 8-pole machine is 50 Hz, at what speed is the rotor rotating?

3.6. A 10-cm long conductor is placed perpendicularly in a magnetic field, the intensity of which is 1.2 T. If the current in the conductor is 120 A, calculate the force experienced by the conductor.

3.7. A single conductor is being moved perpendicularly to a magnetic field. The magnetic field strength is 0.8 T. The length of the conductor influenced by the magnetic field is 15 cm. If the conductor is moved at a speed of 2 m/s, what is the induced voltage in the conductor? If a 0.2-Ω resistance is connected across the conductor, what is the force exerted by the field on the conductor?

3.8. A 120-turn coil on the rotor of a 2-pole motor carries a current of 10 A. The flux density per pole is 1.2 T. If the core diameter is 25 cm and its length is 10 cm, determine (a) the force per conductor, (b) the force acting on the coil, and (c) the torque acting on the rotor.

3.5 A Coil in a Time-Varying Field

Among numerous possibilities of time-varying magnetic fields, we focus our attention on a sinusoidally varying (alternating, or ac) magnetic field. In all the rotating machines using the time-varying magnetic field, the field winding is wound on the stator and the coil is wound on the rotor. When the field winding is connected to an alternating source and the coil is shorted, the machine is called an **induction motor**. Hence, the rotor of an induction motor receives its power (energy) inductively and converts it into useful mechanical power (energy). For that reason, the shorted coil (closed loop) on the rotor is referred to as the **induction winding**. Although we consider only one closed loop to explain the development of torque in an induction motor, there are, in fact, many closed loops on the rotor.

Without going into the winding details at this time, we aim to show in this section the following:

(a) When the stator windings of a polyphase induction machine are connected to a polyphase supply, they produce a **revolving magnetic field**.
(b) The strength of the revolving magnetic field is constant.
(c) The field rotates a distance covered by two poles for each cycle of the input waveform.
(d) The force acting on the conductors of the coil wound on the rotor causes the rotor to rotate in the direction of the revolving field.
(e) The rotor turns at a speed lower than the speed of the revolving field.

The most commonly used polyphase motor is a three-phase motor owing to the worldwide generation and distribution of the three-phase power. Two-phase induction motors are rarely designed, as the two-phase power supply is not directly accessible. However, a single-phase induction motor is designed to simulate primitively a two-phase motor in order to empower it with a self-starting feature. Therefore, it is essential to explore how the revolving fields are developed by both the three-phase and the two-phase motors.

Revolving Field of a Three-Phase Motor

The stator of a three-phase induction motor is wound with identical coils that are interconnected to form three phases. The phase windings are spaced 120° electrical apart (actual placement of coils is discussed in Chapter 5). Figure 3.19 shows the phase-winding arrangement of a 2-pole, three-phase motor. In order to explain the principle, we have shown only one full-pitch coil per phase. The unprimed coils 1, 2, and 3 are for one pole and the primed coils, 1′, 2′, and 3′ are for the other. The numbers 1, 2, and 3 refer to the three phases of the motor.

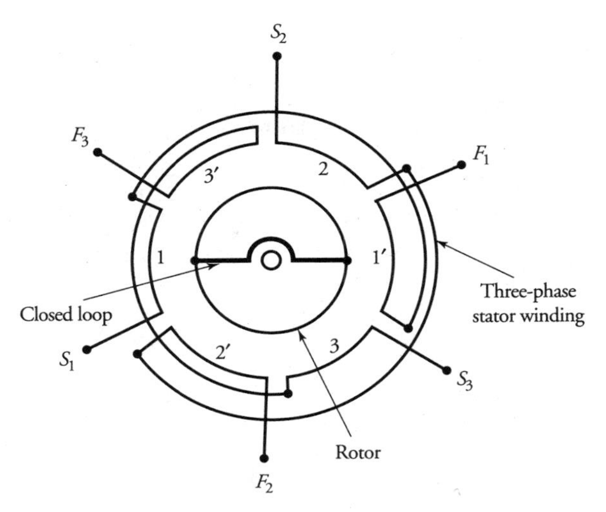

Figure 3.19 Winding arrangement of a three-phase two-pole induction motor.

When the windings are connected to a balanced three-phase source, we expect the currents in all windings to be equal in magnitude but displaced in phase by 120°. The design of each phase winding is such that the spatial distribution of the flux in the air-gap due to that phase winding alone is almost sinusoidal. If we consider the current in phase-1 as the reference, then the currents in the three phases, as shown in Figure 3.20 for a positive phase sequence, are

$$i_1 = I_m \sin \omega t \tag{3.36a}$$

$$i_2 = I_m \sin(\omega t - 120°) \tag{3.36b}$$

$$i_3 = I_m \sin(\omega t + 120°) \tag{3.36c}$$

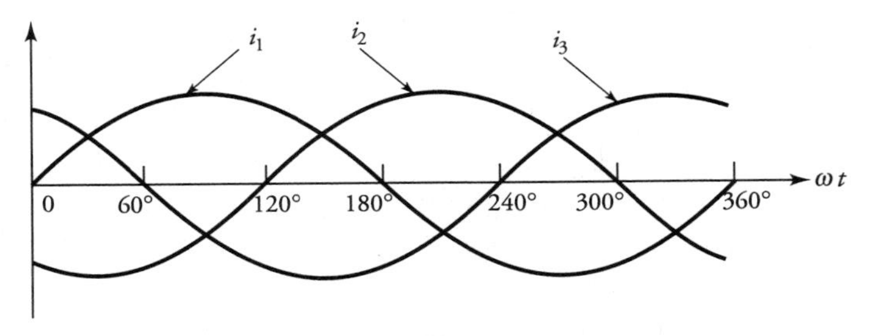

Figure 3.20 The current waveforms in the stator windings of a three-phase motor.

where I_m is the amplitude of each phase current, and $\omega = 2\pi f$ is the angular frequency of the source. Under linear conditions, the flux produced by each current also varies sinusoidally. Therefore, the current waveforms can also be labeled the flux waveforms.

In the discussion that follows we assume that the current flows from S (the starting end) of a coil toward F (the finishing end) during the positive half-cycle. In other words, during the positive half-cycle the current in a phase coil is in the clockwise direction, and it produces a flux that points toward it as viewed from the inside of the stator.

We now consider three instants of time to show that the resultant field is constant in magnitude and revolves around the periphery of the rotor at a constant speed determined by the frequency of the applied source.

Instant I ($\omega t = 0$): At the outset, the phase currents are

$$i_1 = 0$$
$$i_2 = -0.866 I_m$$
$$i_3 = 0.866 I_m$$

The positive direction of currents in the coils is shown in Figure 3.21**a**. If Φ_m is the amplitude of the flux produced by I_m ($\Phi \propto i$), then the corresponding flux magnitudes are

$$\Phi_1 = 0$$
$$\Phi_2 = 0.866 \Phi_m$$
$$\Phi_3 = 0.866 \Phi_m$$

and their directions are as indicated in Figure 3.21**a**. The angle between Φ_2 and Φ_3 is 60°. The resultant flux is

$$\begin{aligned} \Phi_{r1} &= \sqrt{\Phi_2^2 + \Phi_3^2 + 2\Phi_2\Phi_3 \cos(60°)} \\ &= 1.5\Phi_m \end{aligned} \tag{3.37}$$

The resultant flux is directed vertically downward inside the motor. A simple explanation is provided by the direction of the currents in the windings. Because the currents in the two phases are equal in magnitude, the phase coils 2 and 3′ act as one and form a north pole. On the other hand, the phase coils 3 and 2′ join together to form a south pole. These are the two north and south poles of a 2-pole, three-phase motor at $\omega t = 0$. The magnetic axis is, therefore, along the vertical line.

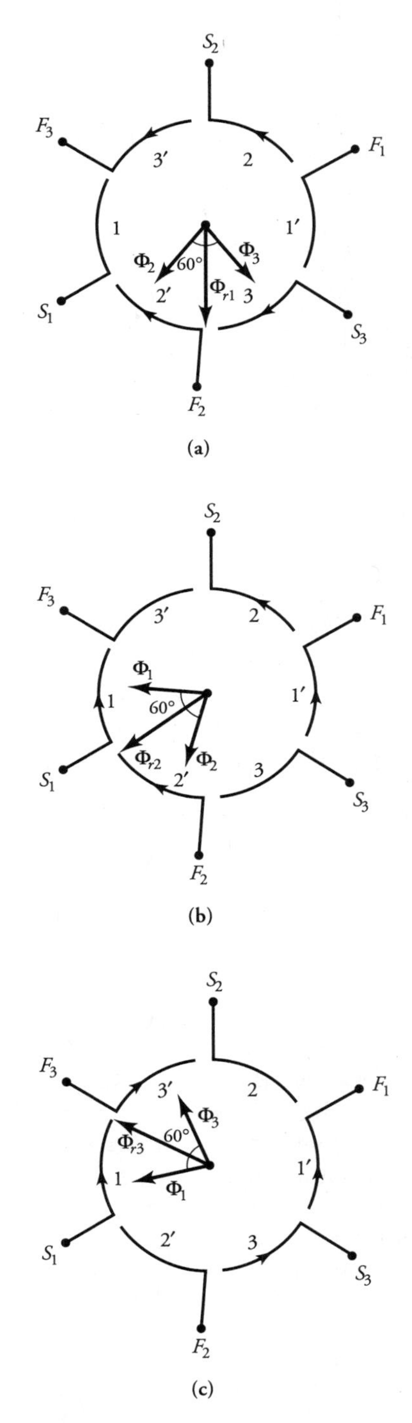

Figure 3.21 Revolving field in a two-pole three-phase induction motor at three instants of time **(a)** $\omega t = 0°$, **(b)** $\omega t = 60°$, and **(c)** $\omega t = 120°$.

Instant II (ωt = 60°): One-sixth of the cycle later, the currents are

$$i_1 = 0.866 I_m$$
$$i_2 = -0.866 I_m$$
$$i_3 = 0$$

The magnitude of the flux created by either i_1 or i_2 is $0.866\Phi_m$. The positive direction of the currents and the fluxes are shown in Figure 3.21**b**. Once again, we have two fluxes, each of magnitude $0.866\Phi_m$ and 60° apart in time phase. Therefore, the resultant flux is

$$\Phi_{r2} = 1.5\Phi_m$$

Note that Φ_{r2} is equal to Φ_{r1} but its direction is 60° clockwise from Φ_{r1}. In other words, a 60° advancement in the time phase of the currents has shifted the resultant flux by 60° electrical in space. Note that phase coils 1′ and 2 act as a single north pole, and phase coils 2′ and 1 behave as a single south pole. It appears as if the two magnetic poles have been rotated by 60° electrical in the clockwise direction.

Instant III (ωt = 120°): One-third of a cycle later, the currents in the phase coils are

$$i_1 = 0.866 I_m$$
$$i_2 = 0$$
$$i_3 = -0.866 I_m$$

The magnitude of the flux produced by either i_1 or i_3 is $0.866\Phi_m$. The positive direction of each flux and the current in each coil are shown in Figure 3.21**c**. By combining the two fluxes, we find that the resultant flux Φ_{r3} is the same in magnitude as Φ_{r1} and is directed 120° clockwise from Φ_{r1}. In other words, a 120° phase shift in time-domain has spatially rotated the flux by 120°. The phase coils 3 and 1′ now act as a north pole while coils 1 and 3′ behave as a south pole. The two poles have now rotated by 120° in the clockwise direction.

Another 60° later we find that coils 2′ and 3 act as a north pole and coils 3′ and 2 as a south pole. The pole positions have now reversed. In other words, the resultant flux with a magnitude of $1.5\Phi_m$ has rotated by 180° (an arc covered by one pole) when each current has undergone a change of one-half cycle. Accordingly, **the resultant field rotates 360° electrical along the periphery of the air-gap when the currents undergo one cycle of a change**.

In a 2-pole motor, the time taken by the flux to complete one revolution along the periphery of the air-gap is the same as the time period T of the input waveform

($T = 1/f$). Since one revolution in a 4-pole machine is equivalent to 720° electrical, the time taken by the flux to complete one rotation is $2T$. The time required by the flux to revolve once around the air-gap is $3T$ for a 6-pole motor. If T_s is the time taken by the flux to revolve along the periphery of a P-pole motor, then

$$T_s = \frac{P}{2} T \tag{3.38}$$

If n_s is the speed of the revolving field in revolutions per second, more aptly referred to as the **synchronous speed**, then

$$n_s = \frac{1}{T_s} = \frac{2f}{P} \tag{3.39a}$$

Equation (3.39a) can also be expressed in revolutions per minute (N_s) or radians per second (ω_s) as

$$N_s = \frac{120f}{P} \tag{3.39b}$$

or

$$\omega_s = \frac{4\pi f}{P} = \frac{2}{P}\omega \tag{3.39c}$$

From the above equation it is evident that **the synchronous speed of the revolving field is constant for a constant frequency source.**

In our discussion we have tacitly assumed that the phase windings are arranged in the clockwise direction and the power supply has a positive phase sequence. This combination gave birth to a uniform revolving field that revolves in the clockwise direction at a constant speed. However, we can force the field to revolve in the counterclockwise direction simply by swapping two of the three supply connections.

As the flux rotates, it induces an emf in the coil. Since the coil forms a closed loop, the induced emf gives rise to a current in the loop as shown in Figure 3.22. A current-carrying conductor immersed in a magnetic field experiences a force that tends to move it. In this case, the force acting on each conductor of the loop thrusts the rotor to rotate in the direction of the revolving field. Under no-load (nothing is coupled to the rotor shaft), the rotor attains a speed slightly less than the synchronous speed. The rotor of an induction motor can never rotate at synchronous speed for the following reason: If the rotor were to rotate at the synchronous speed, the closed loop on the rotor would encounter a constant flux passing through it. Thus, there would be no induced emf in it. In the absence of an induced emf, there would be no current in the closed loop and thereby no force acting on the conductors. Thus, the rotor would tend to slow down as a result of

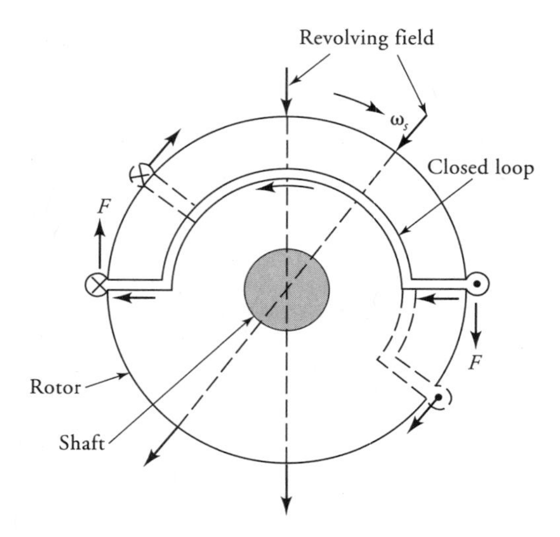

Figure 3.22 Force exerted by the constant revolving field on each conductor of a closed loop.

rotational losses due to friction and windage. As soon as the rotor slows down, it experiences a change in the flux that induces an emf and thereby the current in the closed loop. The current creates a force on the conductors that tends to increase the speed of the rotor. When the steady state is reached, the rotor attains the speed at which the torque developed by it is equal and opposite to the applied torque. As we increase the load on the motor, the rotor speed falls even further. Therefore, **the speed of an induction motor depends upon the load on the motor.**

The difference between the synchronous speed and the rotor speed is called the **slip speed**. If $N_m(\omega_m)$ is the rotor speed, the slip speed is

$$N_r = N_s - N_m \tag{3.40a}$$

or

$$\omega_r = \omega_s - \omega_m \tag{3.40b}$$

If we imagine two points, one on the revolving field and the other on the rotor, then the slip speed is simply the relative speed with which the point on the revolving field is moving ahead of the point on the rotor. Most often, the slip speed is defined in terms of the synchronous speed and is called the **per-unit slip** or **slip**. That is,

$$s = \frac{\omega_r}{\omega_s} = \frac{\omega_s - \omega_m}{\omega_s} \tag{3.41}$$

In terms of the slip and the synchronous speed, the rotor speed, from the above equation, is

$$\omega_m = (1 - s)\omega_s \tag{3.42a}$$

or

$$N_m = (1 - s)N_s \tag{3.42b}$$

EXAMPLE 3.7

Calculate the synchronous speed of a 4-pole, 50-Hz, three-phase induction motor. What is the percent slip if the rotor rotates at a speed of 1200 rpm?

● SOLUTION

$$N_s = \frac{120f}{P} = \frac{120 \times 50}{4} = 1500 \text{ rpm, or } 157.08 \text{ rad/s}$$

$$s = \frac{1500 - 1200}{1500} = 0.2$$

Hence, the slip is 0.2, or the percent slip is 20%.

■

Revolving Field of a Two-Phase Motor

Figure 3.23**a** shows a 2-pole motor whose phase windings are placed in space quadrature. The current waveforms in the two windings when connected to a balanced two-phase source are given in Figure 3.23**b**, where phase-1 has been assumed as a reference. That is,

$$i_1 = I_m \sin \omega t \tag{3.43a}$$

$$i_2 = I_m \sin(\omega t - 90°) = -I_m \cos \omega t \tag{3.43b}$$

The corresponding values of the instantaneous fluxes produced by the two currents are

$$\Phi_1 = \Phi_m \sin \omega t \tag{3.44a}$$

$$\Phi_2 = -\Phi_m \cos \omega t \tag{3.44b}$$

Once again, the positive direction of the current is from S to F. Also, each phase winding is so wound that it produces a south pole when the current in the coil is in the clockwise direction.

At the outset, $\omega t = 0$, the current in coils 1 and 1′ is zero. Therefore, no flux is produced by this winding. The current in the second phase winding is maximum and flows from F_2 toward S_2, as shown in Figure 3.24**a**. The flux established

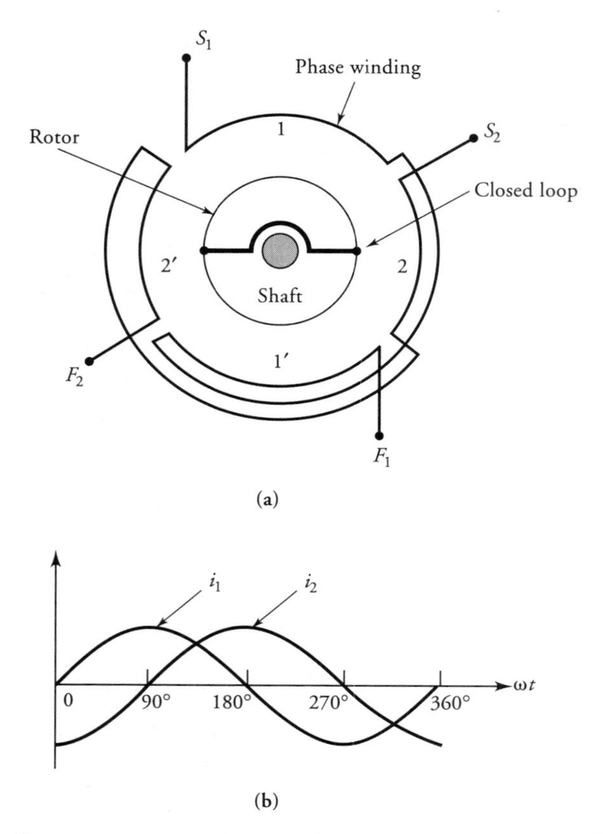

Figure 3.23 **(a)** Winding arrangement of a two-phase, two-pole induction motor. **(b)** The current waveforms in the stator windings.

by this current is maximum, that is, $\Phi_{r1} = \Phi_m$, and is directed from coil 2 toward coil 2′. Thus, the phase coil 2 acts like a north pole while 2′ is a south pole.

One-eighth of a cycle later, $\omega t = 45°$, the currents in the two windings are

$$i_1 = 0.707 I_m$$
$$i_2 = -0.707 I_m$$

Since the magnitude of the currents in the two windings is the same, the magnitude of the flux created by either winding is $0.707\Phi_m$. The directions of the two fluxes are as shown in Figure 3.24**b**. Since the two fluxes are orthogonal, the resultant flux is

$$\Phi_{r2} = \sqrt{(0.707\Phi_m)^2 + (0.707\Phi_m)^2} = \Phi_m \tag{3.45}$$

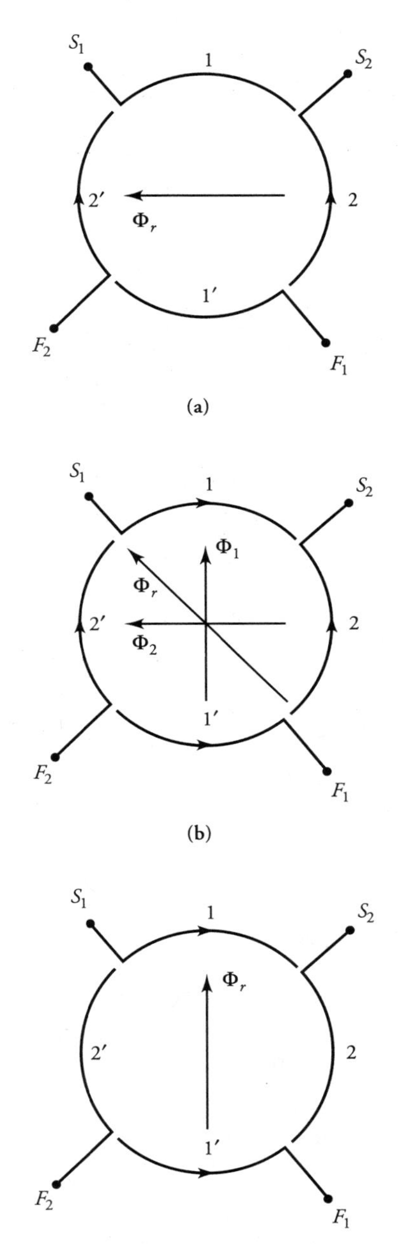

Figure 3.24 Revolving field in a two-pole, two-phase inductor motor at **(a)** $\omega t = 0°$, **(b)** $\omega t = 45°$, and **(c)** $\omega t = 90°$.

Note that Φ_{r2} is equal in magnitude to Φ_{r1} and has advanced spatially by an angle of 45°, as shown in the figure. It is evident from the figure that the coils 1′ and 2 now act as a north pole, and the coils 2′ and 1 form a south pole. A time shift of 45° has rotated the two poles spatially by 45° in the clockwise direction.

A quarter-cycle later, $\omega t = 90°$, the two currents are

$$i_1 = I_m$$

$$i_2 = 0$$

At this instant, only the first-phase winding carries a maximum current and produces a maximum flux, which is directed from coil 1′ toward coil 1, as depicted in Figure 3.24c. The resultant flux, $\Phi_{r3} = \Phi_m$, is the same in magnitude as Φ_{r1} but leads it in the clockwise direction by an angle of 90°. The coil 1′ is now a north pole and the coil 1 is a south pole. A time-phase increment of 90° in the phase currents has rotated the two magnetic poles by 90° electrical spatially. The strength of each pole is Φ_m.

If this process is continued, we find that after one complete cycle of currents in the two phases of a 2-pole motor, the constant flux has completed one revolution around the periphery of the rotor. This situation is similar to what we studied earlier for a three-phase induction motor. Thus, all the equations we formulated for a three-phase motor are also valid for a two-phase motor.

Since the flux established by a single-phase winding pulsates in time along the same magnetic axis, it does not revolve. Therefore, the minimum number of phases must at least be two in order to produce a revolving field of constant magnitude.

From the above discussion we conclude the following:

(1) The magnitude of the revolving field in a two-phase motor is constant and is equal to Φ_m, where Φ_m is the maximum flux produced by either winding.
(2) The magnitude of the revolving field in a three-phase machine is constant and is equal to $1.5\Phi_m$.
(3) In fact, we can show that the magnitude of the revolving field in an n-phase machine is $n/2\Phi_m$.
(4) The revolving field rotates at a synchronous speed determined by the frequency of the applied source and the number of poles of the motor.
(5) The rotor speed can never be equal to the synchronous speed unless the rotor is being driven by an external prime mover.

Exercises

3.9. A 6-pole, three-phase induction motor operates from a supply whose frequency is 60 Hz. Calculate (a) the synchronous speed of the revolving field and (b) the rotor speed if the percent slip is 5%.

3.10. Show that the strength of the revolving field for a six-phase machine is $3\Phi_m$, where Φ_m is the maximum flux produced by each phase winding.

3.11. At no-load, the rotor speed of a three-phase induction motor is 895 rpm. Determine (a) the number of poles, (b) the frequency of the source, (c) the synchronous speed, and (d) the percent slip.

3.6 Synchronous Motor

The stator of a polyphase synchronous motor is wound exactly in the same fashion as that of a polyphase induction motor. However, the rotor of a synchronous motor has two windings. One of the windings is identical to that of an induction motor and is referred to as an **induction winding**. This winding is in the form of closed loops. The other winding, called the **field winding**, is designed to carry a constant (dc) current so as to produce constant flux in the air-gap as shown in Figure 3.25.

When the stator winding is connected to a polyphase power supply, it produces a constant field that revolves around the periphery of the rotor. The field winding is not excited at the time of starting. The revolving field induces emf in the closed loop, which causes a current to flow. A current-carrying conductor immersed in a magnetic field experiences a force that creates the driving torque. In a nutshell, the synchronous motor starts as an induction motor.

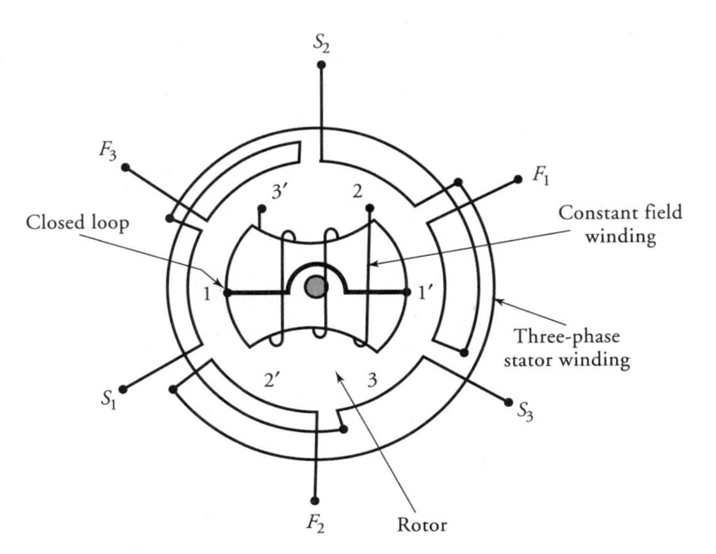

Figure 3.25 Winding arrangement of a three-phase synchronous motor with induction and constant flux windings on the rotor.

As soon as the rotor attains a speed in excess of 75% of the synchronous speed, the constant current is gradually applied to the field winding on the rotor. The magnetic poles thus created on the surface of the rotor "lock up" with the revolving field and enable the rotor to "pull up" to the synchronous speed.

The position of alignment of the magnetic poles depends upon the load on the machine. The angle δ by which the magnetic axis of the rotor lags the magnetic axis of the revolving field is called the **power angle**, or the **torque angle**. We show later that the power (torque) developed by a synchronous motor is proportional to sin δ. Thus, under no-load, δ is almost zero. On the other hand, the torque developed is maximum when $\delta = 90°$. If the applied torque (also known as the **brake torque**) exceeds the maximum torque that a synchronous motor can develop, it causes the rotor to "pull out" of synchronism. Thus, a synchronous motor rotates at its synchronous speed as long as the applied torque is less than the maximum torque developed by it.

It is quite interesting to note that once the rotor starts rotating at the synchronous speed, the induction winding becomes ineffective because there is no induced emf in it. However, it does serve another important purpose: It helps to stabilize the motor whenever there is a sudden change in the load on the machine. We discuss this attribute of the induction winding in detail in Chapter 8.

3.7 Reluctance Motor

The reluctance motor is essentially a synchronous motor whose reluctance changes as a function of angular displacement θ. Owing to its constant speed operation, it is commonly used in electric clocks, record players, and other precise timing devices. This motor is usually of a single-phase type and is available in the fractional horsepower range. A reluctance motor differs from a synchronous motor in that it does not have field winding on the rotor.

Figure 3.26**a** shows an elementary, single-phase, 2-pole reluctance motor. In order to simplify the theoretical development, we assume that all the reluctance of the magnetic circuit is in the air-gap.

When the angular displacement θ between the rotor and the stator magnetic axes is zero (the direct or d-axis position), the effective air-gap is minimum. The reluctance of the magnetic circuit,

$$\mathcal{R}(0) = \frac{2g}{\mu_0 A} \tag{3.46}$$

where g is the effective air-gap and A is the area per pole, is also minimum. Consequently, the inductance of the magnetic circuit is maximum because the inductance is inversely proportional to the reluctance.

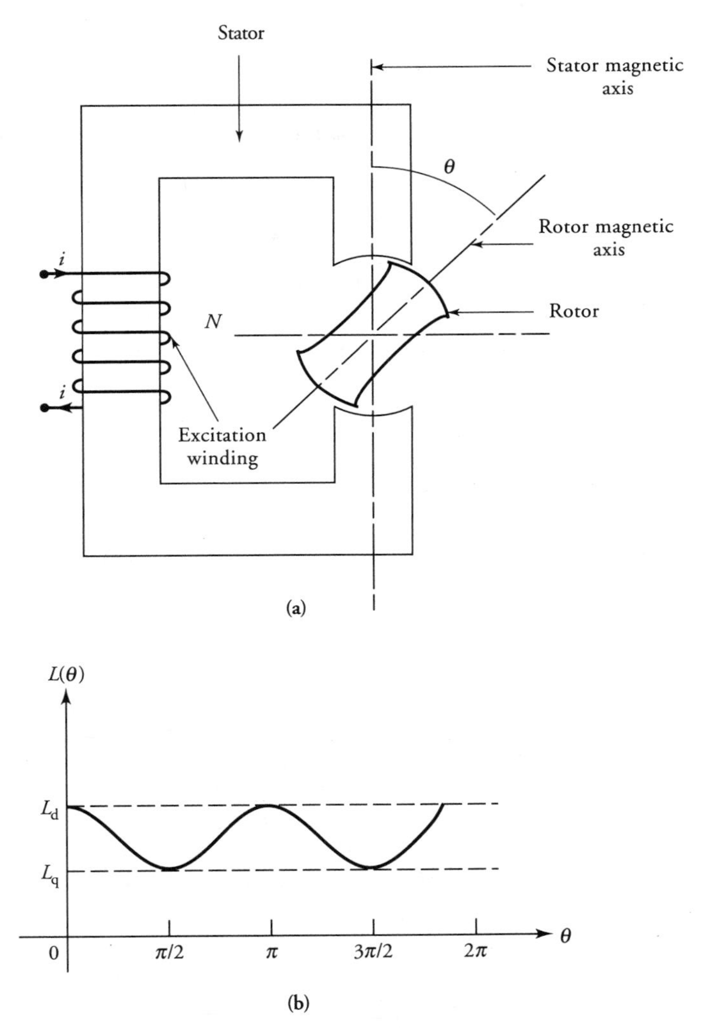

Figure 3.26 (a) A single-phase reluctance motor. **(b)** Variation in the inductance of a reluctance motor as a function of the displacement angle θ.

When the magnetic axes of the rotor and the stator are at right angles to each other (the quadrature or q-axis position), the reluctance is maximum, leading to a minimum inductance. As the rotor rotates with a uniform speed ω_m, the inductance goes through maxima and minima as depicted in Figure 3.26**b**.

The inductance as a function of θ can be expressed as

$$L(\theta) = 0.5(L_d + L_q) + 0.5(L_d - L_q)\cos 2\theta \tag{3.47}$$

The torque developed by a rotating system can be obtained from Eq. (3.23) as

$$T_e = \frac{1}{2} i^2 \frac{\partial L}{\partial \theta} \tag{3.48}$$

Thus, the torque exerted on the rotor of a reluctance motor is

$$T_e = -\frac{1}{2} i^2 (L_d - L_q)\sin 2\theta \tag{3.49}$$

However, we can express θ as

$$\theta = \omega_m t + \delta \tag{3.50}$$

where δ is the initial position of the rotor's magnetic axis with respect to the stator's magnetic axis. The torque experienced by the rotor can now be rewritten as

$$T_e = -\frac{1}{2} i^2 (L_d - L_q)\sin[2(\omega_m t + \delta)] \tag{3.51}$$

It is evident that the initial torque (at $t = 0$) experienced by the rotor is zero if $\delta = 0°$ and is maximum if $\delta = 45°$. The presence of the negative sign in the above equation highlights the fact that the torque tends to align the rotor under the nearest pole of the stator and thereby defines the direction of rotation.

For a sinusoidal variation in the current,

$$i = I_m \cos \omega t$$

the torque developed by the reluctance motor is

$$T_e = -0.5 I_m^2 (L_d - L_q)\cos^2 \omega t \sin(2\omega_m t + 2\delta)$$

Using the following trigonometric identities,

$$2\cos^2\alpha = 1 + \cos 2\alpha$$

and

$$2\sin\alpha\cos\beta = \sin(\alpha + \beta) + \sin(\alpha - \beta)$$

the torque expression becomes

$$T_e = -0.25(L_d - L_q)I_m^2[\sin(2\omega_m t + 2\delta) + 0.5\sin\{2(\omega + \omega_m)t + 2\delta\} + 0.5\sin\{2(\omega - \omega_m)t - 2\delta\}]$$

It is obvious from this expression that the average torque developed by the reluctance motor is zero unless its speed ω_m is equal to ω and $\delta \neq n\pi$, where $n = 0, 1, 2, \ldots$ Thus, for the reluctance motor to develop an average torque, its angular velocity must be equal to the angular frequency of the source. In other words, the motor must rotate at its synchronous speed. The average torque developed at the synchronous speed is

$$T_{e|avg} = -0.125\,(L_d - L_q)\,I_m^2 \sin 2\delta \tag{3.52}$$

which is maximum when $\delta = 45°$.

EXAMPLE 3.8

The current intake of a 2-pole reluctance motor at 60 Hz is 10 A (rms). The minimum and maximum values of the inductances are 2 H and 1 H, respectively. Determine (a) the rotor speed and (b) the average torque developed by the motor.

● SOLUTION

The rotor speed: $\omega_m = \omega = 2\pi f = 2\pi \times 60 \approx 377$ rad/s, or 3600 rpm

The average torque developed by the motor is

$$T_{e|avg} = -0.125(2 - 1)(10 \times \sqrt{2})^2 \sin 2\delta$$
$$= -25 \sin 2\delta$$

The average torque developed is 25 N·m when $\delta = 45°$. ■

Exercises

3.12. The inductance of a 2-pole reluctance motor is given as

$$L(\theta) = 5 + 2\cos 2\theta \text{ H}$$

Determine the torque developed by the motor when the current in the 150-turn coil is 5 A and $\theta = 25°$. Compute the minimum and the maximum inductance of the motor. Sketch the torque developed as a function of angular displacement θ.

3.13. If the current in the above 2-pole reluctance motor is $5 \sin 314t$ A, determine the speed of the motor. Sketch the average torque developed by the motor as a function of the initial position of the rotor (δ).

SUMMARY

In this chapter we have outlined some of the basic energy conversion principles. These are the principles that will be used extensively to examine the energy conversion processes associated with different types of machines. An important point to remember is that the conversion of energy from one form to another satisfies the principle of conservation of energy. The input energy is, therefore, always equal to the sum of the output energy, the increase in the stored energy, and the loss in energy. If the loss in energy is negligible in an energy conversion system, it is said to be a lossless system. In a "real-life" system, there is always some loss of energy, however small it may be.

For the electric energy conversion, we can use either an electric field or the magnetic field as a medium. However, the quantity of energy that can be converted by a device using the electric field as a medium is relatively small. A parallel-plate capacitor is a good example of the energy conversion process that uses an electric field as a medium. If V is the potential difference between the plates and x is the separation between them, the force of attraction between the plates is

$$F_e = \frac{1}{2} V^2 \frac{\partial C}{\partial x}$$

where the capacitance C is a function of the distance x between the two plates.

When a large amount of electric energy is required, the magnetic field is the medium of choice. The magnetic force acting on one part of the magnetic circuit is given in terms of its inductance as

$$F_m = \frac{1}{2} i^2 \frac{\partial L}{\partial x}$$

where the inductance L is a function of linear displacement x.

In a rotational system, the above equation is usually expressed in terms of the torque as

$$T = \frac{1}{2} i^2 \frac{\partial L}{\partial \theta}$$

When a single-turn coil is rotated in a radially directed uniform magnetic field, the induced emf in the coil, generator action, is

$$e = \Phi_p \omega \sin \omega t$$

where

$$\omega = \frac{P}{2} \omega_m$$

is the angular frequency of the induced emf, P is the number of poles, and ω_m is the angular velocity of the rotor. In terms of the frequency of the induced emf, the rotor speed in rpm is

$$N_m = \frac{120f}{P}$$

On the other hand, a current-carrying conductor immersed in a magnetic field experiences a force given by

$$\mathbf{F} = i\mathbf{L} \times \mathbf{B}$$

When a conductor is mounted on the periphery of the rotor, the magnetic force acting on the conductor exerts a torque on the rotor. This torque has a tendency to rotate the rotor (motor action).

When the stator of an induction motor or a synchronous motor is wound with a polyphase winding and is excited from a polyphase source, it establishes a magnetic field that rotates at a synchronous speed. For a P-pole motor, the synchronous speed, in rpm, is

$$N_s = \frac{120f}{P}$$

The strength of the rotating magnetic field is constant and is given as

$$\Phi_r = \frac{n}{2} \Phi_m$$

where n is the number of phases in the motor, and Φ_m is the maximum flux produced by each phase.

A closed loop (induction winding) placed on the periphery of the rotor of a polyphase induction motor causes the rotor to rotate in the direction of the synchronous speed. The rotor, however, can never rotate at synchronous speed. In addition to the induction winding, the rotor of a synchronous motor also has a field winding. When the field winding is excited, it enables the rotor to rotate in synchronism with the revolving field.

For low-torque applications, a reluctance motor is commonly used as a synchronous motor. The motor develops the torque because its inductance is a function of the angular displacement of the rotor.

Review Questions

3.1. When a moving conductor is placed in a magnetic field, explain why the positive charges are forced toward one direction and the negative charges toward the other.

3.2. If a conductor of length L is moving with a velocity $\boldsymbol{v}$ in a uniform magnetic field $\boldsymbol{B}$, show that the induced emf between its two ends is $e = (\boldsymbol{v} \times \boldsymbol{B}) \cdot \boldsymbol{L}$.

3.3. If $\boldsymbol{v}$, $\boldsymbol{B}$, and $\boldsymbol{L}$ are mutually perpendicular, show that the induced emf in a conductor is $e = BLv$.

3.4. If β is the angle between $\boldsymbol{B}$ and $\boldsymbol{v}$, and α is the angle between $\boldsymbol{L}$ and the plane containing $\boldsymbol{v}$ and $\boldsymbol{B}$, show that the induced emf is $e = BLv \cos \alpha \sin \beta$.

3.5. Explain the principle of operation of an ac generator.

3.6. Explain the principle of operation of a dc generator.

3.7. Explain the difference between slip rings and a split ring.

3.8. What is a commutator?

3.9. What is the difference between a generator and a motor?

3.10. State Faraday's law. Is the induced emf in a moving conductor in a magnetic field in accordance with Faraday's law?

3.11. What is the effect of number of poles on the induced emf in an ac generator?

3.12. What is the effect of number of poles on the synchronous speed of the revolving field?

3.13. Why is it necessary to have a dc field winding on the rotor of a synchronous motor?

3.14. Is it possible for a synchronous motor to develop starting torque in the absence of an induction winding?

3.15. What is the difference between an induction winding and the field winding of a synchronous motor?

3.16. If a current-carrying conductor is placed parallel to the magnetic field, will it experience a force? Give reasons.

3.17. When a current-carrying conductor is placed in a uniform magnetic field, what happens to the field in the vicinity of the conductor? Sketch the field lines to explain your answer.

3.18. Why is there no induced emf in the induction winding of a synchronous motor when the motor rotates at its synchronous speed?

3.19. What happens when the speed of a synchronous motor with an induction winding tends to decrease from its synchronous value?
3.20. What happens when the speed of a synchronous motor with an induction winding tends to increase from its synchronous value?
3.21. If the pole-face area of an electromagnet is decreased slightly, what will the effect be on its lifting force?
3.22. If the pole-face area of an electromagnet is increased slightly, what will the effect be on its lifting force?
3.23. If a reluctance motor has a round rotor, will it exert torque on the rotor?
3.24. If the field winding of a reluctance motor carries a constant current, will there be an average torque developed by the motor?
3.25. What is the nature of the current in the armature coil of a dc generator?
3.26. The induced emf e_{ab} between the two open ends a, b of a coil appears as a voltage source v_{ab} for an external circuit that can be connected to the coil. Express v_{ab} in terms of e_{ab}.
3.27. Explain the development of a revolving field in a three-phase induction motor.
3.28. Explain the development of a revolving field in a two-phase induction motor.
3.29. Does the field really revolve in a polyphase machine?
3.30. Explain the principle of operation of a reluctance motor.
3.31. Define slip. Explain per-unit slip and percent slip.
3.32. Explain how the rotor of an induction motor develops the torque.

Problems

3.1. The upper plate of a parallel-plate capacitor is held fixed while the lower plate is free to move. The surface area of each plate is 40 cm^2. The mass of each plate is 8 g. What must be the potential difference between the plates to maintain a separation of 4 mm between them? What is the electric energy stored in the electric field?

3.2. Two parallel plates each having dimensions of 20 cm $\times$ 20 cm are held 5 mm apart. If the force experienced by each plate is 285 μN, what must be the potential difference between them? What is the energy stored in the electric field?

3.3. The energy stored in a parallel-plate capacitor is 30 mJ. The potential difference between the plates is 50 kV. If the separation between the plates is 4 mm, determine (a) the surface area of each plate and (b) the force of attraction between them.

3.4. The area of each plate of a parallel-plate capacitor is 50 cm^2. The potential difference between the plates is 50 kV. The mass of each plate is 100 g. If the upper plate is fixed while the lower plate is free to move, what must be the separation between the two plates to keep them stationary?

3.5. The magnetic flux density in free space is 0.8 T. Compute the energy density in the region.

3.6. The energy stored in a magnetic circuit is given as $W_m = -5 \ln \lambda - \lambda$. What is the relationship between the applied current and the flux linkages? Calculate the inductance of the magnetic circuit when the current is 2 A.

3.7. Determine the force acting on the movable part in Figure P3.7 when the 100-turn coil carries a current of 20 A. The relative permeability of the magnetic material is 1000.

3.8. An electromagnet with a relative permeability of 1000 and a uniform thickness of 10 cm has the dimensions as given in Figure P3.8. What must be the current in the series-connected coils so that the electromagnet is on the verge of lifting a ferromagnetic bar having a permeability of 300 and a mass of 20 kg?

3.9. The magnetic circuit of a plunger is shown in Figure P3.9. Determine (a) the total flux linkages, (b) the energy stored in the magnetic circuit, and (c) the force acting on the plunger. The relative permeability of the magnetic material is 1500.

3.10. The 400-turn coil of the electromagnet as shown in Figure P3.10 carries a current of 150 sin(377t) A. Sketch the force as a function of time experienced by the fixed magnetic bar. What is the average value of the force? The relative permeability of the magnetic material is 800.

3.11. A U-shaped electromagnet is required to lift an iron bar that is at a distance of 1 mm. The cross-sectional area of each pole is 12 cm^2. The flux density

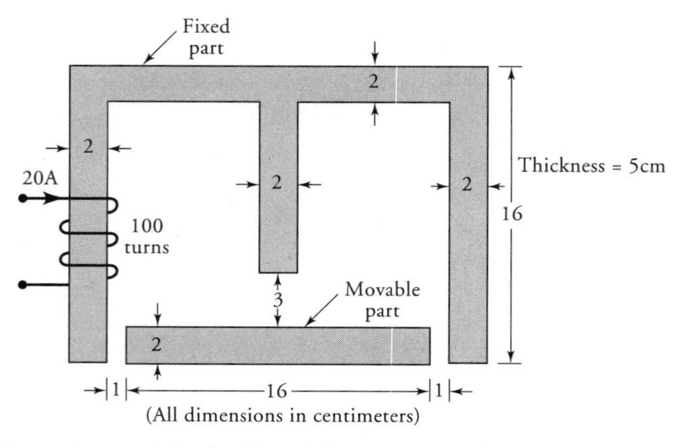

Figure P3.7 Magnetic circuit for Problem 3.7.

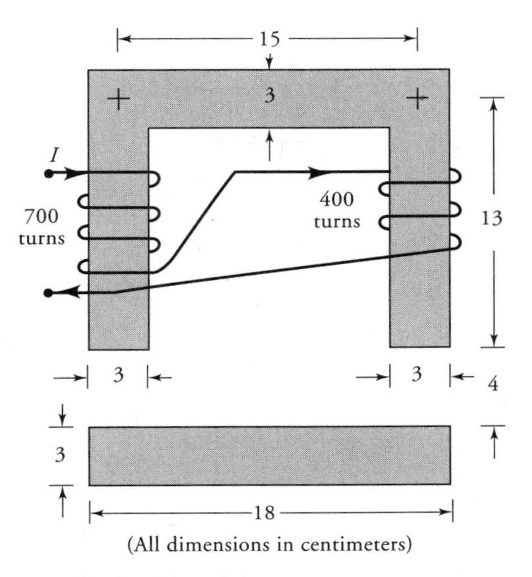

Figure P3.8 Electromagnet for Problem 3.8.

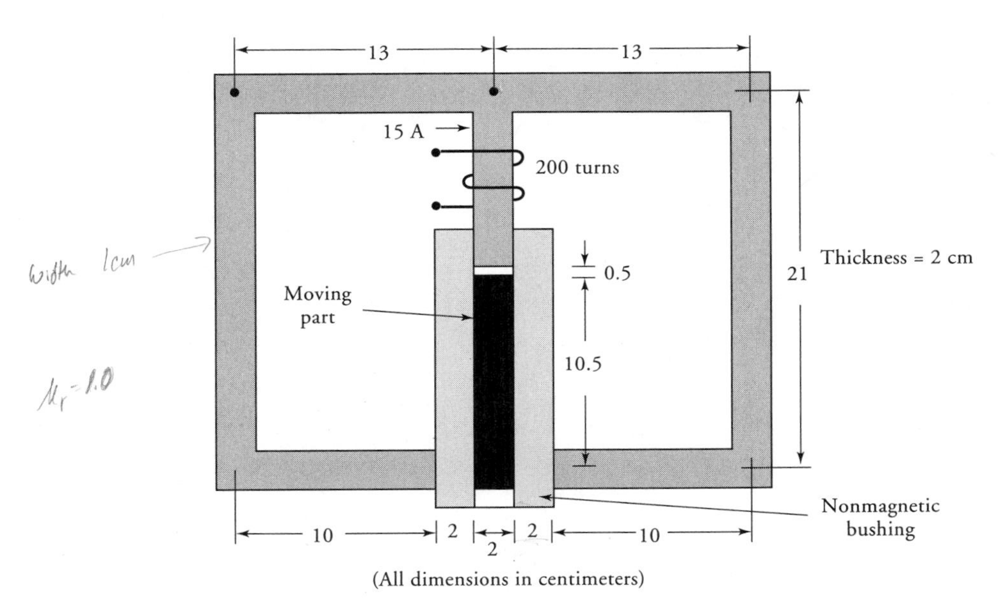

Figure P3.9 Magnetic plunger for Problem 3.9.

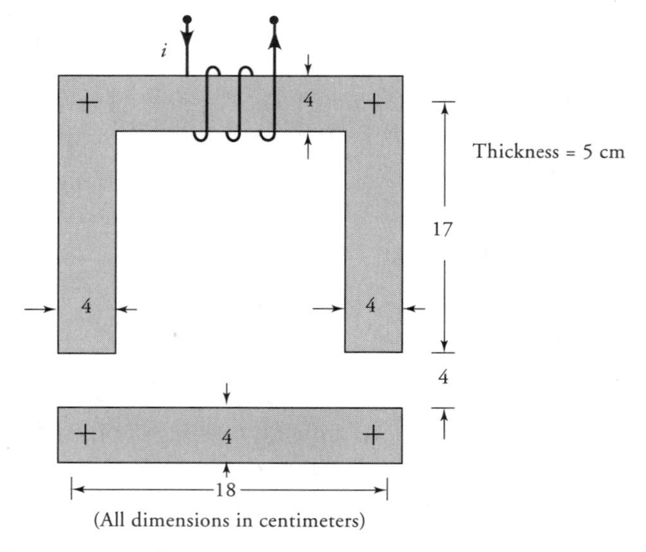

Figure P3.10 Electromagnet for Problem 3.10.

in each air-gap is 0.8 T. If the magnetic material is infinitely permeable, what will be the force exerted by the magnet on the iron bar?

3.12. A magnetic circuit with uniform circular cross-section is shown in Figure P3.12. Determine (a) the force acting on each pole, (b) the energy stored in each air-gap, (c) the energy stored in the magnetic region, and (d) the total energy stored in the magnetic circuit. Verify the total energy stored in the magnetic circuit using the inductance concept.

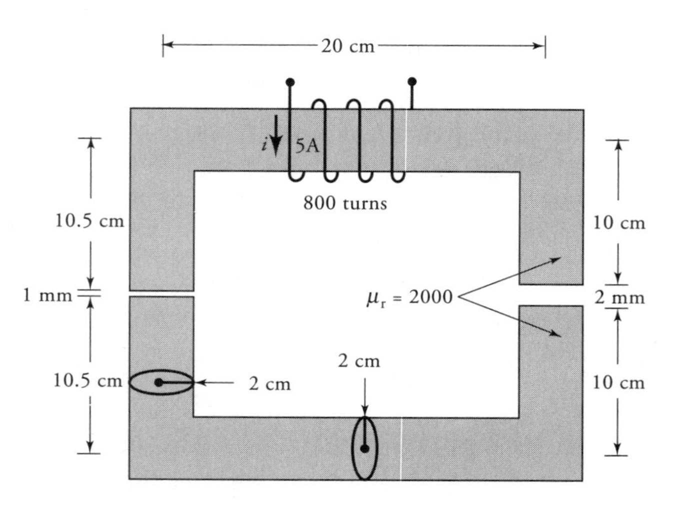

Figure P3.12 Magnetic circuit for Problem 3.12.

3.13. The dimensions of a cylindrical electromagnet in close contact with a 2-cm thick iron disc are given in Figure P3.13. The density of iron is 7.85 g/cm^3, and its relative permeability is 500. The relative permeability of the electromagnet is 2000. What must be the current in the 500-turn coil? Also compute the inductance and the total energy stored in the system.

3.14. Determine the number of poles in an ac generator that operates at a speed of 2400 rpm and the frequency of the induced voltage is 120 Hz.

3.15. Determine the speed of rotation if the frequency of the induced emf in a 12-pole machine is (a) 25 Hz, (b) 50 Hz, (c) 60 Hz, (d) 120 Hz, and (e) 400 Hz.

3.16. A 50-cm long wire passes 20 times a second under the pole-face of a square magnet. If each side of the magnet is 20 cm in length and the flux density is 0.5 T, find the induced emf in the wire.

3.17. A 200-turn coil is rotating at a speed of 3600 rpm in a 4-pole generator. The flux per pole is 2 mWb. Determine (a) the frequency of the induced emf, (b) the maximum value of the induced emf, (c) the rms value of the induced emf in an ac generator, and (d) the average value of the induced emf in a dc generator.

3.18. The average value of the induced emf per conductor in a 4-pole dc generator is 1.5 V. If the average value of the induced emf is 240 V, find the number of turns in the coil. If the coil is rotating at a speed of 800 rpm, what is the flux per pole?

3.19. Determine the force exerted on the rotor of a dc machine by a 10-cm long conductor when it carries a current of 100 A. The magnetic flux density is 0.8 T.

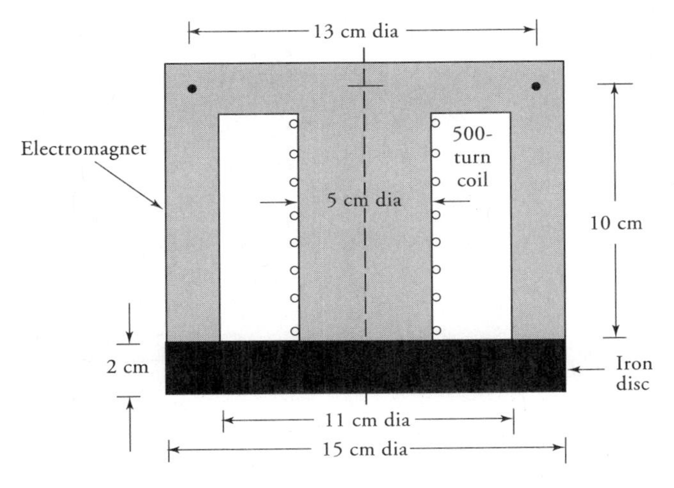

Figure P3.13 Cylindrical electromagnet for Problem 3.13.

3.20. A copper strip of length L pivoted at one end is freely rotating at an angular velocity of ω in a uniform magnetic field. If the field is perpendicular to the plane of the strip and the flux density is B_o, determine the induced emf between the two ends of the strip.

3.21. A copper strip of length $2L$ pivoted at the midpoint is rotating at an angular velocity ω in a uniform magnetic field. The flux density is B_o. The plane of the strip is perpendicular to the magnetic field. Determine the induced emf (a) between its end points and (b) between the midpoint and one of its ends.

3.22. A 4-pole, 230-V, 60-Hz, three-phase induction motor runs at a speed of 1725 rpm when fully loaded. Determine (a) the synchronous speed in rpm, (b) synchronous speed in rps, (c) synchronous speed in rad/s, (d) per-unit slip, and (e) percent slip.

3.23. A synchronous generator (alternator) has 12 poles and is driven at 600 rpm. What is the frequency of the induced emf?

3.24. What must be the number of poles in a synchronous motor so that the revolving field rotates at the maximum speed when the frequency is 50 Hz?

3.25. Repeat Problem 3.24 if the frequency is 60 Hz.

3.26. The rotor of a 6-pole, 50-Hz, 208-V, three-phase induction motor rotates at a speed of 980 rpm. Calculate (a) the synchronous speed, (b) per-unit slip, and (c) percent slip.

3.27. A 12-pole, 460-V, 50-Hz induction motor rotates at a percent slip of 3%. Determine the rotor speed.

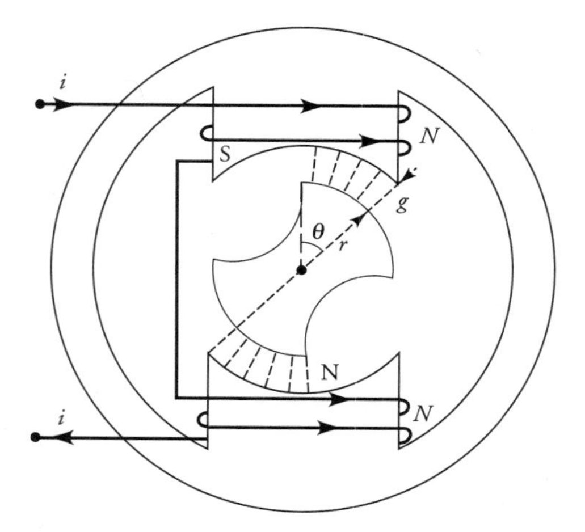

Figure P3.28 Reluctance motor for Problem 3.28.

3.28. Derive an expression for the torque of the 2-pole reluctance motor shown in Figure P3.28. Assume that (a) all the reluctance of the magnetic circuit is in the air-gap, and (b) the effective area of each gap is equal to the area of the overlap. Calculate the torque when the 100-turn coil carries 2 A, the diameter of the rotor is 15 cm, the length of the motor is 10 cm, the displacement angle is 30°, and the air-gap length is 2 mm.

3.29. The reluctance of a 2-pole reluctance motor is given by

$$R(\theta) = 1500 - 850 \cos 2\theta \text{ H}^{-1}$$

Determine the torque developed by the motor when the current in the 150-turn coil is 5 A. Plot the torque as a function of θ.

3.30. The inductance of a 2-pole reluctance motor is given as

$$L(\theta) = 4 + 2 \cos 2\theta + \cos 4\theta \text{ H}$$

Determine the torque developed by the motor when the current in a 200-turn coil is 10 A. Sketch the torque developed by the motor as a function of θ.

CHAPTER 

Transformers

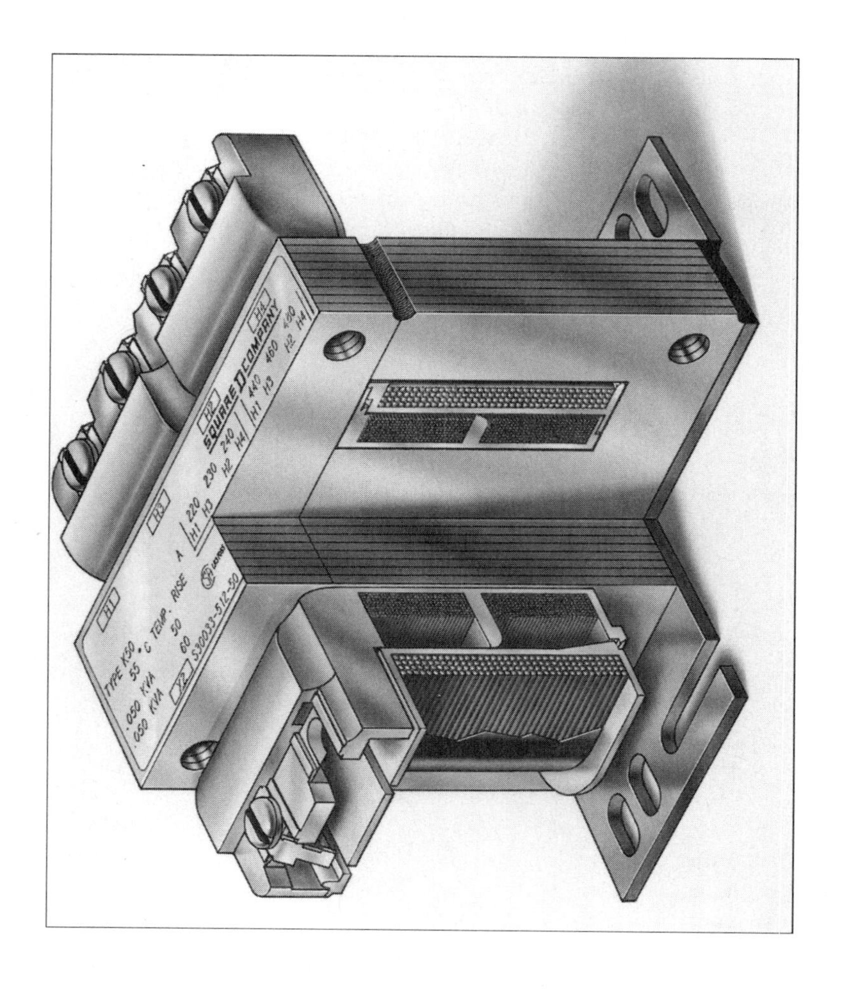

"Cut-away" view of a single-phase transformer. (*Courtesy of Square D Company*)

4.1 Introduction

If we arrange two electrically isolated coils in such a way that the time-varying flux due to one of them causes an electromotive force (emf) to be induced in the other, they are said to form a **transformer**. In other words, a transformer is a device that involves magnetically coupled coils. If only a fraction of the flux produced by one coil links the other, the coils are said to be loosely coupled. In this case, the operation of the transformer is not very efficient.

In order to increase the coupling between the coils, the coils are wound on a common core. When the core is made of a nonmagnetic material, the transformer is called an **air-core transformer**. When the core is made of a ferromagnetic material with relatively high permeability, the transformer is referred to as an **iron-core transformer**. A highly permeable magnetic core ensures that (a) almost all the flux created by one coil links the other and (b) the reluctance of the magnetic path is low. This results in the most efficient operation of a transformer.

In its simplest form, a transformer consists of two coils that are electrically isolated from each other but are wound on the same magnetic core. A time-varying current in one coil sets up a time-varying flux in the magnetic core. Owing to the high permeability of the core, most of the flux links the other coil and induces a time-varying emf (voltage) in that coil. The frequency of the induced emf in the other coil is the same as that of the current in the first coil. If the other coil is connected to a load, the induced emf in the coil establishes a current in it. Thus, the power is transferred from one coil to the other via the magnetic flux in the core.

The coil to which the source supplies the power is called the **primary winding**. The coil that delivers power to the load is called the **secondary winding**. Either winding may be connected to the source and/or the load.

Since the induced emf in a coil is proportional to the number of turns in a coil, it is possible to have a higher voltage across the secondary than the applied voltage to the primary. In this case, the transformer is called a **step-up** transformer. A step-up transformer is used to connect a relatively high-voltage transmission line to a relatively low-voltage generator. On the other hand, a **step-down** transformer has a lower voltage on the secondary side. An example of a step-down transformer is a welding transformer, the secondary of which is designed to deliver a high load current.

When the applied voltage to the primary is equal to the induced emf in the secondary, the transformer is said to have a **one-to-one ratio.** A one-to-one ratio transformer is used basically for the purpose of electrically isolating the secondary side from its primary side. Such a transformer is usually called an **isolation transformer**. An isolation transformer can be utilized for direct current (dc) isolation. That is, if the input voltage on the primary side consists of both dc and alternating currrent (ac) components, the voltage on the secondary side will be purely ac in nature.

4.2 Construction of a Transformer

In order to keep the core loss to a minimum, the core of a transformer is built up of thin laminations of highly permeable ferromagnetic material such as silicon-sheet steel. Silicon steel is used because of its nonaging properties and low magnetic losses. The lamination's thickness varies from 0.014 inch to 0.024 inch. A thin coating of varnish is applied to both sides of the lamination in order to provide high interlamination resistance. The process of cutting the laminations to the proper size results in punching and shearing strains. These strains cause an increase in the core loss. In order to remove the punching and shearing strains, the laminations are subjected to high temperatures in a controlled environment for some time. It is known as the **annealing process**.

Basically two types of construction are in common use for the transformers: **shell type** and **core type**. In the construction of a shell-type transformer, the two windings are usually wound over the same leg of the magnetic core, as shown in Figure 4.1. In a core-type transformer, shown in Figure 4.2, each winding may be evenly split and wound on both legs of the rectangular core. The nomenclature, shell type and core type, is derived from the fact that in a shell-type transformer the core encircles the windings, whereas the windings envelop the core in a core-type transformer.

For relatively low power applications with moderate voltage ratings, the windings may be wound directly on the core of the transformer. However, for high-voltage and/or high-power transformers, the coils are usually form-wound and then assembled over the core.

Both the core loss (hysteresis and eddy-current loss) and the copper loss (electrical loss) in a transformer generate heat, which, in turn, increases the operating temperature of the transformer. For low-power applications, natural air circulation may be enough to keep the temperature of the transformer within an accept-

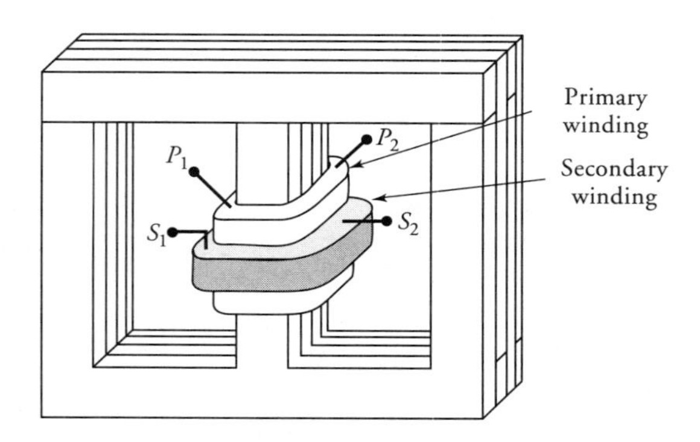

Figure 4.1 Shell-type transformer.

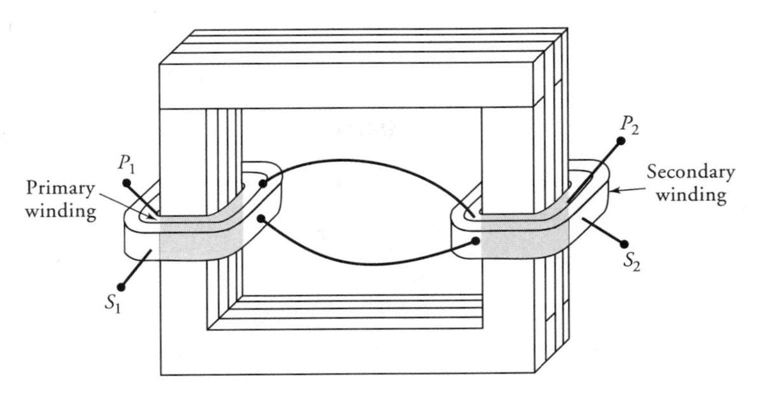

Figure 4.2 Core-type transformer.

able range. If the temperature increase cannot be controlled by natural air circulation, a transformer may be cooled by continuously forcing air through its core and windings. When forced-air circulation is not enough, a transformer may be immersed in a transformer oil, which carries the heat to the walls of the containing tank. In order to increase the radiating surface of the tank, cooling fins may be welded to the tank or the tank may be built from corrugated sheet steel. These are some of the methods used to curb excessive temperature in the transformer.

4.3 An Ideal Transformer

A two-winding transformer with each winding acting as a part of a separate electric circuit is shown in Figure 4.3. Let N_1 and N_2 be the number of turns in the primary and secondary windings. The primary winding is connected to a time-

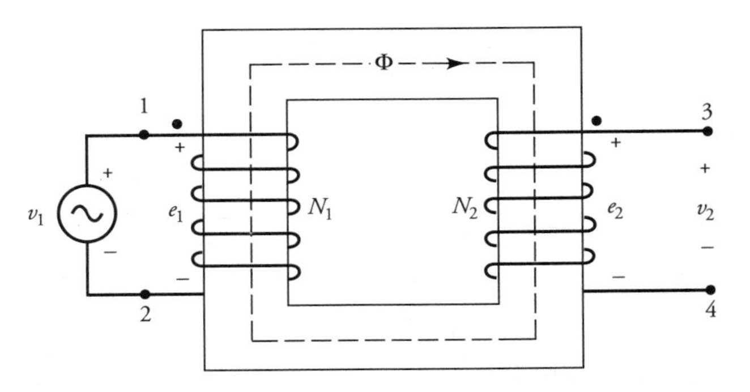

Figure 4.3 An idealized transformer under no load.

varying voltage source v_1, while the secondary winding is left open. For the sake of understanding, let us first consider an idealized transformer in which there are no losses and no leakage flux. In other words, we are postulating the following:

1. The core of the transformer is highly permeable in a sense that it requires vanishingly small magnetomotive force (mmf) to set up the flux Φ as shown in the figure.
2. The core does not exhibit any eddy-current or hysteresis loss.
3. All the flux is confined to circulate within the core.
4. The resistance of each winding is negligible.

According to Faraday's law of induction, the magnetic flux Φ in the core induces an emf e_1 in the primary winding that opposes the applied voltage v_1. For the polarities of the applied voltage and the induced emf, as indicated in the figure for the primary winding, we can write

$$e_1 = N_1 \frac{d\Phi}{dt} \tag{4.1}$$

Similarly, the induced emf in the secondary winding is

$$e_2 = N_2 \frac{d\Phi}{dt} \tag{4.2}$$

with its polarity as indicated in the figure.

In the idealized case assumed, the induced emf's e_1 and e_2 are equal to the corresponding terminal voltages v_1 and v_2, respectively. Thus, from Eqs. (4.1) and (4.2), we obtain

$$\frac{v_1}{v_2} = \frac{e_1}{e_2} = \frac{N_1}{N_2} \tag{4.3}$$

which states that **the ratio of primary to secondary induced emf is equal to the ratio of primary to secondary turns.**

It is a common practice to define the ratio of primary to secondary turns as the ***a*-ratio**, or the **transformation ratio**. That is,

$$\frac{N_1}{N_2} = a \tag{4.4}$$

Let i_2 be the current through the secondary winding when it is connected to a load, as shown in Figure 4.4. The magnitude of i_2 depends upon the load impedance. However, its direction is such that it tends to weaken the core flux Φ

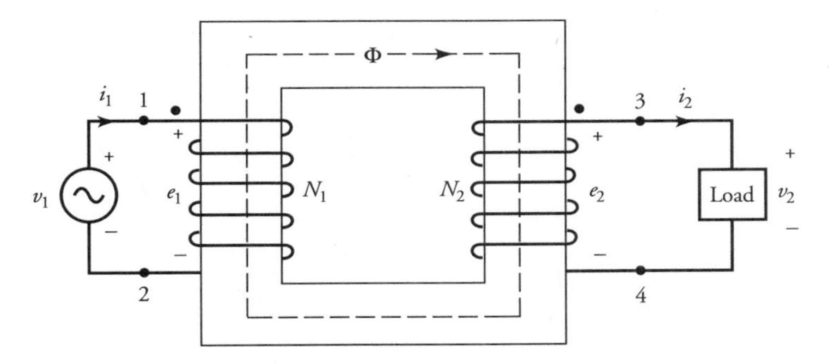

Figure 4.4 An idealized transformer under load.

and to decrease the induced emf in the primary e_1. For an idealized transformer, e_1 must always be equal to v_1. In other words, the flux in the core must always be equal to its original no-load value. In order to restore the flux in the core to its original no-load value, the source v_1 forces a current i_1 in the primary winding, as indicated in the figure. In accordance with our assumptions, the mmf of the primary current $N_1 i_1$ must be equal and opposite to the mmf of the secondary $N_2 i_2$. That is,

$$N_1 i_1 = N_2 i_2$$

or

$$\frac{i_2}{i_1} = \frac{N_1}{N_2} = a \tag{4.5}$$

which states that **the primary and the secondary currents are transformed in the inverse ratio of turns.**

From Eqs. (4.3) and (4.5), it is evident that

$$v_1 i_1 = v_2 i_2 \tag{4.6}$$

This equation simply confirms our assumption of no losses in an idealized transformer. It highlights the fact that, at any instant, **the power output (delivered to the load) is equal to the power input (supplied by the source).**

For sinusoidal variations in the applied voltage, the magnetic flux in the core also varies sinusoidally under ideal conditions. If the flux in the core at any instant t is given as

$$\Phi = \Phi_m \sin \omega t$$

where Φ_m is the amplitude of the flux and $\omega = 2\pi f$ is the angular frequency, then the induced emf in the primary is

$$e_1 = N_1 \omega \Phi_m \cos \omega t$$

The above equation can be expressed in phasor form in terms of its root-mean-square (rms) or effective value as

$$\tilde{E}_1 = \frac{1}{\sqrt{2}} N_1 \omega \Phi_m \underline{/0^\circ} = 4.44 f N_1 \Phi_m \underline{/0^\circ} \tag{4.7}$$

Likewise, the induced emf in the secondary winding is

$$\tilde{E}_2 = 4.44 f N_2 \Phi_m \underline{/0^\circ} \tag{4.8}$$

From Eqs. (4.7) and (4.8), we get

$$\frac{\tilde{V}_1}{\tilde{V}_2} = \frac{\tilde{E}_1}{\tilde{E}_2} = \frac{N_1}{N_2} = a \tag{4.9}$$

where $\tilde{V}_1 = \tilde{E}_1$ and $\tilde{V}_2 = \tilde{E}_2$ under ideal conditions. From the above equation, it is obvious that the induced emfs are in phase. **For an idealized transformer, the terminal voltages are also in phase.**

From the mmf requirements, we can also deduce that

$$\frac{\tilde{I}_2}{\tilde{I}_1} = \frac{N_1}{N_2} = a \tag{4.10}$$

where $\tilde{I}_1$ and $\tilde{I}_2$ are the currents in phasor form through the primary and secondary windings. **The above equation dictates that $\tilde{I}_1$ and $\tilde{I}_2$ must be in phase for an idealized transformer.**

Equation (4.6) can also be expressed in terms of phasor quantities as

$$\tilde{V}_1 \tilde{I}_1^* = \tilde{V}_2 \tilde{I}_2^* \tag{4.11}$$

That is, **the complex power supplied to the primary winding by the source is equal to the complex power delivered to the load by the secondary winding.** In terms of the apparent powers, the above equation becomes

$$V_1 I_1 = V_2 I_2 \tag{4.12}$$

If $\hat{Z}_2$ is the load impedance on the secondary side, then

$$\hat{Z}_2 = \frac{\tilde{V}_2}{\tilde{I}_2} = \frac{1}{a^2}\frac{\tilde{V}_1}{\tilde{I}_1}$$

$$= \frac{1}{a^2}\hat{Z}_1$$

or

$$\hat{Z}_1 = a^2\hat{Z}_2 \tag{4.13}$$

where $\hat{Z}_1 = \tilde{V}_1/\tilde{I}_1$ is the load impedance as referred to the primary side. Equation (4.13) states that **the load impedance as seen by the source on the primary side is equal to a^2 times the actual load impedance on the secondary side.** This equation highlights the fact that transformers can be used for impedance matching. A known impedance can be raised or lowered to match the rest of the circuit for maximum power transfer.

Transformer Polarity

A transformer may have multiple windings that may be connected either in series to increase the voltage rating or in parallel to increase the current rating. Before the connections are made, however, it is necessary that we know the polarity of each winding. By polarity we mean the relative direction of the induced emf in each winding.

Let us examine the transformer shown in Figure 4.4. Let the polarity of the time-varying source connected to the primary winding, at any instant, be as indicated in the figure. Since the induced emf e_1 in the primary of an idealized transformer must be equal and opposite to the applied voltage v_1, terminal 1 of the primary is positive with respect to terminal 2. The wound direction of primary winding as shown in the figure is responsible for the clockwise direction of the flux Φ in the core of the transformer. This flux must induce in the secondary winding an emf e_2 that results in the current i_2 as indicated. The direction of the current i_2 is such that it produces a flux that opposes the change in the original flux Φ. For the wound direction of the secondary winding as depicted in the figure, terminal 3 must be positive with respect to terminal 4. Since terminal 3 has the same polarity as terminal 1, they are said to follow each other. In other words, terminals 1 and 3 are like-polarity terminals. To indicate the like-polarity relationship, we have placed dots at these terminals.

Transformer Ratings

The nameplate of a transformer provides information on the apparent power and the voltage-handling capacity of each winding. From the nameplate data of a 5-kVA, 500/250-V, step-down transformer, we conclude the following:

1. The full-load or nominal power rating of the transformer is 5 kVA. In other words, the transformer can deliver 5 kVA on a continuous basis.
2. Since it is a step-down transformer, the (nominal) primary voltage is $V_1 = 500$ V and the (nominal) secondary voltage is $V_2 = 250$ V.
3. The nominal magnitudes of the primary and the secondary currents at full load are

$$I_1 = \frac{5000}{500} = 10 \text{ A}$$

and

$$I_2 = \frac{5000}{250} = 20 \text{ A}$$

4. Since the information on the number of turns is customarily not given by the manufacturer, we determine the a-ratio from the (nominal) terminal voltages as

$$a = \frac{500}{250} = 2$$

EXAMPLE 4.1

The core of a two-winding transformer, as shown in Figure 4.5**a**, is subjected to a magnetic flux variation as indicated in Figure 4.5**b**. What is the induced emf in each winding?

● SOLUTION

Since the polarities are already marked on the windings, terminals b and c are like-polarity terminals.

For the time interval from 0 to 0.06 s, the magnetic flux increases linearly as

$$\Phi = 0.15\,t \text{ Wb}$$

Thus, the induced emf between terminals a and b is

$$e_{ab} = -e_{ba} = -N_{ab}\frac{d\Phi}{dt} = -200 \times 0.15 = -30 \text{ V}$$

and the induced emf between terminals c and d is

$$e_{cd} = N_{cd}\frac{d\Phi}{dt} = 500 \times 0.15 = 75 \text{ V}$$

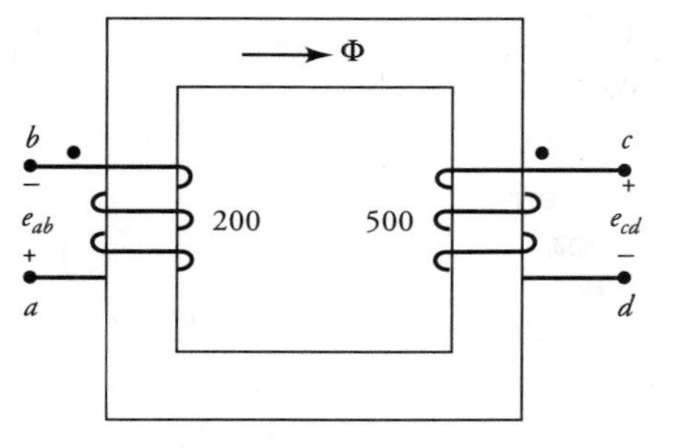

(a) Ideal transformer.

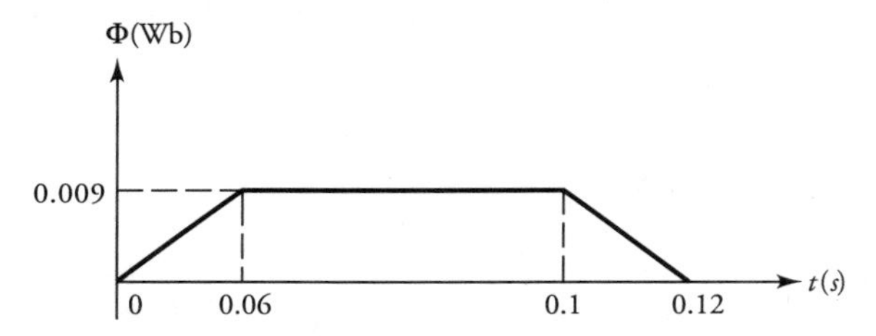

(b) Flux variations in the core of the transformer shown in part (a).

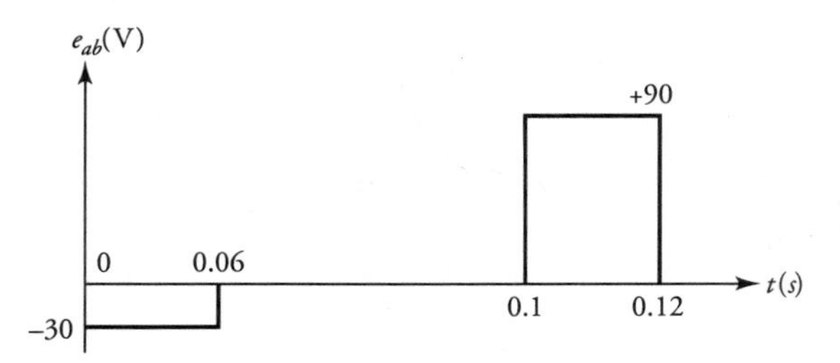

(c) Induced voltage e_{ab} as a function of time.

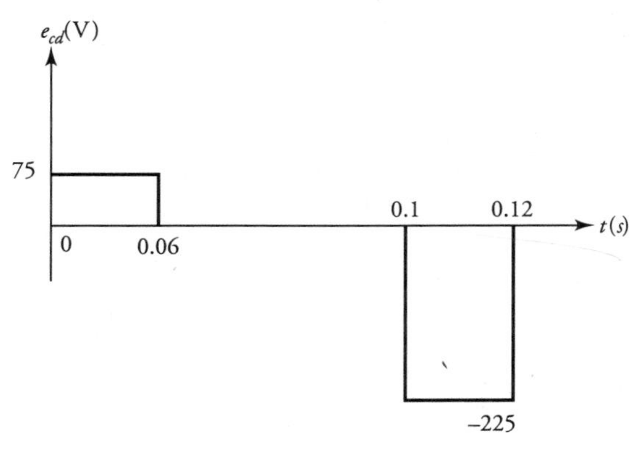

(d) Induced voltage e_{cd} as a function of time.

Figure 4.5 An ideal-transformer and related waveforms for Example 4.1.

From 0.06 to 0.1 s, the induced emfs are zero, as there is no variation in the flux. In the time interval from 0.1 to 0.12 s, the rate of change of the flux is -0.45 Wb/s. Hence, the induced emfs are

$$e_{ab} = -200 \times (-0.45) = 90 \text{ V}$$

and

$$e_{cd} = 500 \times (-0.45) = -225 \text{ V}$$

The waveforms for the induced emfs are depicted in Figures 4.5**c** and **d** for the primary and the secondary windings, respectively.

■

EXAMPLE 4.2

An ideal transformer has a 150-turn primary and 750-turn secondary. The primary is connected to a 240-V, 50-Hz source. The secondary winding supplies a load of 4 A at a lagging power factor (pf) of 0.8. Determine (a) the *a*-ratio, (b) the current in the primary, (c) the power supplied to the load, and (d) the flux in the core.

● SOLUTION

(a) The *a*-ratio: $a = 150/750 = 0.2$

(b) Since $I_2 = 4$ A, the current in the primary is

$$I_1 = \frac{I_2}{a} = \frac{4}{0.2} = 20 \text{ A}$$

(c) The voltage on the secondary side is

$$V_2 = \frac{V_1}{a} = \frac{240}{0.2} = 1200 \text{ V}$$

Thus, the power supplied to the load is

$$P_L = V_2 I_2 \cos\theta = 1200 \times 4 \times 0.8 = 3840 \text{ W}$$

(d) The maximum flux in the core is

$$\Phi_m = \frac{E_1}{4.44 f N_1} = \frac{V_1}{4.44 f N_1} = \frac{240}{4.44 \times 50 \times 150}$$
$$= 7.21 \text{ mWb}$$

■

Exercises

4.1. The magnetic flux density in the core of a 4.4-kVA, 440/4400-V, step-up transformer is 0.8 T (rms). If the induced emf per turn is 10 V, determine (a) the primary and secondary turns, (b) the cross-sectional area of the core, and (c) the full-load current in each winding.

4.2. A 200-turn coil is immersed in a 60-Hz flux with an effective value of 4 mWb. Obtain an expression for the instantaneous value of the induced emf. If a voltmeter is connected between its two ends, what will be the reading on the voltmeter?

4.3. The number of turns in the primary and the secondary of an ideal transformer are 200 and 500, respectively. The transformer is rated at 10 kVA, 250 V, and 60 Hz on the primary side. The cross-sectional area of the core is 40 cm^2. If the transformer is operating at full load with a power factor of 0.8 lagging, determine (a) the effective flux density in the core, (b) the voltage rating of the secondary, (c) the primary and secondary winding currents, and (d) the load impedance on the secondary side and as viewed from the primary side.

4.4 The Real Transformer

In the previous section we placed quite a few restrictions to obtain useful relations for an idealized transformer. In this section, our aim is to lift those restrictions in order to develop an equivalent circuit for a real transformer.

Winding Resistances

However small it may be, each winding has some resistance. Nonetheless, we can replace a real transformer with an idealized transformer by including a lumped resistance equal to the winding resistance in series with each winding. As shown in Figure 4.6, R_1 and R_2 are the winding resistances for the primary and the secondary, respectively. The inclusion of the winding resistances dictates that (a) the power input must be greater than the power output, (b) the terminal voltage is not equal to the induced emf, and (c) the efficiency (the ratio of power output to power input) of a real transformer is less than 100%.

Leakage Fluxes

Not all of the flux created by a winding confines itself to the magnetic core on which the winding is wound. Part of the flux, known as the **leakage flux**, does complete its path through air. Therefore, when both windings in a transformer

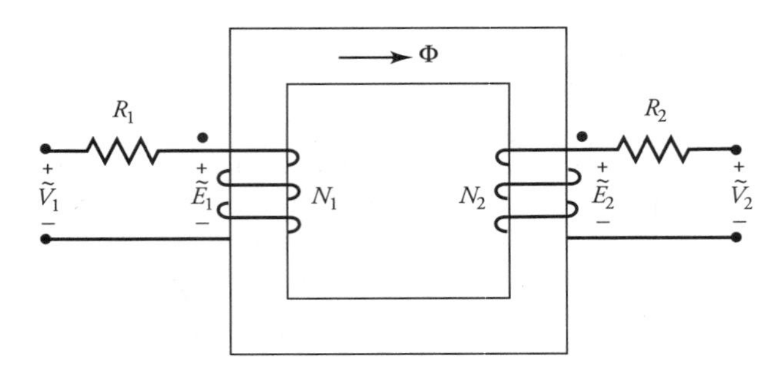

Figure 4.6 An ideal transformer with winding resistances modelled as lumped resistances.

carry currents, each creates its own leakage flux, as illustrated in Figure 4.7. The **primary leakage flux** set up by the primary does not link the secondary. Likewise, the **secondary leakage flux** restricts itself to the secondary and does not link the primary. The common flux that circulates in the core and links both windings is termed the **mutual flux**.

Although a leakage flux is a small fraction of the total flux created by a winding, it does affect the performance of a transformer. We can model a winding as if it consists of two windings: One winding is responsible to create the leakage flux through air, and the other encircles the core. Such hypothetical winding arrangements are shown in Figure 4.8 for a two-winding transformer. The two windings enveloping the core now satisfy the conditions of an idealized transformer.

The leakage flux associated with either winding is responsible for the voltage drop across it. Therefore, we can represent the voltage drop due to the leakage flux by a **leakage reactance**, as discussed in Chapter 2. If X_1 and X_2 are the leakage reactances for the primary and secondary windings, a real transformer can then be represented in terms of an idealized transformer with winding resistances and leakage reactances as shown in Figure 4.9.

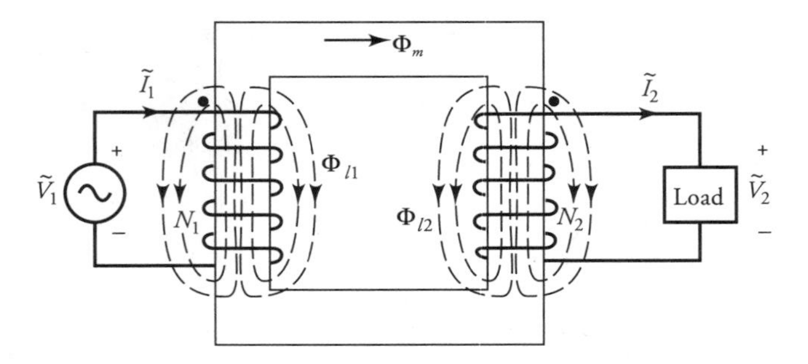

Figure 4.7 Transformer with leakage and mutual fluxes.

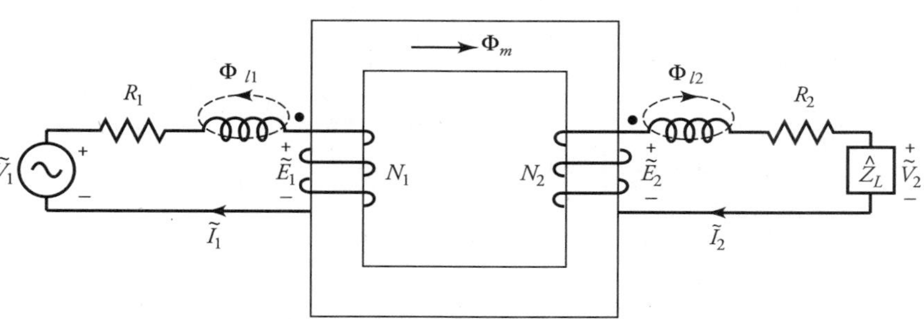

Figure 4.8 Hypothetical windings showing leakage and mutual flux linkages separately.

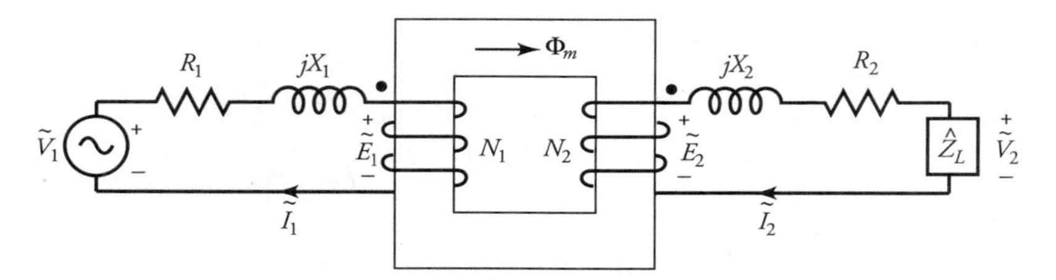

Figure 4.9 A real transformer represented in terms of an ideal transformer with winding resistances and leakage reactances.

EXAMPLE 4.3

A 23-kVA, 2300/230-V, 60-Hz, step-down transformer has the following resistance and leakage-reactance values: $R_1 = 4\ \Omega$, $R_2 = 0.04\ \Omega$, $X_1 = 12\ \Omega$, and $X_2 = 0.12\ \Omega$. The transformer is operating at 75% of its rated load. If the power factor of the load is 0.866 leading, determine the efficiency of the transformer.

● SOLUTION

Since the transformer is operating at 75% of its rated load, the effective value of the secondary winding current is

$$I_2 = \frac{23000}{230} \times 0.75 = 75\ \text{A}$$

Assuming the load voltage as a reference, the load current at a leading power factor of 0.866, in phasor form, is

$$\tilde{I}_2 = 75\underline{/30^\circ}\ \text{A}$$

The secondary winding impedance is

$$\hat{Z}_2 = R_2 + jX_2 = 0.04 + j0.12\ \Omega$$

The induced emf in the secondary winding is

$$\tilde{E}_2 = \tilde{V}_2 + \tilde{I}_2\hat{Z}_2 = 230 + (75\underline{/30°})(0.04 + j0.12)$$
$$= 228.287\underline{/2.33°}\ \text{V}$$

Since the transformation ratio is

$$a = \frac{2300}{230} = 10$$

we can determine the induced emf and the current on the primary side as

$$\tilde{E}_1 = a\tilde{E}_2 = 2282.87\underline{/2.33°}\ \text{V}$$
$$\tilde{I}_1 = \frac{\tilde{I}_2}{a} = 7.5\underline{/30°}\ \text{A}$$

The primary winding impedance is

$$\hat{Z}_1 = R_1 + jX_1 = 4 + j12\ \Omega$$

Hence, the source voltage must be

$$\tilde{V}_1 = \tilde{E}_1 + \tilde{I}_1\hat{Z}_1 = 2282.87\underline{/2.33°} + (7.5\underline{/30°})(4 + j12)$$
$$= 2269.587\underline{/4.7°}\ \text{V}$$

The power supplied to the load is

$$P_o = \text{Re}[\tilde{V}_2\tilde{I}_2^*] = \text{Re}[230 \times 75\underline{/-30°}] = 14{,}938.94\ \text{W}$$

The power input is

$$P_{in} = \text{Re}[\tilde{V}_1\tilde{I}_1^*] = \text{Re}[(2269.587\underline{/4.7°})(7.5\underline{/-30°})]$$
$$= 15{,}388.94\ \text{W}$$

The efficiency of the transformer is

$$\eta = \frac{P_o}{P_{in}} = \frac{14{,}938.94}{15{,}388.94} = 0.971, \text{ or } 97.1\%$$

■

Finite Permeability

The core of a real transformer has finite permeability and core loss. Therefore, even when the secondary is left open (no-load condition) the primary winding draws some current, known as the **excitation current**, from the source. It is a common practice to assume that the excitation current, $\tilde{I}_\Phi$, is the sum of two currents: the **core-loss current**, $\tilde{I}_c$, and the **magnetizing current**, $\tilde{I}_m$. That is,

$$\tilde{I}_\Phi = \tilde{I}_c + \tilde{I}_m \tag{4.14}$$

The core-loss component of the excitation current accounts for the magnetic loss (the hysteresis loss and the eddy-current loss) in the core of the transformer. If $\tilde{E}_1$ is the induced emf on the primary side and R_{c1} is the equivalent core-loss resistance, then the core-loss current, $\hat{I}_c$, is

$$\tilde{I}_c = \frac{\tilde{E}_1}{R_{c1}} \tag{4.15}$$

The magnetizing component of the excitation current is responsible to set up the mutual flux in the core. Since a current-carrying coil forms an inductor, the magnetizing current, $\hat{I}_m$, gives rise to a **magnetizing reactance**, X_{m1}. Thus,

$$X_{m1} = \frac{\tilde{E}_1}{j\tilde{I}_m} \tag{4.16}$$

We can now modify the equivalent circuit of Figure 4.9 to include the core-loss resistance and the magnetizing reactance. Such a circuit is shown in Figure 4.10.

When we increase the load on the transformer, the following sequence of events takes place:

(a) The secondary winding current increases.

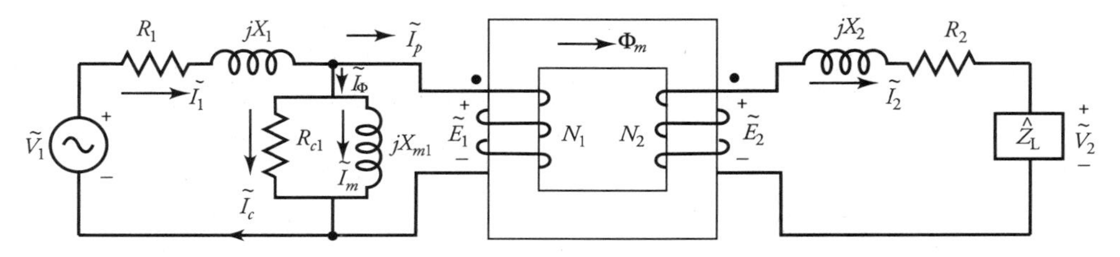

Figure 4.10 Equivalent circuit of a transformer including winding resistances, leakage reactance, core-loss resistance, magnetizing reactance, and an ideal transformer.

(b) The current supplied by the source increases.
(c) The voltage drop across the primary winding impedance $\hat{Z}_1$ increases.
(d) The induced emf $\tilde{E}_1$ drops.
(e) Finally, the mutual flux decreases owing to the decrease in the magnetizing current.

However, in a well-designed transformer, the decrease in the mutual flux from no load to full load is about 1% to 3%. Therefore, for all practical purposes, we can assume that $\tilde{E}_1$ remains substantially the same. In other words, the mutual flux is essentially the same under normal loading conditions and thereby there is no appreciable change in the excitation current.

In the equivalent circuit representation of a transformer, the core is rarely shown. Sometimes, parallel lines are drawn between the two windings to indicate the presence of a magnetic core. We will use such an equivalent circuit representation. If the parallel lines between the two windings are missing, our interpretation is that the core is nonmagnetic. With that understanding, the exact equivalent circuit of a real transformer is given in Figure 4.11. In this figure, a dashed box is also drawn to show that the circuit enclosed by it is the so-called ideal transformer. All the ideal-transformer relationships can be applied to this circuit. The load current $\tilde{I}_2$ on the secondary side is represented on the primary side as $\tilde{I}_p$.

Since the excitation current is supplied by the source, the difference between the mmfs of the primary and secondary windings must be equal to the mmf required for the excitation of the transformer. That is,

$$\tilde{I}_\Phi N_1 = \tilde{I}_1 N_1 - \tilde{I}_2 N_2$$

which yields

$$\tilde{I}_\Phi = \tilde{I}_1 - \frac{\tilde{I}_2}{a} = \tilde{I}_1 - \tilde{I}_p \tag{4.17}$$

It is possible to represent a transformer by an equivalent circuit that does not employ an ideal transformer. Such equivalent circuits are drawn with reference

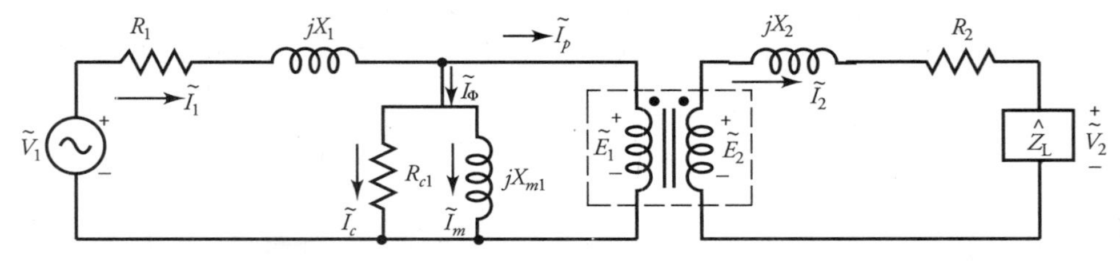

Figure 4.11 An exact equivalent circuit of a real transformer. The coupled-coiled in the dashed box represent an ideal transformer with a magnetic core.

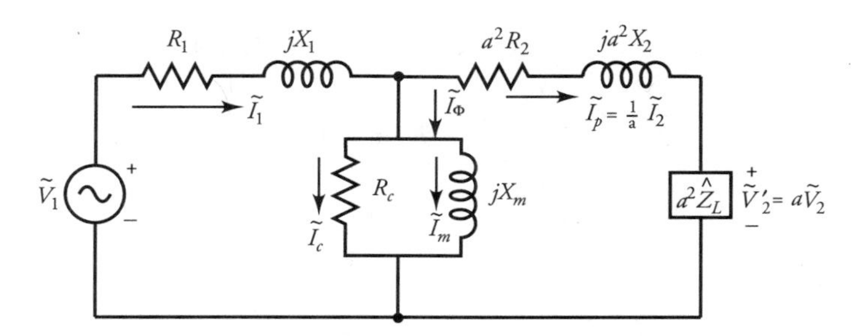

Figure 4.12 The exact equivalent circuit as viewed from the primary side of the transformer.

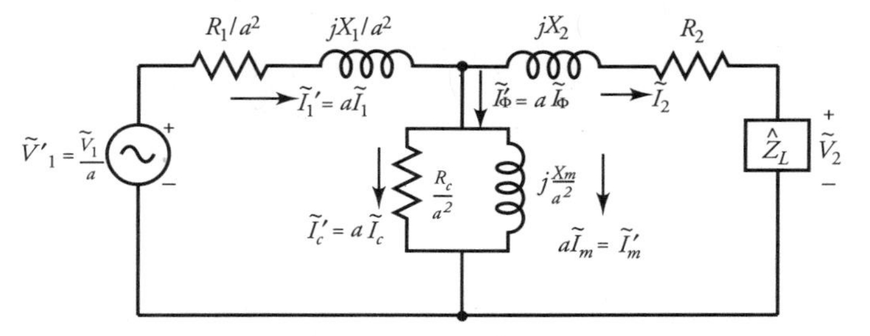

Figure 4.13 An exact equivalent circuit as viewed from the secondary side of the transformer.

to a given winding. Figure 4.12 shows such an equivalent circuit as viewed from the primary side. Note that the circuit elements that were on the secondary side in Figure 4.11 have been transformed to the primary side in Figure 4.12. Figure 4.13 shows the equivalent circuit of the same transformer as referred to the secondary side.

EXAMPLE 4.4

The equivalent core-loss resistance and the magnetizing reactance on the primary side of the transformer discussed in Example 4.3 are 20 kΩ and 15 kΩ, respectively. If the transformer delivers the same load, what is its efficiency?

● SOLUTION

From Example 4.3, we have

$$\tilde{V}_2 = 230 \text{ V} \qquad \tilde{I}_2 = 75\underline{/30^\circ} \text{ A} \qquad \tilde{E}_2 = 228.287\underline{/2.33^\circ} \text{ V}$$

$$a = 10 \qquad \tilde{E}_1 = 2282.87\underline{/2.33^\circ} \text{ V}, \qquad \tilde{I}_p = 7.5\underline{/30^\circ} \text{ A}$$

$$P_o = 14{,}938.94 \text{ W}$$

The core-loss, magnetizing, and excitation currents are

$$\tilde{I}_c = \frac{\tilde{E}_1}{R_{c1}} = \frac{2282.87\underline{/2.33^\circ}}{20{,}000} = 0.114\underline{/2.33^\circ}\text{ A}$$

$$\tilde{I}_m = \frac{\tilde{E}_1}{jX_{m1}} = \frac{2282.87\underline{/2.33^\circ}}{j15{,}000} = 0.152\underline{/-87.67^\circ}\text{ A}$$

$$\tilde{I}_\phi = \tilde{I}_c + \tilde{I}_m = 0.114\underline{/2.33^\circ} + 0.152\underline{/-87.67^\circ} = 0.19\underline{/-50.8^\circ}\text{ A}$$

Thus,

$$\tilde{I}_1 = \tilde{I}_p + \tilde{I}_\phi = 7.5\underline{/30^\circ} + 0.19\underline{/-50.8^\circ} = 7.53\underline{/28.57^\circ}\text{ A}$$

$$\tilde{V}_1 = \tilde{E}_1 + \tilde{I}_1\hat{Z}_1 = 2282.87\underline{/2.33^\circ} + (7.53\underline{/28.57^\circ})(4 + j12)$$

$$= 2271.9\underline{/4.71^\circ}\text{ V}$$

The power input is

$$P_{in} = \text{Re}[\tilde{V}_1\tilde{I}_1^*] = 15{,}651.48\text{ W}$$

The efficiency of the transformer is

$$\eta = \frac{P_o}{P_{in}} = \frac{14{,}938.94}{15{,}651.48} = 0.955,\text{ or } 95.5\%$$

■

Phasor Diagram

When a transformer operates under steady-state conditions, an insight into its currents, voltages, and phase angles can be obtained by sketching its phasor diagram. Even though a phasor diagram can be developed by using any phasor quantity as a reference, we use the load voltage as a reference because quite often it is a known quantity.

Let $\tilde{V}_2$ be the voltage across the load impedance $\hat{Z}_L$ and $\tilde{I}_2$ be the load current. Depending upon $\hat{Z}_L$, $\tilde{I}_2$ may be leading, in phase with, or lagging $\tilde{V}_2$. In this particular case, let us assume that $\tilde{I}_2$ lags $\tilde{V}_2$ by an angle θ_2. We first draw a horizontal line from the origin of magnitude V_2 to represent the phasor $\tilde{V}_2$, as shown in Figure 4.14. The current I_2 is now drawn lagging V_2 by θ_2.

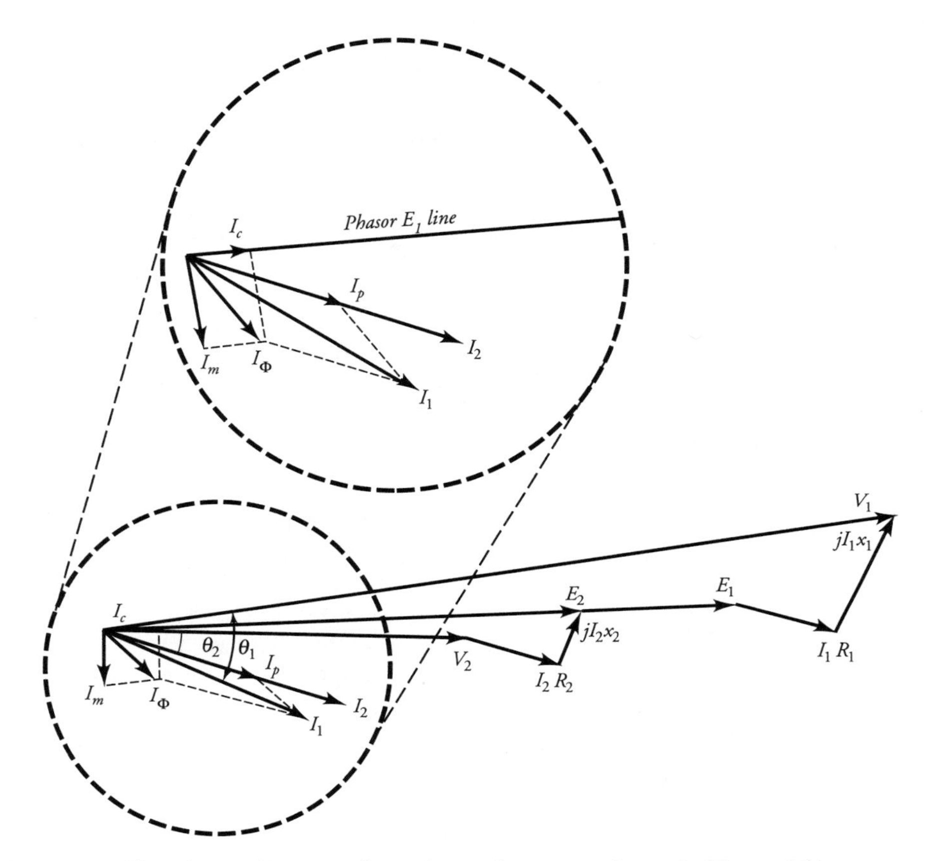

Figure 4.14 The phasor diagram of a real transformer as shown in Figure 4.11.

From the equivalent circuit, Figure 4.11, we have

$$\tilde{E}_2 = \tilde{V}_2 + \tilde{I}_2 R_2 + j\tilde{I}_2 X_2$$

We now proceed to construct the phasor diagram for $\underset{\sim}{E}_2$. Since the voltage drop $\tilde{I}_2 R_2$ is in phase with $\tilde{I}_2$ and it is to be added to $\tilde{V}_2$, we draw a line of magnitude $I_2 R_2$ starting at the tip of V_2 and parallel to I_2. The length of the line from the origin to the tip of $I_2 R_2$ represents the sum of $\tilde{V}_2$ and $\tilde{I}_2 R_2$. We can now add the voltage drop $j\tilde{I}_2 X_2$ at the tip of $I_2 R_2$ by drawing a line equal to its magnitude and leading $\tilde{I}_2$ by 90°. A line from the origin to the tip of $jI_2 X_2$ represents the magnitude of $\tilde{E}_2$. This completes the phasor diagram for the secondary winding.

Since $\tilde{E}_1 = a\tilde{E}_2$, the magnitude of the induced emf on the primary side depends

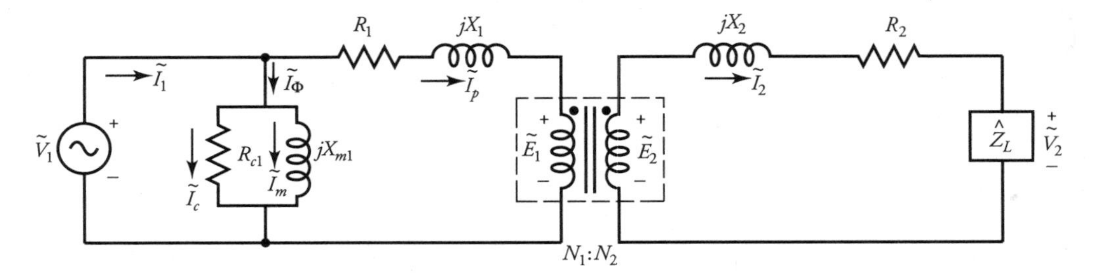

Figure 4.15 An approximate equivalent circuit of a transformer embodying an ideal transformer.

upon the a-ratio. Let us assume that the a-ratio is greater than unity. In that case, $\tilde{E}_1$ is greater than $\tilde{E}_2$ and can be represented by extending E_2 as shown.

The current $\tilde{I}_c$ is in phase with $\tilde{E}_1$, and $\tilde{I}_m$ lags $\tilde{E}_1$ by 90°. These currents are drawn from the origin as shown. The sum of these currents yields the excitation current $\tilde{I}_\Phi$. The source current I_1 is now constructed using the currents I_Φ and I_2/a, as illustrated in the figure. The voltage drop across the primary-winding impedance $\hat{Z}_1 = R_1 + jX_1$ is now added to obtain the phasor $\tilde{V}_1$. The phasor diagram is now complete. The angle between V_1 and I_1 is the power factor angle θ_1 for the input power.

Phasor diagrams for the exact equivalent circuits as shown in Figures 4.12 and 4.13 can also be drawn and are left as exercises for the reader.

Approximate Equivalent Circuits

In a well-designed transformer, the winding resistances, the leakage reactances, and the core loss are kept as low as possible. The low core loss implies high core-loss resistance. The high permeability of the core ensures high magnetizing reac-

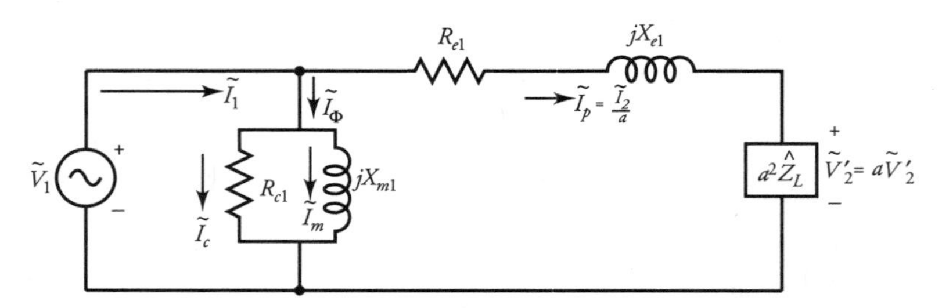

Figure 4.16 Approximate equivalent circuit of a transformer as viewed from the primary side.

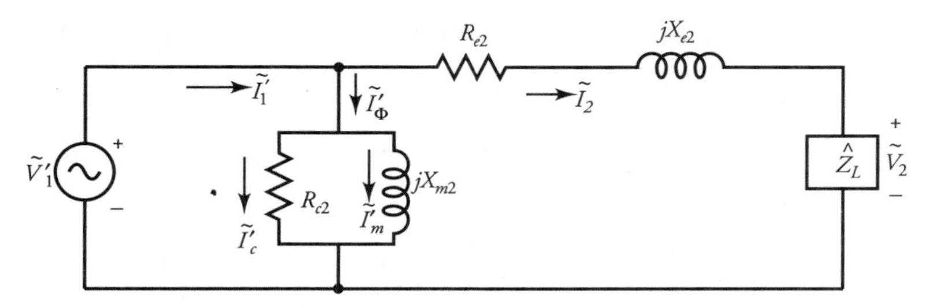

Figure 4.17 An approximate equivalent circuit of a transformer as viewed from the secondary side.

tance. Thus, the impedance of the so-called parallel branch (R_{c1} in parallel with jX_{m1}) across the primary is very high compared with $\hat{Z}_1 = R_1 + jX_1$ and $\hat{Z}_2 = R_2 + jX_2$. The high impedance of the parallel branch assures low excitation current. In the analysis of complex power systems, a great deal of simplification can be achieved by neglecting the excitation current.

Since $\hat{Z}_1$ is kept low, the voltage drop across it is also low in comparison with the applied voltage. Without introducing any appreciable error in our calculations, we can assume that the voltage drop across the parallel branch is the same as the applied voltage. This assumption allows us to move the parallel branch as indicated in Figure 4.15 for the equivalent circuit of a transformer embodying an ideal transformer. This is referred to as the **approximate equivalent circuit of a transformer.**

The approximate equivalent circuit as viewed from the primary side is given in Figure 4.16, where

$$\hat{Z}_{e1} = R_{e1} + jX_{e1} \tag{4.18a}$$

$$R_{e1} = R_1 + a^2R_2 \tag{4.18b}$$

and

$$X_{e1} = X_1 + a^2X_2 \tag{4.18c}$$

Similarly, Figure 4.17 shows the approximate equivalent circuit as referred to the secondary side of the transformer. In this figure,

$$\hat{Z}_{e2} = R_{e2} + jX_{e2} \tag{4.19a}$$

$$R_{e2} = R_2 + \frac{R_1}{a^2} \tag{4.19b}$$

$$X_{e2} = X_2 + \frac{X_1}{a^2} \tag{4.19c}$$

$$R_{c2} = \frac{R_{c1}}{a^2} \tag{4.19d}$$

and

$$X_{m2} = \frac{X_{m1}}{a^2} \tag{4.19e}$$

EXAMPLE 4.5

Analyze the transformer discussed in Examples 4.3 and 4.4 using the approximate equivalent circuit as viewed from the primary side. Also sketch its phasor diagram.

● SOLUTION

$$\tilde{V}_2' = a\tilde{V}_2 = 10 \times 230\angle 0^\circ = 2300\angle 0^\circ \text{ V}$$
$$\tilde{I}_p = 7.5\angle 30^\circ \text{ A}$$
$$R_{e1} = R_1 + a^2R_2 = 4 + (10^2)(0.04) = 8\ \Omega$$
$$X_{e1} = X_1 + a^2X_2 = 12 + (10^2)(0.12) = 24\ \Omega$$
$$\hat{Z}_{e1} = R_{e1} + jX_{e1} = 8 + j24\ \Omega$$

Thus, $\tilde{V}_1 = \tilde{V}_2' + \tilde{I}_p\hat{Z}_{e1} = 2300\angle 0^\circ + (7.5\angle 30^\circ)(8 + j24) = 2269.59\angle 4.7^\circ$ V

The core-loss and magnetizing currents are

$$\tilde{I}_c = \frac{2269.59\angle 4.7^\circ}{20{,}000} = 0.113\angle 4.7^\circ \text{ A}$$
$$\tilde{I}_m = \frac{2269.59\angle 4.7^\circ}{j15{,}000} = 0.151\angle -85.3^\circ \text{ A}$$

Thus, $$\tilde{I}_1 = \tilde{I}_p + \tilde{I}_c + \tilde{I}_m = 7.5\angle 30^\circ + 0.113\angle 4.7^\circ + 0.151\angle -85.3^\circ$$
$$= 7.54\angle 28.6^\circ \text{ A}$$

Hence, the power output, the power input, and the efficiency are

$$P_o = \text{Re}[(2300\angle 0^\circ)(7.5\angle -30^\circ)] = 14{,}938.94 \text{ W}$$
$$P_{in} = \text{Re}[(2269.59\angle 4.7^\circ)(7.54\angle -28.6^\circ)] = 15{,}645.36 \text{ W}$$

$$\eta = \frac{14{,}938.94}{15{,}645.36} = 0.955 \quad \text{or} \quad 95.5\%$$

The corresponding phasor diagram is shown in Figure 4.18.

The reader is encouraged to compare the above results with those obtained in Example 4.4 in order to have some awareness of the errors introduced as a result of the approximations we have made.

■

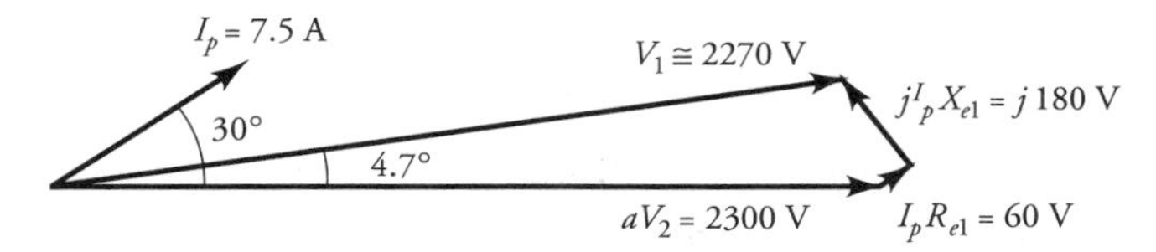

Figure 4.18 The phasor diagram of a transformer discussed in Example 4.5.

Exercises

4.4. A 2.4-kVA, 2400/240-V, 50-Hz, step-down transformer has the following parameters: $R_1 = 1.5\ \Omega$, $X_1 = 2.5\ \Omega$, $R_2 = 0.02\ \Omega$, $X_2 = 0.03\ \Omega$, $R_{c1} = 6\ \text{k}\Omega$, and $X_{m1} = 8\ \text{k}\Omega$. It is operating at 80% of its load at unity power factor. Using the exact equivalent circuit embodying the ideal transformer, determine the efficiency of the transformer. Also sketch its phasor diagram.

4.5. If the transformer in Exercise 4.4 delivers the rated load at 0.8 pf lagging, determine its efficiency using the approximate equivalent circuit as viewed from the secondary side. Also sketch its phasor diagram.

4.6. If the transformer in Exercise 4.4 operates at 50% of its rated load at a power factor of 0.5 leading, compute its efficiency using both exact and approximate equivalent circuits as referred to the primary side. Tabulate the percent errors in the currents, voltages, powers, and efficiency caused by the approximation.

4.5 Voltage Regulation

Consider a transformer whose primary winding voltage is adjusted so that it delivers the rated load at the rated secondary terminal voltage. If we now remove the load, the secondary terminal voltage changes because of the change in the voltage drops across the winding resistances and leakage reactances. A quantity of interest is the net change in the secondary winding voltage from no load to full load for the same primary winding voltage. When the change is expressed as a percentage of its rated voltage, it is called the **voltage regulation (*VR*)** of the transformer. As a percent, it may be written as

$$VR\% = \frac{V_{2NL} - V_{2FL}}{V_{2FL}} \times 100 \qquad (4.20)$$

where V_{2NL} and V_{2FL} are the effective values of no-load and full-load voltages at the secondary terminals.

The voltage regulation is like the figure-of-merit of a transformer. For an ideal transformer, the voltage regulation is zero. The smaller the voltage regulation, the better the operation of the transformer.

The expressions for the percent voltage regulation for the approximate equivalent circuits as viewed from the primary and the secondary sides are

$$VR\% = \frac{V_1 - aV_2}{aV_2} \times 100 \tag{4.21}$$

$$VR\% = \frac{\dfrac{V_1}{a} - V_2}{V_2} \times 100 \tag{4.22}$$

where V_1 is the full-load voltage on the primary side and V_2 is the rated voltage at the secondary.

EXAMPLE 4.6

A 2.2-kVA, 440/220-V, 50-Hz, step-down transformer has the following parameters as referred to the primary side: $R_{e1} = 3\ \Omega$, $X_{e1} = 4\ \Omega$, $R_{c1} = 2.5\ \text{k}\Omega$, and $X_{m1} = 2\ \text{k}\Omega$. The transformer is operating at full load with a power factor of 0.707 lagging. Determine the efficiency and the voltage regulation of the transformer.

● SOLUTION

From the given data,

$$a = \frac{440}{220} = 2 \qquad \tilde{V}_2 = 220\text{ V} \qquad S = 2200\text{ VA} \qquad I_2 = \frac{2200}{220} = 10\text{A}$$

For a lagging power factor of 0.707, $\theta = -45°$
Using load voltage as a reference, $\tilde{I}_2 = 10\underline{/-45°}$ A.
Referring to the equivalent circuit, Figure 4.16, we have

$$\tilde{I}_p = \frac{\tilde{I}_2}{a} = 5\underline{/-45°}\text{ A}$$

$$\tilde{V}_2' = a\tilde{V}_2 = 440\underline{/0°}\text{ V}$$

Thus,

$$\tilde{V}_1 = \tilde{V}_2' + \tilde{I}_p(R_{e1} + jX_{e1}) = 440 + (5\underline{/-45°})(3 + j4)$$

$$= 464.762\underline{/0.44°}\text{ V}$$

The core-loss and magnetizing currents are

$$\tilde{I}_c = \frac{464.762\angle 0.44^\circ}{2500} = 0.186\angle 0.44^\circ \text{ A}$$

$$\tilde{I}_m = \frac{464.762\angle 0.44^\circ}{j2000} = 0.232\angle -89.56^\circ \text{ A}$$

The current supplied by the source is

$$\tilde{I}_1 = \tilde{I}_p + \tilde{I}_c + \tilde{I}_m = 5\angle -45^\circ + 0.186\angle 0.44^\circ + 0.232\angle -89.56^\circ$$
$$= 5.296\angle -45.33^\circ \text{ A}$$

The power output, the power input, and the efficiency are

$$P_o = \text{Re}[(440)(5\angle 45^\circ)] = 1555.63 \text{ W}$$
$$P_{in} = \text{Re}[(464.762\angle 0.44^\circ)(5.296\angle 45.33^\circ)] = 1717.04 \text{ W}$$

$$\eta = \frac{1555.63}{1717.04} = 0.906 \quad \text{or} \quad 90.6\%$$

The voltage regulation is

$$VR\% = \frac{464.762 - 440}{440} \times 100 = 5.63\%$$

■

Exercises

4.7. Determine the voltage regulation of a step-down transformer whose parameters are given in Exercise 4.4 at full load and 0.8 pf leading. What is the full-load efficiency? Use exact equivalent circuit.

4.8. If the transformer discussed in Example 4.6 operates at a power factor of 0.8 leading, what is its full-load efficiency? What is the voltage regulation?

4.9. A 100-kVa, 13.2/2.2-kV, 50-Hz, step-down transformer has a core-loss equivalent resistance of 8 kΩ and a magnetization reactance of 7 kΩ. The equivalent winding impedance as referred to the primary side is $3 + j12$ Ω. If the transformer delivers the rated load at a power factor of 0.707 lagging, determine its voltage regulation and efficiency.

4.6 Maximum Efficiency Criterion

As defined earlier, the efficiency is simply the ratio of power output to power input. In a real transformer, the efficiency is always less than 100% owing to the two types of losses: the magnetic loss and the copper loss.

The magnetic loss, which is commonly referred to as the **core loss**, consists of eddy-current loss and hysteresis loss. For a given flux density and the frequency of operation, the eddy-current loss can be minimized by using thinner laminations. On the other hand, the hysteresis loss depends upon the magnetic characteristics of the type of steel used for the magnetic core. Since the flux, Φ_m, in the core of a transformer is practically constant for all conditions of load, the core (magnetic) loss, P_m, is essentially constant. Thus, the core loss is referred to as the **fixed loss**.

The copper loss (also known as I^2R loss or the electric-power loss) comprises the power dissipated by the primary and secondary windings. The copper loss, P_{cu}, varies as the square of the current in each winding. Therefore, as the load increases, so does the copper loss. For that reason, the copper loss is termed the **variable loss**.

The power output can be obtained simply by subtracting the core loss and the copper loss from the power input. This implies that we can also obtain power input by adding the core loss and the copper loss to the power output. We can highlight the power flow from the input toward the output by a single-line diagram, called the **power-flow diagram**. A power-flow diagram of a transformer (see Figure 4.11) is shown in Figure 4.19.

The efficiency of a transformer is zero at no load. It increases with the increase in the load and rises to a maximum value. Any further increase in the load actually forces the efficiency of a transformer to drop off. Therefore, there exists a definite load for which the efficiency of a transformer is maximum. We now proceed to determine the criterion for the maximum efficiency of a transformer.

Let us consider the approximate equivalent of a transformer as viewed from the primary side (see Figure 4.16). The equivalent load current and the load voltage on the primary side are $I_p\underline{/\theta}$ and aV_2. The power output is

$$P_o = aV_2I_p \cos \theta$$

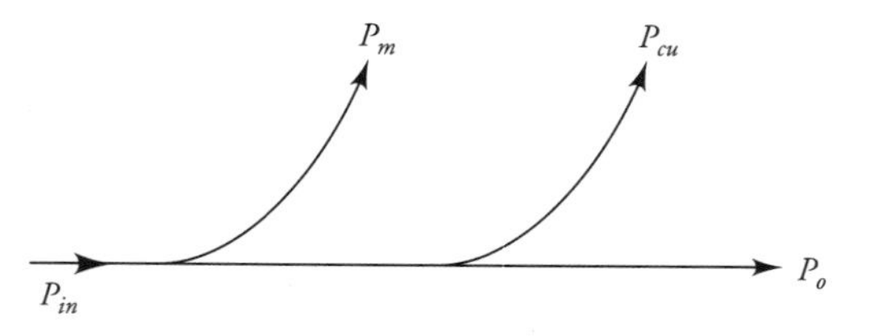

Figure 4.19 Power-flow diagram of a transformer.

and the copper loss is

$$P_{cu} = I_p^2 R_{e1}$$

If the core loss is P_m, then the power input is

$$P_{in} = aV_2 I_p \cos\theta + P_m + I_p^2 R_{e1}$$

Hence the efficiency of the transformer is

$$\eta = \frac{aV_2 I_p \cos\theta}{aV_2 I_p \cos\theta + P_m + I_p^2 R_{e1}}$$

The only variable in the above equation is the load current I_p. Therefore, if we differentiate it with respect to I_p and set it equal to zero, we obtain

$$I_{p\eta}^2 R_{e1} = P_m \tag{4.23a}$$

where $I_{p\eta}$ is the load current on the primary side at maximum efficiency. The above equation states that **the efficiency of a transformer is maximum when the copper loss is equal to the core (magnetic) loss**. In other words, a transformer operates at its maximum efficiency when the copper-loss curve intersects the core-loss curve, as depicted in Figure 4.20.

We can rewrite Eq. (4.23a) as

$$I_{p\eta} = \sqrt{\frac{P_m}{R_{e1}}} \tag{4.23b}$$

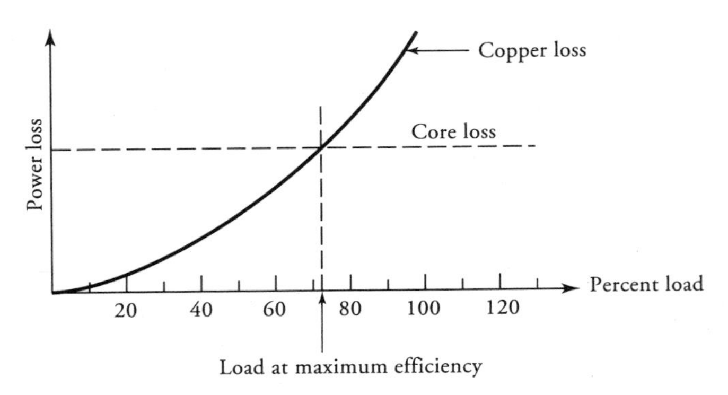

Figure 4.20 Losses in a transformer.

If I_{pfl} is the full-load current on the primary side, then the above equation can also be written as

$$I_{p\eta} = I_{pf1}\sqrt{\frac{P_m}{I_{pf1}^2 R_{e1}}}$$

or

$$I_{p\eta} = I_{pf1}\sqrt{\frac{P_m}{P_{cuf1}}} \tag{4.24}$$

where $P_{cuf1} = I_{pf1}^2 R_{e1}$ is the copper loss at full load.

Multiplying both sides of Eq. (4.24) by the rated load voltage on the primary side (aV_2), we can obtain the rating of the transformer at maximum efficiency in terms of its nominal rating as

$$VA|_{\text{max. eff.}} = VA|_{\text{rated}}\sqrt{\frac{\text{magnetic loss}}{\text{full-load copper loss}}} \tag{4.25}$$

The volt-ampere rating at maximum efficiency can actually be determined by performing tests on the transformer as explained in the ensuing section.

EXAMPLE 4.7

A 120-kVA, 2400/240-V, step-down transformer has the following parameters: $R_1 = 0.75\ \Omega$, $X_1 = 0.8\ \Omega$, $R_2 = 0.01\ \Omega$, $X_2 = 0.02\ \Omega$. The transformer is designed to operate at maximum efficiency at 70% of its rated load with 0.8 pf lagging. Determine (a) the kVA rating of the transformer at maximum efficiency, (b) the maximum efficiency, (c) the efficiency at full load and 0.8 pf lagging, and (d) the equivalent core-loss resistance.

● SOLUTION

Let us use the approximate equivalent circuit of the transformer as shown in Figure 4.16. The rated voltage across the load as viewed from the primary side is 2400 V. Thus, the rated load current is

$$I_p = \frac{120{,}000}{2400} = 50\ \text{A}$$

The load current at maximum efficiency is

$$I_{p\eta} = 0.7 I_p = 0.7 \times 50 = 35\ \text{A}$$

(a) The kVA rating of the transformer at maximum efficiency is

$$\text{kVA}|_{\text{max. eff.}} = \frac{35 \times 2400}{1000} = 84 \text{ kVA}$$

Thus, the copper loss at maximum efficiency is

$$P_{cu\eta} = I^2(R_1 + a^2R_2) = 35^2(0.75 + 10^2 \times 0.01)$$
$$= 2143.75 \text{ W}$$

and the core loss is

$$P_m = P_{cu\eta} = 2143.75 \text{ W}$$

(b) The power output, the power input, and the efficiency when the transformer delivers the load at maximum efficiency are

$$P_o = 2400 \times 35 \times 0.8 = 67{,}200 \text{ W}$$
$$P_{in} = P_o + P_m + P_{cu\eta}$$
$$= 67{,}200 + 2{,}143.75 + 2{,}143.75 = 71{,}487.5 \text{ W}$$
$$\eta = \frac{67{,}200}{71{,}487.5} = 0.94 \quad \text{or} \quad 94\%$$

(c) The power output, the copper loss, and the efficiency at full load are

$$P_o = 2400 \times 50 \times 0.8 = 96{,}000 \text{ W}$$
$$P_{cu} = 50^2 \times (0.75 + 10^2 \times 0.01) = 4{,}375 \text{ W}$$

$$\eta = \frac{96{,}000}{96{,}000 + 4{,}375 + 2{,}143.75} = 0.936 \quad \text{or} \quad 93.6\%$$

(d) The equivalent core-loss resistance at no load is

$$R_c = \frac{2400^2}{2143.75} = 2686.88 \ \Omega$$

■

Exercises

4.10. A 2.4-kVA, 2400/240-V, 50-Hz, step-down transformer has $R_1 = 25\ \Omega$, $X_1 = 35\ \Omega$, $R_2 = 250\ \text{m}\Omega$, $X_2 = 350\ \text{m}\Omega$, and $X_m = 3\ \text{k}\Omega$. The efficiency of the transformer is maximum when it operates at 80% of its rated load and 0.866 pf lagging. Determine (a) its kVA rating at maximum efficiency, (b) the maximum efficiency, (c) the efficiency at full load and 0.866 pf lagging, and (d) the equivalent core-loss resistance.

4.11. A 200-kVA transformer has a core loss of 10 kW and a copper loss of 40 kW at full load. What is the kVA rating at the maximum efficiency of the transformer? Express the load current at maximum efficiency as a percent of the rated current. If the transformer operates at a power factor of 0.8 lagging, determine (a) the maximum efficiency and (b) the full-load efficiency of the transformer.

4.12. A 24-kVA, 120/480-V, step-up transformer performs at its maximum efficiency at 75% of its rated load with a unity power factor. The core loss is 1.2 kW. What is the equivalent winding resistance as referred to the primary side? What is the efficiency of the transformer if it delivers the rated load at a power factor of 0.9 lagging?

4.7 Determination of Transformer Parameters

The equivalent circuit parameters of a transformer can be determined by performing two tests: the open-circuit test and the short-circuit test.

The Open-Circuit Test

As the name implies, one winding of the transformer is left open while the other is excited by applying the rated voltage. The frequency of the applied voltage must be the rated frequency of the transformer. Although it does not matter which side of the transformer is excited, it is safer to conduct the test on the **low-voltage side**. Another justification for performing the test on the low-voltage side is the availability of the low-voltage source in any test facility.

Figure 4.21 shows the connection diagram for the open-circuit test with ammeter, voltmeter, and wattmeter inserted on the low-voltage side. If we assume that the power loss under no load in the low-voltage winding is negligible, then the corresponding approximate equivalent circuit as viewed from the low-voltage side is given in Figure 4.22. From the approximate equivalent of the transformer as referred to the low-voltage side (Figure 4.22), it is evident that the source supplies the excitation current under no load. One component of the excitation current is responsible for the core loss, whereas the other is responsible to establish the required flux in the magnetic core. In order to measure these values exactly, the

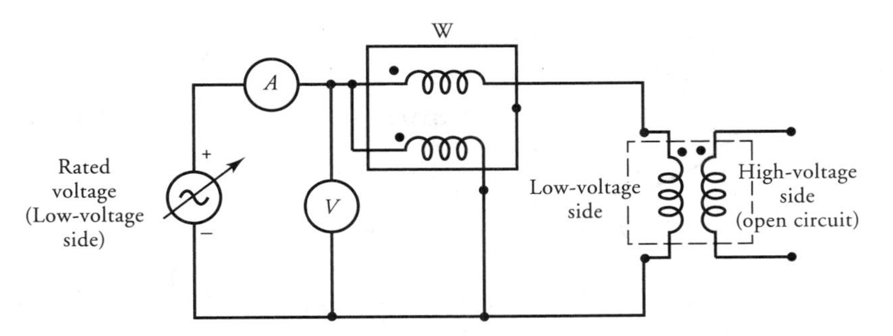

Figure 4.21 A two-winding transformer wired with instruments for open-circuit test.

source voltage must be adjusted carefully to its rated value. Since the only power loss in Figure 4.22 is the core loss, **the wattmeter measures the core loss in the transformer.**

The core-loss component of the excitation current is in phase with the applied voltage while the magnetizing current lags the applied voltage by 90°, as shown in Figure 4.23. If V_{oc} is the rated voltage applied on the low-voltage side, I_{oc} is the excitation current as measured by the ammeter, and P_{oc} is the power recorded by the wattmeter, then the apparent power at no-load is

$$S_{oc} = V_{oc} I_{oc}$$

at a lagging power-factor angle of

$$\phi_{oc} = \cos^{-1}\left[\frac{P_{oc}}{S_{oc}}\right]$$

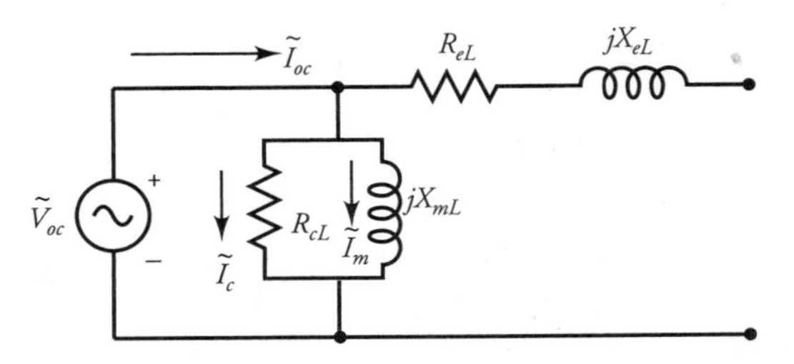

Figure 4.22 The approximate equivalent circuit of a two-winding transformer under open-circuit test.

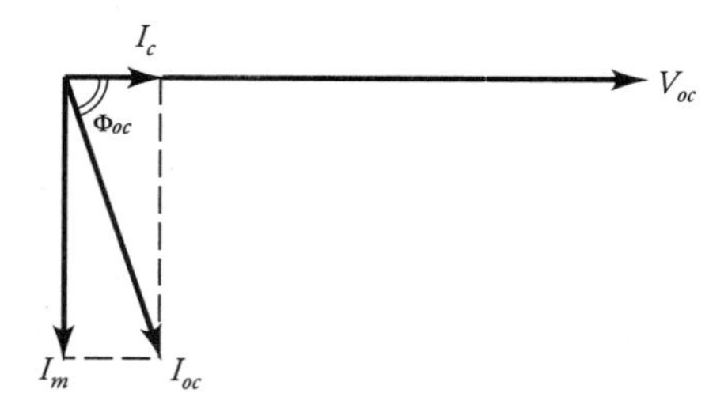

Figure 4.23 The phasor diagram of a two-winding transformer under open-circuit test.

The core-loss and magnetizing currents are

$$I_c = I_{oc} \cos(\phi_{oc})$$

and

$$I_m = I_{oc} \sin(\phi_{oc})$$

Thus, the core-loss resistance and the magnetizing reactance as viewed from the low-voltage side are

$$R_{cL} = \frac{V_{oc}}{I_c} = \frac{V_{oc}^2}{P_{oc}} \tag{4.26}$$

and

$$X_{mL} = \frac{V_{oc}}{I_m} = \frac{V_{oc}^2}{Q_{oc}} \tag{4.27}$$

where

$$Q_{oc} = \sqrt{S_{oc}^2 - P_{oc}^2}$$

The Short-Circuit Test

This test is designed to determine the winding resistances and leakage reactances. The short-circuit test is conducted by placing a short circuit across one winding and exciting the other from an alternating-voltage source of the frequency at which the transformer is rated. The applied voltage is carefully adjusted so that each winding carries a rated current. The rated current in each winding ensures a proper simulation of the leakage flux pattern associated with that winding. Since the short circuit constrains the power output to be zero, the power input to the transformer is low. The low power input at the rated current implies that **the applied voltage is a small fraction of the rated voltage. Therefore, extreme care must be exercised in performing this test.**

Once again, it does not really matter on which side this test is performed. However, the measurement of the rated current suggests that, for safety purposes, the test be performed on the high-voltage side. The test arrangement with all instruments inserted on the high-voltage side with a short circuit on the low-voltage side is shown in Figure 4.24.

Since the applied voltage is a small fraction of the rated voltage, both the core-loss and the magnetizing currents are so small that they can be neglected. In other words, the core loss is practically zero and the magnetizing reactance is almost infinite. The approximate equivalent circuit of the transformer as viewed from the high-voltage side is given in Figure 4.25. **In this case, the wattmeter records the copper loss at full load.**

If V_{sc}, I_{sc}, and P_{sc} are the readings on the voltmeter, ammeter, and wattmeter, then

$$R_{eH} = \frac{P_{sc}}{I_{sc}^2} \tag{4.28}$$

is the total resistance of the two windings as referred to the high-voltage side. The magnitude of the impedance as referred to the high-voltage side is

$$Z_{eH} = \frac{V_{sc}}{I_{sc}} \tag{4.29}$$

Therefore, the total leakage reactance of the two windings as referred to the high-voltage side is

$$X_{eH} = \sqrt{Z_{eH}^2 - R_{eH}^2} \tag{4.30}$$

If we define the a-ratio as

$$a = \frac{N_H}{N_L}$$

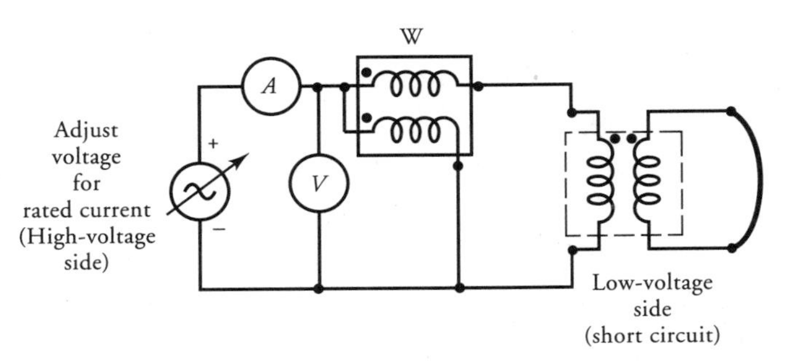

Figure 4.24 A two-winding transformer wired for short-circuit test.

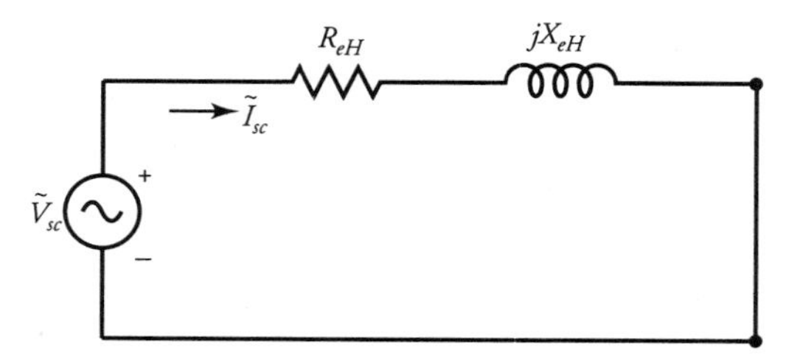

Figure 4.25 An approximate equivalent circuit of a two-winding transformer under short-circuit condition.

then

$$R_{eH} = R_H + a^2 R_L \tag{4.31}$$

and

$$X_{eH} = X_H + a^2 X_L \tag{4.32}$$

where R_H is the resistance of the high-voltage winding, R_L is the resistance of the low-voltage winding, X_H is the leakage reactance of the high-voltage winding, and X_L is the leakage reactance of the low-voltage winding.

If the transformer is available, we can measure R_H and R_L and verify Eq. (31). However, there is no simple way to separate the two leakage reactances. The same is also true for the winding resistances if the transformer is unavailable. If we have to segregate the resistances, we will assume that the transformer has been designed in such a way that the power loss on the high-voltage side is equal to the power loss on the low-voltage side. That is

$$I_H^2 R_H = I_L^2 R_L$$

which yields

$$R_H = a^2 R_L = 0.5 R_{eH} \tag{4.33}$$

Similarly, we can asume that

$$X_H = a^2 X_L = 0.5 X_{eH} \tag{4.34}$$

EXAMPLE 4.8

The following data were obtained from testing a 48-kVA, 4800/240-V, step-down transformer:

	Voltage (V)	Current (A)	Power (W)
Open-circuit test:	240	2	120
Short-circuit test:	150	10	600

Determine the equivalent circuit of the transformer as viewed from (a) the high-voltage side and (b) the low-voltage side.

● SOLUTION

Since the open-circuit test must be conducted at the rated terminal voltage, the above data indicate that it is performed on the low-voltage side. Thus, the equivalent core-loss resistance as referred to the low-voltage side is

$$R_{cL} = \frac{240^2}{120} = 480\ \Omega$$

The apparent power under no load: $S_{oc} = V_{oc}I_{oc} = 240 \times 2 = 480$ VA
Thus, the reactive power is

$$Q_{oc} = \sqrt{480^2 - 120^2} = 464.76 \text{ VAR}$$

Hence, the magnetization reactance as referred to the low-voltage side is

$$X_{mL} = \frac{240^2}{464.76} = 123.94\ \Omega$$

The core-loss resistance and the magnetization reactance as referred to the high-voltage side can be obtained as follows:

$$a = 4800/240 = 20$$

$$R_{cH} = a^2R_{cL} = (20^2)(480) = 192\ \text{k}\Omega$$

$$X_{mH} = a^2X_{mL} = (20^2)(123.94) = 49.58\ \text{k}\Omega$$

Since the short-circuit current is 10 A, the short-circuit test is performed on the high-voltage side. Thus,

$$R_{eH} = \frac{P_{sc}}{I_{sc}^2} = \frac{600}{10^2} = 6\ \Omega$$

$$Z_{eH} = \frac{V_{sc}}{I_{sc}} = \frac{150}{10} = 15\ \Omega$$

$$X_{eH} = \sqrt{15^2 - 6^2} = 13.75\ \Omega$$

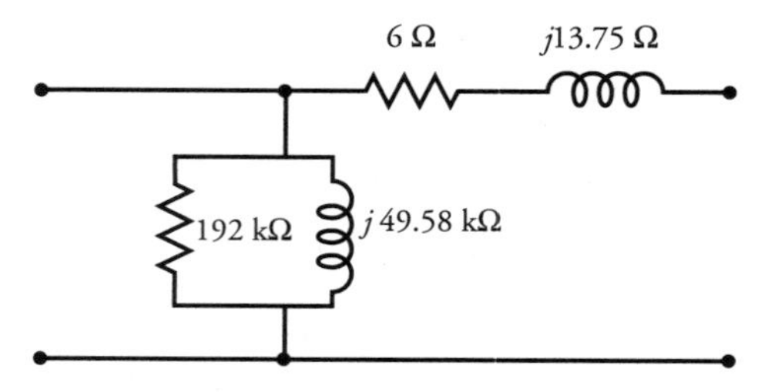

Figure 4.26 The approximate equivalent circuit as viewed from the high-voltage side for Example 4.8.

The winding parameters as referred to the low-voltage side are

$$R_{eL} = \frac{R_{eH}}{a^2} = \frac{6}{20^2} = 0.015\ \Omega \qquad \text{or} \qquad 15\ \text{m}\Omega$$

$$X_{eL} = \frac{X_{eH}}{a^2} = \frac{13.75}{20^2} = 0.034\ \Omega \qquad \text{or} \qquad 34\ \text{m}\Omega$$

The approximate equivalent circuits as viewed from the high-voltage and the low-voltage sides are given in Figures 4.26 and 4.27, respectively. In order to draw an exact equivalent circuit, we can segregate the winding resistances and leakage reactances using Eqs. (4.33) and (4.34). That is,

$$R_H = 0.5R_{eH} = 3\ \Omega$$

$$X_H = 0.5X_{eH} = 6.88\ \Omega$$

$$R_L = \frac{0.5R_{eH}}{a^2} = 0.0075\ \Omega, \qquad \text{or} \qquad 7.5\ \text{m}\Omega$$

$$X_L = \frac{0.5X_{eH}}{a^2} = 0.017\ \Omega, \qquad \text{or} \qquad 17\ \text{m}\Omega$$

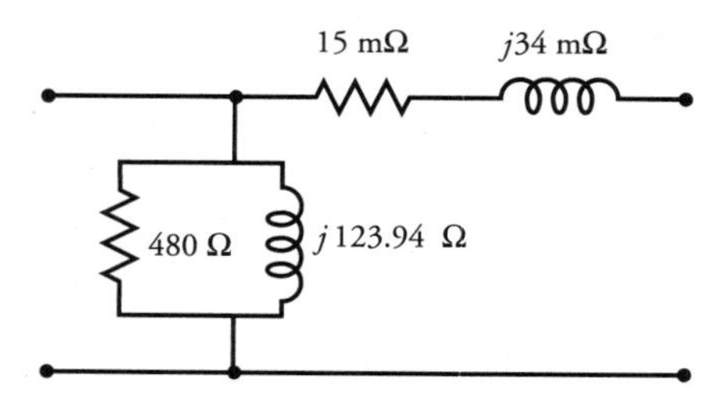

Figure 4.27 The approximate equivalent circuit as viewed from the low-voltage side for Example 4.8.

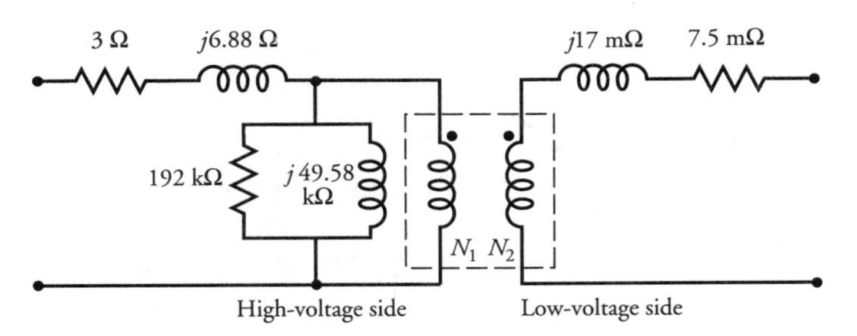

Figure 4.28 The exact equivalent circuit of a transformer for Example 4.8.

The exact equivalent circuit incorporating an ideal transformer is shown in Figure 4.28. ■

Exercises

4.13. The following data were obtained when a 25-kVA, 2300/460-V, 50-Hz transformer was tested:

	Voltage (V)	Current (A)	Power (W)
Open-circuit test:	460	1.48	460
Short-circuit test:	108.7	10.87	709

Determine the approximate equivalent circuit as viewed from (a) the high-voltage side and (b) the low-voltage side. Also draw the exact equivalent circuit.

4.14. A 25-kVA, 4000/400-V, 60-Hz transformer has the following parameters: $R_1 = 18\ \Omega$, $X_1 = 25\ \Omega$, $R_2 = 180\ \text{m}\Omega$, $X_2 = 250\ \text{m}\Omega$, $R_{cH} = 15\ \text{k}\Omega$, and $X_{mH} = 25\ \text{k}\Omega$. If the open-circuit and short-circuit tests are performed on this transformer, what are the readings of the instruments in each case?

4.15. From the data given in Exercise 4.14, determine the rating of the transformer at maximum efficiency. What is the maximum efficiency of the transformer at unity power factor? What is the efficiency at full load and unity power factor?

4.8 Per-Unit Computations

When an electric machine is designed or analyzed using the actual values of its parameters, it is not immediately obvious how its performance compares with another similar-type machine. However, if we express the parameters of a machine as a per-unit (pu) of a base (or reference) value, we will find that the per-unit values of machines of the same type but widely different ratings lie within a narrow range. This is one of the main advantages of a per-unit system.

An electric system has four quantities of interest: voltage, current, apparent power, and impedance. If we select base values of any two of them, the base values of the remaining two can be calculated. If S_b is the apparent base power and V_b is the base voltage, then the base current and base impedance are

$$I_b = \frac{S_b}{V_b} \tag{4.35}$$

$$Z_b = \frac{V_b}{I_b} \tag{4.36}$$

The actual quantity can now be expressed as a decimal fraction of its base value by using the following equation.

$$\text{Quantity, pu} = \frac{\text{actual quantity}}{\text{its base value}} \tag{4.37}$$

Since the power rating of a transformer is the same on both sides, we can use it as one of the base quantities. However, we have to select two base voltages—one for the primary side and the other for the secondary side. The two base voltages must be related by the *a*-ratio. That is,

$$a = \frac{V_{bH}}{V_{bL}} \tag{4.38}$$

where V_{bH} and V_{bL} are the base voltages on the high- and low-voltage sides of a transformer, respectively.

Since the base voltages have been transformed, the currents and impedances are also transformed. In other words, the *a*-ratio is unity when the parameters of a transformer are expressed in terms of their per-unit values. The following example illustrates the analysis of a transformer on a per-unit basis.

EXAMPLE 4.9

A single-phase generator with an internal impedance of $23 + j92$ mΩ is connected to a load via a 46-kVA, 230/2300-V, step-up transformer, a short transmission line, and a 46-kVA, 2300/115-V, step-down transformer. The impedance of the transmission line is $2.07 + j4.14$ Ω. The parameters of step-up and step-down transformers are:

	R_H	X_H	R_L	X_L	R_{cH}	X_{mH}
Step-up:	2.3 Ω	6.9 Ω	23 mΩ	69 mΩ	13.8 kΩ	6.9 kΩ
Step-down:	2.3 Ω	6.9 Ω	5.75 mΩ	17.25 mΩ	11.5 kΩ	9.2 kΩ

Determine (a) the generator voltage, (b) the generator current, and (c) the overall efficiency of the system at full load and 0.866 pf lagging.

● SOLUTION

The exact equivalent circuit of the system incorporating ideal transformers is given in Figure 4.29. The entire system is divided into three regions—A, B, and C, as shown.
Region A: $V_{bA} = 230$ V, and $S_{bA} = 46{,}000$ VA
Thus, $I_{bA} = 46{,}000/230 = 200$ A, and $Z_{bA} = 230/200 = 1.15\ \Omega$
The per-unit impedance of the generator is

$$\hat{Z}_{g,pu} = \frac{0.023 + j0.092}{1.15} = 0.02 + j0.08$$

The per-unit parameters on the low-voltage winding of the step-up transformer are

$$R_{L,pu} = \frac{0.023}{1.15} = 0.02$$

$$X_{L,pu} = \frac{0.069}{1.15} = 0.06$$

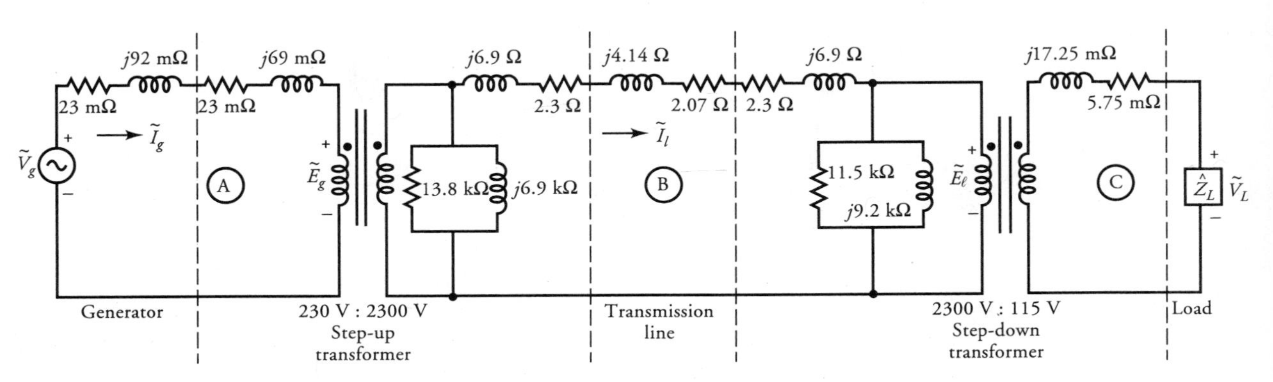

Figure 4.29 A power distribution system of Example 4.9.

Region B: $V_{bB} = 2300$ V, and $S_{bB} = 46{,}000$ VA
Thus, $I_{bB} = 46{,}000/2300 = 20$ A, and $Z_{bB} = 2300/20 = 115\ \Omega$
The per-unit parameters on the high-voltage side of the step-up transformer are

$$R_{H,pu} = \frac{2.3}{115} = 0.02$$

$$X_{H,pu} = \frac{6.9}{115} = 0.06$$

$$R_{cH,pu} = \frac{13{,}800}{115} = 120$$

$$X_{mH,pu} = \frac{6900}{115} = 60$$

The per-unit impedance of the transmission line is

$$\hat{Z}_{\ell,pu} = \frac{2.07 + j4.14}{115} = 0.018 + j0.036$$

The per-unit parameters on the high-voltage side of the step-down transformer are

$$R_{H,pu} = \frac{2.3}{115} = 0.02$$

$$X_{H,pu} = \frac{6.9}{115} = 0.06$$

$$R_{cH,pu} = \frac{11{,}500}{115} = 100$$

$$X_{mH,pu} = \frac{9200}{115} = 80$$

Region C: $V_{bc} = 115$ V, and $S_{bC} = 46{,}000$ VA
Thus, $I_{bC} = 46{,}000/115 = 400$ A, and $Z_{bC} = 115/400 = 0.2875\ \Omega$
Finally, the per-unit parameters on the low-voltage side of the step-down transformer are

$$R_{L,pu} = \frac{0.00575}{0.2875} = 0.02$$

$$X_{L,pu} = \frac{0.01725}{0.2875} = 0.06$$

At full load and 0.866 pf lagging, we have

$$\tilde{V}_{L,pu} = 1\angle 0^\circ \qquad \text{and} \qquad \tilde{I}_{L,pu} = 1\angle -30^\circ$$

Referring to the per-unit equivalent circuit of the system as given in Figure 4.30, we can write the following set of equations to determine the generator voltage and overall efficiency of the system.

$$\tilde{E}_{\ell,pu} = 1\angle 0^\circ + (0.02 + j0.06)(1\angle -30^\circ) = 1.048\angle 2.29^\circ$$

$$\tilde{I}_{\ell,pu} = 1\angle -30^\circ + 1.048\angle 2.29^\circ \left[\frac{1}{100} + \frac{1}{j80}\right] = 1.016\angle -30.31^\circ$$

$$\begin{aligned}\tilde{E}_{g,pu} &= 1.048\angle 2.29^\circ \\ &\quad + (1.016\angle -30.01^\circ)(0.02 + j0.06 + 0.018 + j0.036 + 0.02 + j0.06) \\ &= 1.188\angle 7.21^\circ\end{aligned}$$

$$\tilde{I}_{g,pu} = 1.016\angle -30.31^\circ + 1.188\angle 7.21^\circ \left[\frac{1}{120} + \frac{1}{j60}\right] = 1.036\angle -30.84^\circ$$

$$\begin{aligned}\tilde{V}_{g,pu} &= 1.188\angle 7.21^\circ + (1.036\angle -30.84^\circ)(0.02 + j0.08 + 0.02 + j0.06) \\ &= 1.313\angle 11.08^\circ\end{aligned}$$

(a) Hence the generator voltage is

$$\tilde{V}_g = V_{bA}\tilde{V}_{g,pu} = 230 \times 1.313\angle 11.08^\circ = 301.89\angle 11.08^\circ \text{ V}$$

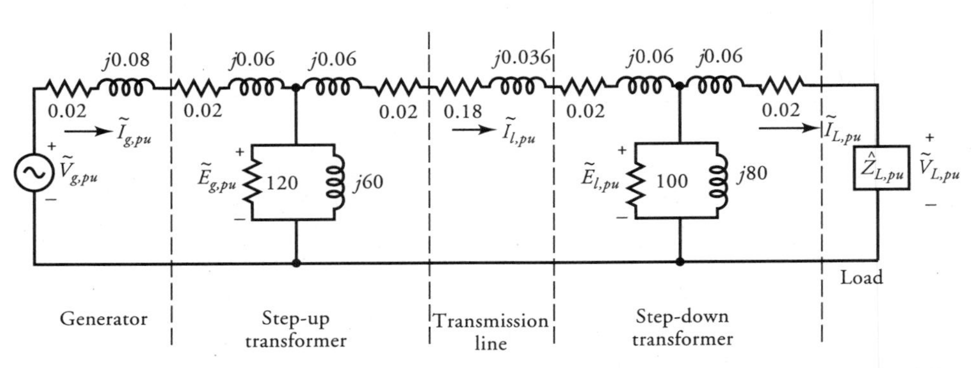

Figure 4.30 Representation of a power distribution system of Figure 4.29 in terms of the per-unit parameters.

(b) The current supplied by the generator is

$$\tilde{I}_g = I_{bA}\tilde{I}_{g,pu} = 200 \times 1.036\underline{/-30.84^\circ} = 207.17\underline{/-30.84^\circ} \text{ A}$$

(c) On a per-unit basis, the rated power output at a 0.866 pf lagging is

$$P_{o,pu} = 0.866$$

The per-unit power supplied by the generator is

$$P_{in,pu} = \text{Re}[(1.313\underline{/11.08^\circ})(1.036\underline{/30.84^\circ})] = 1.012$$

Hence, the efficiency is

$$\eta = \frac{0.866}{1.012} = 0.856 \quad \text{or} \quad 85.6\%$$

■

Exercises

4.16. If $V_{nL,pu}$ is the per-unit output voltage under no load, show that the voltage regulation of a transformer is $VR = V_{nL,pu} - 1$.
4.17. Calculate the voltage regulation of the entire system of Example 4.9.
4.18. Repeat Exercise 4.4 using the per-unit system. What is the voltage regulation of the transformer?

4.9 The Autotransformer

In the two-winding transformer we have considered thus far, the primary winding is electrically insulated from the secondary winding. The two windings are coupled together magnetically by a common core. Thus, the principle of magnetic induction is responsible for the energy transfer from the primary to the secondary.

When the two windings of a transformer are interconnected electrically, it is called an **autotransformer**. An autotransformer may have a single continuous winding that is common to both the primary and the secondary. Alternatively, we can connect two or more distinct coils wound on the same magnetic core to form an autotransformer. The principle of operation is the same in either case.

The direct electrical connection between the windings ensures that a part of the energy is transferred from the primary to the secondary by **conduction**. The magnetic coupling between the windings guarantees that some of the energy is also delivered by **induction**.

Autotransformers may be used for almost all applications in which we use a two-winding transformer. The only disadvantage in doing so is the loss of electrical isolation between the high- and low-voltage sides of the autotransformer. Listed below are some of the advantages of an autotransformer compared with a two-winding transformer.

1. It is cheaper in first cost than a conventional two-winding transformer of a similar rating.
2. It delivers more power than a two-winding transformer of similar physical dimensions.
3. For a similar power rating, an autotransformer is more efficient than a two-winding transformer.
4. An autotransformer requires lower excitation current than a two-winding transformer to establish the same flux in the core.

We begin our discussion of an autotransformer by connecting an ideal two-winding transformer as an autotransformer. In fact, there are four possible ways to connect a two-winding transformer as an autotransformer, as shown in Figure 4.31.

Let us consider the circuit shown in Figure 4.31**a**. The two-winding transformer is connected as a step-down autotransformer. Note that the secondary winding of the two-winding transformer is now the common winding for the autotransformer. Under ideal conditions,

$$\tilde{V}_{1a} = \tilde{E}_{1a} = \tilde{E}_1 + \tilde{E}_2$$

$$\tilde{V}_{2a} = \tilde{E}_{2a} = \tilde{E}_2$$

$$\frac{\tilde{V}_{1a}}{\tilde{V}_{2a}} = \frac{\tilde{E}_{1a}}{\tilde{E}_{2a}} = \frac{\tilde{E}_1 + \tilde{E}_2}{\tilde{E}_2} = \frac{N_1 + N_2}{N_2} = 1 + a = a_T \tag{4.39}$$

where $a = N_1/N_2$ is the a-ratio of a two-winding transformer, and $a_T = 1 + a$ is the a-ratio of the autotransformer under consideration. The a-ratio for the other connections should also be computed in the same way. Note that a_T is not the same for all connections.

In an ideal autotransformer, the primary mmf must be equal and opposite to the secondary mmf. That is,

$$(N_1 + N_2)I_{1a} = N_2 I_{2a}$$

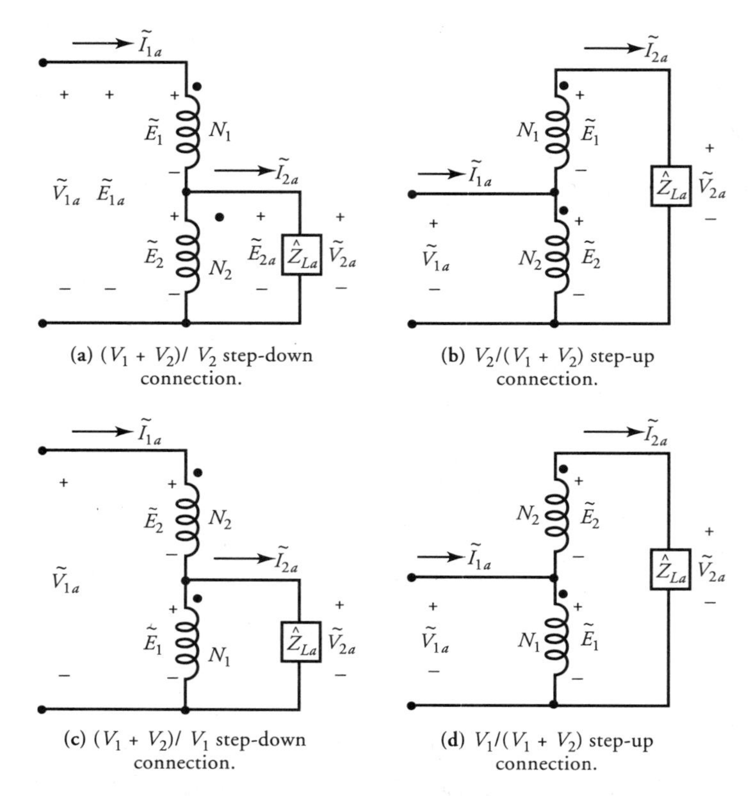

(a) $(V_1 + V_2)/V_2$ step-down connection.

(b) $V_2/(V_1 + V_2)$ step-up connection.

(c) $(V_1 + V_2)/V_1$ step-down connection.

(d) $V_1/(V_1 + V_2)$ step-up connection.

Figure 4.31 Possible ways to connect a two-winding transformer as an autotransformer.

From this equation, we obtain

$$\frac{I_{2a}}{I_{1a}} = \frac{N_1 + N_2}{N_2} = 1 + a = a_T \tag{4.40}$$

Thus, the apparent power supplied by an ideal transformer to the load, S_{oa}, is

$$\begin{aligned} S_{oa} &= V_{2a} I_{2a} \\ &= \left[\frac{V_{1a}}{a_T}\right][a_T I_{1a}] \\ &= V_{1a} I_{1a} \\ &= S_{ina} \end{aligned} \tag{4.41}$$

where S_{ina} is the apparent power input to the autotransformer. The above equation simply highlights the fact that the power input is equal to the power output under ideal conditions.

Let us now express the apparent output power in terms of the parameters of a two-winding transformer. For the configuration under consideration,

$$V_{2a} = V_2$$

and

$$I_{2a} = a_T I_{1a} = (a + 1)I_{1a}$$

However, for the rated load, $I_{1a} = I_1$. Thus,

$$S_{oa} = V_2 I_1 (a + 1)$$

$$= V_2 I_2 \frac{a + 1}{a} = S_o \left[1 + \frac{1}{a}\right]$$

where $S_o = V_2 I_2$ is the apparent power output of a two-winding transformer. This power is associated with the common winding of the autotransformer. This, therefore, is the power transferred to the load by induction in an autotransformer. The rest of the power, S_o / a in this case, is conducted directly from the source to the load and is called the **conduction power**. Hence, a two-winding transformer delivers more power when connected as an autotransformer.

EXAMPLE 4.10

A 24-kVA, 2400/240-V distribution transformer is to be connected as an autotransformer. For each possible combination, determine (a) the primary winding voltage, (b) the secondary winding voltage, (c) the ratio of transformation, and (d) the nominal rating of the autotransformer.

● SOLUTION

From the given information for the two-winding transformer, we conclude

$$V_1 = 2400 \text{ V}, \quad V_2 = 240 \text{ V}, \quad S_o = 24 \text{ kVA} \quad I_1 = 10 \text{ A}, \quad \text{and} \quad I_2 = 100 \text{ A}$$

(a) For the autotransformer operation shown in Figure 4.31**a**,

$$V_{1a} = 2400 + 240 = 2640 \text{ V}$$

$$V_{2a} = 240 \text{ V}$$

$$a_T = \frac{2640}{240} = 11$$

$$S_{oa} = V_{2a} I_{2a} = V_{1a} I_{1a} = V_{1a} I_1$$

$$= 2640 \times 10 = 26{,}400 \text{ VA} \quad \text{or} \quad 26.4 \text{ kVA}$$

Thus, the nominal rating of the autotransformer is 26.4 kVA, 2640/240 V.

(b) For the autotransformer connection shown in Figure 4.31**b**,

$$V_{1a} = 240 \text{ V}$$

$$V_{2a} = 2400 + 240 = 2640 \text{ V}$$

$$a_T = \frac{240}{2640} = 0.091$$

$$S_{oa} = V_{2a}I_{2a} = V_{2a}I_1$$

$$= 2640 \times 10 = 26{,}400 \text{ VA} \quad \text{or} \quad 26.4 \text{ kVA}$$

The nominal rating of the autotransformer is 26.4 kVA, 240/2640 V.

(c) If an autotransformer connection is as shown in Figure 4.31**c**,

$$V_{1a} = 240 + 2400 = 2640 \text{ V}$$

$$V_{2a} = 2400 \text{ V}$$

$$a_T = \frac{2640}{2400} = 1.1$$

$$S_{oa} = V_{2a}I_{2a} = V_{1a}I_{1a} = V_{1a}I_2$$

$$= 2640 \times 100 = 264{,}000 \text{ VA} \quad \text{or} \quad 264 \text{ kVA}$$

The nominal rating of the autotransformer is 264 kVA, 2640/2400 V.

(d) Finally, if the autotransformer connection is as shown in Figure 4.31**d**,

$$V_{1a} = 2400 \text{ V}$$

$$V_{2a} = 2400 + 240 = 2640 \text{ V}$$

$$a_T = \frac{2400}{2640} = 0.91$$

$$S_{oa} = V_{2a}I_{2a} = V_{2a}I_2$$

$$= 2640 \times 100 = 264{,}000 \text{ VA} \quad \text{or} \quad 264 \text{ kVA}$$

The nominal rating of the autotransformer is 264 kVA, 2400/2640 V. Note that the power rating of a two-winding transformer, when connected as an autotransformer (Figure 4.31**c** or **d**), has an 11-fold increase.

■

A Real Autotransformer

An equivalent circuit of a real autotransformer can be obtained by including the winding resistances, the leakage reactances, the core-loss resistance, and the magnetizing reactance, as depicted in Figure 4.32. We can analyze each circuit in Figure 4.32 as it is drawn or derive an equivalent circuit as viewed from the primary or the secondary side by following the techniques used earlier. The equivalent circuit as viewed from the primary of the autotransformer for each connection is illustrated in Figure 4.33. The reader is urged to verify each circuit in order to gain confidence in making the transformations.

EXAMPLE 4.11

A 720-VA, 360/120-V, two-winding transformer has the following constants: $R_H = 18.9\ \Omega$, $X_H = 21.6\ \Omega$, $R_L = 2.1\ \Omega$, $X_L = 2.4\ \Omega$, $R_{cH} = 8.64\ \text{k}\Omega$, and $X_{mH} = 6.84\ \text{k}\Omega$. The transformer is connected as a 120/480-V, step-up autotransformer. If the autotransformer delivers the full load at 0.707 pf leading, determine its efficiency and voltage regulation.

SOLUTION

The equivalent circuit of a 120/480-V, step-up transformer is shown in Figure 4.34. Its ratio of transformation is

$$a_T = \frac{120}{480} = 0.25$$

The ratio of transformation of a two-winding transformer is

$$a = \frac{360}{120} = 3$$

Thus, the equivalent core-loss resistance and the magnetizing reactance on the low-voltage side are

$$R_{cL} = \frac{8640}{3^2} = 960\ \Omega$$

$$X_{mL} = \frac{6840}{3^2} = 760\ \Omega$$

At full load, the load current is

$$I_{2a} = I_H = \frac{720}{360} = 2\ \text{A}$$

Hence,

$$\tilde{I}_{2a} = 2\angle 45^\circ\ \text{A}$$

$$\tilde{I}_{pa} = \frac{\tilde{I}_{2a}}{a_T} = \frac{2\angle 45^\circ}{0.25} = 8\angle 45^\circ\ \text{A}$$

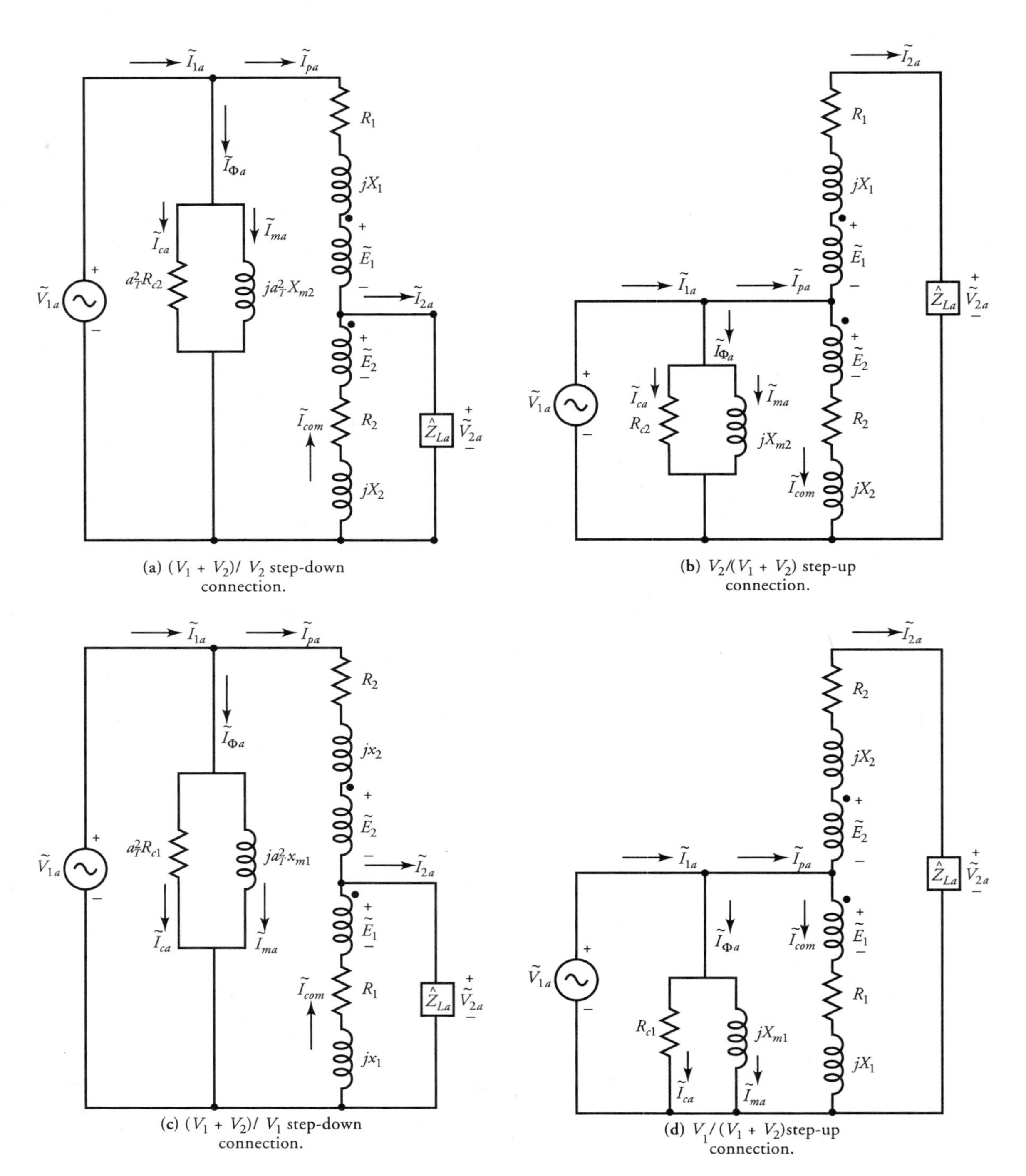

(a) $(V_1 + V_2)/\ V_2$ step-down connection.

(b) $V_2/(V_1 + V_2)$ step-up connection.

(c) $(V_1 + V_2)/\ V_1$ step-down connection.

(d) $V_1/(V_1 + V_2)$step-up connection.

Figure 4.32 A real two-winding transformer connected as an autotransformer.

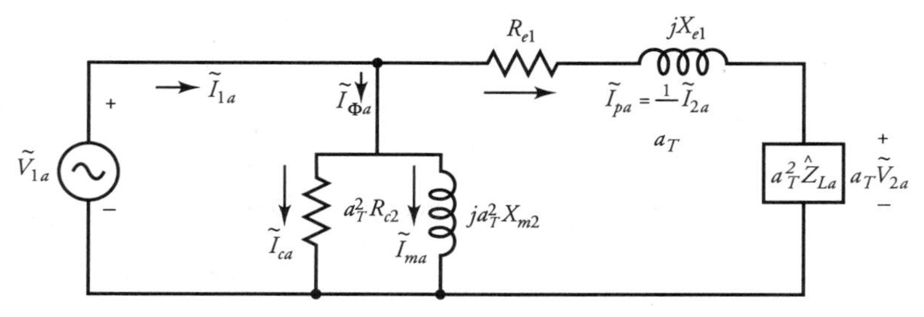

(a) The equivalent circuit of a $(V_1 + V_2)/V_2$ step-down autotransformer as viewed from the primary side.

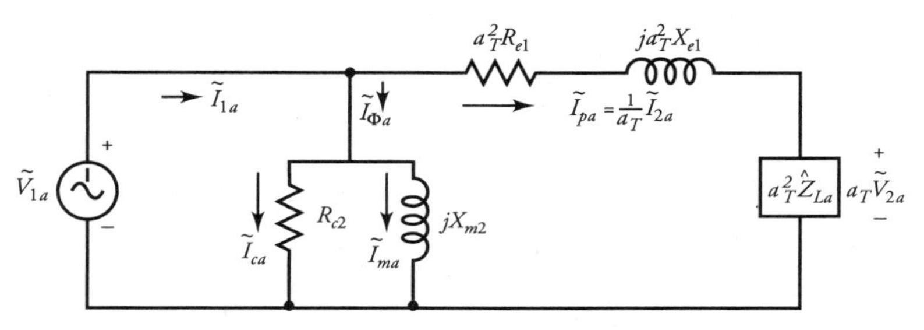

(b) The equivalent circuit of a $V_2/(V_1 + V_2)$ step-up autotransformer as viewed from the primary side.

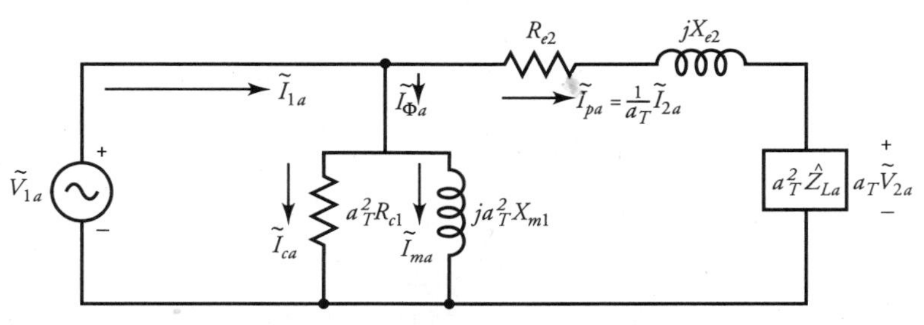

(c) The equivalent circuit of a $(V_1 + V_2)/V_1$ step-dowm autotransformer as viewed from the primary side.

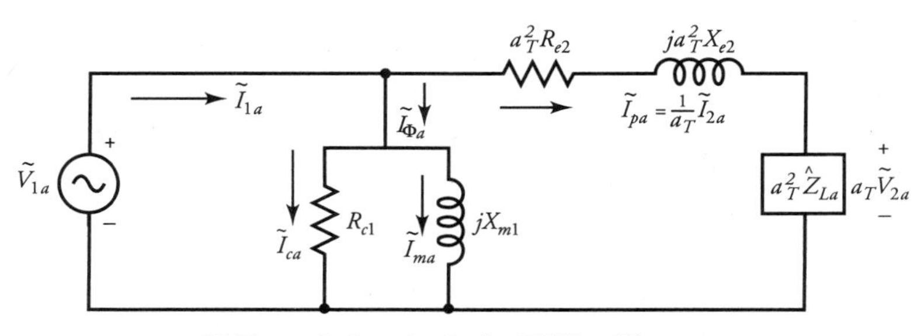

(d) The equivalent circuit of a $V_1/(V_1 + V_2)$ step-up autotransformer as viewed from the primary side.

Figure 4.33 The equivalent circuits of an autotransformer.

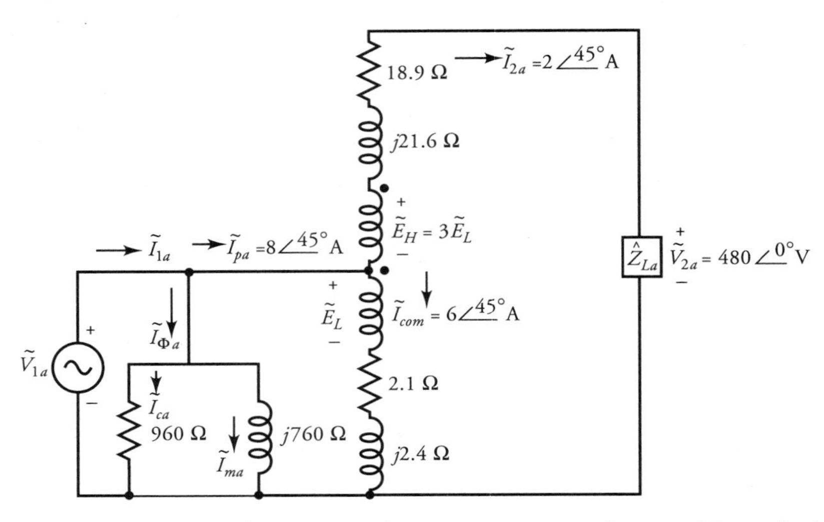

Figure 4.34 An exact equivalent circuit of a step-up autotransformer of Example 4.11.

The current through the common winding is

$$\tilde{I}_{\text{com}} = \tilde{I}_{pa} - \tilde{I}_{2a} = 6\underline{/45°}$$

In addition, $\tilde{E}_H = a\tilde{E}_L = \tilde{E}_L$.
We can now obtain $\tilde{E}_L$ by applying KVL to the output loop. That is,

$$4\tilde{E}_L = \tilde{I}_{2a}(R_H + jX_H) + \tilde{V}_{2a} - \tilde{I}_{\text{com}}(R_L + jX_L)$$
$$= 2\underline{/45°}(18.9 + j21.6) + 480 - 6\underline{/45°}(2.1 + j2.4)$$

or

$$\tilde{E}_L = 119.745\underline{/4.57°} \text{ V}$$

Thus,

$$\tilde{V}_{1a} = \tilde{E}_L + \tilde{I}_{\text{com}}(R_L + jX_L)$$
$$= 119.745\underline{/4.57°} + 6\underline{/45°}(2.1 + j2.4)$$
$$= 121.514\underline{/13.63°} \text{ V}$$

The core-loss, magnetizing, and excitation currents are

$$\tilde{I}_{ca} = \frac{\tilde{V}_{1a}}{R_{cL}} = \frac{121.514\underline{/13.63°}}{960} = 0.127\underline{/13.63°} \text{ A}$$

$$\tilde{I}_{ma} = \frac{\tilde{V}_{1a}}{jX_{mL}} = \frac{121.514\underline{/13.63°}}{j760} = 0.160\underline{/-76.37°}$$

$$\tilde{I}_{\phi a} = \tilde{I}_{ca} + \tilde{I}_{ma} = 0.127\underline{/13.63°} + 0.160\underline{/-76.37°}$$
$$= 0.204\underline{/-38°} \text{ A}$$

Hence,

$$\tilde{I}_{1a} = \tilde{I}_{pa} + \tilde{I}_{\phi a} = 8\underline{/45^\circ} + 0.204\underline{/-38^\circ} = 8.027\underline{/43.56^\circ} \text{ A}$$

$$P_o = \text{Re}[\tilde{V}_{2a}\tilde{I}^*_{2a}] = \text{Re}[480 \times 2\underline{/-45^\circ}] = 678.82 \text{ W}$$

$$P_{in} = \text{Re}[\tilde{V}_{1a}\tilde{I}^*_{1a}] = \text{Re}[(121.514\underline{/13.63^\circ})(8.027\underline{/-43.56^\circ})] = 845.4 \text{ W}$$

$$\eta = \frac{678.82}{845.4} = 0.803 \quad \text{or} \quad 80.3\%.$$

If we now remove the load, the no-load voltage at the secondary of the autotransformer is

$$\tilde{V}_{2anL} = \frac{\tilde{V}_{1a}}{a_T} = \frac{121.514\underline{/13.63^\circ}}{0.25} = 486.055\underline{/13.63^\circ} \text{ V}$$

We can now compute the voltage regulation as

$$VR\% = \frac{V_{2anL} - V_{2a}}{V_{2a}} \times 100 = \frac{486.055 - 480}{480} \times 100 = 1.26\%$$

■

Exercises

4.19. Repeat Example 4.11 if the two-winding transformer is connected as a 480/120-V, step-down autotransformer.

4.20. Repeat Example 4.11 if the two-winding transformer is connected as a 480/360-V, step-down autotransformer.

4.21. Repeat Example 4.11 if the two-winding transformer is connected as a 360/480-V, step-up transformer.

4.10 Three-Phase Transformers

Since most of the power generated and transmitted over long distances is of the three-phase type, we can use three exactly alike single-phase transformers to form a single three-phase transformer. For economic reasons, however, a three-phase transformer is designed to have all six windings on a common magnetic core. A common magnetic core, three-phase transformer can also be either a core type (Figure 4.35) or a shell type (Figure 4.36).

Since the third harmonic flux created by each winding is in phase, a shell-type transformer is preferred because it provides an external path for this flux. In other

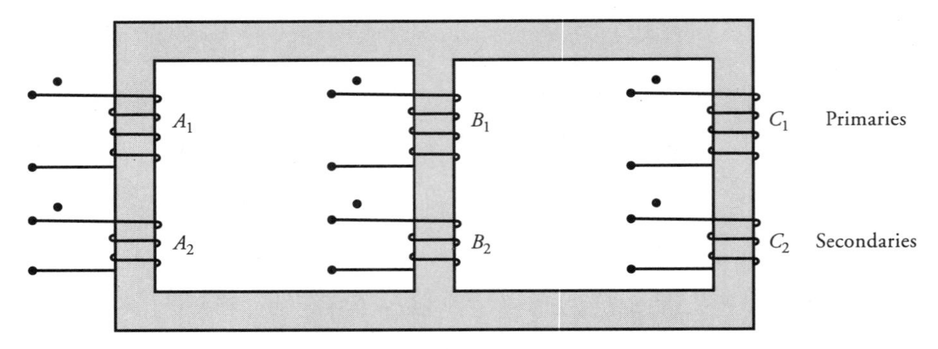

Figure 4.35 A core type three-phase transformer.

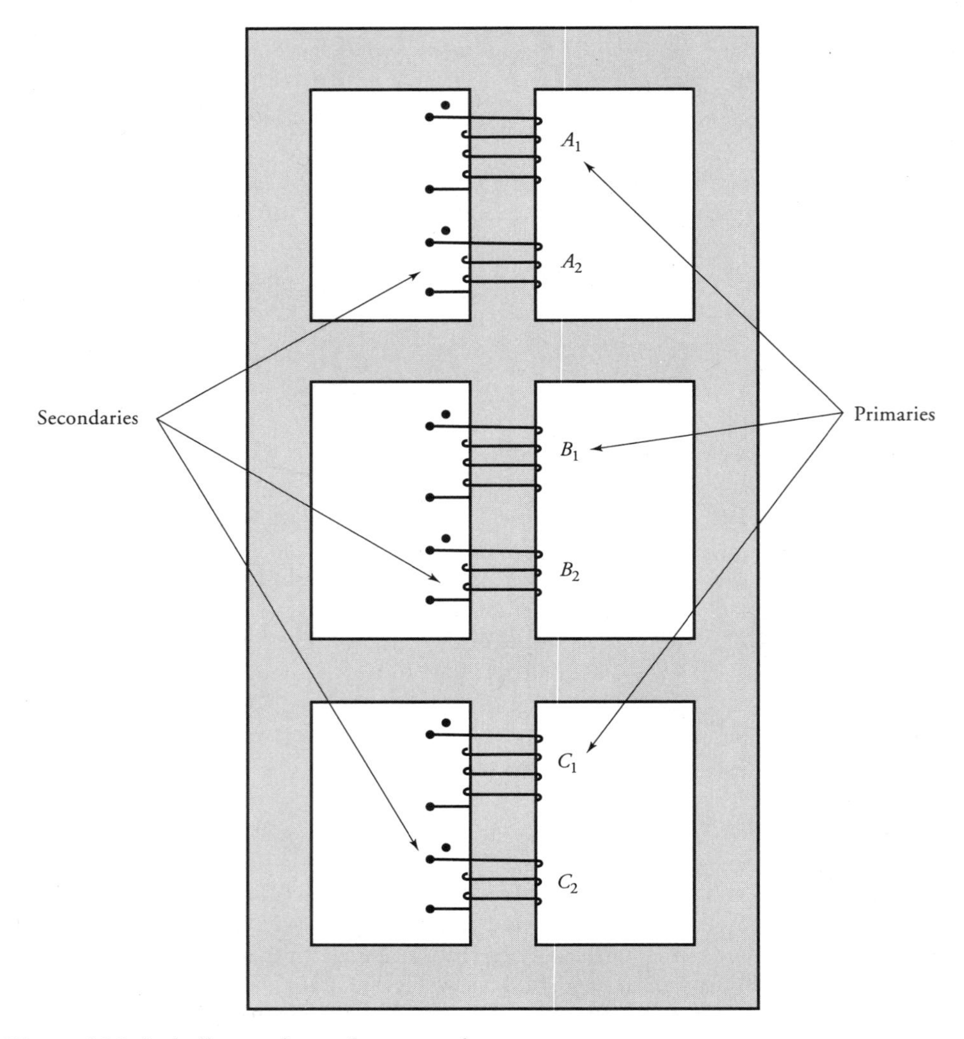

Figure 4.36 A shell-type three-phase transformer.

words, the voltage waveforms are less distorted for a shell-type transformer than for a core-type transformer of similar ratings.

The three windings on either side of a three-phase transformer can be connected either in wye (Y) or in delta (Δ). Therefore, a three-phase transformer can be connected in four possible ways: Y/Y, Y/Δ, Δ/Y, and Δ/Δ. Some of the advantages and drawbacks of each connection are highlighted below.

Y/Y Connection

A Y/Y connection for the primary and secondary windings of a three-phase transformer is depicted in Figure 4.37. The line-to-line voltage on each side of the three-phase transformer is $\sqrt{3}$ times the nominal voltage of the single-phase transformer. The main advantage of a Y/Y connection is that we have access to the neutral terminal on each side and it can be grounded if desired. Without grounding the neutral terminals, the Y/Y operation is satisfactory only when the three-phase load is balanced. The electrical insulation is stressed only to about 58% of the line voltage in a Y-connected transformer.

Since most of the transformers are designed to operate at or above the knee of the curve, such a design causes the induced emfs and currents to be distorted. The reason is as follows: Although the excitation currents are still 120° out of phase with respect to each other, their waveforms are no more sinusoidal. These currents, therefore, do not add up to zero. If the neutral is not grounded, these currents are forced to add up to zero. Thus, they affect the waveforms of the induced emfs.

Δ/Δ Connection

Figure 4.38 shows the three transformers with the primary and secondary windings connected as Δ/Δ. The line-to-line voltage on either side is equal to the cor-

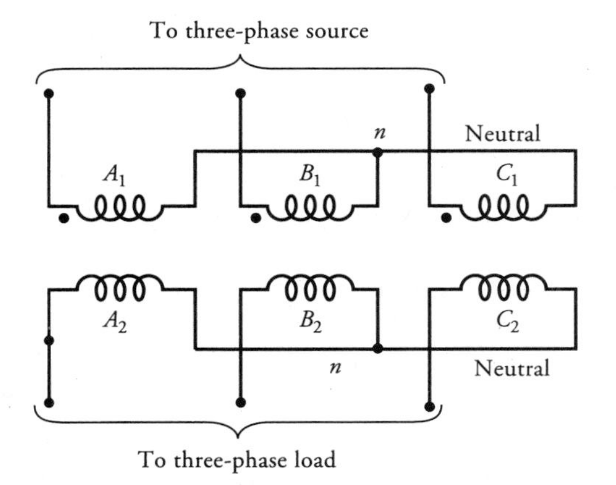

Figure 4.37 A Y/Y-connected three-phase transformer.

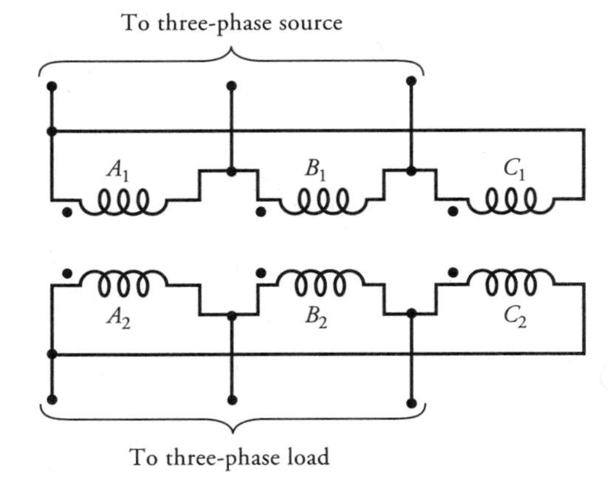

Figure 4.38 A Δ/Δ-connected three-phase transformer.

responding phase voltage. Therefore, this arrangement is useful when the voltages are not very high. The advantage of this connection is that even under unbalanced loads the three-phase load voltages remain substantially equal. The disadvantage of the Δ/Δ connection is the absence of a neutral terminal on either side. Another drawback is that the electrical insulation is stressed to the line voltage. Therefore, a Δ-connected winding requires more expensive insulation than a Y-connected winding for the same power rating.

A Δ/Δ connection can be analyzed theoretically by transforming it into a simulated Y/Y connection using Δ-to-Y transformations.

Y/Δ Connection

This connection, as shown in Figure 4.39, is very suitable for step-down applications. The secondary winding current is about 58% of the load current. On the primary side the voltages are from line to neutral, whereas the voltages are from line to line on the secondary side. Therefore, the voltage and the current in the primary are out of phase with the voltage and the current in the secondary. In a Y/Δ connection, the distortion in the waveform of the induced voltages is not as drastic as it is in a Y/Y-connected transformer when the neutral is not connected to the ground. The reason is that the distorted currents in the primary give rise to a circulating current in the Δ-connected secondary. The circulating current acts more like a magnetizing current and tends to correct the distortion.

Δ/Y Connection

This connection, as depicted in Figure 4.40, is proper for a step-up application. However, this connection is now being exploited to satisfy the requirements of

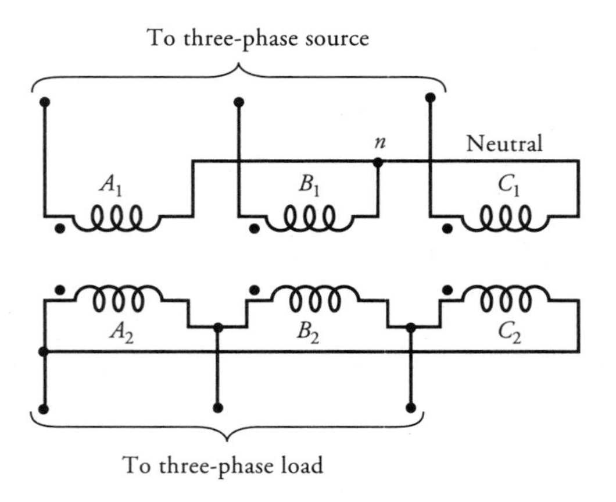

Figure 4.39 A Y/Δ-connected three-phase transformer.

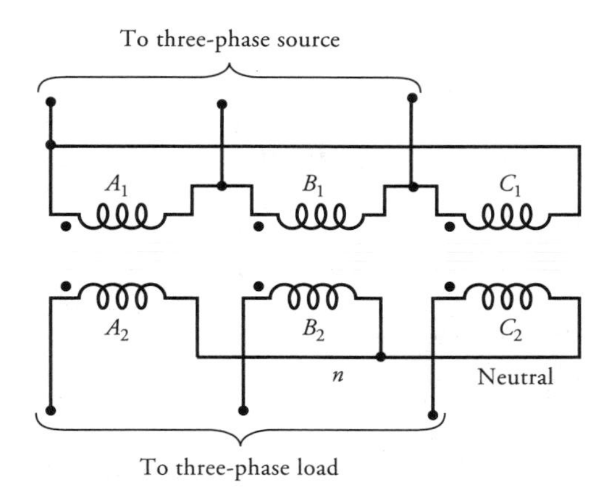

Figure 4.40 A Δ/Y-connected three-phase transformer.

both the three-phase and the single-phase loads. In this case, we use a four-wire secondary. The single-phase loads are taken care of by the three line-to-neutral circuits. An attempt is invariably made to distribute the single-phase loads almost equally among the three phases.

Analysis of a Three-Phase Transformer

Under steady-state conditions, a single three-phase transformer operates exactly the same way as three single-phase transformers connected together. Therefore, in our discussion that follows we assume that we have three identical single-phase transformers connected to form a three-phase transformer. Such an understanding helps us to develop the per-phase equivalent circuit of a three-phase transformer.

We also assume that the three-phase transformer delivers a balanced load, and the waveforms of the induced emfs are purely sinusoidal. In other words, the magnetizing currents are not distorted and there are no third harmonics.

We can analyze a three-phase transformer in the same way we have analyzed a three-phase circuit. That is, we can employ the per-phase equivalent circuit of a transformer. A Δ-connected winding of a three-phase transformer can be replaced by an equivalent Y-connected winding using Δ-to-Y transformation. If $\hat{Z}_\Delta$ is the impedance in a Δ-connected winding, the equivalent impedance $\hat{Z}_Y$ in a Y-connected winding is

$$\hat{Z}_Y = \frac{\hat{Z}_\Delta}{3} \tag{4.42}$$

On the other hand, if $\tilde{V}_L$ is the line-to-line voltage in a Y-connected winding, the line-to-neutral voltage $\tilde{V}_n$ is

$$\tilde{V}_n = \frac{\tilde{V}_L}{\sqrt{3}} \angle \pm 30° \tag{4.43}$$

where the plus sign is for the negative phase-sequence (counterclockwise) and the minus sign is for the positive phase-sequence (clockwise).

The following examples show how to analyze a system of balanced three-phase transformers. Unless it is specified otherwise, we assume that the impressed voltages on the primary side follow the positive phase-sequence.

EXAMPLE 4.12

A three-phase transformer is assembled by connecting three 720-VA, 360/120-V, single-phase transformers. The constants for each transformer are $R_H = 18.9\ \Omega$, $X_H = 21.6\ \Omega$, $R_L = 2.1\ \Omega$, $X_L = 2.4\ \Omega$, $R_{cH} = 8.64\ \text{k}\Omega$, and $X_{mH} = 6.84\ \text{k}\Omega$. For each of the four configurations, determine the nominal voltage and power ratings of the three-phase transformer. Draw the winding arrangements and the per-phase equivalent circuit for each configuration.

● SOLUTION

The power rating of a three-phase transformer for each connection is

$$S_{3\phi} = 3 \times 720 = 2160 \text{ VA} \quad \text{or} \quad 2.16 \text{ kVA}$$

(a) For a Y/Y connection, the nominal values of the line voltages on the primary and the secondary sides are

$$V_{1L} = \sqrt{3} \times 360 = 623.54 \text{ V}$$
$$V_{2L} = \sqrt{3} \times 120 = 207.85 \text{ V}$$

Thus, the nominal ratings of a three-phase transformer are

2.16-kVA 624/208-V Y/Y connection

The winding arrangement and the per-phase equivalent circuit are shown in Figures 4.41**a** and **b**, respectively.

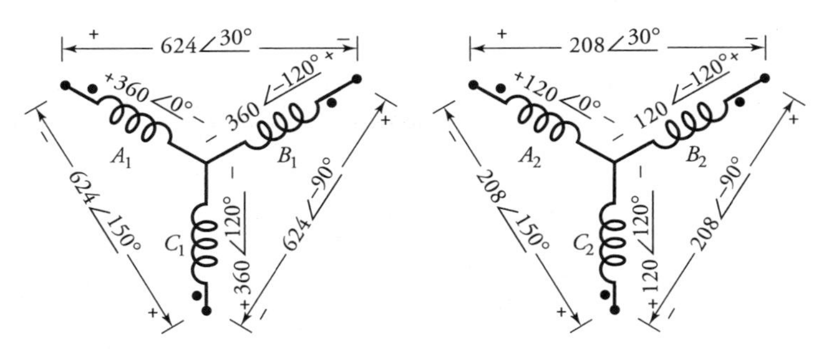

(a) Winding arrangement, phase and line voltages.

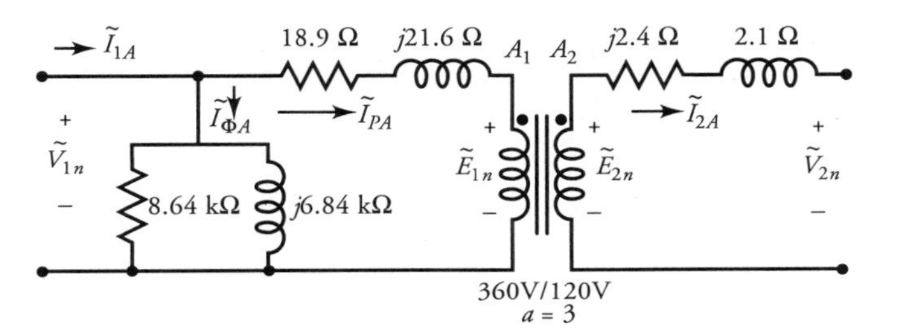

(b) Per-phase equivalent circuit.

Figure 4.41 Winding arrangement and per-phase equivalent circuit of a three-phase Y/Y-connected transformer.

(b) The nominal values of the line voltages, for Δ/Δ connection, are

$$V_{1L} = 360 \text{ V}$$
$$V_{2L} = 120 \text{ V}$$

Hence, the nominal ratings of a three-phase transformer are

2.16-kVA 360/120-V Δ/Δ connection

The winding connections, the Y/Y equivalent representation, and the per-phase equivalent circuit are illustrated in Figures 4.42**a**, **b**, and **c**, respectively.

(c) When the windings form a Y/Δ connection, the nominal ratings are

2.16-kVA 624/120-V Y/Δ connection

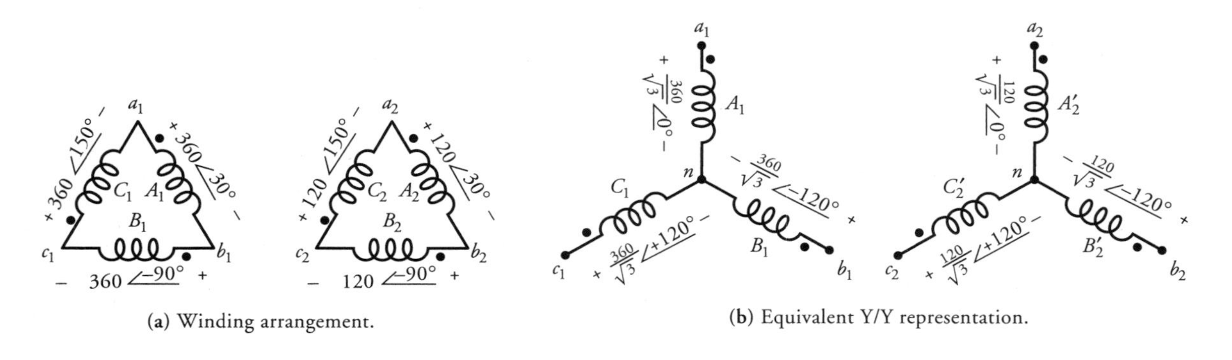

(a) Winding arrangement.

(b) Equivalent Y/Y representation.

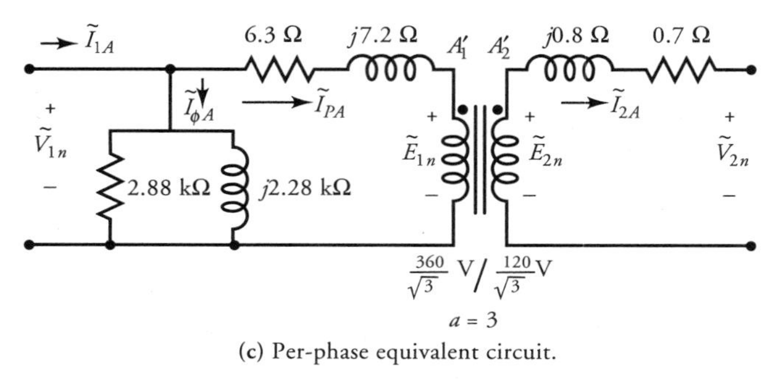

(c) Per-phase equivalent circuit.

Figure 4.42 The winding connections, equivalent Y/Y representation, and per-phase equivalent circuit of a Δ/Y connected three-phase transformer.

The winding connections, an equivalent Y/Y representation, and the per-phase equivalent circuit are depicted in Figures 4.43**a**, **b**, and **c**, respectively. Note that the secondary winding voltages lead the corresponding primary winding voltages by 30°.
cm(d)Finally, the nominal ratings for a Δ/Y connection are

$$2.16\text{-kVA} \qquad 360/208\text{-V} \qquad \Delta/\text{Y connection}$$

The winding connections, the equivalent Y/Y representation, and the per-phase equivalent circuit are drawn in Figures 4.44**a**, **b**, and **c**, respectively. In this case, the primary winding voltages lead the corresponding secondary winding voltages by 30°.

■

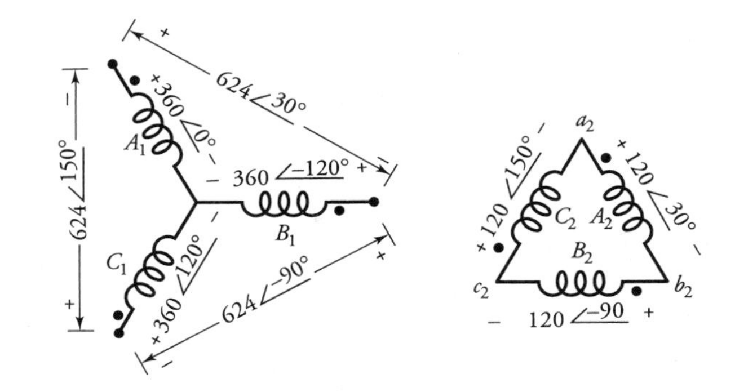

(a) Winding arrangement.

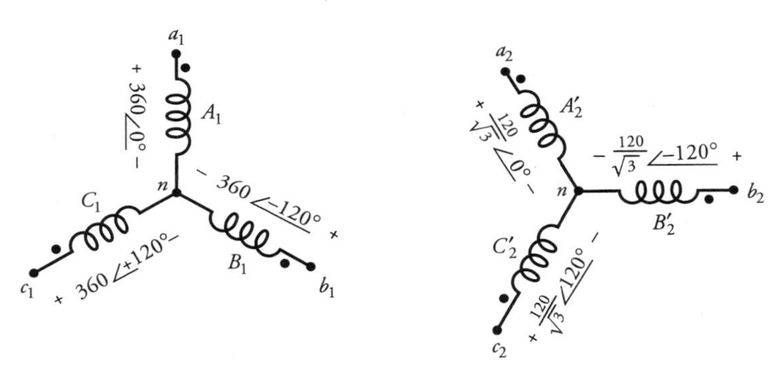

(b) Equivalent Y/Y representation

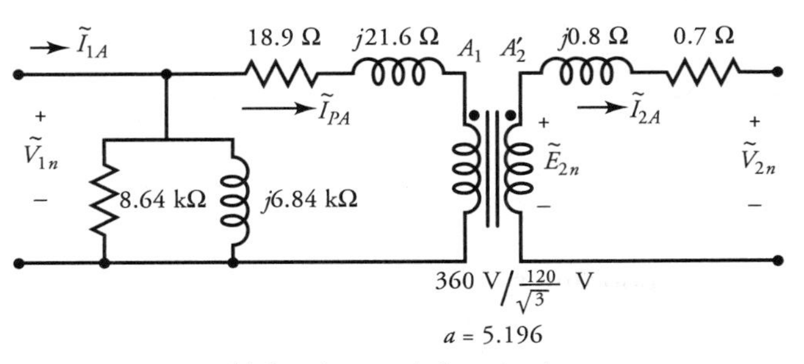

(c) Per-phase equivalent circuit

Figure 4.43 The winding connections, equivalent Y/Y representation, and per-phase equivalent circuit of a Y/Δ three-phase transformer.

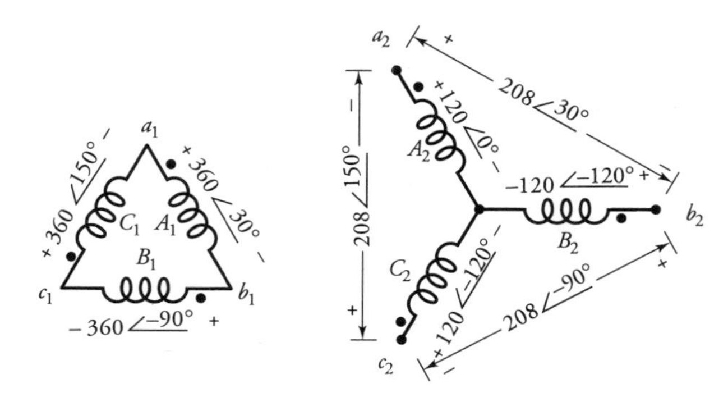

(a) Winding arrangement.

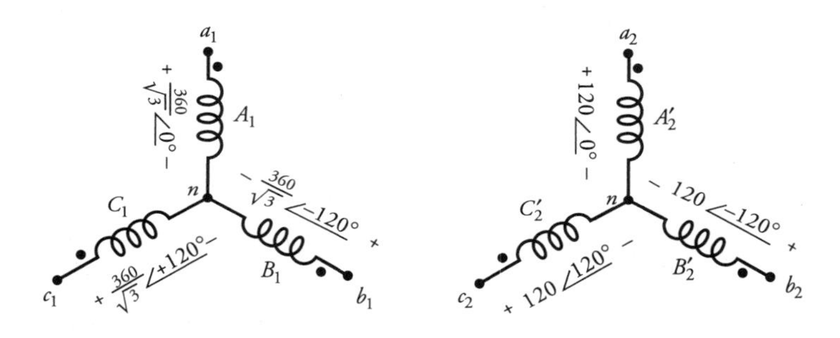

(b) Equivalent Y/Y representation.

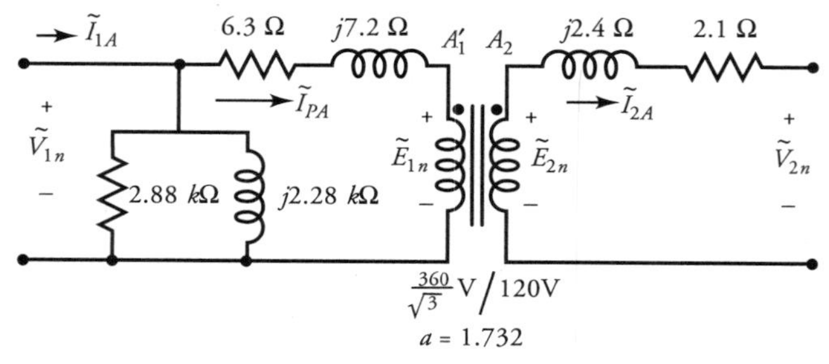

(c) Per-phase equivalent circuit.

Figure 4.44 The winding connections, equivalent Y/Y representation, and per-phase equivalent circuit of a Δ/Y three-phase transformer.

EXAMPLE 4.13

A 30-kVA, 400/120-V, Y/Δ-connected, three-phase, step-down transformer has the following parameters on a per-phase basis: $R_H = 50\ \text{m}\Omega$, $X_H = 100\ \text{m}\Omega$, $R_L = 12\ \text{m}\Omega$, $X_L = 30\ \text{m}\Omega$, $R_{cH} = 400\ \Omega$, and $X_{mH} = 250\ \Omega$. The transformer delivers the rated load at a 0.8 pf lagging. Calculate (a) the line voltage on the primary side and (b) the efficiency of the transformer.

● SOLUTION

The apparent power and the nominal voltages on a per-phase basis for a simulated Y/Y connection are

$$S_\phi = 10\ \text{kVA}$$

$$V_{1n} = \frac{400}{\sqrt{3}} = 230.94\ \text{V}$$

$$V_{2n} = \frac{120}{\sqrt{3}} = 69.282\ \text{V}$$

Thus, the *a*-ratio is

$$a = \frac{230.94}{69.282} = 3.333$$

The per-phase equivalent circuit is given in Figure 4.45.
The rated load current is

$$I_{2A} = \frac{10{,}000}{69.282} = 144.338\ \text{A}$$

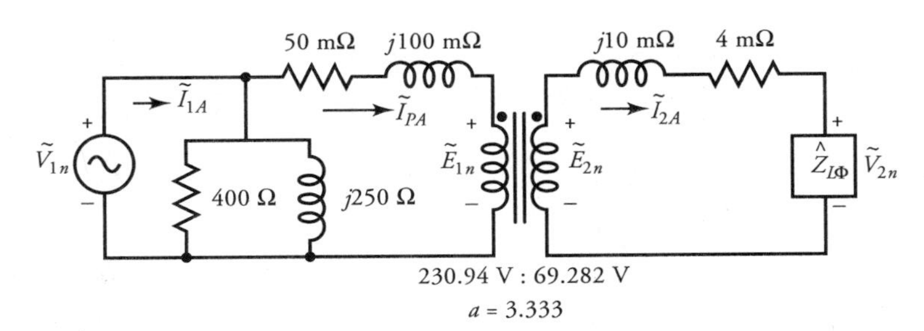

Figure 4.45 Per-phase equivalent circuit of a three-phase transformer for Example 4.13.

Assuming V_{2n} as the reference, the load current in the phasor form for a 0.8 lagging power factor is

$$\tilde{I}_{2A} = 144.338\angle{-36.87^\circ} \text{ A}$$

The primary winding current is

$$\tilde{I}_{pA} = \frac{\tilde{I}_{2A}}{a} = 144.338\angle{-36.87^\circ}/3.333 = 43.301\angle{-36.87^\circ} \text{ A}$$

The induced emf in the simulated (Y-connected) secondary winding is

$$\begin{aligned}\tilde{E}_{2n} &= \tilde{V}_{2n} + \tilde{I}_{2A}(0.004 + j0.01) \\ &= 69.282 + (144.338\angle{-36.87^\circ})(0.004 + j0.01) = 70.615\angle{0.66^\circ} \text{ V}\end{aligned}$$

Assuming a positive phase-sequence, the induced emf in the actual Δ-connected secondary winding is

$$\tilde{E}_{2L} = \sqrt{3}\tilde{E}_{2n}\angle{30^\circ} = \sqrt{3} \times 70.615\angle{30.66^\circ} = 122.31\angle{30.66^\circ} \text{ V}$$

The induced emf in the Y-connected primary winding is

$$\tilde{E}_{1n} = a\tilde{E}_{2n} = 3.333 \times 70.615\angle{0.66^\circ} = 235.382\angle{0.66^\circ} \text{ V}$$

Thus, the induced emf between lines a_1 and b_1 is

$$\tilde{E}_{1L} = \sqrt{3}\tilde{E}_{1n}\angle{30^\circ} = \sqrt{3} \times 235.382\angle{30.66^\circ} = 407.69\angle{30.66^\circ} \text{ V}$$

The per-phase applied voltage is

$$\begin{aligned}\tilde{V}_{1n} &= \tilde{E}_{1n} + \tilde{I}_{pa}(0.05 + j0.1) \\ &= 235.382\angle{0.66^\circ} + (43.301\angle{-36.87^\circ})(0.05 + j0.1) = 239.746\angle{1.16^\circ} \text{ V}\end{aligned}$$

(a) Hence the effective value of the line voltage is

$$V_{1L} = \sqrt{3}V_{1n} = \sqrt{3} \times 239.746 = 415.25 \text{ V}$$

The current supplied by the source is

$$\begin{aligned}\tilde{I}_{1A} &= \tilde{I}_{pA} + \tilde{V}_{1n}\left[\frac{1}{R_{cH}} - \frac{j}{X_{mH}}\right] \\ &= 43.301\angle{-36.87^\circ} + 239.746\angle{1.16^\circ}\left[\frac{1}{400} - \frac{j}{250}\right] \\ &= 44.366\angle{-37.37^\circ} \text{ A}\end{aligned}$$

The power supplied by the three-phase transformer is

$$P_o = 3\ \text{Re}[\tilde{V}_{2n}\tilde{I}^*_{2A}] = 3\ \text{Re}[69.282 \times 144.338\underline{/36.87°}] = 24{,}000\ \text{W}$$

The power input to the three-phase transformer is

$$P_{\text{in}} = 3\ \text{Re}[\tilde{V}_{1n}\tilde{I}^*_{1A}] = 3\ \text{Re}[(239.746\underline{/1.16°})(44.366\underline{/37.37°})]$$
$$= 24{,}962\ \text{W}$$

(b) The efficiency of the three-phase transformer is

$$\eta = \frac{P_o}{P_{in}} = \frac{24{,}000}{24{,}962} = 0.961 \quad \text{or} \quad 96.1\%$$

■

Exercises

4.22. A 300-kVA, 460-V, balanced three-phase load is supplied by three single-phase transformers connected in Y/Δ to form a three-phase transformer. The primary winding of the three-phase transformer is connected to a 4.8-kV, three-phase transmission line. Determine (a) the primary voltage, (b) the secondary voltage, and (c) the power rating of each transformer.

4.23. Three 2.2-kVA, 440/220-V transformers are connected to form a three-phase transformer. Each transformer has the following parameters: $R_H = 1.2\ \Omega$, $X_H = 2\ \Omega$, $R_L = 0.3\ \Omega$, $X_L = 0.5\ \Omega$, $R_{cH} = 2.2\ \text{k}\Omega$, and $X_{mH} = 1.8\ \text{k}\Omega$. Determine the primary winding voltage and current, the secondary winding voltage and current, and the efficiency at full load at 0.707 pf leading for each of the four connections.

4.11 The Constant-Current Transformer

Thus far we have devoted our attention to the study of constant-potential transformers, for which the load voltage is essentially constant and the current varies with the load. A transformer of this type is designed to operate at or just above the knee of the magnetization curve in order to ensure relatively high permeability and low initial cost. The two windings are wound on top of each other to reduce the leakage fluxes.

A constant-current transformer, on the other hand, is designed to satisfy the constant-current requirement while the voltage drop varies with the load. In other words, the secondary voltage varies directly with the load in a constant-current transformer.

Figure 4.46 is an illustration of a constant-current transformer with a fixed primary winding and a movable secondary winding. The transformer is designed to operate at a relatively high flux density so that the core is highly saturated. A saturated core assures low permeability which, in turn, implies high leakage flux.

The constant-current transformer operates on the principles that (a) a current-carrying coil manifests itself as an electromagnet and (b) like polarity poles of a magnet exhibit a force of repulsion.

In a constant-current transformer, the current in the primary winding induces a current in the secondary winding such that the mmf produced by one winding opposes the mmf of the other. In other words, at any instant, the two coils act like electromagnets with like polarity poles facing each other. Hence, a force of repulsion is always present between the two windings that makes the secondary winding move up or down.

Let us suppose that the secondary winding is delivering some current to the resistive load, as indicated in the figure. Correspondingly, some voltage is induced

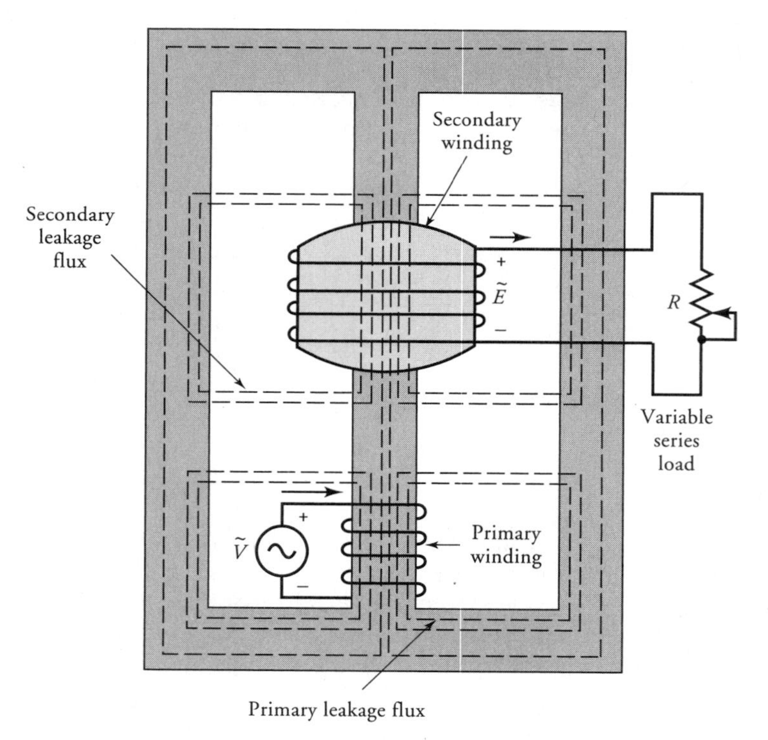

Figure 4.46 A constant-current transformer feeding a variable resistive load.

in the secondary coil. If we now decrease the resistance, the following sequence of events takes place.

1. The induced emf in the secondary at its present position results in an increase in the current through the resistance.
2. An increase in the secondary winding current increases its mmf and, therefore, causes the current in the primary winding to increase.
3. An increase in mmf boosts the flux associated with each winding, which, in turn, makes each winding act as a stronger magnet.
4. The increase in the magnetic strength of each magnet causes the secondary winding to move up.
5. As the secondary winding tends to move up, the flux linking the winding decreases because of high leakage flux.
6. The decrease in the flux linking the secondary winding decreases the induced emf in the winding.
7. As the induced emf decreases, so does the current in the secondary winding.
8. The secondary winding continues to move up until the secondary current drops to its original value. At that time the forces are once again in equilibrium.

Following the same logic, we can say that the secondary winding moves closer to the primary if the load on the transformer is increased. At no load, the secondary winding is supported by a mechanical suspension system to rest on the primary winding.

4.12 Instrument Transformers

Instrument transformers are designed to facilitate the measurements of high currents and voltages in a power system with standard low-range ammeters and voltmeters. They also provide the needed safety in making these measurements, as the primary and the secondary windings are electrically isolated. Instrument transformers are of two kinds: **current transformers** and **potential transformers**.

The Current Transformer

The current transformer, as the name suggests, is designed to measure high current in a power system. The primary winding has few turns of heavy wire, whereas the secondary has many turns of very fine wire. In a clamp-on type current transformer, the current-carrying conductor itself acts as a one-turn primary. The wiring arrangements for a wound primary and clamp-on current transformers are shown in Figure 4.47. It is evident from the figure that a current transformer is merely a well-designed step-up transformer. As the voltage is stepped up, the current is stepped down.

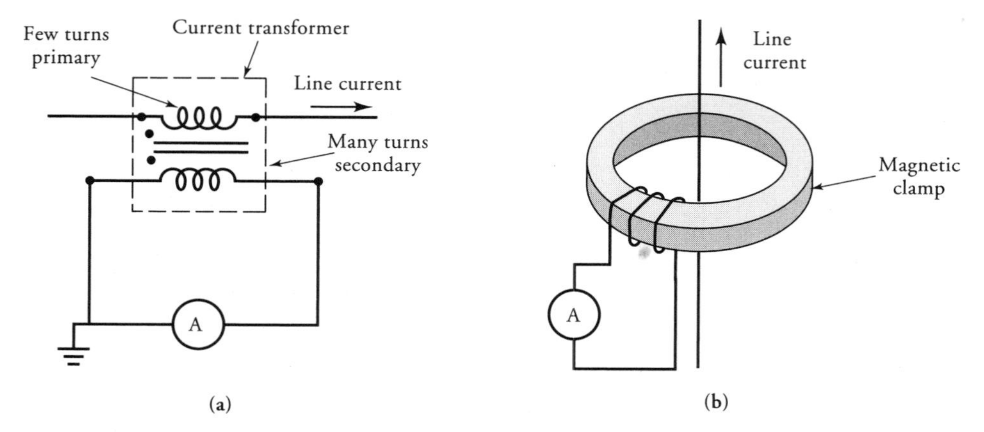

Figure 4.47 **(a)** Current transformer with wound primary. **(b)** Clamp-on type current transformer.

The low-range ammeter is connected across the secondary winding. Because the internal resistance of an ammeter is almost negligible compared with the winding resistance of the secondary, the ammeter can be treated as a short circuit. Therefore, the current transformers are always designed to operate under short-circuit conditions. The magnetizing current is almost negligible, and the flux density in the core is relatively low. Consequently, the core of a current transformer never saturates under normal operating conditions.

The secondary winding of a current transformer should never be left open. Otherwise, the current transformer may lose its calibration and yield inaccurate reading. The reason is that the primary winding is still carrying current, and no secondary current is present to counteract its mmf. The primary winding current acts like a magnetizing current and increases the flux in the core. The increased flux may saturate and magnetize the core. When the secondary is closed again, the hysteresis loop may not be symmetric around the origin but is displaced in the direction of residual flux in the core. The increase in the flux causes an increase in the magnetizing current, which, in turn, invalidates its calibration. In addition, the primary current can produce excessive heat over a period of time and may destroy the insulation. Furthermore, the saturation may result in excessively high voltage across the secondary.

A current transformer is usually given a designation like 100:1. This simply means that if the ammeter measures 1 A, the current in the primary is 100 A. If an ammeter connected to a 100:5 transformer registers 2 A, the line current is 40 A.

The Potential Transformer

As the name suggests, a potential transformer is used to measure high potential difference (voltage) with a standard low-range voltmeter. A potential transformer,

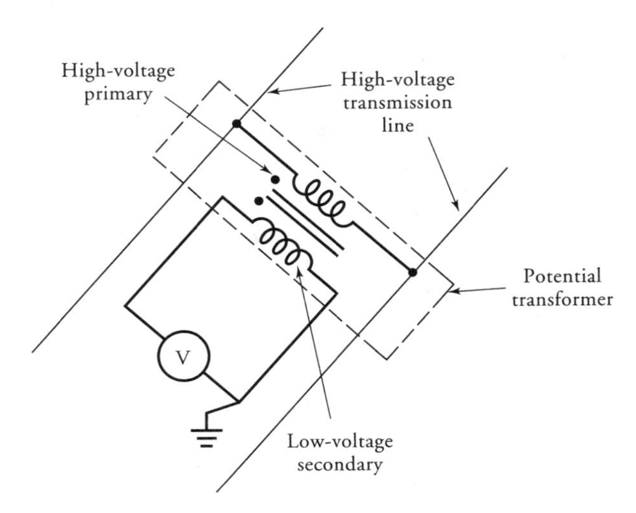

Figure 4.48 A potential transformer connected to measure high voltage with standard low-range voltmeter.

therefore, must be of the step-down type. The primary winding has many turns and is connected across the high-voltage line. The secondary winding has few turns and is connected to a voltmeter. The magnetic core of a potential transformer usually has a shell-type construction for better accuracy. In order to provide adequate protection to the operator, one end of the secondary winding is usually grounded as illustrated in Figure 4.48.

Since a voltmeter behaves more like an open circuit, the power rating of a potential transformer is low. Other than that, a potential transformer operates like any other constant-potential transformer. The a-ratio is simply the ratio of transformation. For example, if the voltmeter on a 100:1 potential transformer reads 120 V, the line voltage is 12,000 V. Some common ratios of transformation are 10:1, 20:1, 100:1, and 120:1.

The insulation between the winding presents a major problem in the design of potential transformers. The primary winding may, in fact, be wound in layers. Each layer is then insulated in order to avoid insulation breakdown. Some of the insulation materials used in potential transformers are oil, oil-impregnated paper, gaseous dielectrics such as sulfur hexafluoride, and epoxy resins.

EXAMPLE 4.14

A typical application employing a 100:1 potential transformer and an 80:5 current transformer is shown in Figure 4.49. If the ammeter, voltmeter, and wattmeter register 4 A, 110 V, and 352 W, respectively, determine (a) the line current, (b) the line voltage, and (c) the power on the transmission line.

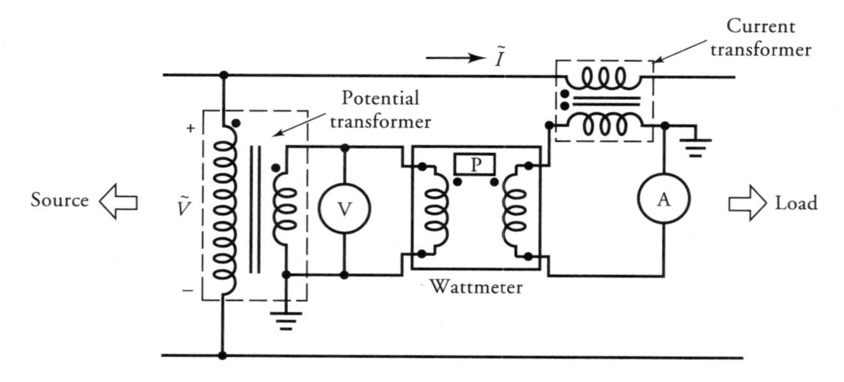

Figure 4.49 Voltage, current, and power measurements using current and potential transformers.

● SOLUTION

(a) A designation of 80:5 indicates that the current transformation ratio is 16:1. A current of 4 A on the ammeter translates into a current of 64 A (16 × 4) in the transmission line.
(b) The voltage transformation ratio is 100:1. Therefore, a voltmeter reading of 110 V signifies that the line voltage is 110 × 100 = 11,000 V.
(c) The power transformation is 100 × 16 = 1600. Thus, the power supplied by the transmission line is 563.2 kW (1600 × 352 W).

■

Exercises

4.24. What must be the readings on the voltmeter, ammeter, and wattmeter, if the designation of the current transformer is changed to 100:5 and the voltage transformation ratio is changed to 120:1 in Example 4.14?

4.25. Refer to Figure 4.49. The current transformer is 100:5. The potential transformer is 80:1. If the readings on the instruments are 4.8 A, 60 V, and 260 W, calculate (a) the line current, (b) the line voltage, and (c) the power on the transmission line.

SUMMARY

We began our discussion by defining different types of transformers that are being used in power systems. The most commonly used transformer is a constant-potential transformer. A constant-potential transformer delivers power to substantially constant voltage under normal conditions of loading.

To develop the theory for a real transformer, we first defined an ideal transformer and then derived the following equations:

$$\tilde{E}_1 = a\tilde{E}_2$$
$$\tilde{I}_2 = a\tilde{I}_1$$
$$\hat{Z}_1 = a^2\hat{Z}_2$$

By including the winding resistances, the leakage reactances, the equivalent resistance to represent the core loss, and the equivalent reactance to represent the mmf drop in the core, we were able to represent a real transformer by an exact equivalent circuit. The exact equivalent circuit included the ideal transformer.

By transforming impedance from one side to the other, we were also able to develop exact equivalent circuits as viewed from the primary and secondary sides.

Based on the fact that the flux in the core stays almost the same under normal conditions of loading, we were able to develop the approximate equivalent circuits. These approximate circuits were used to experimentally determine the parameters of a transformer.

The open-circuit test was conducted by applying the rated voltage on the low-voltage side while the high-voltage side was left open. This test allowed us to measure the current necessary to account for the hysteresis and the eddy-current losses and the current needed to create the magnetic flux in the core. From the measurements of power, current, and voltage, we were able to determine

$$R_{cL} = \frac{V_{oc}^2}{P_{oc}}$$

$$X_{mL} = \frac{V_{oc}^2}{Q_{oc}}$$

where $$Q_{oc} = \sqrt{S_{oc}^2 - P_{oc}^2}$$

and $$S_{oc} = V_{oc}I_{oc}$$

The short-circuit test was conducted at the rated current on the high-voltage side by shorting the low-voltage side. Because the applied voltage was a fraction of the rated voltage, we assumed that the excitation current was almost negligible. From the measured values of power, voltage, and current, we were able to calculate the equivalent resistance and leakage reactance as referred to the high-voltage side. The following equations were used to do so.

$$Z_{eH} = \frac{V_{sc}}{I_{sc}}$$

$$R_{eH} = \frac{P_{sc}}{I_{sc}^2}$$

$$X_{eH} = \sqrt{Z_{eH}^2 - R_{eH}^2}$$

We have also shown that a transformer operates at its maximum efficiency when the copper loss is equal to the core (magnetic) loss. That is

$$I_{p\eta}^2 R_{e1} = P_m$$

We also analyzed a transformer on a per-unit basis and highlighted its importance. The per-unit values of machines of the same type and widely different ratings fall within a narrow range.

When we connected the two windings, the transformer was referred to as an autotransformer. We explained the four possible ways of interconnecting the two windings. An equivalent circuit was developed for each connection. We derived expressions to show that a two-winding transformer delivers more power when it is connected as an autotransformer. The equations for an ideal autotransformer are

$$\tilde{V}_{1a} = a_T \tilde{V}_{2a}$$

$$\tilde{I}_{2a} = a_T \tilde{I}_{1a}$$

$$\hat{Z}_{1a} = a_T^2 \hat{Z}_{2a}$$

To handle three-phase power, we depicted four possible ways we can connect the primary and secondary windings of a three-phase transformer. A three-phase transformer can either be designed as a single unit or formed by connecting three single-phase transformers. We shed some light on the advantages and drawbacks of each connection. We stated that a three-phase transformer can be analyzed on a per-phase basis by simulating it as a Y/Y connection.

We gave some insight into the constant-current transformer and the instrument transformers. The constant-current transformer is designed to supply constant current to a variable load connected in series. The transformer is designed to operate in a highly saturated state. The instrument transformers are, in fact, very carefully designed constant-potential transformers. They allow us to measure high currents and high voltages by using standard low-range meters.

Review Questions

4.1. What is a transformer?
4.2. Can a transformer be used to transform direct voltage and direct current?
4.3. In a well-designed transformer, what is the coefficient of coupling between the two windings?
4.4. Why is the frequency of the induced emf in the secondary winding of a transformer the same as that of the impressed voltage on the primary winding?
4.5. What is an ideal transformer?
4.6. What differentiates a core-type transformer from a shell-type transformer?
4.7. What are the drawbacks of using a solid magnetic core for a transformer?
4.8. What is a potential transformer?
4.9. What is the difference between a distribution transformer and a power transformer?
4.10. What is the a-ratio, or ratio of transformation? How can the a-ratio be determined experimentally?
4.11. In a transformer, the primary current is twice as much as the secondary current. Is this a step-up or step-down transformer?
4.12. Explain why the primary mmf must be equal and opposite to the secondary mmf in an ideal transformer.
4.13. Why does a real transformer draw some current when the secondary is open?
4.14. A 22-kVA, 2200/1100-V, step-down ideal transformer delivers a rated load at a leading power factor of 0.5. Determine (a) the secondary winding current, (b) the primary winding current, (c) the impedance on the secondary side as a parallel combination of resistance and reactance, and (d) the impedance on the primary side as a series combination of resistance and reactance.
4.15. Distinguish between the excitation current, the core-loss current, and the magnetizing current.
4.16. What is a leakage flux? How can the leakage flux be minimized? Is it possible to have no leakage flux?
4.17. Why is an open circuit test conducted at the rated voltage?
4.18. Outline carefully the procedure for performing the open-circuit test.
4.19. What is the advantage of performing the open-circuit test on the low-voltage side?
4.20. Why is a short-circuit test performed at the rated current?
4.21. Outline carefully the procedure for performing the short-circuit test.
4.22. Why do we prefer to perform the short-circuit test on the high-voltage side?
4.23. Explain why the core loss is not affected by the load.
4.24. Explain why the copper loss varies with the load.
4.25. What is the criterion for maximum efficiency of a transformer?

4.26. In a distribution transformer, the maximum efficiency occurs at 80% of the load. What does this mean? If the operating frequency is increased, what happens to the load current at maximum efficiency?
4.27. What is an autotransformer? List its advantages and drawbacks.
4.28. Draw sketches showing how a 22-kVA, 2200/1100-V, two-winding transformer can be connected as an autotransformer. Determine (a) the voltage rating, (b) the power rating, (c) the power transferred by conduction, and (d) the power transferred by induction.
4.29. What is the significance of a per-unit system?
4.30. Do we have to use the nominal ratings of a transformer as the base quantities?
4.31. Using the apparent power and the terminal voltages as the base quantities in a 44-kVA, 1100/250-V transformer, determine the other base quantities. What is the a-ratio on a per-unit basis?
4.32. What is the importance of polarity markings in a multiwinding transformer?
4.33. Explain voltage regulation.
4.34. When a transformer operates at half load, the secondary winding voltage is 250 V. At no load, it is 270 V. What is the voltage regulation?
4.35. The no-load voltage is lower than the full-load voltage on the secondary side of a transformer. Under what conditions can this happen?
4.36. Write equations to justify that the third harmonics in a three-phase transformer must be in phase.
4.37. Sketch the four possible ways of connecting three single-phase transformers as a three-phase transformer. State the advantages and drawbacks of each connection.
4.38. Why is a shell-type three-phase transformer better than a core-type?
4.39. Explain the procedure for performing the open-circuit and short-circuit tests on a three-phase transformer.
4.40. Explain the principle of operation of a constant-current transformer.
4.41. What is the advantage of a clamp-on current transformer?

Problems

4.1. A magnetic core carries a flux of $5 \sin 314t$ mWb. Determine the rms value of the induced emf in a 100-turn coil wound on the core using (a) Faraday's law and (b) the transformer equation. If the cross-sectional area of the core is 25 cm^2, what is the effective flux density in the core?

4.2. The effective flux density in the core of a 220/110-V ideal transformer is 1.2 T. The cross-sectional area of the core is 80 cm^2, and the stacking factor is 0.93. Calculate the number of turns in the primary and secondary windings if the frequency of oscillations is (a) 25 Hz, (b) 50 Hz, (c) 60 Hz, and (d) 400 Hz.

4.3. The cross-sectional area of the core of a 1.2-kVA, 120/208-V, 50-Hz ideal transformer is 100 cm^2. The primary has 60 turns. Calculate (a) the induced emf per turn, (b) the number of turns in the secondary, and (c) the effective flux density in the core.

4.4. A 12-kVA, 480/120-V, step-down ideal transformer is operating at its full load with a 70.7% pf leading. The transformer is designed such that 2 V per turn are induced in its windings. Determine (a) the number of turns in its primary and secondary windings, (b) the primary and secondary winding currents, (c) the power output, and (d) the load impedance on the secondary side. Draw the equivalent circuit by representing the load on the primary side.

4.5. A 25-kVA transformer has 500 turns on the primary and 50 turns on the secondary. If the primary is rated at 2.5 kV, find the rating of the secondary. What is the flux in the core at no load if the operating frequency is 50 Hz? If the transformer delivers 80% of its full load at 0.8 pf leading, determine (a) the primary and the secondary winding currents, (b) the power output, (c) the load impedance on the secondary side, and (d) the load impedance as referred to the primary side.

4.6. A power amplifier can be represented by a current source in parallel with a 200-Ω resistance. An 8-Ω loudspeaker is connected to the power amplifier via an audio transformer so that maximum power is transferred to the loudspeaker. What must be the a-ratio of the transformer?

4.7. A 48-kVA, 4800/480-V, 60-Hz, step-down transformer is supplying a load of 18 kW at the rated voltage and 0.5 pf lagging. Determine the extra load that must be connected to the transformer so that it delivers full load at the rated voltage with unity power factor. If each load consists of a series combination of R, L, and/or C, what are their values?

4.8. An ac generator can be modelled as a voltage source in series with an impedance of $0.5 + j10\ \Omega$. It is connected to a transmission line having an equivalent series impedance of $5 + j12\ \Omega$ via an ideal step-up transformer with an a-ratio of 0.05. An ideal step-down transformer with an a-ratio of 25 connects the other end of the transmission line to a load impedance of $30 + j40\ \Omega$. If the load voltage is 240 V, determine (a) the power supplied to the load, (b) the power dissipated by the transmission line, (c) the generator voltage, and (d) the power supplied by the generator.

4.9. A 240-kVA, 480/4800-V, step-up transformer has the following constants: $R_H = 2.5\ \Omega$, $X_H = 5.75\ \Omega$, $R_L = 25\ \text{m}\Omega$, $X_L = 57.5\ \text{m}\Omega$. The transformer is operating at 50% of its rated load. If the load is purely resistive, determine (a) the applied voltage on the primary side, (b) the secondary current, (c) the primary current, and (d) the efficiency of the transformer.

4.10. Repeat Problem 4.9 if the core-loss resistance and the magnetizing reactance on the high-voltage side are 18 kΩ and 12 kΩ, respectively. What is the effective value of the excitation current?

4.11. A 100-kVA, 2500/125-V, 50-Hz, step-down transformer has the following parameters: $R_H = 1.5\ \Omega$, $X_H = 2.8\ \Omega$, $R_L = 15\ \text{m}\Omega$, $X_L = 20\ \text{m}\Omega$, $R_{cH} = 3\ \text{k}\Omega$, and $X_{mH} = 5\ \text{k}\Omega$. The transformer delivers 85% of the rated load at a terminal voltage of 110 V and a power factor of 0.866 lagging. Determine (a) the core loss, (b) the copper loss, and (c) the efficiency of the transformer.

4.12. The parameters of a 12-kVA, 120/480-V, 60-Hz, two-winding, step-up transformer are: $R_H = 0.6\ \Omega$, $X_H = 1.2\ \Omega$, $R_L = 0.\ 1\ \Omega$, $X_L = 0.3\ \Omega$, $R_{cH} = 3.2\ \text{k}\Omega$, $X_{mH} = 1.2\ \text{k}\Omega$. The transformer is operating at 80% of its load and 0.866 pf lagging. Determine the copper losses, the core loss, and the efficiency of the transformer.

4.13. Repeat Problem 4.12 if the power factor of the load is 0.866 leading.

4.14. Repeat Problem 4.12 if the power factor of the load is unity.

4.15. A 230-kVA, 2300/230-V, 60-Hz, step-down, two-winding transformer has the following parameters: $R_H = 1.2\ \Omega$, $X_H = 3\ \Omega$, $R_L = 12\ \text{m}\Omega$, $X_L = 30\ \text{m}\Omega$, $R_{cH} = 2\ \text{k}\Omega$, $X_{mH} = 1.8\ \text{k}\Omega$. If the transformer operates at half load with a unity power factor, what is its efficiency?

4.16. For the transformer of Problem 4.9, determine the voltage regulation when the power factor of the load is (a) unity, (b) 0.8 lagging, and (c) 0.8 leading. Use the equivalent circuit as referred to the high-voltage side.

4.17. Determine the voltage regulation for the transformer of Problem 4.11 when the power factor of the load is (a) unity, (b) 0.707 lagging, and (c) 0.707 leading. Use the approximate equivalent circuit as referred to the high-voltage side.

4.18. For the transformer of Problem 4.15, draw approximate equivalent circuits as viewed from (a) the primary side and (b) the secondary side. In each case, determine (a) the efficiency and (b) the regulation if the load has a lagging power factor of 0.8.

4.19. The transformer of Problem 4.9 is operating at full load and 0.866 pf lagging. Determine its voltage regulation and efficiency using the per-unit quantities.

4.20. The transformer of Problem 4.11 is operating at full load and 0.85 pf leading. Determine its voltage regulation using the per-unit quantities.

4.21. The transformer of Problem 4.15 operates at full load and 0.95 pf leading. Determine its voltage regulation and efficiency using the per-unit quantities.

4.22. The transformer of Problem 4.9 operates at a frequency of 60 Hz. Assuming that the transformer operates in a linear region, what must be its efficiency at full load and a power factor of 0.9 lagging? What is its voltage regulation? Use the per-unit system.

4.23. A 320-kVA, 240/4800-V, 60-Hz transformer yielded the following information when tested:

	Voltage (V)	Current (A)	Power (W)
Open-circuit test:	240	39.5	1200
Short-circuit test:	195	66.67	3925

Find the equivalent circuit of the transformer as viewed from (a) the low-voltage side and (b) the high-voltage side.

4.24. The following information was obtained when a 60-VA, 120/208-V, 60-Hz transformer was tested:

	Voltage (V)	Current (mA)	Power (W)
Open-circuit test:	120	25.07	2
Short-circuit test:	16.85	300	4.7

Determine the equivalent circuit of the transformer as viewed from (a) the low-voltage side and (b) the high-voltage side. Draw the exact equivalent circuit embodying an ideal transformer using equal power-loss criterion.

4.25. The following data were taken for a 46-kVA, 2300/230-V, 60-Hz transformer:

	Voltage (V)	Current (A)	Power (W)
Open-circuit test:	2300	0.9	1680
Short-circuit test:	95	20	670

Determine the equivalent circuit of the transformer as viewed from (a) the low-voltage side and (b) the high-voltage side.

4.26. The following data were obtained from the tests on a 12-kVA, 480/120-V, 60-Hz, two-winding transformer:

	Voltage (V)	Current (A)	Power (W)
Open-circuit test:	120	1.71	72
Short-circuit test:	73	25	937.5

Sketch the equivalent circuit of the transformer as viewed from (a) the low-voltage side and (b) the high-voltage side.

4.27. Using only the experimental data of Problem 4.23, determine the efficiency of the transformer when the power factor at full load is unity. Determine the maximum efficiency of the transformer.

4.28. Using only the experimental data of Problem 4.24, determine the efficiency of the transformer when the power factor at full load is 0.8 leading. Calculate the maximum efficiency of the transformer.

4.29. Use the information given in Problem 4.25 to find the efficiency of the transformer when the power factor at full load is 0.8 lagging.

4.30. Using the information given in Problem 4.26, determine the efficiency of the transformer when the power factor at full load is 0.866 lagging. What is the maximum efficiency of the transformer?

4.31. If the open-circuit and short-circuit tests are performed on the transformer of Problem 4.10, what must be the readings on the meters in each test?

4.32. If the open-circuit and short-circuit tests are performed on the transformer of Problem 4.11, what must be the readings on the meters in each test?

4.33. If the open-circuit and short-circuit tests are performed on the transformer of Problem 4.12, what must be the readings on the meters in each test?

4.34. An 11-kVA, 220/110-V, 60-Hz, two-winding transformer is connected as a step-up autotransformer. Sketch each possibility and determine (a) the nominal power rating, (b) the nominal voltage rating, and (c) the power transferred by conduction.

4.35. The transformer of Problem 4.34 is connected as a step-down autotransformer. Sketch each possibility and determine (a) the nominal power rating, (b) the nominal voltage rating, and (c) the power transferred by conduction.

4.36. The transformer of Problem 4.9 is connected as a 120/600-V, step-up autotransformer. It is operating at 80% of its load with unity power factor. Determine the efficiency and the voltage regulation of the autotransformer.

4.37. The two-winding transformer of Problem 4.12 is connected as a 480/600-V, step-up autotransformer. It delivers its rated load at a power factor of 0.866 leading. Determine its efficiency and voltage regulation.

4.38. The two-winding transformer of Problem 4.12 is connected as a 600/480-V, step-down autotransformer. Find its efficiency at rated load and 0.866 pf leading. What is its voltage regulation?

4.39. The two-winding transformer of Problem 4.12 is connected as a 600/120-V, step-down autotransformer. It delivers its rated load at a power factor of 0.866 leading. Determine its efficiency and voltage regulation.

4.40. The two-winding transformer of Problem 4.12 is connected as a 120/600-V, step-up autotransformer. It delivers its rated load at 0.866 pf leading. Determine its efficiency and voltage regulation.

4.41. The two-winding transformer of Problem 4.15 is connected to operate as 230/2530-V autotransformer. Determine the voltage regulation and efficiency at rated load when the power factor is 0.8 lagging. Use the equivalent circuit as referred to the primary side of the autotransformer.

4.42. The two-winding transformer of Problem 4.15 is connected to operate as a 2300/2530-V autotransformer. Determine the voltage regulation and efficiency at rated load when the power factor is 0.8 lagging. Use the equivalent circuit as referred to the primary side of the autotransformer.

4.43. The two-winding transformer of Problem 4.15 is connected to operate as a 2530/230-V autotransformer. Determine the voltage regulation and efficiency at rated load with 0.8 pf leading. Use the equivalent circuit as referred to the primary side of the autotransformer.

4.44. The two-winding transformer of Problem 4.15 is connected to operate as a 2530/2300-V autotransformer. Determine the voltage regulation and efficiency at rated load with 0.8 pf leading. Use the equivalent circuit as referred to the primary side of the autotransformer.

4.45. Three 60-VA, 120/208-V, single-phase transformers are to be used to form a three-phase transformer. For each of the four connections, determine (a) the phase voltages, (b) the phase currents, (c) the a-ratio, and (d) the power rating.

4.46. Three identical transformers are needed to connect a 6-kVA, 120-V, three-phase load to a 4800-V, three-phase transmission line. For a Y/Δ connection, determine (a) the power rating, (b) the voltage rating, and (c) the current rating of each transformer. What is the a-ratio?

4.47. A 150-kVA, 2080/208-V, 60-Hz, Y/Y-connected, three-phase, step-down transformer consists of three identical single-phase transformers. Each transformer has the following parameter values: $R_H = 0.45\ \Omega$, $X_H = 2.2\ \Omega$, $R_L = 4.5\ \text{m}\Omega$, $X_L = 22\ \text{m}\Omega$, $R_{cH} = 10\ \text{k}\Omega$, and $X_{mH} = 8\ \text{k}\Omega$. The load on the transformer is 120 kVA, 90 kW (lagging) at the rated terminal voltage. Determine (a) the primary voltage, (b) the efficiency, and (c) the voltage regulation of the three-phase transformer.

4.48. If the three transformers of Problem 4.47 are connected in Y/Δ, what are the nominal voltage and power rating of the three-phase transformer? When the transformer delivers the rated load at 0.8 pf lagging, determine its efficiency and voltage regulation.

4.49. If the three transformers of Problem 4.47 are connected in Δ/Y, determine its efficiency and voltage regulation when the load on the transformer is 96 kW, 72 kVAR at the rated voltage. What are the nominal voltage and power rating of the three-phase transformer?

4.50. If the three transformers of Problem 4.47 are connected in Δ/Δ, what are the nominal voltage and power rating of the three-phase transformer? For a load of 100 kVA, 70.7 kW (lagging) at the rated voltage, determine (a) the primary voltage, (b) the primary and the secondary winding currents, (c) the efficiency, and (d) the voltage regulation.

4.51. A three-phase generator is connected to a three-phase load via a 230/2300-V, Δ/Y-connected, three-phase, step-up transformer, a short transmission line, and a 2300/230-V, Y/Δ-connected, step-down transformer. The per-phase winding resistance and the leakage reactance as referred to the high-voltage side for each transformer are 1.2 Ω and 4.8 Ω, respectively. The impedance of the transmission line is $2.5 + j2.1\ \Omega$. The per-phase impedance of the generator is $0.3 + j1.2\ \Omega$. If the load is 60 kVA at a 230 V and a lagging power factor of 0.9, determine the generator voltage and the efficiency of the system.

4.52. A 120:1 potential transformer is used for the measurement of the high voltage on a transmission line. If the voltmeter reading is 85 V, what is the transmission line voltage?

4.53. An 80:5 current transformer is used for the measurement of the current on a transmission line. If the ammeter records 3.5 A, what is the line current?

4.54. Determine the designation of a current transformer so that an ammeter with a maximum deflection of 2.5 A can measure a current of 100 A.

4.55. Find the a-ratio of a potential transformer so that a voltmeter with a maximum deflection of 100 V can measure a high voltage of 23 kV.

CHAPTER

5

Direct-Current Generators

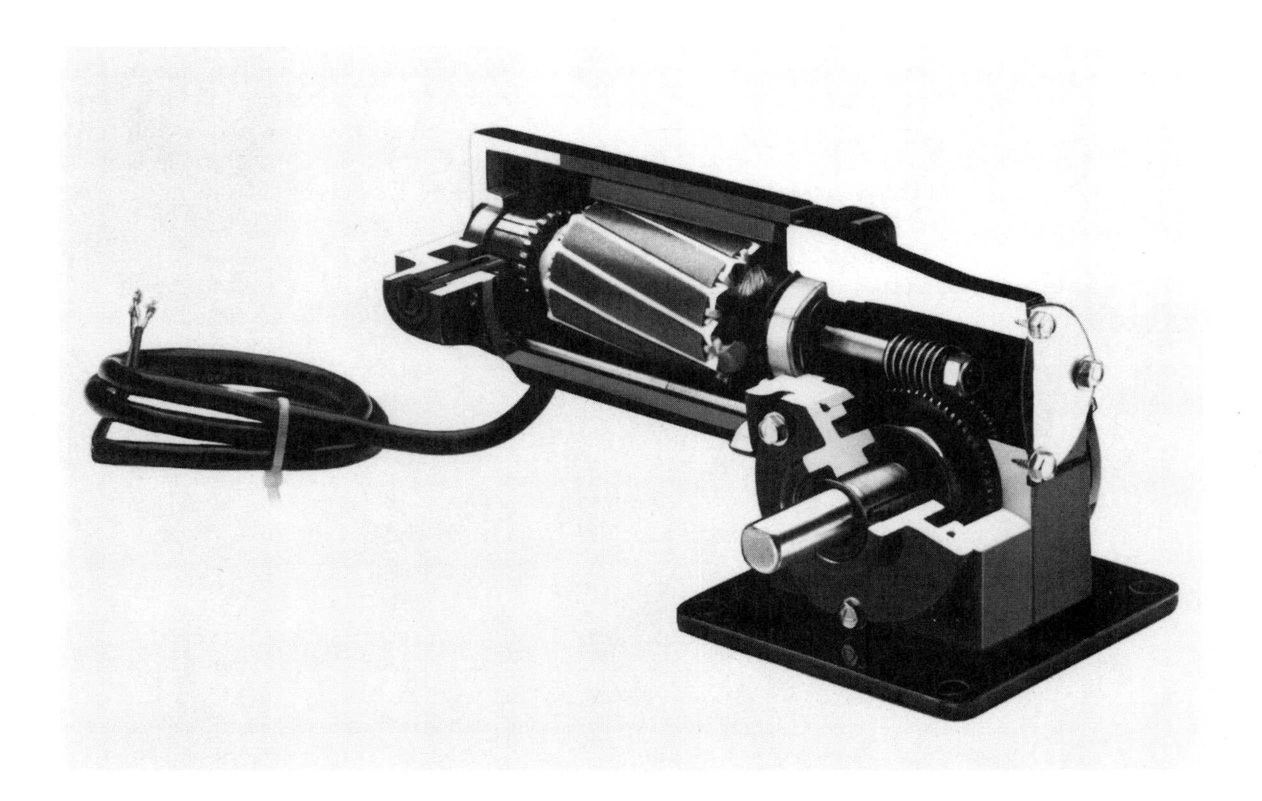

Sectional view of a dc machine showing the commutator and armature arrangement. (*Courtesy of Bodine Electric Company*)

5.1 Introduction

From our discussion in Chapter 3, it should be obvious that there are two types of rotating electric energy conversion machines—the direct-current (dc) machines and the alternating-current (ac) machines. When a rotating machine converts electric energy into mechanical energy, it is called a **motor**. A **generator**, on the other hand, converts mechanical energy into electric energy. Hence, there are dc motors, dc generators, ac motors, and ac generators. A significant portion of this book is devoted to the study of these machines.

The word *machine* is commonly used to explain features that are common to both the motor and the generator. Quite often, a given machine can be operated as either a motor or a generator without making any modifications. This is especially true for all dc machines.

We commit this chapter entirely to the study of dc generators and reserve discussion of dc motors for the next chapter. In addition, we limit our discussion to the steady-state performance characteristics of the machine. Transient behavior is analyzed in a later chapter.

In a dc machine, the uniform magnetic flux is established by fixed poles mounted on the inside of the stationary member called the **stator**. We may either use permanent magnets as the poles or wind the **field windings** (**excitation coils**) around the poles. One of the major advantages of a wound machine is that we can control the flux in the machine by regulating the direct current in the field winding. The winding in which the electromotive force (emf) is induced is wound on the rotating member called the **armature** and is called the **armature winding**. The armature is mechanically supported and aligned inside the stator by the end bells as shown in Figure 5.1. Before we go any further, let us first discuss the construction of a dc machine.

5.2 Mechanical Construction

A cross-section of a 4-pole dc machine is shown in Figure 5.2. Only the main components of the machine have been identified and are discussed below.

Stator

The stator of a dc machine provides the mechanical support for the machine and consists of the **yoke** and the **poles** (or **field poles**). The yoke serves the basic function of providing a highly permeable path for the magnetic flux. For small permanent-magnet (PM) machines, it can be a rolled-ring structure welded at its ends. For small wound machines, the field poles and the yoke are punched as one unit from thin steel laminations. For large machines, the yoke is built using cast steel sections.

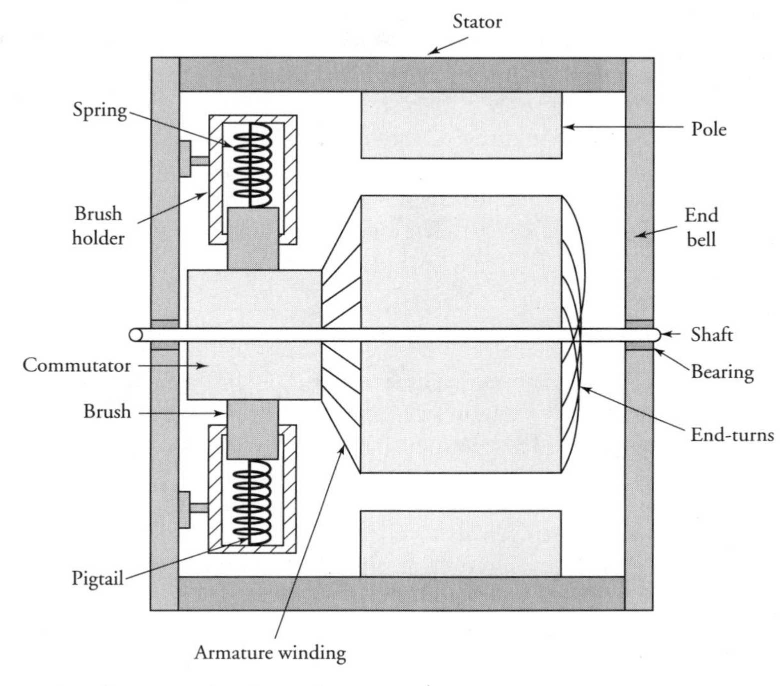

Figure 5.1 Main features of a dc machine.

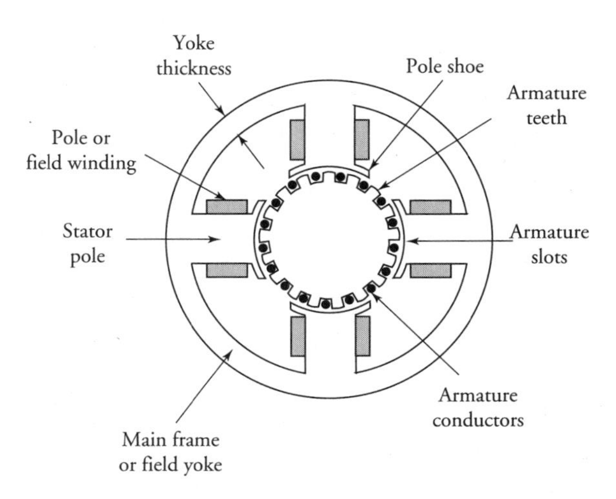

Figure 5.2 The cross-sectional view of a 4-pole dc machine.

The poles are mounted inside the yoke and are properly designed to accommodate the field windings. Most often, the field poles are made of thin laminations stacked together. This is done to minimize the magnetic losses due to the poles' proximity to the armature flux. For large machines, the field poles are built separately and then bolted to the yoke. A typical field pole and a field winding are shown in Figure 5.3. Note that the cross-sectional area of the field pole is smaller than that of the pole shoe. This is done (a) to provide sufficient room for the field winding and (b) to decrease the mean turn length of the wire and thereby reduce its weight and cost. The pole shoe helps to spread the flux in the air-gap region.

Field Winding

The field coils are wound on the poles in such a way that the poles alternate in their polarity. There are two types of field windings—a **shunt field winding** and a **series field winding**. The shunt field winding has many turns of fine wire and derives its name from the fact that it is connected in parallel with the armature winding. A series field winding, as the name implies, is connected in series with the armature winding and has comparatively fewer turns of heavy wire. A dc machine may have both field windings wound on the same pole.

A machine with a shunt field winding is called a **shunt machine**. A **series machine** is wired only with series field winding. A **compound machine** has both windings. When both field windings in a compound machine produce fluxes in the same direction, the machine is said to be of the **cumulative** type. The machine

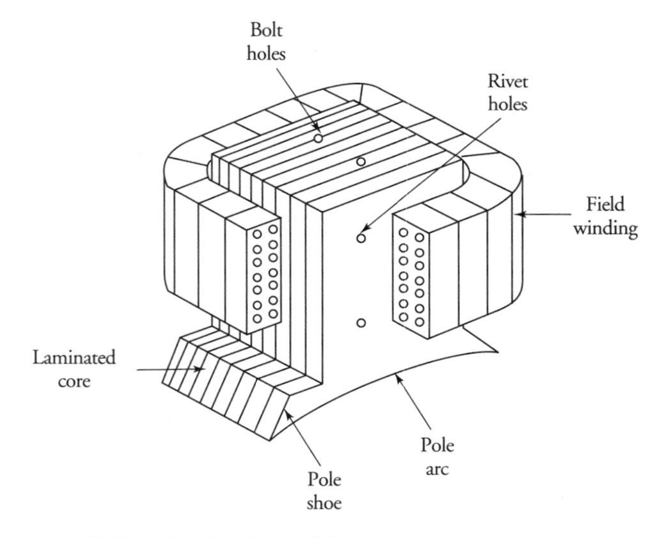

Figure 5.3 A wound field pole of a dc machine.

is of the **differential** type when the field set up by the shunt field winding is opposed by the field established by the series field winding.

Since the field winding carries a constant current, it dissipates power. By using permanent magnets instead of a shunt field winding, we can eliminate the power loss and thereby improve the efficiency of the machine. For the same power rating, a PM machine is smaller and lighter than a wound machine. The disadvantage of a PM machine, of course, is its constant flux.

Armature

The rotating part of a dc machine, which is shrouded by the fixed poles on the stator, is called the armature. The effective length of the armature is usually the same as that of the pole. Circular in cross-section, it is made of thin, highly permeable, and electrically insulated steel laminations that are stacked together and rigidly mounted on the shaft. High permeability ensures a low reluctance path for the magnetic flux, and electrical insulation reduces the eddy currents in the armature core. The laminations have axial slots on their periphery to house the **armature coils** (**armature winding**). Usually an insulated copper wire is used for the armature coils owing to its low resistivity.

Commutator

The commutator is made of wedge-shaped, hard-drawn copper segments as shown in Figure 5.4. It is also rigidly mounted on the shaft as depicted in Figure 5.1. The copper segments are insulated from one another by sheets of mica. One end of two armature coils is electrically connected to a copper segment of the commutator. How each coil is connected to the commutator segment defines the type of armature winding. There are basically two types of armature windings—the **lap winding** and the **wave winding**. The armature winding is the heart of a dc machine. This is the winding in which the emf is induced (generator action) and the torque is developed (motor action). The armature winding, therefore, warrants a detailed discussion. That, in fact, is the subject of the next section.

The commutator is a very well conceived device that serves the function of a rectifier. It converts the alternating emf induced in the armature coils into a unidirectional voltage.

Brushes

Brushes are held in a fixed position on the commutator by means of brush holders. An adjustable spring inside the brush holder exerts a constant pressure on the brush in order to maintain a proper contact between the brush and the commutator. The brush pressure should be just right. If the pressure is low, the contact

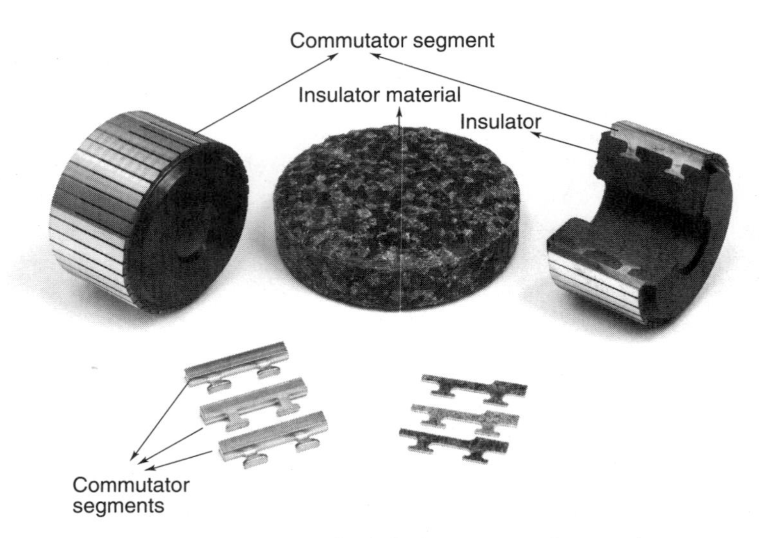

Figure 5.4 Commutator structure. (*Courtesy of Toledo Commutator Company*)

between the brush and the commutator is poor. The poor contact results in excessive sparking and burning of the commutator. On the other hand, too high a pressure results in excessive wear of the brush and overheating of the commutator through friction.

There are many different brush grades, depending upon their composition. A brush may be made of carbon, carbon graphite, or a copper-filled carbon mixture. The graphite in a brush provides self-lubrication between the brush and the commutator.

Although the brush holders are mounted on the end bell, they are electrically insulated from it. A brush is electrically connected to the brush holder by braided copper wire called the **pigtail**. Through these brush holders we can establish an electrical connection between the external circuit and the armature coils.

5.3 Armature Windings

As mentioned in the previous section, the outer periphery of the armature has a plurality of slots into which the coils are either placed or wound. The armature slots are usually insulated with "fish" paper to protect the windings. For small machines, the coils are directly wound into the armature slots using automatic winders. For large machines, the coils are preformed and then inserted into the armature slots. Each coil may have many turns of enamel-covered (insulated) copper wire.

We mentioned in Chapter 3 that the maximum emf is induced in a **full-pitch** coil, that is, when the distance between the sides of a coil is 180° electrical. A full-pitch coil, in other words, implies that when one side is under the center of a south pole, the other side must be under the center of the adjacent north pole. For 2-pole machines, it is quite tedious to place full-pitch coils. In these machines, a **fractional-pitch** coil (coil span less than 180° electrical) is usually employed. Another advantage of a fractional-pitch coil is that it uses less copper than the full-pitch coil. However, the induced emf is reduced by a factor called the pitch factor.

The most commonly used winding is a **two-layer** winding. The number of coils for a two-layer winding is equal to the number of armature slots. Thus, each armature slot has two sides of two different coils. The automatic winders wind the two sides of a coil either at the bottom half or the top half of the two slots. However, when we place the preformed coils in the slots, one side of a coil is placed at the bottom half of a slot and the other side at the top half. This method not only results in the symmetric placement of the coils but also ensures that all coils are electrically equivalent.

When the number of slots is not exactly divisible by the number of poles, it is not even possible to wind a full-pitch coil. In that case, the maximum possible pitch may be used as the fractional pitch of the coil. The maximum pitch of the coil can be determined from the following equation

$$y = \text{integer value of } \left(\frac{S}{P}\right) \tag{5.1}$$

where y is the coil pitch in slots, S is the number of slots in the armature, and P is the number of poles in the machine. This equation yields the pitch as an integer value of the slots per pole. If we place one side of the coil in slot m, the other side must be inserted in slot $m + y$.

EXAMPLE 5.1

The armature of a dc machine has ten slots. Calculate the coil pitch for a (a) 2-pole winding and (b) 4-pole winding.

● SOLUTION

A 10-slot armature employing two-layer winding requires ten coils.

(a) For a 2-pole machine, the slots per pole are

$$S_p = \frac{10}{2} = 5 \text{ slots}$$

Thus, $y = 5$

Since there are five slots per pole and a pole spans 180° electrical, the angle from the center of one slot to the next (**slot span**) is 180/5 = 36° electrical. In this case, it is possible to use a full-pitch coil; that is, if we place one side of a coil in slot 1, then the other side can be placed in slot 6 as shown in Figure 5.5**a**. The second coil goes in slots 2 and 7, the third in slots 3 and 8, and so on. Since the number of slots is equal to the number of teeth, each coil spans five teeth. Quite often, it is easier to count the teeth than the slots.

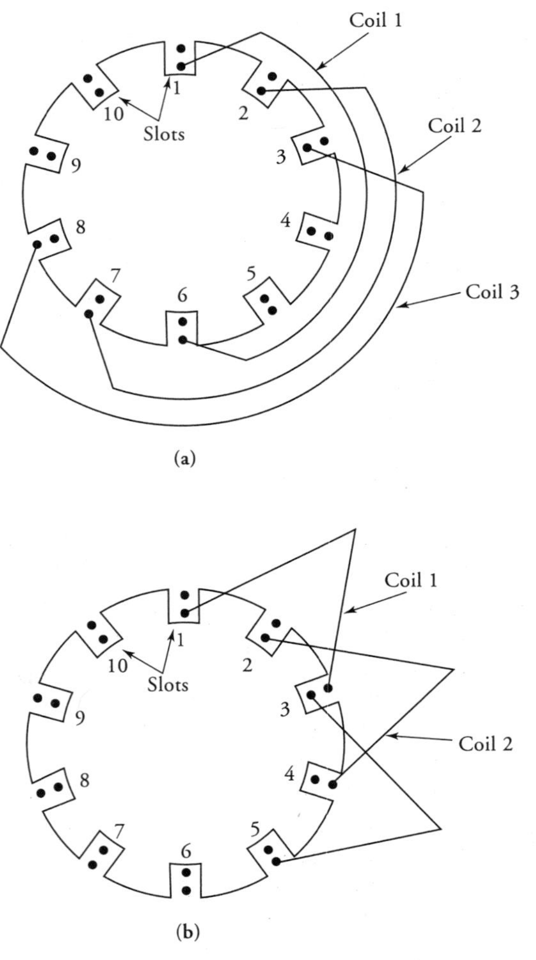

Figure 5.5 Placement of coils in a 10-slot armature of a **(a)** 2-pole and **(b)** 4-pole dc machine.

(b) For a 4-pole machine, the slots per pole are

$$S_p = \frac{10}{4} = 2.5 \qquad \text{and} \qquad y = 2$$

The slot span in this case is $180/2.5 = 72°$ electrical. The maximum number of slots that the coil can span is two. Hence, the coil pitch is 144° electrical. The coils must be inserted in slots 1 and 3, 2 and 4, 3 and 5, etc. as shown in Figure 5.5**b**.

■

How we connect the armature winding to the commutator describes the type of winding. There are two general types of windings—the **lap winding** and the **wave winding**. The lap winding is used for low-voltage and high-current machines. The wave winding, on the other hand, is employed to satisfy the requirements of high voltage and low current.

Each winding is further classified as simplex, duplex, triplex, and so on. We limit our discussion to simplex-lap and simplex-wave windings and refer to them simply as lap and wave windings. There is no difference between the two types for a 2-pole machine. Both require two brushes and have two parallel paths. The number of parallel paths of a lap-wound machine is equal to the number of poles. However, a wave-wound machine always has two parallel paths, regardless of the number of poles.

Lap Winding

In a lap-wound machine the two ends of a coil are connected to adjacent commutator segments. Let us assume that C coils are to be connected to C segments of the commutator. If we connect coil 1 to commutator segments 1 and 2, then coil 2 can be connected to commutator segments 2 and 3. As viewed from the commutator segments 1 and 3, the two coils are now connected in series. We can now connect coil 3 to commutator segments 3 and 4. Following this procedure, we end up connecting coil C to commutator segments C and 1. All the windings are now connected in series and form a closed loop. **The winding is said to close upon itself**. A polar diagram of a 6-pole, 12-coil dc machine with 12 commutator segments is shown in Figure 5.6 with a coil pitch of 2.

For a clockwise rotation, coil 1 is leaving the north pole and the flux linking the coil is decreasing. The indicated direction of the current in coil 1 ensures that the flux created by it opposes the decrease in the flux in accordance with Faraday's law of induction. On the other hand, the flux linking coil 12 is increasing as it moves under the north pole. The direction of the current in this coil must create a flux that opposes the increase. Progressing in this manner from one coil to another, we determine the directions of the currents in all coils. Note that at segments 1, 5, and 7,

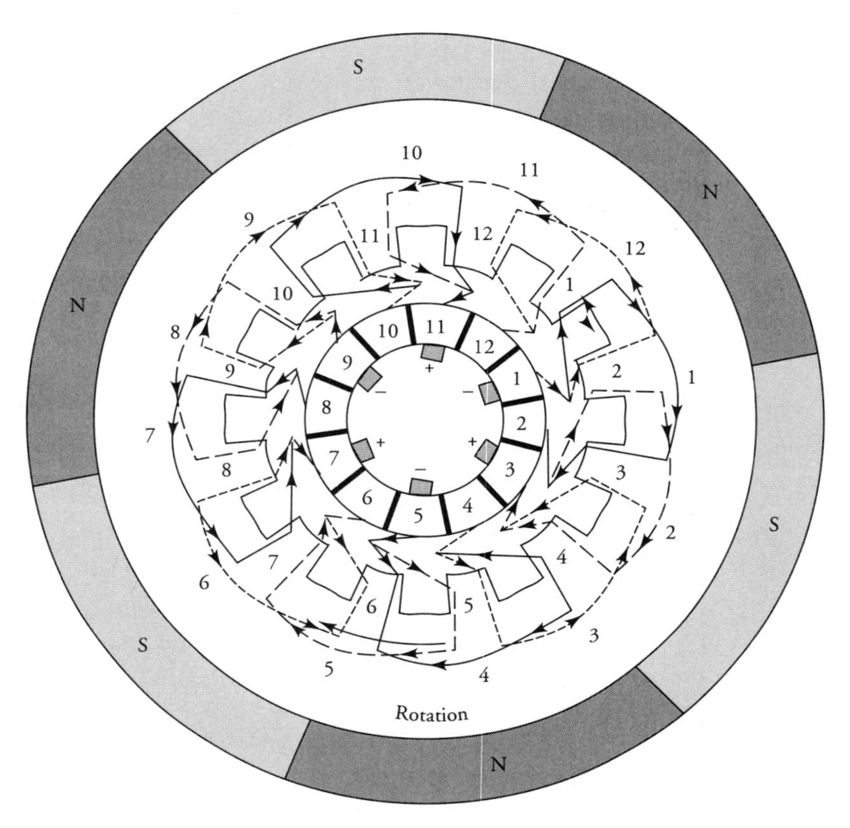

Figure 5.6 A polar-winding diagram of a 6-pole, 12-coil, lap-wound dc machine with 12 commutator segments.

both currents are directed away from the commutator. For a dc generator, these segments mark the placement of brushes having negative polarity. Segments 3, 7, and 11 have both currents directed toward them. They, therefore, represent the placement of brushes having positive polarity. The potential difference between a positive brush at commutator segment 3 and a negative brush at commutator segment 1 is equal to the induced emf in coils 1 and 2. In fact, only two coils are contributing to the potential difference between a positive and a negative brush. Thus, the three negative brushes can be electrically connected together to form a single connection. Likewise, we can connect the three positive brushes to form a single connection. Such an arrangement is shown in Figure 5.7, where each coil is depicted by a single loop. It must be borne in mind, however, that each loop represents the two sides of a coil that are properly placed in armature slots. Note that there are six parallel paths for a 6-pole, lap-wound machine. Each path contributes to one-sixth of the armature current. As shown in Figure 5.7, when the armature supplies a current of 12 A, the current in each coil is 2 A.

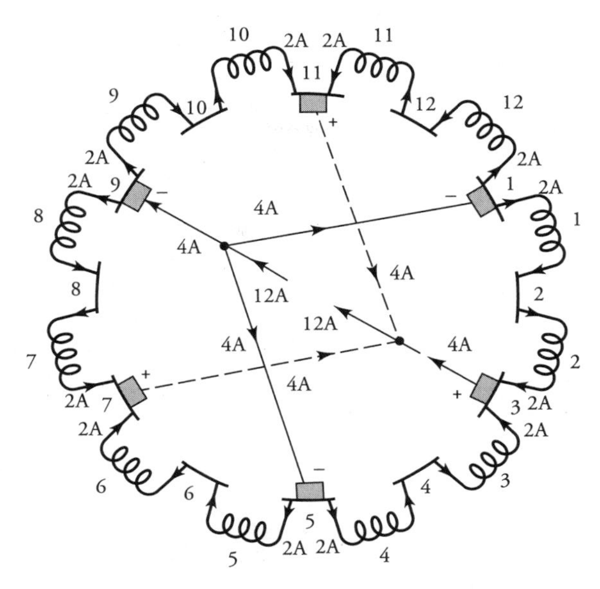

Figure 5.7 Brush connections and currents in six parallel paths of a lap-wound machine of Figure 5.6 when operated as a dc generator.

If we view each side of a coil as a conductor, then we can determine how these conductors are connected in the front (commutator side) and in the back (opposite side to the commutator). A part of the polar diagram of Figure 5.6 is shown in Figure 5.8. The coil sides have been numbered in the clockwise direction beginning with sides in slot 1. For example, the sides of coil 1 have been numbered 1 and 6, and those of coil 2, 3 and 8. Since sides 1 and 6 are connected on the back side, the **back pitch** (y_b) is 5. Side 6 of coil 1 and side 3 of coil 2 are connected to commutator segment 2. Thus, the **front pitch** (y_f) is 3. **The front pitch and the back pitch must both be odd for the winding to be properly placed in armature slots. The difference between the two pitches is always 2.** The winding is said to be **progressive** when $y_b = y_f + 2$. If $y_f = y_b + 2$, the winding is said to be **retrogressive**. A (progressive/retrogressive) winding advances in the (clockwise/counterclockwise) direction when viewed from the commutator side. In our example, the winding is progressive.

As viewed from the winding connections made to the commutator segments, we observe that the winding advances one commutator segment for each coil. Hence, the commutator pitch (y_c) is 1.

It is left to the reader to verify that a 4-pole, lap-wound machine needs four brushes and has four parallel paths. In summary, **the number of brushes and the parallel paths in a lap-wound machine are equal to its number of poles.**

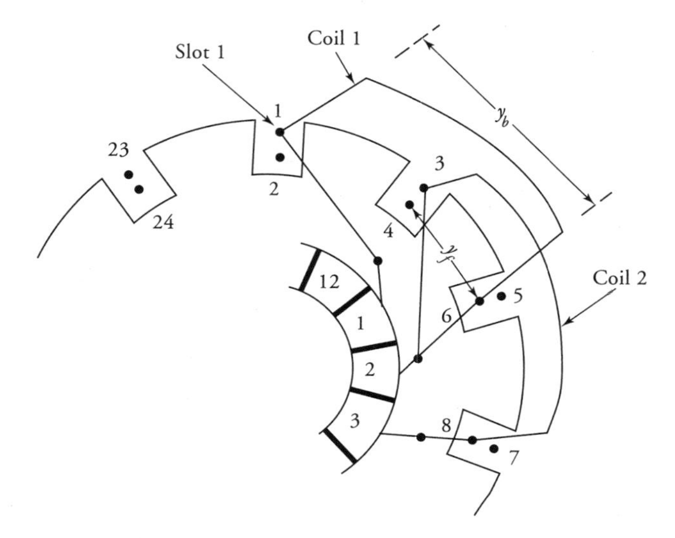

Figure 5.8 Connections of conductors in a lap-wound machine of Figure 5.6 showing front pitch and back pitch.

Wave Winding

The wave winding differs from the lap winding only in the way the coils are connected to the commutator segments. In the lap winding, the two ends of a coil are connected to adjacent commutator segments ($y_c = 1$). In the wave winding, the two ends of a coil are connected to those segments of the commutator that are approximately (but not exactly) 360° electrical apart (2-pole pitches). This is done to ensure that the entire winding closes onto itself only once. By making connections almost 2-pole pitches apart, we are connecting in series only those coils that are under the same polarity poles. That is, a coil under one north pole is connected to another coil comparably placed under the next north pole, and so on.

For the simplex-wave winding, the number of commutator segments per pole should be such that the following are true:

1. The commutator pitch can be either little more or less than 360° electrical.
2. After passing once around the commutator, the last coil should be either one segment ahead (progressive) or behind (retrogressive) the starting segment.

The above requirements dictate that the number of commutator segments per pair of poles should not be an integer. Since the commutator pitch must be an integer, the number of commutator segments for a simplex-wave winding must be determined from the following equation:

$$C = y_c\left(\frac{P}{2}\right) \pm 1 \tag{5.2}$$

where C is the total number of commutator segments, y_c is the commutator pitch (an integer), and P is the number of poles. The (plus/minus) sign is for the (progressive/retrogressive) winding.

The above equation can also be written as

$$y_c = \frac{C \pm 1}{\frac{P}{2}} \tag{5.3}$$

EXAMPLE 5.2

The commutator of a 6-pole machine has 35 segments. Determine the commutator pitch. Can the coils be connected using both retrogressive and progressive windings?

● SOLUTION

From Eq. (5.3), the commutator pitch is

$$y_c = \frac{35 \pm 1}{3} = 12,\ 11.33$$

Since the commutator pitch is an integer only when 1 is added to the commutator segments, the coils can be connected only using progressive windings.

■

In order to understand the placement of the coils and their connections using two-layer, simplex-wave winding, let us consider a 9-slot, 4-pole armature with a nine-segment commutator as shown in Figure 5.9. The coil pitch is 2 slots. The commutator pitch can be 4 segments (320° electrical for a retrogressive winding) or 5 segments (400° electrical for a progressive winding). We have selected a commutator pitch of 4 segments in Figure 5.9. This winding arrangement yields a back pitch y_b of 3 and a front pitch y_f of 5. The average pitch is the same as the commutator pitch.

For a dc generator rotating in the clockwise direction, the direction of the currents in the coils is as shown in the figure. The commutator segment 5 marks the position of a positive brush and the commutator segment 7, the negative brush. Therefore, we need only two brushes to make connections between the external circuit and the armature winding. When we trace the winding, we find that coil 1 is connected to commutator segments 1 and 5. Coil 1 can be depicted as a single loop by drawing segments 1 and 5 adjacent to each other as shown in Figure 5.10. It is now obvious that there are only two parallel paths: coils 1, 6, 2, and 7 form

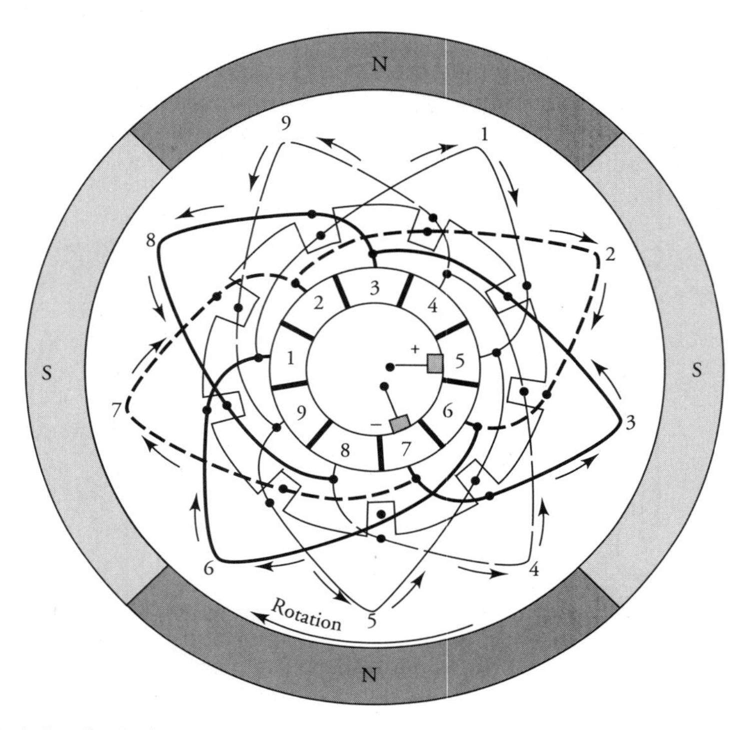

Figure 5.9 A 4-pole, 9-slot, wave-wound dc machine operating as a generator.

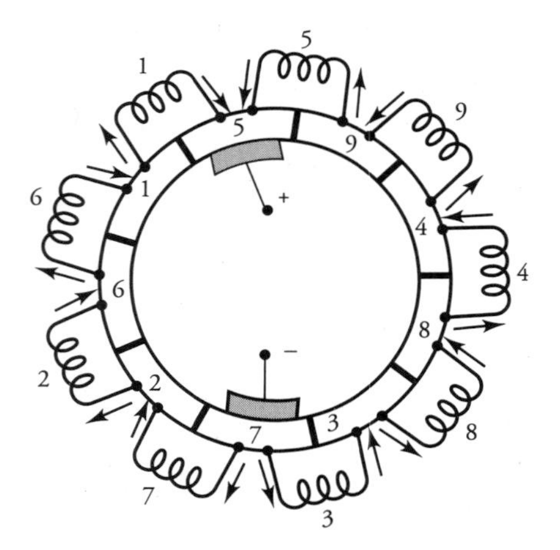

Figure 5.10 Rearrangement of commutator segments to show the number of parallel paths and the coils connected in series in each path.

one parallel path and coils 5, 9, 4, 8, and 3 constitute the other parallel path. We can, therefore, make a general statement for the wave-wound machine as follows: **Regardless of the number of poles, a wave winding requires only two brushes and has two parallel paths.**

Exercises

5.1. An armature of a 4-pole, lap-wound machine has 12 slots. What are the back pitch and the front pitch? Sketch the windings and show the brush positions. Also indicate the direction of the currents in all coils.

5.2. The commutator of a 6-pole machine has 34 segments. Determine the commutator pitch, the front pitch, and the back pitch. Can the coils be connected using both retrogressive and progressive wave windings?

5.3. The commutator of an 8-pole machine has 35 segments. Determine the commutator pitch, the front pitch, and the back pitch. Can the coils be connected using both retrogressive and progressive wave windings?

5.4. An armature of a 6-pole, wave-wound machine has 11 slots. What are the back pitch, the front pitch, and the commutator pitch? Sketch the windings and show the brush positions.

5.4 Induced emf Equation

When a single-turn coil rotates in a uniform magnetic field, the average value of the induced emf, from Section 3.4, is

$$E_c = \frac{P}{\pi} \Phi_P \omega_m \tag{5.4}$$

where P is the number of poles in a dc machine, Φ_P is the flux per pole, and ω_m is the angular velocity of the armature. We also obtained an expression for the frequency of the induced voltage in the coil as

$$f = \frac{P}{4\pi} \omega_m \tag{5.5}$$

From the above equation, we can obtain an expression for the armature speed in revolutions per minute (rpm) as

$$N_m = \frac{120f}{P} \tag{5.6}$$

From our discussion in the preceding sections, it must be obvious not only that a dc machine has many coils properly wound on the armature but that each coil usually has several turns. In addition, the coils are connected in parallel paths: two for the wave winding, and P for the lap winding.

Let N_c be the turns per coil, C the total number of coils (slots for a two-layer winding), and a the number of parallel paths ($a = 2$ for wave winding, or $a = P$ for the lap winding); then the total turns per parallel path are

$$N_a = \frac{C}{a} N_c \tag{5.7}$$

Note that N_a represents the turns connected in series between a negative brush and a positive brush. Consequently, the average value of the total emf induced between the terminals of the two brushes is

$$E_a = \frac{PC}{\pi a} N_c \omega_m \Phi_P \tag{5.8}$$

Since there are two conductors per turn, the total number of conductors, Z, in all the slots of the armature is

$$Z = 2CN_c \tag{5.9}$$

Expressing Eq. (5.8) in terms of the total conductors in the armature slots, we obtain

$$E_a = \frac{PZ}{2\pi a} \Phi_P \omega_m$$

The above equation for the induced emf in the armature winding is traditionally written as

$$E_a = K_a \Phi_P \omega_m \tag{5.10}$$

where

$$K_a = \frac{ZP}{2\pi a} \tag{5.11}$$

is a constant quantity for a given machine and is routinely referred to as the **machine constant**. Equation (5.10) is valid for both dc generators and motors. In the case of a dc generator, E_a is known as the **generated emf**, or **generated voltage**. It is called the **back emf** when the dc machine operates as a motor.

In the above development, we have tacitly assumed that the magnetic poles cover the entire armature periphery. That is, a pole arc subtends an angle of 180° electrical. This is a virtual impossibility, especially for a wound machine. However, we can design machines in which the poles cover 60% to 80% of the armature periphery. This fact can be taken into consideration when computing the flux per pole.

If the armature of a dc generator supplies a constant current I_a to an external load, the electrical power developed by the generator is

$$P_d = E_a I_a = K_a \Phi_P \omega_m I_a \tag{5.12}$$

An equivalent power must be supplied by the mechanical system (the prime mover coupled to the armature). If T_d is the average mechanical torque developed by the armature of a dc generator, the prime mover must supply an equal amount of torque in the opposite direction to keep the armature rotating at a constant speed ω_m. Since in a mechanical system the power developed is

$$P_d = T_d \omega_m \tag{5.13}$$

the torque developed, from Eq. (5.12), is

$$T_d = K_a \Phi_P I_a \tag{5.14}$$

Equation (5.13) is also valid for a dc motor in which the electrical power supplied to the armature (P_d) must be balanced by the mechanical power ($T_d\omega_m$) delivered to the load. Equation (5.13), therefore, symbolizes the transition from mechanical power to electrical power in a dc generator, or vice versa in a dc motor. In the next section, we arrive at the torque expression, Eq. (5.14), from a different perspective.

EXAMPLE 5.3

A 24-slot, 2-pole dc machine has 18 turns per coil. The average flux density per pole is 1 T. The effective length of the machine is 20 cm, and the radius of the armature is 10 cm. The magnetic poles are designed to cover 80% of the armature periphery. If the armature angular velocity is 183.2 rad/s, determine (a) the induced emf in the armature winding, (b) the induced emf per coil, (c) the induced emf per turn, and (d) the induced emf per conductor.

● SOLUTION

For a two-layer winding, the number of coils is the same as the number of armature slots. That is, $C = 24$. N_c is given to be 18.

Thus, the total armature conductors: $Z = 2 \times 24 \times 18 = 864$
For a 2-pole machine: $a = 2$
The actual pole area is

$$A_P = \frac{2\pi r L}{P} = \frac{2\pi \times 0.1 \times 0.2}{2} = 0.063 \text{ m}^2$$

and the effective pole area is

$$A_e = 0.063 \times 0.8 = 0.05 \text{ m}^2$$

Thus, the effective flux per pole is

$$\Phi_P = BA_e = 1 \times 0.05 = 0.05 \text{ Wb}$$

(a) From Eq. (5.11), the machine constant is

$$K_a = \frac{2 \times 864}{2\pi \times 2} = 137.51$$

Hence, the induced emf in the armature winding, from Eq. (5.10), is

$$E_a = 137.51 \times 0.05 \times 183.2 = 1259.6 \text{ V}$$

(b) Since there are two parallel paths, the number of coils in each path is $24/2 = 12$. Thus, the induced emf per coil is

$$E_{\text{coil}} = \frac{1259.6}{12} = 104.97 \text{ V}$$

(c) As there are 18 turns in each coil, the induced emf per turn is

$$E_{\text{turn}} = \frac{104.97}{18} = 5.83 \text{ V}$$

(d) Finally, the induced emf per conductor is

$$E_{\text{cond}} = \frac{5.83}{2} = 2.92 \text{ V}$$

■

Exercises

5.5. A 6-pole dc machine has 360 conductors in its armature slots. Each magnetic pole subtends an arc of 20 cm and has a length of 20 cm. The flux density per pole is 0.8 T. The armature speed is 900 rpm. Determine the induced emf in the armature if the machine has (a) a lap winding and (b) a wave winding.

5.6. The induced emf in a PM machine is 440 V at 1600 rpm. What is the induced emf if the speed is changed to (a) 800 rpm? (b) 1200 rpm? (c) 2000 rpm? (d) 3200 rpm?

5.7. The armature of a 4-pole, lap-wound dc generator has 28 slots with ten conductors in each slot. The flux per pole is 0.03 Wb and the armature speed is 1200 rpm. Calculate (a) the frequency of the induced emf in each armature conductor, (b) the induced emf in the armature, (c) the induced emf per coil, (d) the induced emf per turn, and (e) the induced emf per conductor.

5.8. When the generator of Exercise 5.7 is connected to the load, the current in each conductor is 2 A. What is the armature current? What are the power and torque developed by the generator?

5.5 Developed Torque

In Section 3.4, we obtained an expression for the torque experienced by a single-turn, current-carrying coil in a uniform magnetic field as

$$T_e = 2BiLr \tag{5.15}$$

where B is the uniform flux density, i is the current in the coil, L is the effective length of each conductor of the coil that is exposed to the magnetic field, and r is the radius at which each conductor is located. In the case of a dc machine, L is the length (stack height) of the armature and r is its radius.

Although the current in the coil varies sinusoidally, the current in a conductor under a magnetic pole is of the full-wave rectified type. If I_c is the average value of the current, then the average torque acting on a single-turn coil is

$$T_c = 2BLrI_c \tag{5.16}$$

Since a dc machine with P poles has C coils connected in parallel paths and each coil has N_c turns, the total number of turns is CN_c. If I_a is the average dc current, then I_a/a is the current in each turn. Hence, the average torque developed by a dc machine is

$$T_d = \text{Torque developed by one turn} \times \text{total turns}$$

$$= \left[2BLr\frac{I_a}{a}\right][CN_c]$$

$$= \frac{BLrZ}{a} I_a$$

If A_P is the area of each pole, that is,

$$A_P = \frac{2\pi rL}{P}$$

then the torque developed by the dc machine becomes

$$T_d = \frac{PZ}{2\pi a} BA_P I_a$$

$$= K_a \Phi_P I_a \tag{5.17}$$

where

$$K_a = \frac{PZ}{2\pi a}$$

is the machine constant and $\Phi_P = BA_P$ is the total flux per pole. Equation (5.17) is exactly the same as Eq. (5.14), which was obtained from the power developed point of view.

EXAMPLE 5.4

If the armature current of the machine in Example 5.3 is 25 A, determine (a) the current in each conductor, (b) the torque developed, and (c) the power developed.

● SOLUTION

(a) Since there are two parallel paths, the (average) current per conductor is 12.5 A.

(b) The (average) torque developed by the machine, from Eq. (5.17), is

$$T_d = 137.51 \times 0.05 \times 25 = 171.89 \text{ N·m}$$

(c) The (average) power developed is

$$P_d = E_a I_a = 1259.6 \times 25 = 31{,}490 \text{ W} \quad \text{or} \quad 31.49 \text{ kW}$$

The power developed can also be computed as

$$P_d = T_d \omega_m = 171.89 \times 183.2 \approx 31{,}490 \text{ W} \quad \text{or} \quad 31.49 \text{ kW}$$

■

Exercises

5.9. Determine the torque developed by a 6-pole dc machine with 300 conductors arranged in a lap winding. The flux per pole is 0.3 Wb. The conductor current is 12 A. If the armature speed is 600 rpm, what is the power developed? What is the induced emf?

5.10. The armature of a 6-pole dc machine has 126 slots and is wound with five turns per coil using lap winding. The induced emf is 440 V at 120 rad/s. Determine the flux per pole. If the conductor current is 5 A, what is the torque developed? Also determine the average power developed by the machine.

5.11. The wave-wound armature of an 8-pole dc generator has 95 coils. Each coil has five turns. The flux per pole is 0.5 Wb. The armature speed is 600 rpm. The current per conductor is 20 A. Determine (a) the power developed and (b) the torque developed.

5.6 Magnetization Characteristic of a DC Machine

The induced emf in the armature winding of a dc machine is directly proportional to (a) the flux per pole and (b) the armature speed. Let us assume that the field winding of a dc machine is connected to a variable dc source that can supply a desired amount of constant current. If the armature circuit is left open and the armature is rotated at the rated speed of the machine, then the induced emf in the machine can be expressed as

$$E_a = K_1 \Phi_P \tag{5.18}$$

where $K_1 = K_a \omega_m$ is a constant quantity. In other words, the induced emf is directly proportional to the flux in the machine.

The flux per pole Φ_P depends upon the magnetomotive force (mmf) provided by the current I_f in the field winding. Since the number of turns per pole is fixed, the flux per pole is a function of the field current I_f. That is,

$$\Phi_P = k_f I_f \tag{5.19}$$

where k_f is a constant of proportionality.

The induced emf can now be written as

$$E_a = K_1 k_f I_f \tag{5.20}$$

Since the magnetic circuit of a dc machine consists of both linear regions (air-gaps) and nonlinear regions (magnetic material for the stator and the armature), the value of k_f changes with the change in the flux (or flux density) in the machine. To be precise, its value decreases as the flux in the machine increases. Simply stated, the induced emf E_a does not vary linearly with the field current I_f.

The relationship between E_a and I_f can, however, be determined by measuring E_a at different values of I_f at a constant (usually rated) armature speed. When E_a is plotted as a function of I_f (Figure 5.11), the curve is known as the **no-load characteristic** owing to the fact that the armature is not loaded. Since E_a is an indirect measure of flux (or flux density) per pole and I_f is a measure of the applied mmf (ampere-turns per pole), the curve is bound to be similar to the *B-H* curve of the magnetic material. For that reason, the no-load curve is frequently referred to as the **magnetization curve (characteristic) of a dc machine**.

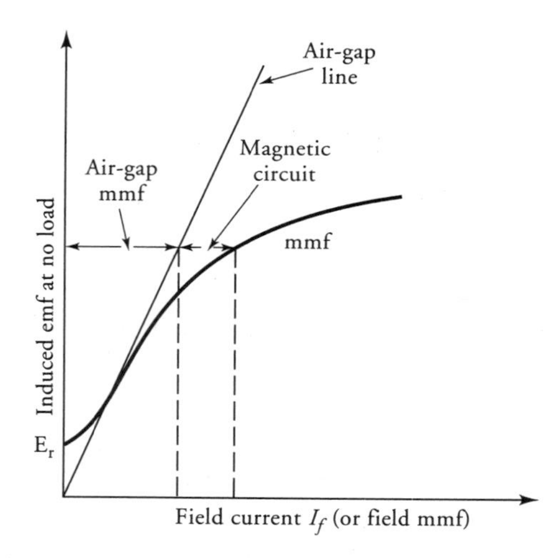

Figure 5.11 Magnetization (no-load) characteristic of a dc machine.

Magnetization curves can be experimentally determined for both the increasing (ascending) and decreasing (descending) values of the field current. The two curves should not be expected to overlap owing to hysteresis. In other words, for any given value of the field current, the flux in the machine depends upon whether the current was increasing or decreasing to reach the value in question. Therefore, **during the experiment the field current should be varied continuously in one direction only.** In order to simplify theoretical calculations involving the use of the magnetization curve, it is an accepted practice to take the mean of the two curves and refer to it as the magnetization curve.

The induced emf does not start at zero when the field current is zero but at some value somewhat greater owing to the residual magnetism from the previous operation of the machine. This value of the induced emf is called the **residual emf**, E_r. Except for that, the lower part of the magnetization curve is practically a straight line. This is due to the fact that the mmf required for the magnetic material is almost negligible at small values of flux density in the machine. Stated differently, most of the reluctance of the magnetic circuit is in the air-gap. The upper part of the curve shows the aftermath of saturation of the magnetic material when the flux density in the machine is high. The bending of the curve to the right provides evidence that part of the applied mmf is being consumed by the reluctance of the magnetic material. The rest of the mmf is, of course, necessary to set up the flux in the air-gap. By drawing an air-gap line, a line passing through the origin and tangent to the magnetization curve, we can always determine the needed mmf of the air-gap to set up a given flux (to induce a given emf). The remainder of the mmf is required by the magnetic material.

Since the induced emf is directly proportional to the armature speed, we can plot the magnetization curve at any speed by making use of the magnetization curve at the rated speed. If E_{a1} is the induced emf at a field current of I_f when the armature is rotating at a speed of ω_{m1}, then the induced emf E_{a2} at the same field current but at a speed of ω_{m2} is

$$E_{a2} = \frac{\omega_{m2}}{\omega_{m1}} E_{a1} \tag{5.21}$$

For the operation of all self-excited generators, we will show that the saturation of the magnetic material is, in fact, a blessing in disguise. In other words, the bending of the magnetization curve to the right is necessary for the successful operation of a self-excited generator.

5.7 Theory of Commutation

For the successful operation of a dc machine, the induced emf in each conductor under a pole must have the same polarity. If the armature winding is carrying current, the current in each conductor under a pole must be directed in the same direction. It implies that as the conductor moves from one pole to the next, there must be a reversal of the current in that conductor. The conductor and thereby the coil in which the current reversal is taking place are said to be **commutating**. The process of reversal of current in a commutating coil is known as **commutation**.

Ideally, the process of commutation should be instantaneous, as indicated in Figure 5.12**a**. This can, however, be achieved only if the brush width and the commutator segments are infinitesimally small. In practice, not only do the brush and the commutator have finite width but the coil also has a finite inductance. Therefore, it takes some time for the current reversal to take place, as illustrated in Figure 5.12**b**.

Figure 5.13**a** shows a set of eight coils connected to the commutator segments of a 2-pole dc generator. The coils g, h, a, and b are under the north pole and form one parallel path, while the coils c, d, e, and f are under the south pole and form another parallel path. The current in the coils under the north pole, therefore, is in the opposite direction to the current in the coils under the south pole. However, the magnitude of the current in each coil is I_c. The width of each brush is assumed to be equal to the width of the commutator segment. Brush A is riding on commutator segment 3 while brush B is on segment 7. The current through each brush is $2I_c$. As the commutator rotates in the clockwise direction, the leading tip of

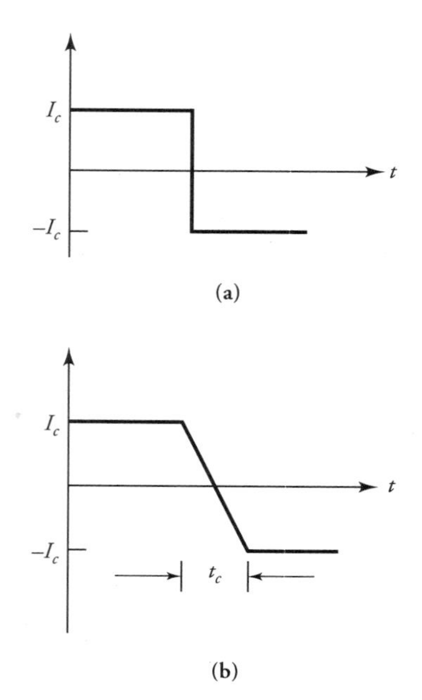

Figure 5.12 Current reversal in a coil undergoing commutation with brushes of **(a)** infinitesimal thickness and **(b)** finite thickness.

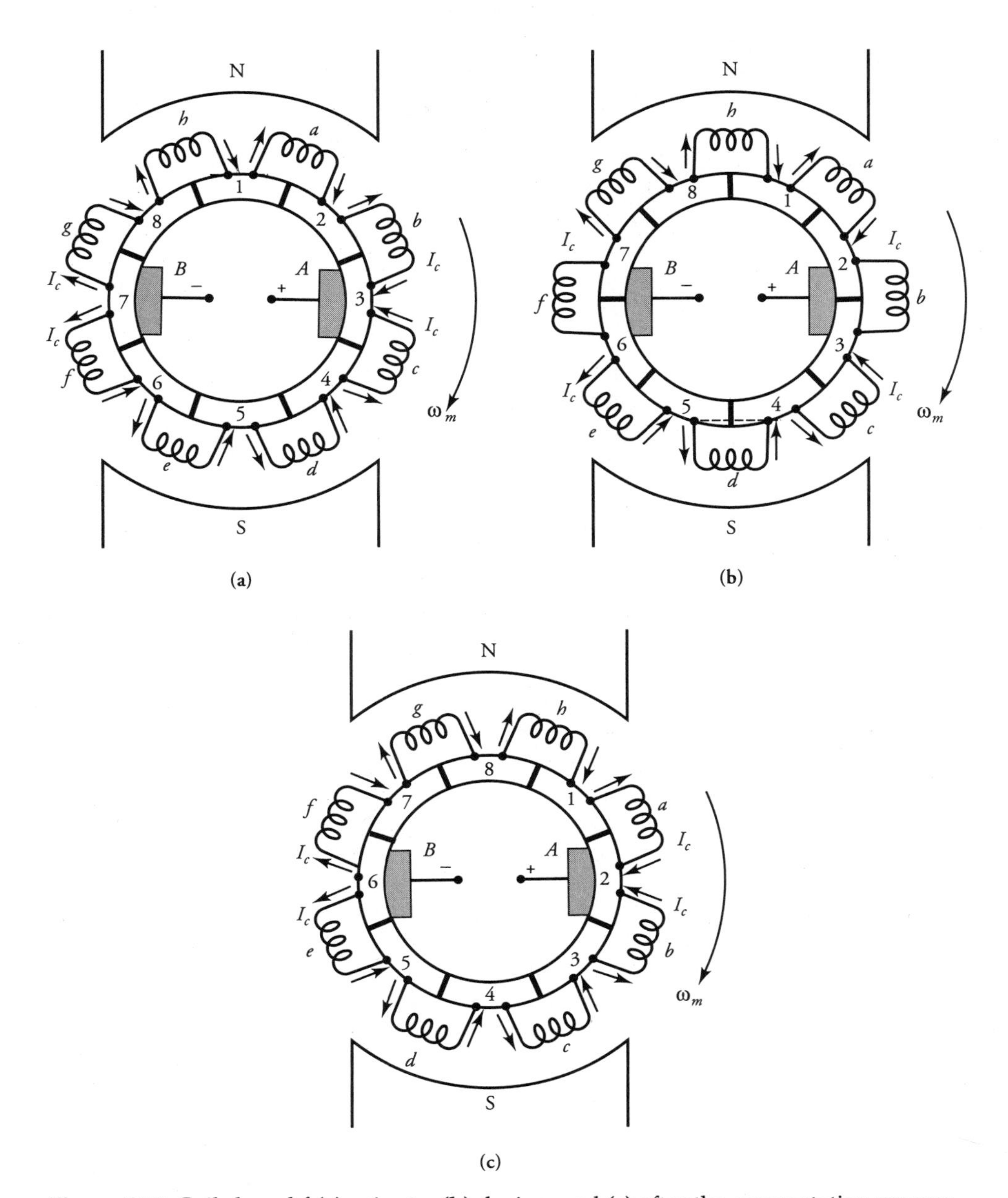

Figure 5.13 Coils b and f **(a)** prior to, **(b)** during, and **(c)** after the commutation process.

brush A comes in contact with commutator segment 2 and shorts coil b as indicated in Figure 5.13**b**. Similarly, the coil f is also short-circuited by brush B. The coils b and f are now undergoing commutation. From Figure 5.13**b** it is evident that the current through each brush is still $2I_c$. At this instant, the induced emfs in coils b and f are zero because each lies in the plane perpendicular to the flux. However, an instant later, the contacts of brushes A and B with commutator segments 3 and 7 are broken as depicted in Figure 5.13**c**. Coil b is now a part of the parallel path formed by coils c, d, and e, and the direction of the current in the coil has reversed. Similarly, coil f has become a part of the parallel path formed by coils a, h, and g, and its current has also reversed its direction. The commutation process for coils b and f is complete. Coils a and e are now ready for commutation. In a multipole machine, the number of coils undergoing commutation at any instant is equal to the number of parallel paths when the brush width is the same as the width of the commutator segment.

For a commutation process to be perfect, the reversal of current from its value in one direction to an equal value in the other direction must take place during the time interval t_c. Otherwise, the excess current (difference of the currents in coils b and c) prompts a flashover from commutator segment 3 to the trailing tip of brush A. Likewise, a flashover also takes place from commutator segment 7 to the trailing tip of brush B.

When the current reverses its direction during commutation in a straight-line fashion as illustrated in Figure 5.12**b**, the commutation process is said to be linear. A **linear commutation** process is considered to be ideal in the sense that no flashover occurs from the commutator segments to the trailing tips of the brushes. In the next section, we discuss methods that allow us to improve commutation under varying loads.

5.8 Armature Reaction

When there is no current in the armature winding (a no-load condition), the flux produced by the field winding is uniformly distributed over the pole faces as shown in Figure 5.14**a** for a 2-pole dc machine. The induced emf in a coil that lies in the **neutral plane**, a plane perpendicular to the field-winding flux, is zero. This, therefore, is the neutral position under no load where the brushes must be positioned for proper commutation.

Let us now assume that the 2-pole dc machine is driven by a prime mover in the clockwise direction and is, therefore, operating as a generator. The direction of the currents in the armature conductors under load is shown in Figure 5.14**b**. The armature flux distribution due to the armature mmf is shown in Figure 5.14**b**. The flux distribution due to the field winding is suppressed in order to highlight the flux distribution due to the armature mmf. Note that the magnetic axis of the armature flux (the quadrature, or q-axis) is perpendicular to the magnetic axis of the field-winding flux (direct, or d-axis). Since both fluxes exist at the same time

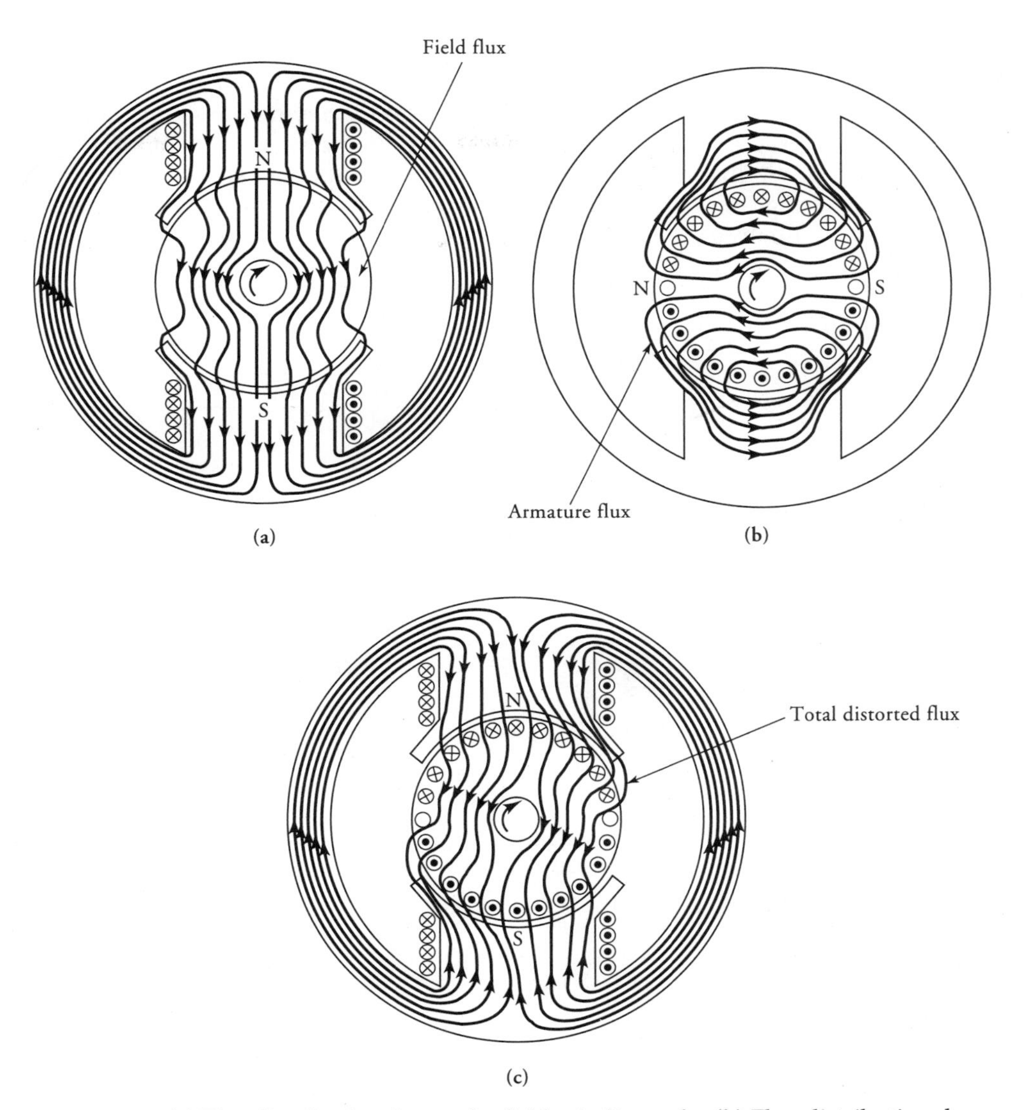

Figure 5.14 **(a)** Flux distribution due to the field-winding only. **(b)** Flux distribution due to armature mmf only. **(c)** Flux distribution due to the field-winding and the armature mmf.

when the armature is loaded, the resultant flux is distorted as indicated in Figure 5.14**c**. The armature flux has weakened the flux in one-half of the pole and has strengthened it in the other half. The armature current has, therefore, displaced the magnetic-field axis of the resultant flux in the direction of rotation of the generator. As the neutral plane is perpendicular to the resultant field, it has also advanced. The effect of the armature mmf upon the field distribution is called the **armature reaction**. We can get a better picture of what is taking place in the generator by looking at its developed diagram.

The developed diagram of the flux per pole under no load is shown in Figure 5.15**a**. In order to simplify the discussion, let us assume that the conductors are uniformly distributed over the surface of the armature. Then the ar-

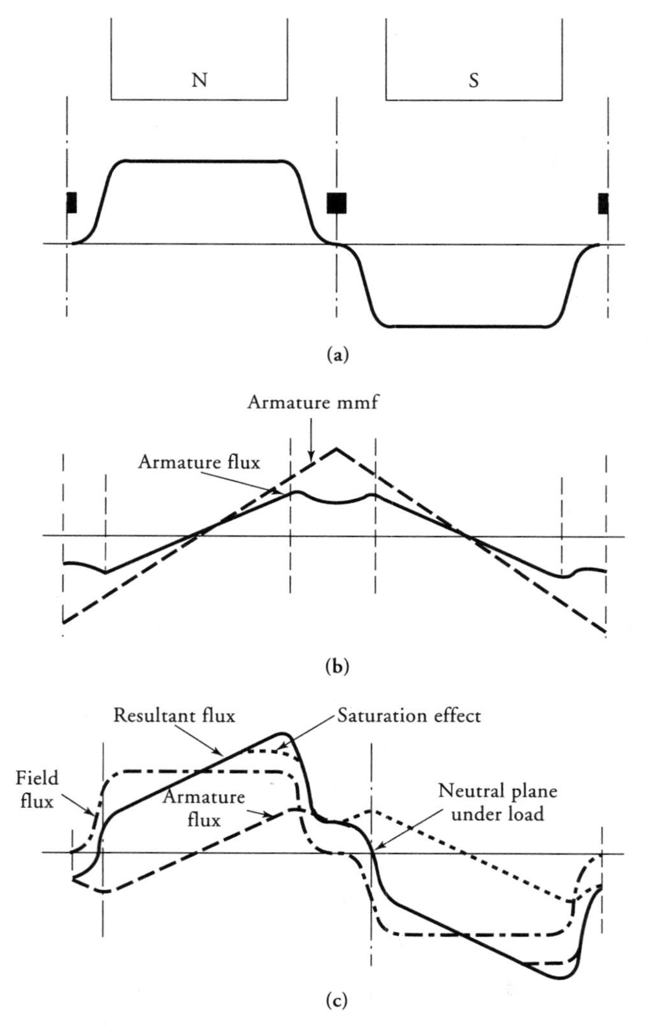

Figure 5.15 (a) Flux per pole under no load. **(b)** MMF and the flux due to armature reaction. **(c)** Resultant flux.

mature mmf under load has a triangular waveform as depicted in Figure 5.15**b**. The flux distribution due to the armature mmf is also a straight line under the pole. If the pole arc is less than 180° electrical, the armature flux has a saddle-shaped curve in the interpolar region due to its higher reluctance. The resultant (total) flux distribution is shown in Figure 5.15**c**. The distortion in the flux distribution at load compared with that at no load is evident. If saturation is low, the decrease in the flux in one-half of the pole is accompanied by an equal increase in the flux in the other half. The net flux per pole, therefore, is the same under load as at no load. On the other hand, if the poles were already close to the saturation point under no load, the increase in the flux is smaller than the decrease, as indicated by the dotted line in Figure 5.15**c**. In this case, there is a net loss in the total flux. For a constant armature speed, the induced emf in the armature winding decreases owing to the decrease in the flux when the armature is loaded.

As mentioned earlier, the neutral plane in the generator moves in the direction of rotation as the armature is loaded. Because the neutral zone is the ideal zone for the coils to undergo commutation, the brushes must be moved accordingly. Otherwise, the forced commutation results in excessive sparking. As the armature flux varies with the load, so does its influence on the flux set up by the field winding. Hence, the shift in the neutral plane is a function of the armature current.

The armature reaction has a demagnetizing effect on the machine. The reduction in the flux due to armature reaction suggests a substantial loss in the applied mmf per pole of the machine. In large machines, the armature reaction may have a devastating effect on the machine's performance under full load. Therefore, techniques must be developed to counteract its demagnetization effect. Some of the measures that are being used to combat armature reaction are summarized below:

1. The brushes may be advanced from their neutral position at no load (geometrical neutral axis) to the new neutral plane under load. This measure is the least expensive. However, it is useful only for constant-load generators.
2. **Interpoles**, or **commutating poles** as they are sometimes called, are narrow poles that may be located in the interpolar region centered along the mechanical neutral axis of the generator. The interpole windings are permanently connected in series with the armature to make them effective for varying loads. The interpoles produce flux that opposes the flux due to the armature mmf. When the interpole is properly designed, the net flux along the geometric neutral axis can be brought to zero for any load. Because the interpole winding carries armature current, we need only a few turns of comparatively heavy wire to provide the necessary interpole mmf. Figure 5.16 shows the armature flux distribution with interpoles.
3. Another method to nullify the effect of armature reaction is to make use of **compensating windings**. These windings, which also carry the armature current, are placed in the shallow slots cut in the pole faces as shown in

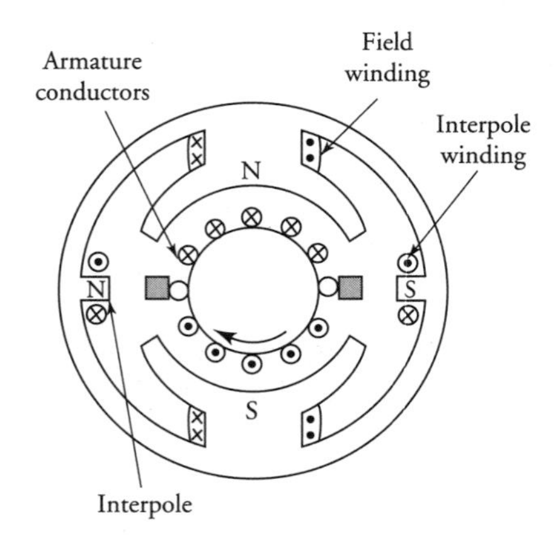

Figure 5.16 Interpole windings of a dc generator.

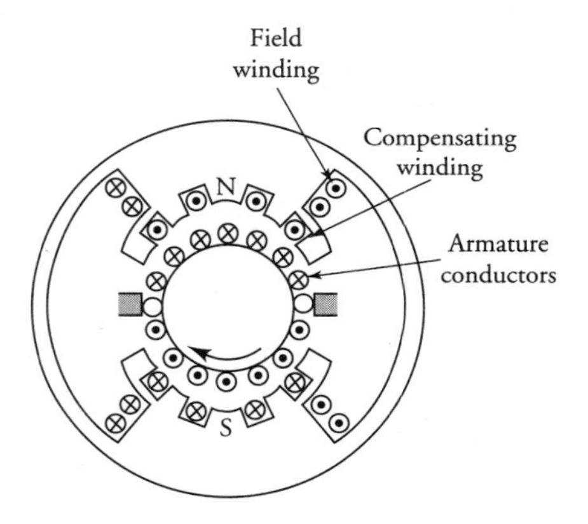

Figure 5.17 Compensating winding of a dc generator.

Figure 5.17. Once again, the flux produced by the compensating winding is made equal and opposite to the flux established by the armature mmf.

5.9 Types of DC Generators

Based upon the method of excitation, dc generators can be divided into two classes: Separately excited generators and self-excited generators. A PM generator can be considered a separately excited generator with constant magnetic flux.

The field (excitation) current in a separately excited generator is supplied by an external source of power. However, a self-excited generator provides its own excitation current. According to the method of connection of the field winding(s), a self-excited generator can be further classified as (a) a **shunt generator** if its field winding, called the shunt field winding, is connected across the armature terminals, (b) a **series generator** when its field winding, called the series field winding, is connected in series with the armature, and (c) a **compound generator** with both shunt and series field windings.

We investigate the operation of each type of dc generator by examining its external characteristics under steady-state conditions. The **external characteristic** of a dc generator is the variation of the load voltage (**terminal voltage**) with load current. This information enables us to highlight the application for which each generator is most suitable.

During our discussion of dc generators we should keep in mind the following facts:

1. The generator is driven by a prime mover, such as a synchronous motor, at a constant speed.
2. The induced emf in the armature winding is proportional to the flux in the machine; that is, $E_a = K_a \Phi_P \omega_m$.
3. The armature terminals are connected to the load. Thus, the armature current supplies the load current.
4. The armature winding has finite resistance, however small it may be. Therefore, the armature terminal voltage is bound to be lower than the induced emf.
5. If the generator is not compensated for the armature reaction, there is less overall flux in the machine under load than at no load. Thus, the induced emf is lower under load than at no load. This results in a further decrease of the terminal voltage.
6. The torque developed by the armature conductors, $T_d = K_a \Phi_P I_a$, is equal and opposite to the torque applied by the prime mover. That is, the torque developed **opposes** the armature rotation (Lenz's law).
7. The voltage drop between the brushes and the commutator segments is known as the **brush-contact** drop. If needed and not specified, it may be assumed to be about 2 V.
8. If the pertinent information regarding the adverse effect of armature reaction on generator performance is not known, we assume that either the armature reaction is negligible or the generator is appropriately compensated for it.
9. We commonly use a term called **load** in electrical machines to signify the load-current. Thus, "no load" means an open circuit and "full load" implies the rated load-current.

5.10 Voltage Regulation

As the load-current increases, the terminal voltage decreases owing to the increase in the voltage drop across the armature-winding resistance as well as the demagnetization effect of the armature reaction. The voltage regulation is a measure of the terminal voltage drop at full load. If V_{nL} is the no-load terminal voltage and V_{fL} is the full-load terminal voltage, the **voltage regulation** is defined as

$$VR\% = \frac{V_{nL} - V_{fL}}{V_{fL}} \times 100 \tag{5.22}$$

where $VR\%$ is the percent voltage regulation. For an ideal (constant-voltage) generator, the voltage regulation should be zero. The voltage regulation is considered positive when the terminal voltage at no load is higher than at full load. A negative voltage regulation indicates that the terminal voltage at full load is higher than that at no load.

5.11 Losses in DC Machines

Once again, we are using the term **machine** in the discussion of power losses owing to the fact that no distinction need be made between the losses in the dc generator and the motor. The law of conservation of energy dictates that the input power must always be equal to the output power plus the losses in the machine. There are three major categories of losses: mechanical losses, copper losses, and magnetic losses.

Mechanical Losses

Mechanical losses are the result of (a) the friction between the bearings and the shaft, (b) the friction between the brushes and the commutator, and (c) the drag on the armature caused by air enveloping the armature (windage loss).

The bearing-friction loss depends upon the diameter of the shaft at the bearing, the shaft's peripheral speed, and the coefficient of friction between the shaft and the bearing. To reduce the coefficient of friction, the bearings are usually lubricated.

The brush-friction loss depends upon the peripheral speed of the commutator, the brush pressure, and the coefficient of friction between the brush and the commutator. The graphite in the brush helps provide lubrication to lessen the coefficient of friction.

The windage loss depends upon the peripheral speed of the armature, the number of slots on its periphery, and its length.

Mechanical losses can be determined by rotating the armature of an unexcited machine at its rated speed by coupling it to a calibrated motor. Becuase there is no power output, the power supplied to the armature is the mechanical loss. The mechanical loss is denoted P_{fw}.

Magnetic Loss

Since the induced emf in the conductors of the armature alternates with a frequency determined by the speed of rotation and the number of poles, a magnetic loss (hysteresis and the eddy-current) exists in the armature. Let us denote the magnetic loss P_m.

The hysteresis loss depends upon the frequency of the induced emf, the area of the hysteresis loop, the magnetic flux density, and the volume of the magnetic material. The area of the hysteresis loop is smaller for soft magnetic materials than for hard magnetic materials. This is one of the reasons why soft magnetic materials are used for electrical machines.

Although the armature is built using thin laminations, the eddy currents do appear in each lamination and produce eddy-current loss. The eddy-current loss depends upon the thickness of the lamination, the magnetic flux density, the frequency of the induced emf, and the volume of the magnetic material.

A considerable reduction in the magnetic loss can be obtained by operating the machine in the linear region at a low flux density but at the expense of its size and initial cost.

Rotational Losses

In the analysis of a dc machine, it is a common practice to lump the mechanical loss and the magnetic loss together. The sum of the two losses is called the **rotational loss, P_r**. That is, $P_r = P_{fw} + P_m$.

The rotational loss of a dc machine can be determined by running the machine as a separately excited motor (to be discussed later) under no load. The armature winding voltage should be so adjusted that the induced emf in the armature winding is equal to its rated value, E_a. If V_t is the terminal voltage and R_a is the armature-winding resistance, then the voltage that must be applied to the armature terminals is

$$V_a = V_t + I_a R_a \tag{5.23a}$$

for the generator and

$$V_a = V_t - I_a R_a \tag{5.23b}$$

for the motor.

Apply V_a across the armature terminals and adjust the field excitation until the machine rotates at its rated speed. Then measure the armature current. Because the armature current under no load is a small fraction of its rated value and the armature winding resistance is usually very small, we can neglect the power loss in the armature winding. As there is no power output, the power supplied to the motor, $V_a I_a$, must be equal to the rotational loss in the machine. By subtracting the mechanical loss, we can determine the magnetic loss in the machine.

Copper Losses

Whenever a current flows in a wire, a copper loss, P_{cu}, is associated with it. The copper losses, also known as electrical or I^2R losses, can be segregated as follows:

1. Armature-winding loss
2. Shunt field–winding loss
3. Series field–winding loss
4. Interpole field–winding loss
5. Compensating field–winding loss

Stray-Load Loss

A machine always has some losses that cannot be easily accounted for; they are termed **stray-load losses.** It is suspected that the stray-load losses in dc machines are the result of (a) the distorted flux due to armature reaction and (b) short-circuit currents in the coils undergoing commutation. As a rule of thumb, the stray-load loss is assumed to be 1% of the power output in large machines (above 100 horsepower) and can be neglected in small machines.

Power-Flow Diagram

In a dc generator, the mechanical energy supplied to the armature by the prime mover is converted into electric energy. At the outset, some of the mechanical energy is lost as the rotational loss. The mechanical power that is available for conversion into electrical power is the difference between the power supplied to the shaft and the rotational loss. We refer to the available power as **developed power**. Subtract all the copper losses in the machine from the developed power to obtain the power output. If T_s is the torque at the shaft and ω_m is the angular velocity of rotation, then the power output, P_o, for a generator is

$$P_o = T_s \omega_m - P_r - P_{cu} \tag{5.24}$$

A typical power-flow diagram for the generator is shown in Figure 5.18.

Efficiency

The efficiency of a machine is simply the ratio of its power output to the power input. In the case of a separately excited machine, the power lost in the field

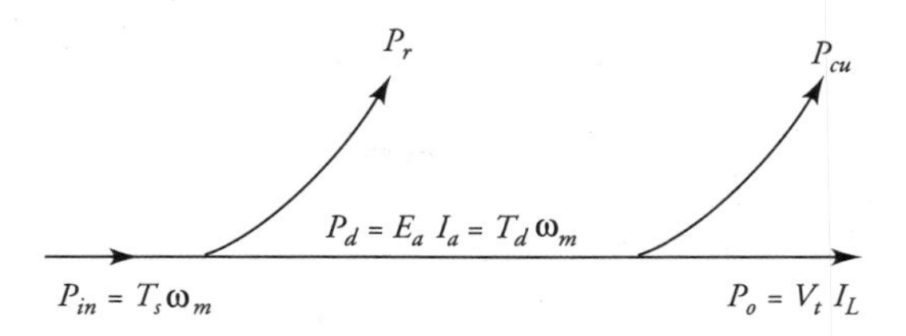

Figure 5.18 Power-flow diagram of a dc generator.

winding may also be included in the input power when computing the efficiency of the machine.

5.12 A Separately Excited DC Generator

As the name suggests, a separately excited dc generator requires an external source of power for the field winding and for that reason is used mainly in (a) laboratory and commercial testing and (b) special regulation sets. The external power source can be another dc generator, a controlled or uncontrolled rectifier, or simply a battery.

The equivalent circuit representation under steady-state condition of a separately excited dc generator is given in Figure 5.19. The steady-state condition implies that no appreciable change occurs in either the armature current or the armature speed for a given load. In other words, there is essentially no change in the mechanical energy or the magnetic energy of the system. Therefore, there is no need to include the inductance of each winding and the inertia of the system as part of the equivalent circuit. We include these effects when we discuss the dynamics of electric machines in Chapter 11.

In the equivalent circuit, E_a is the induced emf in the armature winding, R_a is the effective armature-winding resistance which may also include the resistance of each brush, I_a is the armature current, V_t is the terminal output voltage, I_L is

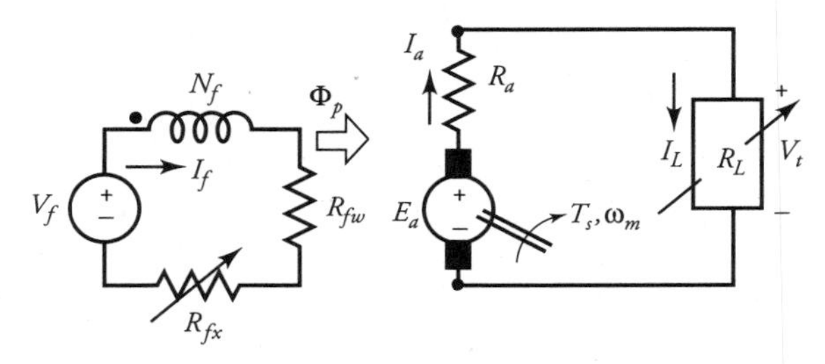

Figure 5.19 An equivalent circuit of a separately excited dc generator.

the load current, I_f is the field-winding current, R_{fw} is the field-winding resistance, R_{fx} is an external resistance added in series with the field winding to control the field current, N_f is the number of turns per pole for the field winding, and V_f is the source voltage of an external source.

The defining equations under steady-state operation are

$$V_f = I_f(R_{fw} + R_{fx}) = I_f R_f \tag{5.25}$$

$$E_a = V_t + I_a R_a \tag{5.26}$$

and

$$I_L = I_a \tag{5.26a}$$

where $R_f = R_{fw} + R_{fx}$ is the total resistance in the shunt field–winding circuit.

From Eq. (5.26), the terminal voltage is

$$V_t = E_a - I_a R_a \tag{5.27}$$

When the field current is held constant and the armature is rotating at a constant speed, the induced emf in an ideal generator is independent of the armature current, as shown by the dotted line in Figure 5.20. As the load current I_L increases, the terminal voltage V_t decreases, as indicated by the solid line. In the absence of the armature reaction, the decrease in V_t should be linear and equal to the voltage drop across R_a. However, if the generator is operating near its saturation and is not properly compensated for the armature reaction, the armature reaction causes a further drop in the terminal voltage.

A plot of the terminal voltage versus the load current is called the **external (terminal) characteristic** of a dc generator. The external characteristic can be obtained experimentally by varying the load from no load to as high as 150% of the

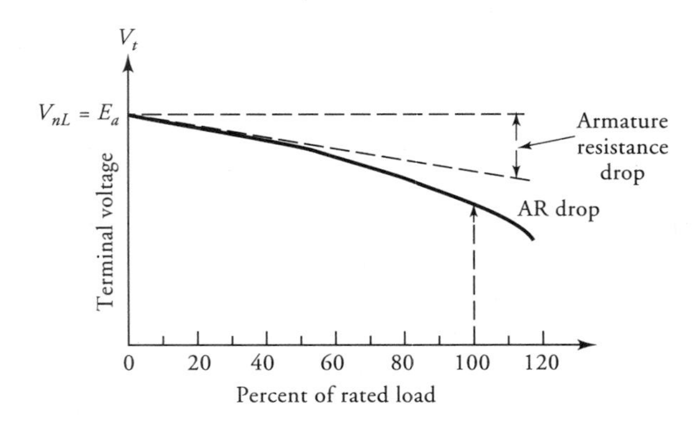

Figure 5.20 The external characteristic of a separately excited dc generator.

rated load. The terminal voltage at no load, V_{nL}, is simply E_a. If we draw a line tangent to the curve at no load, we obtain the terminal characteristic of the machine with no armature reaction. The difference between the no-load voltage and the tangent line yields $I_a R_a$ drop. Since I_a is known, we can experimentally determine the effective armature-winding resistance. The term "effective" signifies that it is not only the resistance of the armature winding but also includes the brush-contact resistance.

EXAMPLE 5.5

A 240-kW, 240-V, 6-pole, 600-rpm, separately excited generator is delivering the rated load at the rated voltage. The generator has $R_a = 0.01\ \Omega$, $R_{fw} = 30\ \Omega$, $V_f = 120$ V, $N_f = 500$ turns per pole, and $P_r = 10$ kW. The demagnetizing mmf due to armature reaction is 25% of the armature current. Its magnetization curve is given in Figure 5.21. Determine (a) the induced emf at full load, (b) the power developed, (c) the torque developed, (d) the applied torque, (e) the efficiency, (f) the external resistance in the field winding, and (g) the voltage regulation.

● SOLUTION

Since $P_o = 240$ kW and $V_t = 240$ V, the full-load current is

$$I_L = I_a = \frac{240{,}000}{240} = 1000 \text{ A}$$

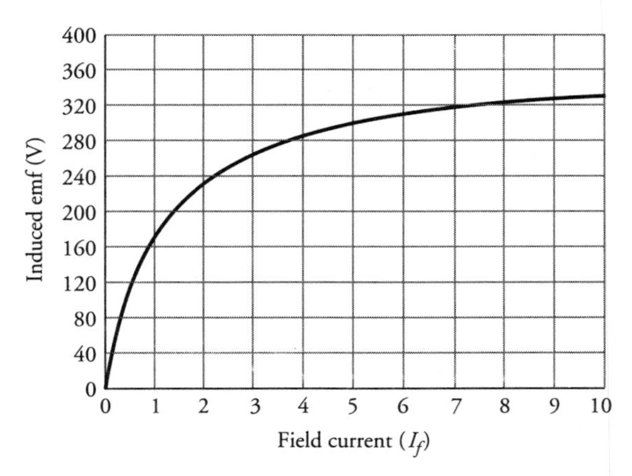

Figure 5.21 Magnetization curve of a dc machine at 600 rpm ($N_F = 500$ turns per pole)

(a) At full load, the induced emf is

$$E_{afl} = V_t + I_a R_a = 240 + 1000 \times 0.01 = 250 \text{ V}$$

(b) Hence, the power developed is

$$P_d = E_{afl} I_a = 250 \times 1000 = 250{,}000 \text{ W} \quad \text{or} \quad 250 \text{ kW}$$

$$\omega_m = \frac{2\pi \times 600}{60} = 20\pi \text{ rad/s}$$

(c) The torque developed at full load is

$$T_d = \frac{P_d}{\omega_m} = \frac{250{,}000}{20\pi} = 3978.87 \text{ N·m}$$

(d) The mechanical power input at full load must be

$$P_{inm} = P_d + P_r = 250 + 10 = 260 \text{ kW}$$

Thus, the applied torque is

$$T_s = \frac{260{,}000}{20\pi} = 4138.03 \text{ N·m}$$

(e) From the magnetization curve, the **effective field current** at full load is 2.5 A. This is the field current that must circulate in the field winding when there is no demagnetization effect of the armature reaction. The corresponding mmf is $500 \times 2.5 = 1250$ A·t/pole. The demagnetizing mmf due to the armature reaction is $0.25 \times 1000 = 250$ A·t/pole. Hence, the total mmf that must be provided by the field winding is $1250 + 250 = 1500$ A·t/pole. Thus, the actual field current at full load is

$$I_f = \frac{1500}{500} = 3 \text{ A}$$

The power loss in the field winding is $V_f I_f = 120 \times 3 = 360$ W. The total power input is

$$P_{in} = P_{inm} + V_f I_f = 260 + 0.36 = 260.36 \text{ kW}$$

The efficiency of the generator is

$$\eta = \frac{240}{260.36} = 0.922 \qquad \text{or} \qquad 92.2\%$$

(f) The total resistance in the field-winding circuit is

$$R_f = \frac{120}{3} = 40\ \Omega$$

Hence,

$$R_{fx} = 40 - 30 = 10\ \Omega$$

(g) At no load, the armature reaction is zero. Thus, the field mmf is 1500 A·t/pole. From the magnetization curve, the induced emf at no load is approximately 266 V. Hence, the voltage regulation is

$$\text{VR\%} = \frac{266 - 240}{240} \times 100 = 10.83\%$$

■

Exercises

5.12. A 240-V, 40-A, PM dc generator is rated at a speed of 2000 rpm. The armature-winding resistance is 0.4 Ω. The rotational loss is 10% of the power developed by the generator at full load. If the generator is operating in the linear range, determine (a) the no-load voltage, (b) the voltage regulation, (c) the applied torque, and (d) the efficiency of the generator.

5.13. The separately excited generator of Example 5.5 is operating at no load with maximum field current. What is the induced emf in the armature? If it is loaded gradually until it supplies the rated load, what is the terminal voltage? What is the voltage regulation? Also calculate its efficiency and the applied torque.

5.13 A Shunt Generator

When the field winding of a separately excited generator is connected across the armature, the dc generator is called the **shunt generator**. In this case, the terminal voltage is also the field-winding (field) voltage. Under no load, the armature current is equal to the field current. When loaded, the armature current supplies the load current and the field current as shown in Figure 5.22. Since the terminal voltage can be very high, the resistance of the field circuit must also be high in

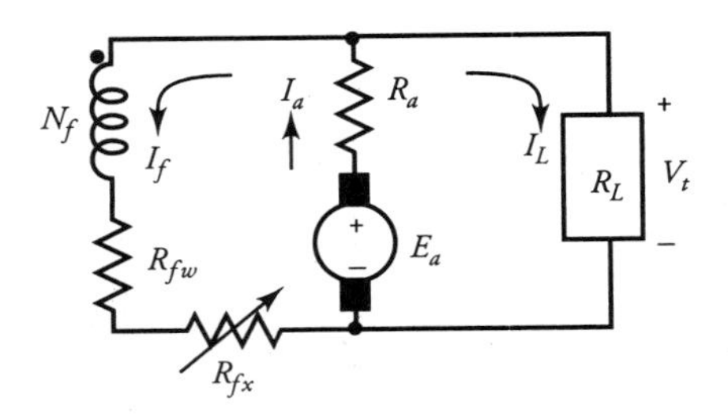

Figure 5.22 An equivalent circuit of a shunt generator.

order to keep its power loss to a minimum. Thus, the shunt field winding has a large number of turns of relatively fine wire.

As long as some residual flux remains in the field poles, the shunt generator is capable of **building up** the terminal voltage. The process of **voltage buildup** is summarized below.

When the generator is rotating at its rated speed, the residual flux in the field poles, however small it may be (but it must be there), induces an emf E_r in the armature winding as shown in Figure 5.23. Because the field winding is connected across the armature, the induced emf sends a small current through the field winding. If the field winding is properly connected, its mmf sets up a flux that aids the residual flux. The total flux per pole increases. The increase in the flux per pole increases the induced emf which, in turn, increases the field current. The

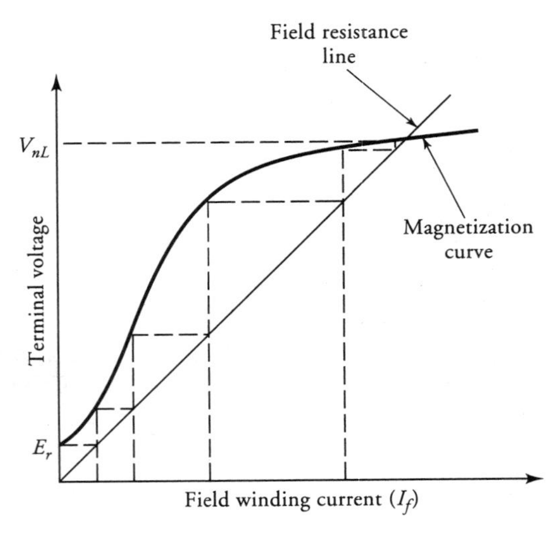

Figure 5.23 Voltage buildup in a shunt generator.

action is therefore cumulative. Does this action continue forever? The answer, of course, is no for the following reason:

The induced emf follows the nonlinear magnetization curve. The current in the field winding depends upon the total resistance in the field-winding circuit. The relation between the field current and the field voltage is linear, and the slope of the curve is simply the resistance in the field-winding circuit. The straight line is known as the **field-resistance** line. The shunt generator continues to build up voltage until the point of intersection of the field-resistance line and the magnetic saturation curve. This voltage is known as the **no-load voltage**.

It is very important to realize that the saturation of the magnetic material is a blessing in the case of a self-excited generator. Otherwise, the voltage buildup would continue indefinitely. We will also show that the saturation is necessary for the generator to deliver load.

If the field winding is connected in such a way that the flux produced by its mmf opposes the residual flux, a voltage **build-down** will occur. This problem can be corrected by either reversing the direction of rotation or interchanging the field-winding connection to the armature terminals, but not both.

The value of no-load voltage at the armature terminals depends upon the field-circuit resistance. A decrease in the field-circuit resistance causes the shunt generator to build up faster to a higher voltage as shown in Figure 5.24. By the same token, the voltage buildup slows down and the voltage level falls when the field-circuit resistance is increased. The value of the field-circuit resistance that makes the field-resistance line tangent to the magnetization curve is called the **critical (field) resistance**. The generator voltage will not build up if the field-circuit resistance is greater than or equal to the critical resistance. The speed at which the field circuit resistance represents the critical resistance is called the **critical speed**.

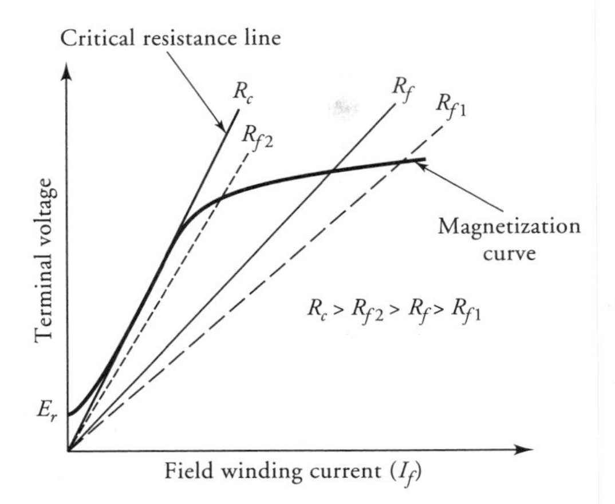

Figure 5.24 Voltage buildup for various values of field-circuit resistances.

Hence, the voltage buildup will take place in a shunt generator if (a) a residual flux exists in the field poles, (b) the field-winding mmf produces the flux that aids the residual flux, and (c) the field-circuit resistance is less than the critical resistance.

The equations that govern the operation of a shunt generator under steady-state are

$$I_a = I_L + I_f \tag{5.28}$$

$$V_t = I_f(R_{fw} + R_{fx}) = I_f R_f \tag{5.29}$$

and

$$V_t = I_L R_L = E_a - I_a R_a \tag{5.30}$$

The External Characteristic

Under no load, the armature current is equal to the field current, which is usually a small fraction of the load current. Therefore, the terminal voltage under no-load V_{nL} is nearly equal to the induced emf E_a owing to the negligible $I_a R_a$ drop. As the load current increases, the terminal voltage decreases for the following reasons:

1. The increase in $I_a R_a$ drop
2. The demagnetization effect of the armature reaction
3. The decrease in the field current due to the drop in the induced emf

The effect of each of these factors is shown in Figure 5.25.

For a successful operation, the shunt generator must operate in the saturated region. Otherwise, the terminal voltage under load can fall to zero for the following reason:

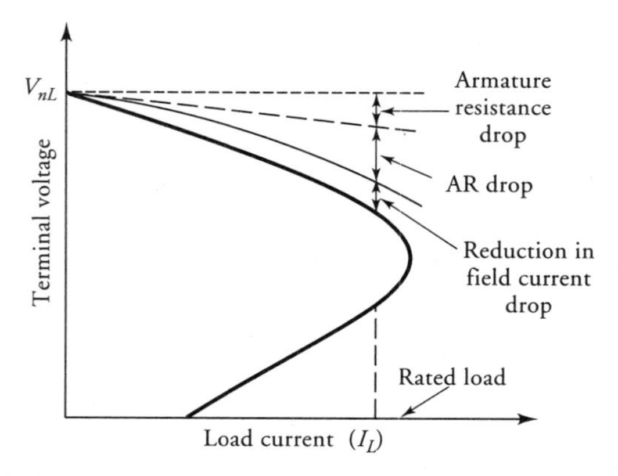

Figure 5.25 External characteristic of a shunt generator.

Suppose the generator is operating in the linear region and there is a 10% drop in the terminal voltage as soon as the load draws some current. A 10% drop in the terminal voltage results in a 10% drop in the field current which, in turn, reduces the flux by 10%. A 10% reduction in the flux decreases the induced emf by 10% and causes the terminal voltage to fall even further, and so on. Soon the terminal voltage falls to a level (almost zero) that is not able to supply any appreciable load. The saturation of the magnetic material comes to the rescue. When the generator is operating in the saturated region, a 10% drop in the field current may result in only a 2% or 3% drop in the flux, and the system stabilizes at a terminal voltage somewhat lower than V_{nL} but at a level suitable for successful operation.

As the generator is loaded, the load current increases to a point called the **break-down point** with the decrease in the load resistance. Any further decrease in the load resistance results in a decrease in the load current owing to a very rapid drop in the terminal voltage. When the load resistance is decreased all the way to zero (a short circuit), the field current goes to zero and the current through the short circuit is the ratio of the residual voltage and the armature-circuit resistance.

EXAMPLE 5.6

The magnetization curve of a shunt generator at 1200 rpm is given in Figure 5.26. The other parameters are $R_{fw} = 40\ \Omega$, $R_a = 0.2\ \Omega$, $N_f = 200$ turns/pole, $P_r = 1200$ W, and the demagnetizing mmf per pole due to armature reaction is 50% of the load current. The external field resistance, R_{fx}, is adjusted to give a no-load voltage of 170 V. What is R_{fx}? If the generator supplies a rated load of 100 A,

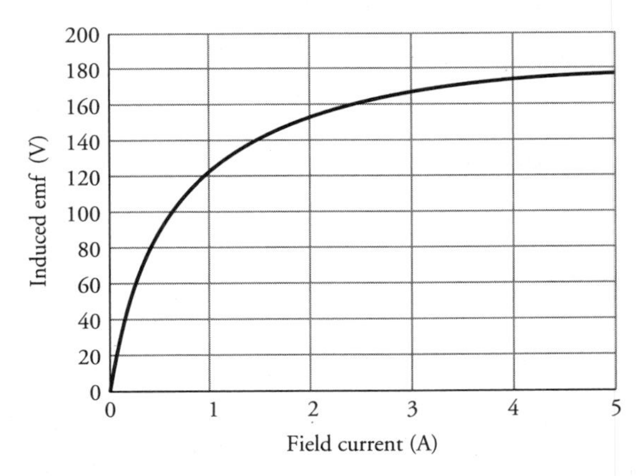

Figure 5.26 The magnetization curve of a dc machine at 1200 rpm (200 turns per pole)

determine (a) the terminal voltage, (b) the voltage regulation, and (c) the efficiency.

● SOLUTION

From the magnetization curve, the field current corresponding to the no-load voltage of 170 V is 3.5 A. The total field-circuit resistance is

$$R_f = \frac{170}{3.5} = 48.57\ \Omega$$

Thus, the external resistance in the field-winding circuit is

$$R_{fx} = R_f - R_{fw} = 48.57 - 40 = 8.57\ \Omega$$

The nonlinear behavior of the induced emf suggests the use of iterations to determine the terminal voltage at the rated current of $I_L = 100$ A. The demagnetizing mmf is $0.5 \times 100 = 50$ A·t/pole.

First iteration:

Assuming $E_a = 170$ V and the field current is 3.5 A, the terminal voltage at the rated load is

$$V_{t1} = 170 - 103.5 \times 0.2 = 149.3 \text{ A}$$

Second iteration:

The actual and the effective field currents are

$$I_{f2} = \frac{149.3}{48.57} = 3.07 \text{ A}$$

$$I_{fe2} = \frac{(200 \times 3.07 - 50)}{200} = 2.82 \text{ A}$$

From the graph, the induced emf is $E_{a2} \approx 165$ V. Thus,

$$V_{t2} = 165 - 103.07 \times 0.2 = 144.4 \text{ V}$$

Third iteration:

$$I_{f3} = \frac{144.4}{48.57} = 2.97 \text{ A}$$

$$I_{fe3} = \frac{(200 \times 2.97 - 50)}{200} = 2.72$$

From the graph, $E_{a3} \approx 163$ V. Hence,

$$V_{t3} = 163 - 102.97 \times 0.2 = 142.4 \text{ V}$$

The difference between V_{t2} and V_{t3} is so small (less than 2%) that no further iteration is necessary.

(a) Thus, $V_t = 142.4$ V and $I_f = 142.4/48.57 = 2.93$ A when $I_L = 100$ A. Thus, $I_{fe} = (200 \times 2.93 - 50)/200 = 2.68$ A. From the magnetization curve, $E_a \approx 163$ V.

(b) The voltage regulation is

$$VR\% = \frac{170 - 142.4}{142.4} \times 100 = 19.38\%$$

(c) The power output: $P_o = 142.4 \times 100 = 14{,}240$ W
The electrical power loss is

$$P_{cu} = (102.93)^2 \times 0.2 + 2.93^2 \times 48.57 = 2535.89 \text{ W}$$

The power developed: $P_d = 14{,}240 + 2535.89 = 16{,}775.89$ W

The power input: $P_{in} = 16{,}775.89 + 1200 = 17{,}975.89$ W

Hence, the efficiency: $\eta = \dfrac{14{,}240}{17{,}975.89} = 0.7922$ or 79.22%

Note that we could also have calculated the power developed as

$$P_d = E_a I_a = 163 \times 102.93 = 16{,}777.59 \text{ W}$$

■

Exercises

5.14. The field-circuit resistance of the shunt generator of Example 5.6 is adjusted to 64 Ω. What is the no-load voltage at its rated speed? If the generator operates at 50% of its rated load, determine (a) the terminal voltage, (b) the efficiency, and (c) the torque applied.

5.15. A 50-kW, 120-V shunt generator has $R_a = 0.09\ \Omega$, $R_{fw} = 30\ \Omega$, $R_{fx} = 15\ \Omega$, $N_m = 900$ rpm, and $P_r = 5$ kW. The generator is delivering the rated load at the rated terminal voltage. Determine (a) the generated emf, (b) the torque applied, and (c) the efficiency. Neglect the armature reaction.

5.14 A Series Generator

As the name suggests, the field winding of a series generator is connected in series with the armature and the external circuit. Because the series field winding has to carry the rated load current, it usually has few turns of heavy wire.

The equivalent circuit of a series generator is given in Figure 5.27. A variable resistance R_d, known as the **series field diverter**, may be connected in shunt with the field winding to control the current through and thereby the flux produced by it.

When the generator is operating under no load, the flux produced by the series field winding is zero. Therefore, the terminal voltage of the generator is equal to the induced emf due to the residual flux, E_r. As soon as the generator supplies a load current, the mmf of the series field winding produces a flux that aids the residual flux. The induced emf, E_a, in the armature winding is higher at load than that at no load. However, the terminal voltage, V_t, is lower than the induced emf due to (a) the voltage drops across the armature resistance, R_a, the series field–winding resistance, R_s, and (b) the demagnetization action of the armature reaction. Since voltage drops across the resistances and the armature reaction are functions of the load current, the induced emf and thereby the terminal voltage depend upon the load current.

The magnetization curve for the series generator is obtained by separately exciting the series field winding. The terminal voltage corresponding to each point on the magnetization curve is less by the amount equal to voltage drops across R_a and R_s when the armature reaction is zero. The terminal voltage drops even further when the armature reaction is also present as illustrated in Figure 5.28. Once the load current pushes the generator into the saturated region, any further increase in its value makes the armature reaction so great as to cause the terminal

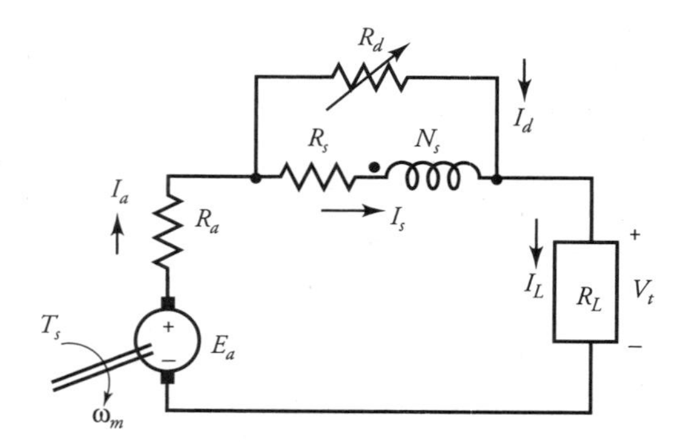

Figure 5.27 An equivalent circuit of a series generator.

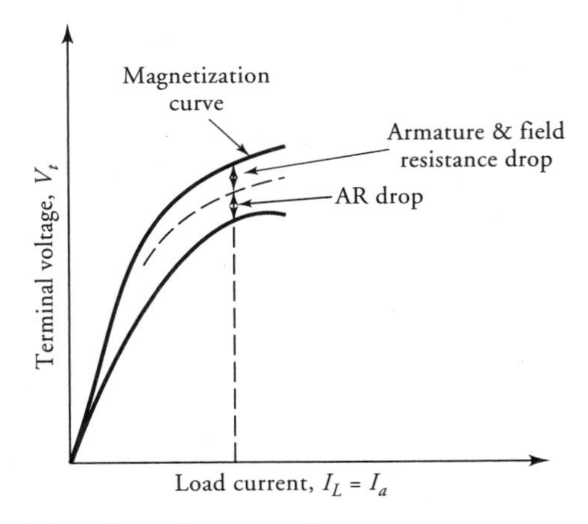

Figure 5.28 Characteristics of a series generator.

voltage to drop sharply. In fact, if driven to its extreme, the terminal voltage drops to zero.

The rising characteristic of a series generator makes it suitable for voltage-boosting purposes. Another clear distinction between a shunt generator and a series generator is that the shunt generator tends to maintain a **constant terminal voltage** while the series generator has a tendency to supply **constant load current**. In Europe, the Thury system of high-voltage direct-current power transmission requires several series generators connected in series and transmitting at constant current.

The basic equations that govern its steady-state operation are

$$V_t = E_a - I_a R_a - I_s R_s \tag{5.31}$$

$$I_s R_s = I_d R_d \tag{5.32}$$

$$I_a = I_L = I_s + I_d \tag{5.33}$$

where I_s is the current in the series field winding, R_s is the resistance of the series field winding, and I_d is the current in the series field–diverter resistance, R_d.

EXAMPLE 5.7

The induced emf in the linear region of a series generator is given as $E_a = 0.4I_s$. The generator has $R_s = 0.03\ \Omega$ and $R_a = 0.02\ \Omega$. It is used as a booster between a 240-V station bus-bar and a feeder of 0.25-Ω resistance. Calculate the voltage between the far end of the feeder and the bus-bar at a current of 300 A.

● SOLUTION

In the absence of R_d, $I_a = I_s = 300$ A. The induced emf is

$$E_a = 0.4 \times 300 = 120 \text{ V}$$

The voltage drop across R_a, R_s, and the feeder is

$$V_d = 300 \times (0.02 + 0.03 + 0.25) = 90 \text{ V}$$

Hence, the voltage between the far end and the bus-bar is

$$V_t = 240 + 120 - 90 = 270 \text{ V}$$

The net increase of 30 V may be beyond the desired limit. The placement of field diverter resistance may be necessary to regulate the far-end terminal voltage. ■

Exercises

5.16. Determine the field diverter resistance that will limit the voltage rise to 5 volts in Example 5.7.

5.17. The magnetization curve of Figure 5.26 was obtained for a dc shunt generator with 200 turns per pole in its shunt field winding. A series field winding with 5 turns per pole is added and is operated as a series generator. The series field resistance is 0.05 Ω, and the armature resistance is 0.2 Ω. The rotational power loss is 1500 W. The generator supplies a current of 140 A to a bank of lamps connected in parallel. If the armature reaction is equivalent to a 6% reduction in the flux, determine (a) the terminal voltage, (b) the power output, (c) the torque applied, and (d) the efficiency.

5.15 Compound Generators

The drooping characteristic of a shunt generator and the rising characteristic of a series generator provide us enough motivation to theorize the possibility of a better external characteristic by fusing the two generators into one. In fact, under certain constraints, putting the two generators together is like transforming two contentiously different generators into a single well-behaved generator. This is done by winding both series and shunt field windings on each pole of the generator. When the mmf of the series field is aiding the mmf of the shunt field, it is referred to as a **cumulative compound generator** (Figure 5.29**a**). Otherwise, it is termed a **differential compound generator** (Figure 5.29**b**).

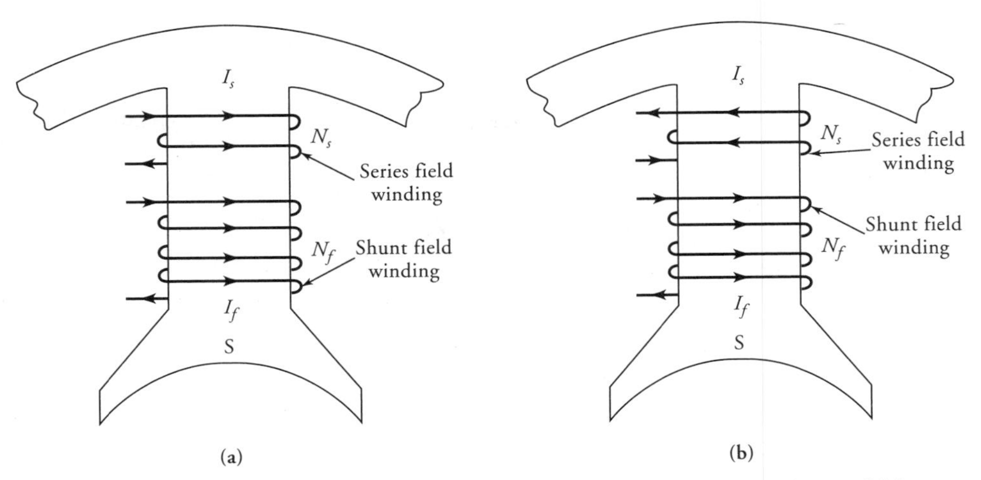

Figure 5.29 Current distributions in the series and the shunt field windings of **(a)** cumulative and **(b)** differential compound generators.

When the shunt field winding is connected directly across the armature terminals, it is called a **short-shunt generator**. In a short-shunt generator (Figure 5.30), the series field winding carries the load current in the absence of a field diverter resistance. A generator is said to be **long shunt** when the shunt field winding is across the load as indicated in Figure 5.31. We hasten to add that most

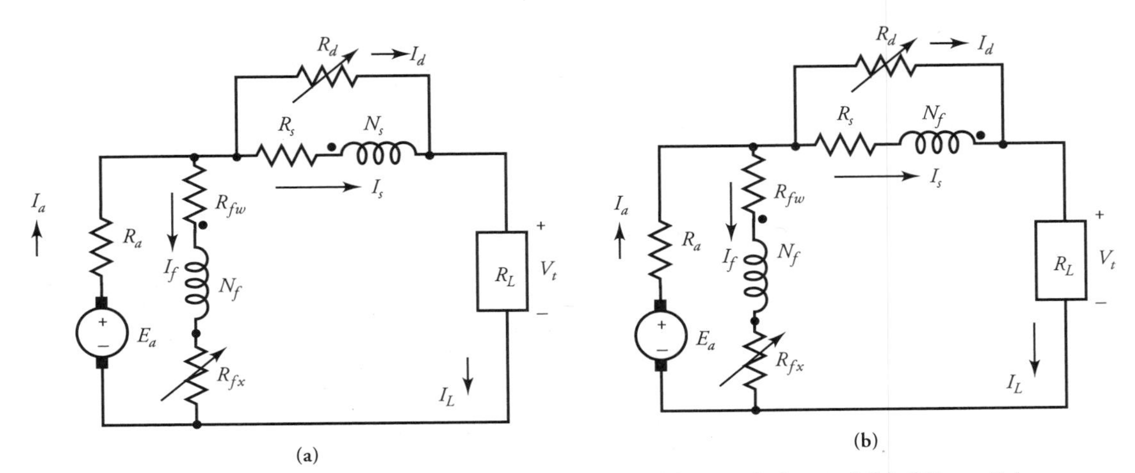

Figure 5.30 Equivalent circuits of short-shunt **(a)** cumulative and **(b)** differential compound generators.

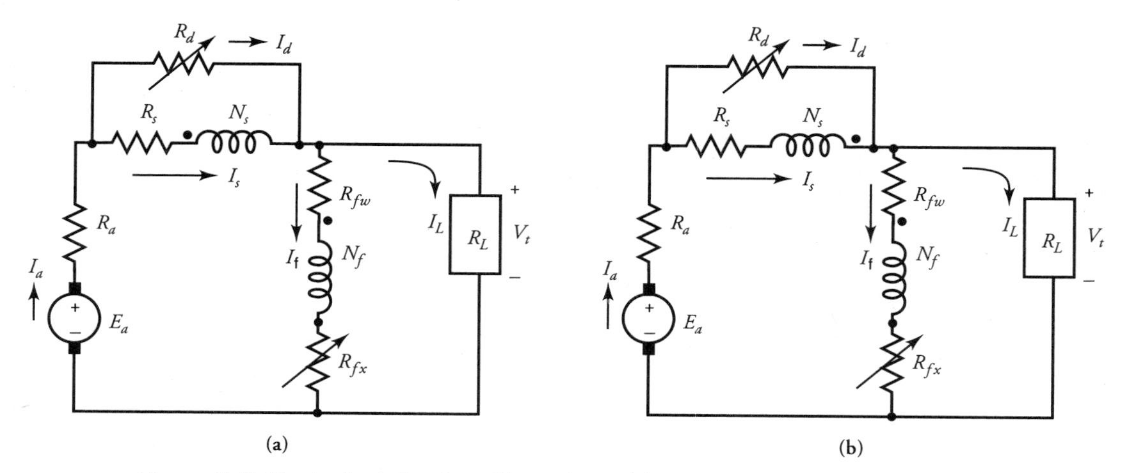

Figure 5.31 Equivalent circuits of long-shunt **(a)** cumulative and **(b)** differential compound generators.

of the flux is created by the shunt field. The series field mainly provides a control over the total flux. Therefore, different levels of compounding can be obtained by limiting the current through the series field, as explained in Section 5.14. Three distinct degrees of compounding that are of considerable interest are discussed below.

Undercompound Generator

When the full-load voltage in a compound generator is somewhat higher than that of a shunt generator but still lower than the no-load voltage, it is called an undercompound generator. The voltage regulation is slightly better than that of the shunt generator.

Flat or Normal Compound Generator

When the no-load voltage is equal to the full-load voltage, the generator is known as a **flat compound generator**. A flat compound generator is used when the distance between the generator and the load is short. In other words, no significant voltage drop occurs on the transmission line (called the **feeder**) connecting the generator to the load.

Overcompound Generator

When the full-load voltage is higher than the no-load voltage, the generator is said to be **overcompound**. An overcompound generator is the generator of choice when the generator is connected to a load via a long transmission line. The long transmission line implies a significant drop in voltage and loss in power over the transmission line.

The usual practice is to design an overcompound generator. The adjustments can then be made by channeling the current away from the series field winding by using the field diverter resistance.

The external characteristics of compound generators are given in Figure 5.32. For comparison purposes, the external characteristics of other generators are also included.

The fundamental equations that govern the steady-state behavior of short-shunt and long-shunt cumulative generators are given below.

Short-shunt:

$$I_a = I_L + I_f \tag{5.34}$$

$$I_s = \frac{R_d}{R_d + R_s} I_L \tag{5.35}$$

$$V_t = E_a - I_a R_a - I_s R_s \tag{5.36}$$

$$V_f = E_a - I_a R_a \tag{5.37}$$

$$\text{mmf} = I_f N_f \pm I_s N_s - \text{mmf}_d \tag{5.38}$$

where the (plus/minus) sign is for the (cumulative/differential) compound generator, and mmf_d is the demagnetizing mmf due to the armature reaction.

Long-shunt:

$$I_a = I_L + I_f \tag{5.39}$$

$$I_s = \frac{R_d}{R_d + R_s} I_a \tag{5.40}$$

$$V_f = V_t = E_a - I_a R_a - I_s R_s \tag{5.41}$$

$$\text{mmf} = I_f N_f \pm I_s N_s - \text{mmf}_d \tag{5.42}$$

EXAMPLE 5.8

A 240-V, short-shunt, cumulative compound generator is rated at 100 A. The shunt field current is 3 A. It has an armature resistance of 50 mΩ, a series field resistance of 10 mΩ, a field diverter resistance of 40 mΩ, and a rotational loss of 2 kW. The

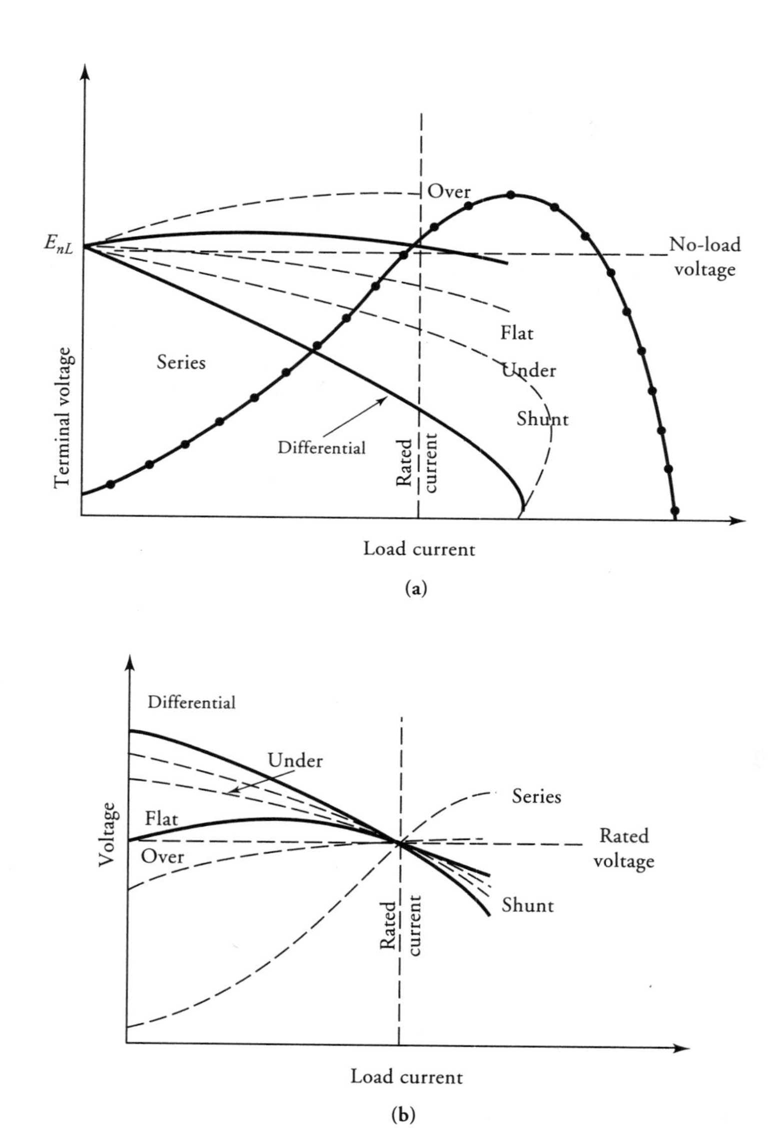

Figure 5.32 **(a)** External characteristics of dc generators. **(b)** Terminal voltage adjusted for rated value.

generator is connected to the load via a feeder, R_{fe}, of 30 mΩ resistance. When the generator is supplying the full load at the rated voltage, determine its efficiency. Draw the power-flow diagram to show the power distribution.

● SOLUTION

$$P_o = 240 \times 100 = 24{,}000 \text{ W}$$

Since $$I_f = 3 \text{ A}, I_a = 100 + 3 = 103 \text{ A}$$

$$I_s = \frac{0.04}{0.04 + 0.01} \times 100 = 80 \text{ A}$$

$$I_d = 100 - 80 = 20 \text{ A}$$

$$E_a = V_t + I_L R_{fe} + I_s R_s + I_a R_a$$

$$= 240 + 100 \times 0.03 + 80 \times 0.01 + 103 \times 0.05 = 248.95 \text{ V}$$

$$V_f = E_a - I_a R_a$$

$$= 248.95 - 103 \times 0.05 = 243.8 \text{ V}$$

Hence, $$R_f = \frac{243.8}{3} = 81.267 \ \Omega$$

Copper losses:

Armature: $I_a^2 R_a = 103^2 \times 0.05 = 530.45$ W

Series field: $I_s^2 R_s = 80^2 \times 0.01 = 64$ W

Shunt field: $I_f^2 R_f = 3^2 \times 81.267 = 731.4$ W

Diverter resistance: $I_d^2 R_d = 20^2 \times 0.04 = 16$ W

Feeder resistance: $I_L^2 R_{fe} = 100^2 \times 0.03 = 300$ W

Total copper loss: $P_{cu} = 530.45 + 64 + 731.4 + 16 + 300$

$= 1641.85$ W

Thus, the power developed is

$$P_d = P_o + P_{cu} = 24{,}000 + 1{,}641.85 = 25{,}641.85 \text{ W}$$

However, the power developed can also be computed as

$$P_d = E_a I_a = 248.95 \times 103 = 25{,}641.85 \text{ W}$$

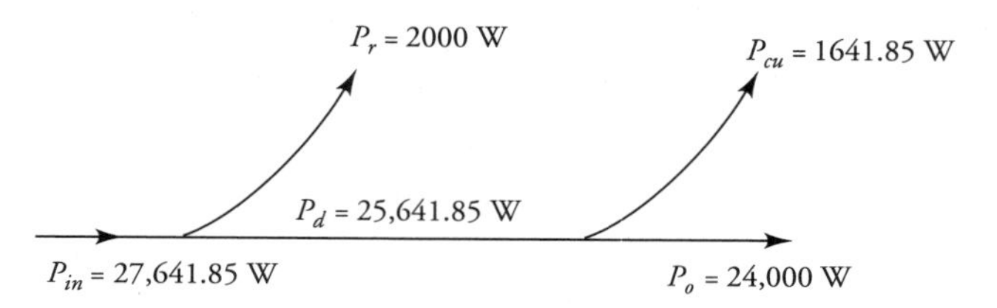

Figure 5.33 Power-flow diagram of a short-shunt compound generator, Example 5.8.

The power input is

$$P_{in} = P_d + P_r = 25{,}641.85 + 2{,}000 = 27{,}641.85 \text{ W}$$

Hence, the efficiency is

$$\eta = \frac{P_o}{P_{in}} = \frac{24{,}000}{27{,}641.85} = 0.8682 \quad \text{or} \quad 86.82\%$$

The power-flow diagram is given in Figure 5.33.

■

Exercises

5.18. The magnetization curve of a short-shunt, flat compound generator is given in Figure 5.26. The other parameters of the generator are $R_a = 80\ \text{m}\Omega$, $R_s = 20\ \text{m}\Omega$, $N_m = 900$ rpm, and $N_f = 200$ turns. If the generator delivers its rated load current of 100 A at its rated terminal voltage of 160 V, determine the number of turns per pole in its series field. Neglect the armature reaction.

5.19. Repeat Exercise 5.18 if the armature reaction causes a 2% reduction in the flux per pole. Assume that R_s is constant. Explain how R_s can be held constant.

5.20. What is the efficiency of the generator in Exercise 5.18 if the rotational loss is 1.5 kW? Sketch its power-flow diagram.

5.16 Maximum Efficiency Criterion

Either by performing an actual load test on a given machine or just by calculating its performance at various loads, we can obtain an efficiency versus load curve. As is evident from the curve, the efficiency increases with load up to some point where it becomes maximum. Any increase in the load thereafter causes a decrease in efficiency. It is, therefore, imperative to know at what load the efficiency of the machine is maximum. Operating the machine at its maximum efficiency results in (a) a decrease in the losses in the machine, which, in turn, drops the operating temperature of the machine, and (b) a reduction in the operating cost of the machine.

Prior to deriving an expression that determines the load at which a generator provides maximum efficiency, let us first take another look at the losses. The losses in a generator can be grouped in two categories: fixed losses and variable losses. Fixed losses do not vary with the load when the generator is driven at a constant speed. The rotational loss falls into this category. Although it is not strictly true for self-excited generators, the shunt-field current, I_f, is taken to be constant. Thus, the power loss due to I_f can also be considered a part of the fixed loss. The variable loss, on the other hand, is the loss that varies with the load current.

For a PM dc generator, the power output is

$$P_o = V_t I_L$$

and the power input is

$$P_{in} = V_t I_L + I_L^2 R_a + P_r$$

Hence, the efficiency is

$$\eta = \frac{V_t I_L}{V_t I_L + I_L^2 R_a + P_r}$$

For the efficiency to be maximum, the rate of change of η with respect to I_L must be zero when $I_L \rightarrow I_{Lm}$, where I_{Lm} is the load current at maximum efficiency. That is

$$\frac{[V_t I_{Lm} + I_{Lm}^2 R_a + P_r]V_t - V_t I_{Lm}[V_t + 2I_{Lm}R_a]}{[V_t I_{Lm} + I_{Lm}^2 R_a + P_r]^2} = 0$$

From this we obtain

$$I_{Lm}^2 R_a = P_r \tag{5.43}$$

Hence, the load current at maximum efficiency for a PM dc generator is

$$I_{Lm} = \sqrt{\frac{P_r}{R_a}} \tag{5.44}$$

The conditions for the maximum efficiency for other generators are as follows:

Separately excited:

$$I_{Lm}^2 R_a = P_r + I_f^2 R_f \tag{5.45}$$

Shunt:

$$I_{Lm}^2 R_a = P_r + I_f^2 (R_a + R_f) \tag{5.46}$$

Series:

$$I_{Lm}^2 (R_a + R_s) = P_r \tag{5.47}$$

Short-shunt:

$$I_{Lm}^2 (R_a + R_s) = P_r + I_f^2 (R_a + R_f) \tag{5.48}$$

Long-shunt:

$$I_{Lm}^2 (R_a + R_s) = P_r + I_f^2 (R_a + R_s + R_f) \tag{5.49}$$

EXAMPLE 5.9

A 50-kW, 120-V, long-shunt compound generator is supplying a load at its maximum efficiency and the rated voltage. The armature resistance is 50 mΩ, series field resistance is 20 mΩ, shunt field resistance is 40 Ω, and rotational loss is 2 kW. What is the maximum efficiency of the generator?

● SOLUTION

The shunt field current is

$$I_f = \frac{120}{40} = 3 \text{ A}$$

From Eq. (5.49), we have

$$I_{Lm}^2 (0.05 + 0.02) = 3^2 (0.05 + 0.02 + 40) + 2000$$

or

$$I_{Lm} = 183.64 \text{ A}$$

Thus, the power output at maximum efficiency is

$$P_o = 120 \times 183.64 = 22{,}036.8 \text{ W}$$

The total copper loss is

$$P_{cu} = I_a^2(R_a + R_s) + I_f^2R_f$$
$$= 186.64^2 \times 0.07 + 3^2 \times 40 = 2798.41 \text{ W}$$

The power developed at maximum efficiency is

$$P_d = P_o + P_{cu} = 22{,}036.8 + 2798.41 = 24{,}835.21 \text{ W}$$

The power input: $P_{in} = 24{,}835.21 + 2000 = 26{,}835.21$ W
Hence, the maximum efficiency is

$$\eta = \frac{22{,}036.8}{26{,}835.21} = 0.8211 \qquad \text{or} \qquad 82.11\%$$

■

Exercises

5.21. Show that the load current at the maximum efficiency of a separately excited generator is given by Eq. (5.45).

5.22. Verify Eq. (5.46) for the load current at the maximum efficiency of a shunt generator.

5.23. Show that the load current at the maximum efficiency of a series generator is given by Eq. (5.47).

5.24. Derive the expressions for the load current at maximum efficiency of short-shunt and long-shunt compound generators.

SUMMARY

We devoted this chapter entirely to the study of different types of dc generators. Most of the principles we have introduced in this chapter are equally applicable to dc machines. The main components of a dc machine are the stator, the armature, and the commutator.

The induced emf in the armature winding varies sinusoidally with a frequency of

$$f = \frac{P}{120} N_m$$

However, the commutator rectifies it and we get almost constant voltage output. The dc-generated voltage is

$$E_a = K_a \Phi_P \omega_m$$

and the torque developed by the generator is

$$T_d = K_a \Phi_P I_a$$

The prime mover must apply the torque in a direction opposite to the torque developed. The applied mechanical energy is then converted into electrical energy.

The armature is either lap wound or wave wound using a two-layer winding. In a lap winding, the two ends of the coil are connected to the adjacent commutator segments. The two ends are almost 360° electrical apart in the case of a wave winding. The lap winding offers as many parallel paths as there are poles in the machine. There are only two parallel paths in a wave-wound machine.

In order to predict the performance of a machine, we need information on its magnet saturation. This information is usually presented in the form of a magnetization curve. For successful operation of a self-excited generator, the point of operation must be in the saturation region.

The current in the armature winding produces its own flux, which has a demagnetizing effect, called armature reaction, on the machine. Techniques such as interpoles and/or commutating windings are used to minimize the armature reaction. When the machine operates at a constant load, advancing the brushes from their geometric neutral axis to the new neutral axis may do the trick.

There are two basic types of generators: separately excited and self-excited. Self-excited generators are further classified as shunt, series, and compound. Except for the overcompound generator, the terminal voltage of all other generators falls as a function of the load current. The voltage regulation is simply a measure of the change in the terminal voltage at full load.

The losses in a dc machine reduce its efficiency. We have categorized the losses as the rotational loss, the copper loss, and the stray-load loss. The rotational loss accounts for the mechanical power loss due to friction and windage, and the magnetic power loss is due to eddy currents and hysteresis. The copper loss accounts for all the electrical losses in the machine. When a loss cannot be properly accounted for, it is considered as a part of the stray-load loss.

Review Questions

5.1. What is the difference between a generator and a motor?
5.2. Explain the construction of a dc machine.

5.3. What are the advantages and drawbacks of a PM dc machine in comparison with a wound dc machine?

5.4. What is a field winding? Name all the types of field windings.

5.5. Why does a shunt field winding have a large number of turns compared with a series field winding?

5.6. Why is the armature resistance kept small?

5.7. Explain the difference between lap winding and wave winding.

5.8. What is the difference between a progressive and a retrogressive wave winding?

5.9. The armature of a 4-pole dc machine has 30 slots. Is this motor a good candidate for the wave winding? 29 slots? 31 slots?

5.10. The armature of a 6-pole dc machine has 30 slots. Can we employ the wave winding? 29 slots? 31 slots?

5.11. What happens to the existing magnetic field when a current-carrying conductor is placed in it? Draw sketches to explain your answer.

5.12. What must be the minimum cross-sectional area of a field pole in order to have the same flux density as in the yoke section?

5.13. The coil pitch of a machine is 5. If one end of the coil is placed in slot 3, determine the slot in which the other must be placed.

5.14. The induced emf in a 4-pole, lap-wound dc generator is 240 V. The maximum current each armature conductor can supply is 25 A. What is the power rating of the generator?

5.15. If the armature coils of the generator in Question 5.14 are reconnected to form a wave winding, what are the voltage, current, and power rating of the generator?

5.16. In a dc machine the air-gap flux density is 6 kG (10 kG = 1 T), and the pole-face area is 3 cm by 5 cm. Determine the flux per pole in the machine.

5.17. A 10-pole, 500-V, lap-wound dc generator is rated at 100 hp. Determine the voltage, current, and power rating if it were wave wound.

5.18. What is the frequency of the induced emf in the armature winding of an 8-pole generator that operates at a speed of 600 rpm?

5.19. Determine the mmf requirements per pole to set up a flux density of 69,000 lines/in^2 in the 0.16-in air-gap of the machine. Assume that the permeability of the magnetic material is infinite. [1 T = 64.516×10^3 lines/in^2]

5.20. What are the various types of dc generators? Draw a schematic for each generator. Draw a cross-sectional view of each machine and show the actual winding connections.

5.21. A series generator requires 620 A·t/pole to deliver a load of 10 A. Determine the number of turns per pole.

5.22. Calculate the armature current per conductor of a 6-pole, 60-kW, 240-V, separately excited dc generator if it is wound using (a) a lap winding and (b) a wave winding.

5.23. What is commutation?

5.24. Explain armature reaction. Cite methods that are commonly used to compensate the demagnetization effect of armature reaction.
5.25. What is the main cause of sparking at the brushes? Does the sparking take place at the leading edge or the trailing edge of a brush?
5.26. What is the meaning of no load? full load?
5.27. Explain voltage regulation. The voltage regulation of generator A is 10% and that of generator B is 5%. Which is a better generator?
5.28. Explain the following terms: mechanical losses, magnetic losses, rotational loss, copper loss, and stray-load loss.
5.29. How can the magnetic losses be reduced?
5.30. What are the various causes for a failure of a shunt generator to build up? What are the remedies?
5.31. Explain the following terms: critical resistance, magnetization (curve) characteristic, external characteristic (curve), and field-resistance line.
5.32. What is the effect of armature reaction on the external characteristic of a dc generator?
5.33. Draw equivalent circuits of different types of dc generators. Explain the voltage buildup in each case.
5.34. Why is the series generator called a booster?
5.35. Why are the commutating poles or interpoles connected in series with the armature?
5.36. Explain the following terms: long shunt, short shunt, undercompound, flat-compound, overcompound, cumulative, and differential.
5.37. What is the importance of a field diverter resistance?
5.38. State the maximum efficiency criterion. What is its significance?
5.39. The induced emf in a separately excited generator is 135 V at no load, while terminal voltage at full load is 115. What is its voltage regulation?
5.40. The copper loss at full load in a 25-kW, 120-V dc generator is 2.5 kW. If the generator is operating at its maximum efficiency, what is the efficiency of the generator? Draw the power-flow diagram.
5.41. A 25-kW, 125-V shunt generator is delivering half load at its rated voltage. The operating speed is 1725 rpm. Its efficiency is 85%. What are the total losses and the power input to the machine? Draw the power-flow diagram. What is the torque applied by the prime mover?

Problems

5.1. The armature of a 4-pole, lap-wound machine has 18 slots. Determine the slot pitch, the front pitch, and the back pitch. Sketch its polar-winding diagram and highlight the parallel paths. If the armature current is 200 A, what is the current per path?

5.2. The armature of a 4-pole, lap-wound machine has 12 slots. What are the slot pitch, back pitch, and front pitch? Show the armature windings and highlight its four parallel paths. Determine the current per path if the armature current is 40 A.

5.3. Sketch a polar-winding diagram for a 4-pole, 20-slot, lap-wound armature. Calculate the slot pitch, the front pitch, and the back pitch. If each coil is capable of carrying a maximum current of 20 A, what is the maximum current rating of the armature?

5.4. The armature of a 6-pole, lap-wound machine has 18 slots. Determine the slot pitch, the front pitch, and the back pitch. Sketch its polar-winding diagram and highlight the parallel paths. If each coil can carry a current of 12 A, determine the current rating of the armature.

5.5. Draw a complete polar-winding diagram for a 4-pole, 21-slot, wave-wound armature. What are the slot pitch, the front pitch, the back pitch, and the commutator pitch? If the rated current of the armature is 80 A, calculate the current per path.

5.6. Determine the back pitch, the front pitch, and the commutator pitch for a 17-slot, 4-pole, wave-wound armature. Sketch its polar-winding diagram. If the current per path is 25 A, what is the current rating of the armature?

5.7. Determine the back pitch, the front pitch, and the commutator pitch for an 11-slot, 4-pole, wave-wound armature. Sketch its polar-winding diagram. If the current in each coil is limited to 30 A, what is the limitation on the armature current?

5.8. Determine the back pitch, the front pitch, and the commutator pitch for a 19-slot, 4-pole, wave-wound armature. Sketch its polar-winding diagram. If the maximum armature current is 100 A, what is the current rating of each coil?

5.9. The armature of a 4-pole, lap-wound machine has 48 slots. There are 10 conductors per slot. The axial length of the armature is 10.8 cm, and its diameter is 25 cm. The pole arc is 15 cm and the flux density per pole is 1.2 T. If the armature speed is 1200 rpm, determine (a) the induced emf in the armature winding, (b) the induced emf per coil, (c) the induced emf per turn, and (d) the induced emf in each conductor. What is the frequency of the induced emf in the armature winding?

5.10. Repeat Problem 5.9 for a wave-wound machine.

5.11. The armature of a 6-pole, lap-wound machine has 48 coils. Each coil has 12 turns. The armature is enveloped by a ceramic permanent-magnet ring, and the flux density per pole is 0.7 T. The armature is 12 cm in length and 31.6 cm in diameter. If the armature speed is 900 rpm, determine the induced emf (a) in the armature, (b) in each coil, and (c) in each conductor. Determine the frequency of the induced emf in the armature winding.

5.12. Determine the induced emf in the armature winding if the speed of the armature in Problem 5.11 is changed to (a) 600 rpm, (b) 450 rpm, and (c) 1200 rpm. What is the frequency of the induced emf at each speed?

5.13. If the armature current in Problem 5.11 is 30 A, determine (a) the current per path, (b) the power developed, and (c) the torque developed. If 90% of the mechanical power is converted into electrical power, what must be the applied torque of the prime mover?

5.14. The magnetization curve of a 120-kW, 240-V, 4-pole dc machine at its rated speed of 600 rpm is approximated as

$$E_a = \frac{2400 I_f}{7.5 + 6.5 I_f}$$

Plot the magnetization curves for speeds of 600 rpm, 400 rpm, and 300 rpm. The field current is restricted to a maximum of 10 A.

5.15. The following data are taken to obtain the magnetization curve of a 15-kW, 150-V, 6-pole dc machine at its rated speed of 1200 rpm.

I_f (A):	0	0.5	1.0	1.5	2.0	2.5	3.0	3.5	4.0	4.5	5.0	5.5
E_a (V):	6	70	112	130	140	145	149	152	155	157	158	159

Sketch the magnetization curves for speeds of 1200 rpm and 800 rpm.

5.16. A PM, lap-wound dc generator has 8 poles, 95 slots, and 6 turns per coil. The rated current per conductor is 12.5 A. The no-load terminal voltage at 1600 rpm is 120 V. Calculate (a) the required flux per pole and (b) the power developed. If the machine was wave wound, what would be its rated voltage, armature current, and power?

5.17. A wave-wound dc machine has 4 poles, 71 slots, and 2 turns per coil. The armature radius is 25.4 cm, and the length is 20 cm. The pole arc covers 70% of the armature. The flux density per pole is 0.9 T. The armature speed is 560 rpm. For an armature current of 40 A, determine the torque developed by the machine.

5.18. The no-load voltage of a PM dc generator at 1800 rpm is 120 V. The armature circuit resistance is 0.02 Ω. The rotational loss at 1800 rpm is 2.5 kW. If the generator is rated at 200 A and operates in the linear region, determine (a) the terminal voltage, (b) the voltage regulation, (c) the applied torque, and (d) the efficiency.

5.19. The generator of Problem 5.18 is operating at half load when the speed is adjusted to 1200 rpm. If the rotational loss is proportional to the speed, determine the applied torque and the efficiency at half load.

5.20. A 120-V, 1200-rpm, separately excited dc generator has $R_a = 0.05\ \Omega$, $R_{fw} = 50\ \Omega$, $V_f = 120$ V, and $N_f = 100$ turns/pole. The demagnetizing mmf due to armature reaction is 20% of the load current. The induced emf at no load is known to follow the relation

$$E_a = \frac{2000I_f}{5 + 8I_f}$$

The generator delivers its rated load of 50 A at the rated voltage of 120 V. The rotational loss is 850 W. Determine (a) the field-winding resistance, (b) an external resistance that must be added in the field circuit, (c) the efficiency, and (d) the voltage regulation.

5.21. The dc generator of Problem 5.16 has an armature circuit resistance of 0.05 Ω. The rotational loss is 1500 W. Determine (a) the efficiency, (b) the torque applied, and (c) the voltage regulation. Neglect the armature reaction effect.

5.22. The dc machine of Problem 5.17 is connected as a separately excited generator. Its parameters are $R_a = 0.25\ \Omega$, $R_f = 90\ \Omega$, $V_f = 90$ V, and $P_r = 2$ kW. Determine the efficiency and the voltage regulation at the full-load current of 40 A. Neglect the armature reaction.

5.23. A 50-kW, 250-V, 600-rpm, separately excited generator has an armature resistance of 0.05 Ω. The generator supplies the rated load at its rated voltage. The field-winding circuit has a resistance of 100 Ω and is connected across a 200-V source. The rotation loss is 5 kW. Determine the efficiency, the applied torque, and the voltage regulation when the armature reaction is zero.

5.24. A 230-V, 1800-rpm, 4.6-kW shunt generator has an armature-circuit resistance of 0.2 Ω and a field-circuit resistance of 65 Ω. The induced emf due to residual flux is 10 V. If the generator is short-circuited prior to its operation, what is the short-circuit current? Will there be a voltage buildup?

5.25. The generator of Problem 5.20 is connected as a shunt generator. The field current is adjusted so that it delivers the rated load at its rated voltage. Determine (a) the external resistance in the field-winding circuit, (b) the efficiency, and (c) the voltage regulation.

5.26. A 23-kW, 230-V, 900-rpm shunt generator has 240.5 V induced in its armature winding when it delivers the rated load at the rated voltage. The field current is 5 A. What is the armature resistance? What is the field-circuit resistance? The no-load voltage of the generator is 245 V. What is the voltage regulation? Why is the induced emf at the rated load not equal to that at no load?

5.27. Determine the value of the critical field-circuit resistance for the magnetization curve (a) given in Figure 5.21 and (b) given in Figure 5.26.

5.28. The dc machine of Problem 5.15 is delivering the rated load at the rated voltage when connected as a shunt generator at a speed of 1200 rpm. The field-circuit resistance is 37.5 Ω. The rotational loss is 1800 W. The armature reaction is zero. Determine (a) the armature resistance, (b) the voltage regulation, and (c) the efficiency.

5.29. An 8-pole, 600-rpm dc shunt generator with 800 wave-connected armature conductors supplies a load of 20 A at its rated terminal voltage of 240 V. The armature resistance is 0.2 Ω, and the field winding resistance is 240 Ω. The rotational loss is 280 W. Determine (a) the armature current, (b) the induced emf, (c) the flux per pole, (d) the torque applied, and (e) the efficiency.

5.30. A 4-pole dc shunt generator with a shunt field resistance of 100 Ω and an armature resistance of 1 Ω has 400 lap-connected conductors in its armature. The flux per pole is 20 mWb. The rotational loss is 500 W. If a load resistance of 10 Ω is connected across the armature and the generator is driven at 1200 rpm, compute (a) the terminal voltage and (b) the efficiency. Neglect the armature reaction.

5.31. The magnetization curve for a 4-pole, 110-V, 1000-rpm shunt generator is as follows:

I_f (A):	0	0.5	1.0	1.5	2.0	2.5	3.0	3.5	4.0
E_a (V):	8	60	100	116	120	123	125	126	127

The resistance of 140 lap-connected conductors in the armature is 0.1 Ω. Field-winding resistance is 40 Ω. Determine (a) the no-load voltage, (b) the critical resistance, (c) the residual flux, and (d) the speed at which the machine fails to build up.

5.32. The generator of Problem 5.31 is supplying a load at its rated terminal voltage. The rotational loss is equal to the copper loss. If the flux is reduced by 5% owing to the armature reaction, find (a) the load supplied by the generator, (b) the applied torque, (c) the voltage regulation, and (d) the efficiency.

5.33. A 12-kW, 120-V dc series generator has a series field resistance of 0.05 Ω. The generated emf is 135 V when the generator delivers the rated load at its rated terminal voltage. What is the armature resistance? If the load is reduced to 80% of its rated value at the rated voltage, what is the induced emf in the armature? If the generated emf is 130 V and the generator is supplying 4.167 kW, what is the terminal voltage?

5.34. A series generator is operating in the linear region where the flux is directly proportional to the armature current. The armature resistance is 80 mΩ, and the series field–winding resistance is 20 mΩ. At a speed of 1200 rpm, the terminal voltage is 75 V and the armature current is 20 A. What must be the armature current for a terminal voltage of 120 V at 1200 rpm? If it is required to have a terminal voltage of 100 V at an armature current of 25 A, what must be the armature speed?

5.35. The following data were obtained under no load by exciting the 200-turn shunt field winding of a dc generator at a speed of 1200 rpm:

I_f (A):	0	2	4	6	8	10	12	14	16	18
E_a (V):	5	50	90	112	120	125	128	130	131	131.5

The armature resistance is 50 mΩ and the series field–winding resistance is 30 mΩ. When connected as a short-shunt generator, the no-load voltage is 120 V. For a flat compound operation at the rated load current of 100 A, determine the number of turns in the series field winding.

5.36. Repeat Problem 5.35 when the generator is connected as a long shunt.

5.37. Repeat Problem 5.35 when the armature reaction reduces the flux by 2% at the rated load.

5.38. Repeat Problem 5.36 when the armature reaction reduces the flux by 2% at the rated load.

5.39. A 240-V, 48-kW, long-shunt compound generator has a series resistance of 0.04 Ω and a shunt field resistance of 200 Ω. When the generator supplies the rated load at the rated voltage, the power input is 54 kW. The rotational loss is 2.5 kW. Find (a) the armature-circuit resistance and (b) the efficiency.

5.40. A 50-kW, 250-V, 600-rpm, long-shunt compound generator equipped with a field diverter resistance delivers the rated load at the rated voltage. The armature and series field resistances are 0.05 Ω and 0.1 Ω, respectively. There are 100 turns per pole in the shunt field winding and 2 turns per pole in the series field winding. The rotational loss is 4.75 kW. The measured currents through the shunt field winding and the diverter resistance are 5 A and 41 A, respectively. Determine (a) the series field current, (b) the diverter resistance, (c) the efficiency, and (d) the torque applied by the prime mover.

5.41. If the generator of Problem 5.40 is reconnected as a short-shunt compound generator and delivers the rated load at the rated voltage, what sort of adjustments are necessary to maintain the same flux and the power developed?

5.42. The data from the open-circuit test on a 30-kW, 150-V, long-shunt cumulative compound dc generator at 1200 rpm is as follows:

I_f (A):	0.5	1.0	1.5	2.0	2.5	3.0	3.5
E_a (V):	56	112	150	180	200	216	230

The other parameters are $R_a = 0.06\ \Omega$, $R_s = 0.04\ \Omega$, $N_f = 2000$ turns, $N_s = 6$ turns. Determine the armature reaction in terms of an equivalent field current when the generator supplies a load of 24 kW at 120 V. Determine the speed at which the generator must be driven to supply a load of 20 kW at 200 V. Assume that the armature reaction is proportional to the armature current squared.

5.43. A 10-kW, 125-V, long-shunt compound generator has $R_a = 10$ mΩ, $R_s = 20$ mΩ, $R_f = 50$ Ω, $N_s = 5$ turns, $N_f = 500$ turns, and $P_r = 740$ W. Under no-load conditions the terminal voltage reduces to 100 V. To make the generator flat compound, the series field must produce 300 A·t/pole. Calculate (a) the diverter resistance to accomplish the change, (b) the total mmf per pole at full load and at no load, and (c) the efficiency.

5.44. A short-shunt compound generator is designed to supply a load of 75 A at 440 V. The other parameters are $R_a = 0.2$ Ω, $R_s = 0.5$ Ω, $R_d = 1.5$ Ω, $I_f = 2.5$ A, and $P_r = 1$ kW. Calculate (a) the power developed by the armature, (b) all the electrical losses, and (c) the efficiency. Sketch the power-flow diagram.

CHAPTER 

Direct-Current Motors

"Cut-away" view of a PM dc motor. (*Courtesy of Bodine Electric Company*)

6.1 Introduction

In Chapter 5, we stated that a generator is a machine that converts mechanical energy into electric energy. When a machine converts electric energy into mechanical energy, it is called a **motor**. There is no fundamental difference in either the construction or the operation of the two machines. In fact, the same machine may be used as either a motor or a generator.

There are basically two types of motors: alternating-current (ac) motors and direct-current (dc) motors. An ac motor (discussed in Chapters 9 and 10) converts alternating (time-varying) power into mechanical power. When a machine converts time-invariant power into mechanical power, it is said to be a dc motor. This chapter is devoted to the study of dc motors.

When most of the power being generated, transmitted, and consumed is of the ac form, the use of a dc motor requires the installation of extra equipment for converting ac into dc. To justify the additional cost of a commutator on one hand and the installation of ac-to-dc converters on the other, a dc motor is put into service only when its performance is superior to that of an ac motor. By superior performance we simply mean that a dc motor is capable of doing what cannot be easily accomplished with an ac motor. For example, a dc motor can develop starting torque several orders of magnitude higher than a comparable size ac motor. A dc motor can operate at speeds that cannot be attained by an ac motor.

A dc motor is extensively used in control systems as a positioning device because its speed as well as its torque can be precisely controlled over a wide range. A dc motor is, of course, a logical choice when a dc power source is easily available.

The nominal unit to specify the power output of a dc motor is the horsepower (1 hp = 746 W). Direct-current motors are built in sizes ranging from fractional horsepower to over 1000 horsepower. Some of the applications of dc motors include automobiles, boats, airplanes, computers, printers, robots, electric shavers, toys, audio and video cassette recorders, movie cameras, traction motors for railways, subway trains and trams, rolling mills, hoist cranes, punch presses, and forklifts.

6.2 Operation of a DC Motor

Since there is no difference in construction between a dc generator and a dc motor, the three types of dc generators discussed in Chapter 5 can also be used as dc motors. Therefore, there are three general types of dc motors: shunt, series, and compound. The permanent-magnet (PM) motor is a special case of a shunt motor with uniform (constant) flux density. We can also have a separately excited motor if we use an auxiliary source of power for the field winding. Because it is not practical to employ two power sources, one for the field winding and the other for the armature circuit, a separately excited motor is virtually nonexistent. However, a separately excited motor can also be treated as a special case of a shunt motor.

The principle of operation of a dc motor is explained in detail in Section 3.4. A brief review is given here. In a dc motor, a uniform magnetic field is created by its poles. The armature conductors are forced to carry current by connecting them to a dc power source (supply) as shown in Figure 6.1. The current direction in the conductors under each pole is kept the same by the commutator. According to the Lorentz force equation, **a current-carrying conductor when placed in a magnetic field experiences a force that tends to move it**. This is essentially the principle of operation of a dc motor. All the conductors placed on the periphery of a dc motor are subjected to these forces, as shown in the figure. These forces cause the armature to rotate in the clockwise direction. Therefore, **the armature of a dc motor rotates in the direction of the torque developed by the motor**. For this reason, the torque developed by the motor is called the **driving torque**. Note that the torque developed by the conductors placed on the armature of a dc generator is in a direction opposite to its motion. Therefore, it can be labeled the **retarding torque**.

The magnitude of the average torque developed by these forces must be the same in both machines, since it does not matter whether the current is forced

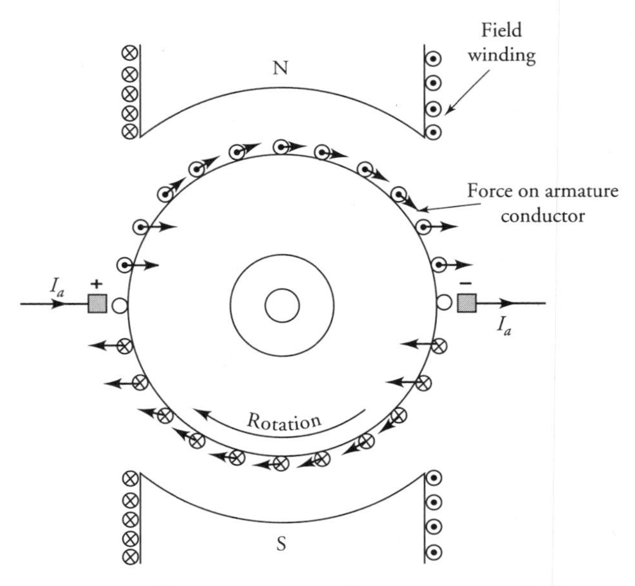

Figure 6.1 Force experienced by armature conductors in a 2-pole dc motor.

through the armature conductors by an external power source or is the result of the induced emf in the conductors. Thus,

$$T_d = K_a \Phi_P I_a \tag{6.1}$$

where $K_a = PZ/2\pi a$ is the machine constant, Φ_P is the flux per pole, and I_a is the armature current.

As the armature rotates, each coil on the armature experiences a change in the flux passing through its plane. Therefore, an electromotive force (emf) is induced in each coil. In accordance with Faraday's law of induction, the induced emf must **oppose** the current entering the armature. In other words, the induced emf opposes the applied voltage. For this reason, we commonly refer to the induced emf in a motor as the **back emf** or **counter emf** of the motor.

The average value of the induced emf at the armature terminals should, however, be the same as that for a dc generator because it does not really matter whether the armature is being driven by a prime mover or by its own driving torque. Thus,

$$E_a = K_a \Phi_P \omega_m \tag{6.2}$$

where ω_m is the angular velocity of the armature (rad/s).

If R is the effective (total) resistance in the armature circuit and V_s is the applied voltage across the armature terminals, then the armature current is

$$I_a = \frac{V_s - E_a}{R} \tag{6.3}$$

This equation can also be written as

$$V_s = E_a + I_a R \tag{6.4}$$

Since the resistance of the armature circuit R is usually very small, the voltage drop across it is also small in comparison with the back emf E_a. Therefore, most of the applied voltage V_s is needed to overcome the back emf E_a. It should also be evident from the above equation that the back emf in the motor is smaller than the applied voltage.

Starting a DC Motor

At the time of starting, the back emf in the motor is zero because the armature is not rotating. For a small value of the armature-circuit resistance R, the starting current in the armature will be extremely high if the rated value of V_s is impressed across the armature terminals. The excessive current can cause permanent damage to the armature windings. Thus, **a dc motor should never be started at its rated voltage.** In order to start a dc motor, an external resistance must be added in series with the armature circuit, as shown in Figure 6.2 for a PM motor. The external resistance is gradually decreased as the armature comes up to speed. Finally, when the armature has attained its normal speed, the external resistance is "cut out" of the armature circuit.

Armature Reaction

The theory of commutation and armature reaction as outlined in Chapter 5 for dc generators is also applicable to dc motors. The only difference is that the direction of the current in a conductor under a pole in a dc motor is opposite to that in a generator for the same direction of rotation. Therefore, the field created by the armature current in a dc motor is opposite in direction to that created by the current in the armature of a dc generator. Since the brushes were advanced to secure a good commutation in a dc generator, the brushes should be retreated in a dc motor.

If a dc machine has interpoles, the polarities of the interpoles for the dc motor must be opposite to those of the dc generator. As the interpoles carry armature current and the armature current is already in the opposite direction, the interpole polarities are automatically reversed. The same is true for the compensating windings. Therefore, no action is needed when a dc generator designed with interpoles or compensating windings is used as a dc motor.

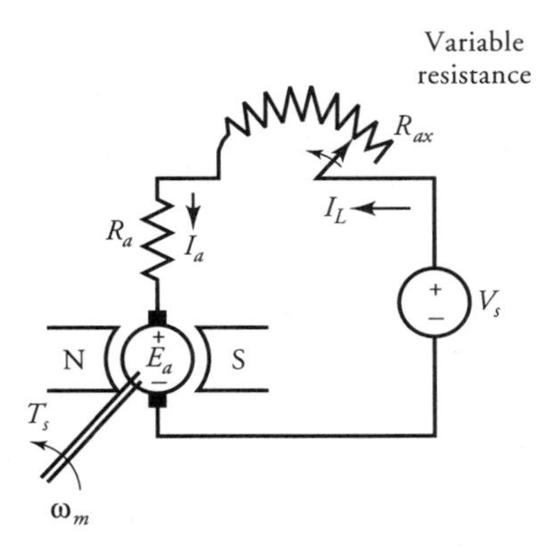

Figure 6.2 Variable resistance inserted in series with the armature circuit during starting of a dc (PM) motor.

6.3 Speed Regulation

The armature current of a motor increases with load. For a constant applied voltage, the increase in the armature current results in a decrease in the back emf. The reduction in the back emf causes a drop in the speed of the motor. The speed regulation is a measure of the change in speed from no load to full load. When the change in speed at full load is expressed as a percent of its full-load speed, it is called the **percent speed regulation ($SR\%$).** In equation form, the percent speed regulation is

$$SR\% = \frac{N_{mnL} - N_{mfL}}{N_{mfL}} \times 100 = \frac{\omega_{mnL} - \omega_{mfL}}{\omega_{mfL}} \times 100 \tag{6.5}$$

where N_{mnL} (ω_{mnL}) is the no-load speed, and N_{mfL} (ω_{mfL}) is the full-load speed of a dc motor.

As we continue our discussion of dc motors, we will observe that (a) a series motor is a variable-speed motor because its speed regulation is very high, (b) a shunt motor is essentially a constant-speed motor because its speed regulation is very small, and (c) a compound motor is a variable-speed motor because its speed regulation is higher than that of the shunt motor.

6.4 Losses in a DC Motor

The power input to a dc motor is electrical and the power output is mechanical. The difference between the power input and the power output is the power loss. A dc motor portrays the same power losses as a dc generator. (For details, refer to Section 5.11.)

When the power is supplied to a motor, a significant portion of that power is dissipated by the resistances of the armature and the field windings as copper loss. The remainder power (the developed power) is converted by the motor into mechanical power. A part of the developed power is consumed by the rotational loss. The difference is the net mechanical power available at the shaft of the motor. A typical power-flow diagram of a dc motor is shown in Figure 6.3.

6.5 A Series Motor

In a series motor the field winding is connected in series with the armature circuit as shown in Figure 6.4. We have also included an external resistance R_{ax} in series with the armature that can be used either to start the motor and then be shorted

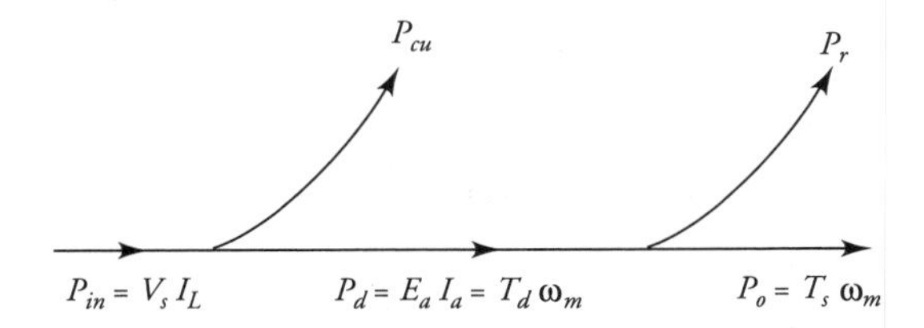

Figure 6.3 A typical power-flow diagram of a dc motor.

or to control the speed of the motor. Since the series field winding carries the rated armature current of the motor, it has few turns of heavy wire. As the armature current changes with the load, so does the flux produced by the field winding. In other words, the flux set up by a series motor is a function of the armature current. If the flux per pole can be expressed as

$$\Phi_P = k_f I_a \tag{6.6}$$

then the back emf is

$$E_a = K_a k_f I_a \omega_m \tag{6.7}$$

and the torque developed by the series motor is

$$T_d = K_a k_f I_a^2 \tag{6.8}$$

From the above equations, it is evident that the back emf in the motor is proportional to the armature current, and the torque developed by a series motor is proportional to the **square** of the armature current as long as the motor is

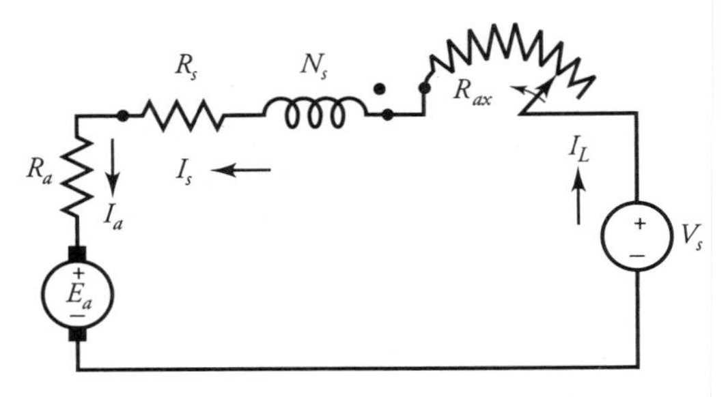

Figure 6.4 An equivalent circuit of a series motor with a variable starting resistance.

operating in the linear region. As the armature current increases, so does the flux produced by it. An increase in the flux enhances the level of saturation in the motor. When the motor is saturated, the flux increases only gradually with further increase in the armature current. Hence, the torque developed is no longer proportional to the square of the current. The torque versus armature current characteristic of a series motor is given in Figure 6.5.

When a series motor operates under no load, the torque developed by the motor is just sufficient to overcome the rotational loss in the machine. Since the rotational loss is only a fraction of the full-load torque, the torque developed by the machine is very small at no load. From Eq. (6.8), the armature current must also be very small. Therefore, the back emf at no load must be nearly equal to the applied voltage V_s. Since the back emf is also proportional to the armature current and the armature current is a small fraction of its rated value, the motor must attain a relatively high speed. In fact, it is possible for a series motor to self-destruct under no load owing to centrifugal action.

As we load the motor, the torque developed by it must increase. The increase in the torque necessitates an increase in the armature current. The increase in the armature current causes an increase in the voltage drop across the armature-circuit resistance, the field-winding resistance, and the external resistance. For a fixed applied voltage, the back emf must decrease with load. Since the back emf is also proportional to the armature current, the speed of the motor must drop. Figure 6.6 shows the torque-speed characteristic of a series motor. We will comment on the nature of the curve shortly. Also shown in the figure is the power developed by a series motor as a function of its speed.

From the equivalent circuit of Figure 6.4, we have

$$\begin{aligned} E_a &= V_s - I_a(R_a + R_s + R_{ax}) \\ &= V_s - I_a R \end{aligned}$$

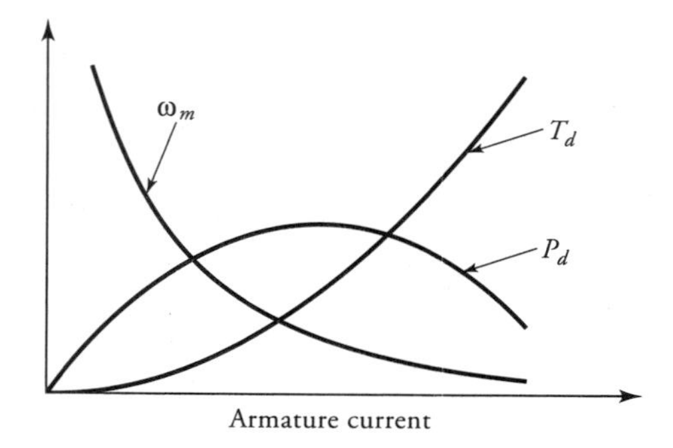

Figure 6.5 Torque developed, power developed, and speed characteristics of a series motor as a function of armature current.

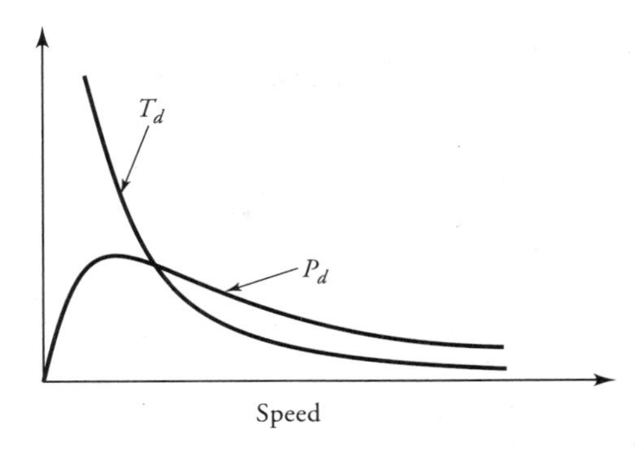

Figure 6.6 Torque and power developed characteristics of a series motor as a function of speed.

where $R = R_a + R_s + R_{ax}$ is the total armature-circuit resistance. Substituting for E_a from Eq. (6.7), we can obtain an expression for the speed of the motor in terms of armature current as

$$\omega_m = \frac{V_s - I_a R}{K_a k_f I_a} \tag{6.9}$$

This equation states that the speed of a series motor is practically inversely proportional to the armature current. The nature of the speed-current characteristic is depicted in Figure 6.5.

Equation (6.9) can also be rewritten as

$$I_a = \frac{V_s}{K_a k_f \omega_m + R} \tag{6.10}$$

The torque developed, from Eq. (6.8), can now be expressed as

$$T_d = \frac{K_a k_f V_s^2}{[R + K_a k_f \omega_m]^2} \tag{6.11}$$

From this equation it is evident that, for all practical purposes, the torque developed by a series motor is inversely proportional to the square of its speed. The nature of the speed-torque characteristic in Figure 6.6 should now be obvious. A series motor provides high torque at low speed and low torque at high speed. For this reason, a series motor is suitable for hoists, cranes, electric trains, and a host of other applications that require large starting torques.

Since the torque developed by a series motor is also proportional to the square of the applied voltage, the torque developed by it can be controlled by controlling the applied voltage. For example, at a given speed, doubling the applied voltage results in quadrupling of the torque. The torque developed as a function of armature current is shown in Figure 6.5.

The power developed by a series motor is

$$P_d = E_a I_a$$
$$= [V_s - I_a R] I_a$$

The power developed by a series motor for a constant applied voltage is maximum when $I_a \rightarrow I_{am}$ and $dP_d/dI_a \rightarrow 0$. Thus, differentiating the above equation with respect to I_a and setting it to zero, we obtain

$$I_{am} = \frac{V_s}{2R} \tag{6.12}$$

The maximum power developed by a series motor is

$$P_{dm} = \frac{V_s^2}{4R} \tag{6.13}$$

For a stable operation, the operating range of a series motor is well below the maximum power developed by it. The power developed as a function of armature current is shown in Figure 6.5.

EXAMPLE 6.1

The magnetization curve of a 10-hp, 220-V series motor is given in Figure 6.7 at 1200 rpm. The other parameters of the series motor are $R_a = 0.75\ \Omega$, $R_s = 0.25\ \Omega$, and $P_r = 1.04$ kW. What is the armature current when the motor delivers its rated

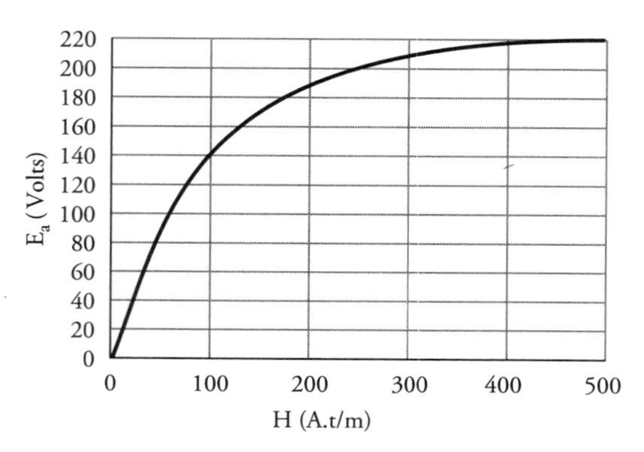

Figure 6.7 The magnetization curve of a dc motor.

load at 1200 rpm? What is the efficiency of the motor at full load? What is the number of turns per pole? When the load is gradually reduced, the armature current decreases to 16.67 A. Determine (a) the new speed of the motor and (b) the driving torque.

● SOLUTION

The power output: $P_o = 10 \times 746 = 7460$ W

The power developed: $P_d = P_o + P_r = 7460 + 1040 = 8500$ W

The armature-circuit resistance: $R = 0.25 + 0.75 = 1\ \Omega$

Since $P_d = E_a I_a = 8500$ W and $E_a = V_s - I_a R$, we have

$$E_a I_a = [V_s - I_a R]\, I_a$$

Substituting the values, we obtain

$$I_a^2 - 220 I_a + 8500 = 0$$

or

$$I_a = 50 \text{ A}$$

The power input: $P_{in} = V_s I_a = 220 \times 50 = 11{,}000$ W

Hence,

$$\eta = \frac{7460}{11{,}000} = 0.6782 \quad \text{or} \quad 67.82\%$$

The induced emf: $E_a = 220 - 50 \times 1 = 170$ V

From the magnetization curve at 1200 rpm, the magnetomotive force (mmf) of the series field winding corresponding to a back emf of 170 V is 150 A·t/pole. Hence, the turns per pole are

$$N_s = \frac{150}{50} = 3 \text{ turns/pole}$$

At the reduced load: $I_{an} = 16.67$ A

$$E_{an} = 220 - 16.67 \times 1 = 203.33 \text{ V}$$

The mmf of the series winding is $16.67 \times 3 = 50$ A·t/pole. From the magnetization curve, the back emf at 1200 rpm is 90 V. In order to obtain a back emf of 203.33 V at the same mmf of 50 A·t/pole, the new speed of the motor is

$$N_{mn} = \frac{203.33}{90} \times 1200 \approx 2711 \text{ rpm} \qquad \text{or} \qquad \omega_{mn} = 283.9 \text{ rad/s}$$

The power developed at reduced load: $P_{dn} = 203.33 \times 16.67 = 3389.51$ W

$$T_{dn} = \frac{3389.51}{283.9} = 11.94 \text{ N·m}$$

■

Exercises

6.1. A series motor develops 20 N·m of torque when the armature current is 40 A. When the load on the motor is gradually increased, the armature current increases to 60 A. If the motor is operating in a linear region, determine the torque developed by the motor.

6.2. A 240-V series motor takes 80 A when driving its rated load at 600 rpm. The other parameters of the motor are $R_a = 0.2\ \Omega$ and $R_s = 0.3\ \Omega$. Calculate the efficiency if 5% of the power developed is lost as rotational loss. What is the horsepower rating of the motor? If a 50% increase in the armature current results in a 20% increase in the flux, determine (a) the speed of the motor when the armature current is 120 A and (b) the torque developed by the motor.

6.3. A series motor operates in the linear region in which the flux is proportional to the armature current. When the armature current is 12 A, the motor speed is 600 rpm. The line voltage is 120 V, the armature resistance is 0.7 Ω, and the series field–winding resistance is 0.5 Ω. What is the torque developed by the motor? For the motor to operate at a speed of 2400 rpm, determine (a) the armature current and (b) the driving torque.

6.6 A Shunt Motor

The equivalent circuit of a shunt motor is shown in Figure 6.8 with a starting resistor in the armature circuit. The field winding is connected directly across the source. For a constant-source voltage, the flux created by the field winding is constant. The torque developed by the motor is

$$T_d = K_a \Phi_P I_a = K I_a \tag{6.14}$$

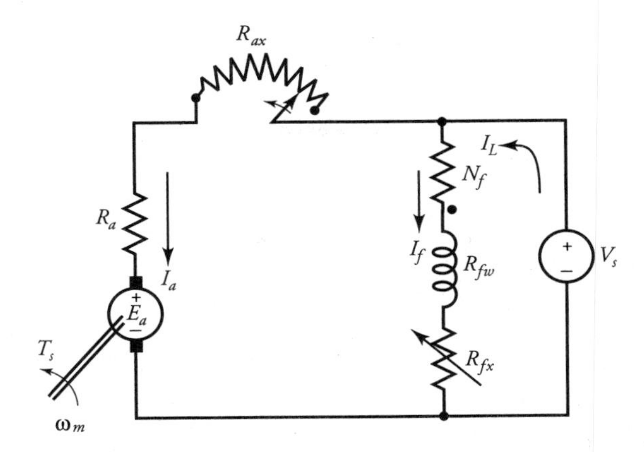

Figure 6.8 An equivalent circuit of a shunt motor with a starting resistor in the armature circuit.

where $K = K_a\Phi_P$ is a constant quantity. Hence the torque developed by a shunt motor is proportional to the armature current as shown in Figure 6.9.

When the shunt motor is driving a certain load, the back emf of the motor is

$$E_a = V_s - I_aR_a \tag{6.15}$$

Since $E_a = K_a\Phi_P\omega_m$, the operating angular velocity (speed) of the motor is

$$\omega_m = \frac{V_s - I_aR_a}{K_a\Phi_P} \tag{6.16}$$

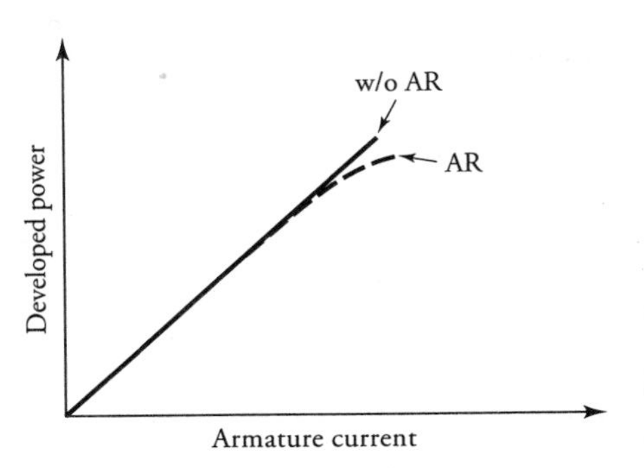

Figure 6.9 Torque developed by a shunt motor as a function of armature current with armature reaction (AR) and without armature reaction (w/o AR).

Although we are developing relationships in terms of the angular velocity ω_m, we still refer to it as the speed of the motor.

When the load on the motor is increased, the following sequence of events takes place:

(a) The armature current I_a increases to satisfy the demand of increased load.
(b) The voltage drop across the armature circuit resistance R_a increases.
(c) For a fixed-source voltage, the back emf E_a decreases.
(d) Since the flux is constant when armature reaction is negligible, the decrease in the back emf of the motor is accompanied by a decrease in its speed as shown in Figure 6.10 (curve d).

With the increase in the armature current, the armature reaction becomes more significant if the motor is not compensated for it. The increase in the armature reaction decreases the flux in the motor, which, in turn, causes an increase in the speed. Depending upon the magnetic saturation of the motor and the severity of the armature reaction, the increase in the speed due to the armature reaction may be less than, equal to, or greater than the drop in the speed due to the increase in the armature current, as depicted in Figure 6.10 by curves a, b, and c, respectively. Curve c is not really desirable, as it may lead to instability of operation. We can overcome this undesirable effect by adding few series turns in the motor. Such a winding, when included as a part of a shunt motor, is called the **stabilizing** winding.

Equation (6.16) can also be written as

$$\omega_m = \omega_{mnL} - \frac{R_a}{K_a\Phi_P} I_a \tag{6.17}$$

where $\omega_{mnL} = V_s/K_a\Phi_P$ is the no-load speed of the motor. This is the speed at which the torque developed by the motor is zero. The actual speed of the motor

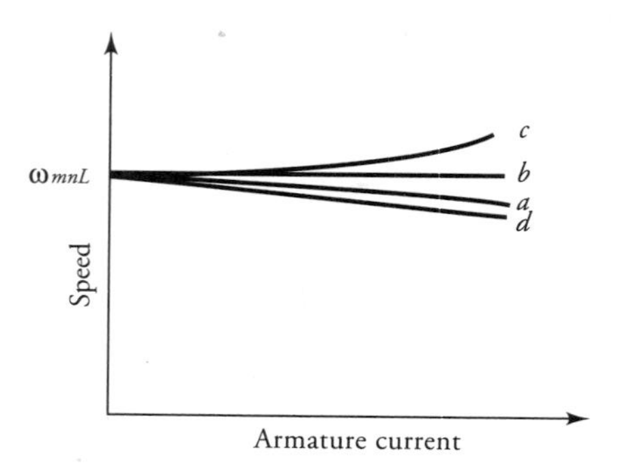

Figure 6.10 Speed versus armature current characteristic of a shunt motor.

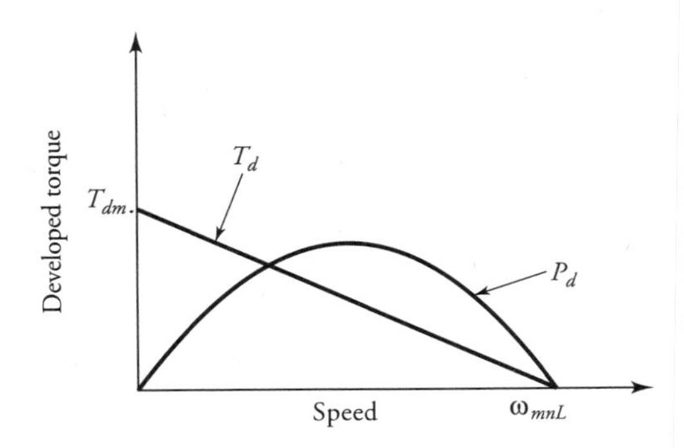

Figure 6.11 Torque developed and power developed as a function of speed of a typical shunt motor.

is bound to be lower than ω_{mnL} owing to the rotational loss. As $R_a \to 0$, the speed of a shunt motor $\omega_m \to \omega_{mnL}$. In other words, **a shunt motor is a constant-speed motor.**

We can also obtain an expression for the torque developed in terms of the speed of the motor from Eqs. (6.14) and (6.16) as

$$T_d = \frac{K_a \Phi_P V_s}{R_a} - \frac{K_a^2 \Phi_P^2}{R_a} \omega_m \tag{6.18}$$

This straight-line relationship, depicted in Figure 6.11, is known as the **torque-speed characteristic** of a shunt motor. It simply states that the speed increases as the load on the motor decreases. Also plotted in Figure 6.11 is the power developed by the machine as a function of its speed. We can show that the power developed by a shunt motor is maximum when its speed is equal to $0.5\omega_{mnL}$. The corresponding expression for the maximum power developed by the machine is

$$P_{dm} = \frac{V_s^2}{4R_a} \tag{6.19}$$

EXAMPLE 6.2

The data for the magnetization characteristic of a 4-pole, 120-V shunt motor at 2400 rpm are

E_a (V):	10	50	78	95	110	120	127	134	138
I_f (A):	0.2	0.4	0.6	0.8	1.0	1.2	1.4	1.6	1.8

The armature-circuit resistance is 0.1 Ω. The shunt winding has 500 turns per pole with a total resistance of 100 Ω. The full-load current is 40 A, and the demagnetization effect of armature reaction is 0.2 A in terms of the shunt field current. The rotational loss is 240 W. Determine (a) the speed of the motor at full load, (b) the speed regulation, (c) the torque available at the shaft, and (d) its efficiency.

● SOLUTION

The magnetization curve is plotted in Figure 6.12. At no load, the field current is

$$I_f = \frac{120}{100} = 1.2 \text{ A}$$

At full load, the effective field current is 1.2 − 0.2 = 1.0 A. The back emf of the motor at 2400 rpm is 110 V from the magnetization characteristic. The full-load armature current is

$$I_a = I_L - I_f = 40 - 1.2 = 38.8 \text{ A}$$

Thus, the back emf of the motor must be

$$E_a = V_s - I_a R_a = 120 - 38.8 \times 0.1 = 116.12 \text{ V}$$

(a) Hence, the motor speed is

$$N_{m2} = \frac{E_{a2}}{E_{a1}} N_{m1} = \frac{116.12}{110} \times 2400 \approx 2534 \text{ rpm} \quad \text{or} \quad 265.3 \text{ rad/s}$$

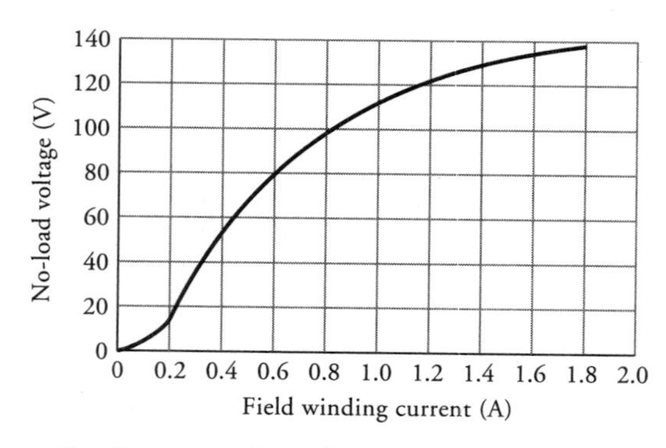

Figure 6.12 The magnetization curve for a shunt motor at 2400 rpm. The shunt field winding has 500 turns per pole.

At no load, the field current is 1.2 A and the corresponding value of the back emf is 120 V. Since the rotational loss is 240 W, the no-load armature current (first iteration) is 240/120 = 2 A. The voltage drop across R_a is 0.2 V. Thus, the back emf must be 119.8 V. The no-load speed of the motor is approximately 2400 rpm.

The armature reaction has resulted in a full-load speed higher than the no-load speed of the motor.

(b) The speed regulation of the motor is

$$SR\% = \frac{2400 - 2534}{2534} \times 100 = -5.3\%$$

The power developed by the motor is

$$P_d = E_a I_a = 116.12 \times 38.8 = 4505.46 \text{ W}$$

The power output is

$$P_o = P_d - P_r = 4505.46 - 240 = 4265.46 \text{ W}$$

(c) The torque available at the shaft is

$$T_s = \frac{4265.46}{265.3} = 16.08 \text{ N·m}$$

(d) The power input is

$$P_{in} = V_s I_L = 120 \times 40 = 4800 \text{ W}$$

Thus, the efficiency of the motor is

$$\eta = \frac{4265.46}{4800} = 0.889 \quad \text{or} \quad 88.9\%$$

■

Exercises

6.4. Verify Eq. (6.19).
6.5. Repeat Example 6.2 if the armature reaction is negligible.
6.6. A 220-V shunt motor draws 10 A at 1800 rpm. The armature-circuit resistance is 0.2 Ω, and the field-winding resistance is 440 Ω. The rotational loss

is 180 W. Determine (a) the back emf, (b) the driving torque, (c) the shaft torque, and (d) the efficiency of the motor.

6.7. If the torque developed by the motor in Exercise 6.6 is 20 N·m, determine its (a) speed, (b) line current, and (c) efficiency. Assume that the armature reaction is zero and the rotational loss is proportional to the speed.

6.7 The Compound Motor

A shunt motor may have an additional series field winding in the same manner as a shunt generator. The series field winding may be connected so that the flux produced by it aids the flux set up by the shunt field winding, in which case the motor is said to be a **cumulative compound motor**. A motor is said to be **differential compound** when the flux of the series field winding opposes the flux of the shunt field winding. In fact, the stabilizing winding discussed in the previous section makes a shunt motor to behave like a cumulative compound motor.

A compound motor may be connected either as a **short-shunt motor** or a **long-shunt motor**. In a long-shunt motor, the shunt field winding is directly connected across the power source as depicted in Figure 6.13**a**. Therefore, the flux created by the shunt field winding is constant under all loading conditions. On the other hand, the shunt field winding of a short-shunt compound motor is connected directly across the armature terminals as shown in Figure 6.13**b**. The flux created by the shunt field winding of a short-shunt motor decreases (somewhat) with an increase in the load owing to the voltage drop across the series field winding.

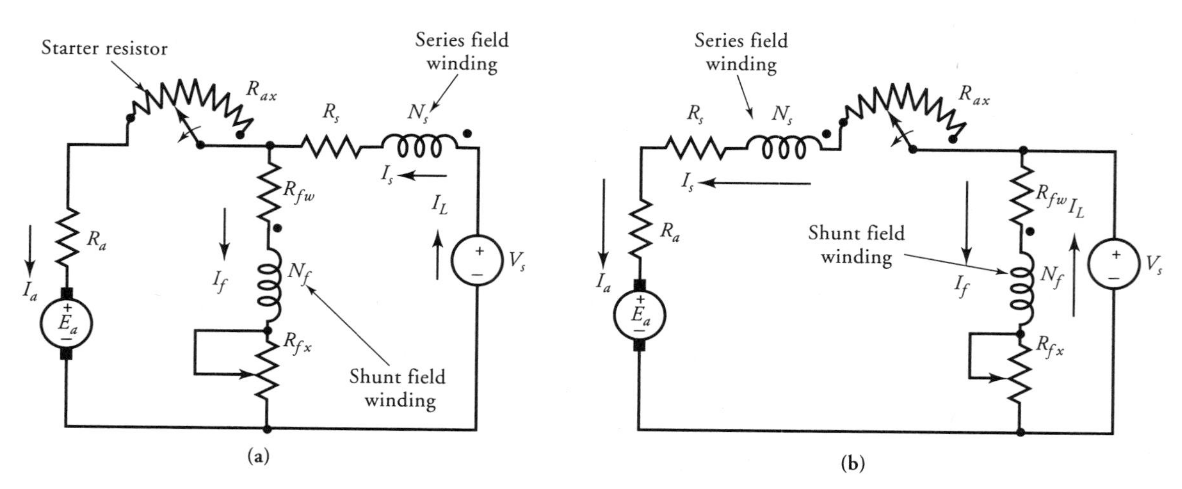

Figure 6.13 **(a)** A short-shunt and **(b)** a long-shunt compound motor.

As you may have guessed, the characteristics of a compound motor are a combination of the characteristics of a shunt motor and a series motor. As we increase the load on a compound motor, the following sequence of events takes place:

1. The total flux (increases/decreases) owing to the increase in the series-winding current for a (cumulative/differential) compound motor. That is,

$$\Phi_P = \Phi_{sh} \pm k_f I_a \tag{6.20}$$

where Φ_P is the total flux in the motor, Φ_{sh} is the flux due to the shunt field winding, and $k_f I_a$ is the flux produced by the series field winding. Note that the minus sign is for the differential compound motor.

2. The (increase/decrease) in the flux with an increase in the armature current causes the torque to (increase/decrease) at a faster rate in a (cumulative/differential) compound motor than in a shunt motor. Thus, the torque developed is

$$T_d = K_a I_a \Phi_{sh} \pm K_a k_f I_a^2 \tag{6.21}$$

The torque versus armature current characteristics of cumulative and differential compound motors are shown in Figure 6.14.

3. The (increase/decrease) in the flux accompanied by an increase in the voltage drops across the armature circuit and the series field–winding resistances causes the motor speed to (decrease/increase) more rapidly than it does in a shunt motor. Hence, the motor speed is

$$\omega_m = \frac{V_s - I_a(R_a + R_s)}{K_a(\Phi_{sh} \pm k_f I_a)} \tag{6.22}$$

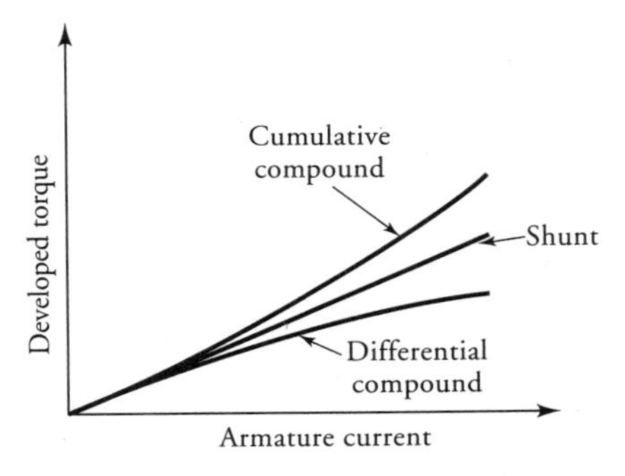

Figure 6.14 Torque-current characteristics of shunt, cumulative compound, and differential compound motors.

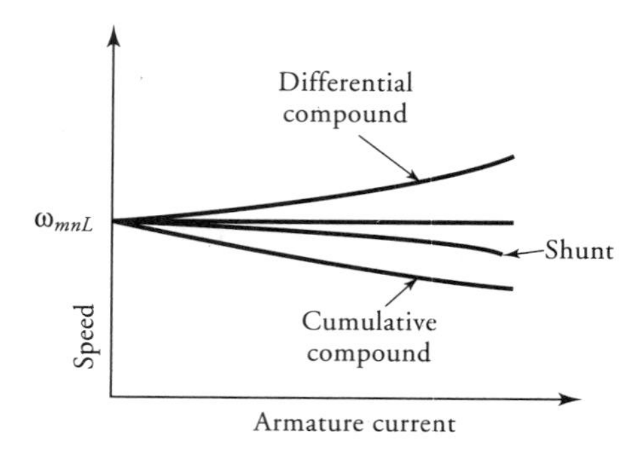

Figure 6.15 Speed-current characteristics of shunt, cumulative compound, and differential compound motors.

The speed versus armature current characteristics of cumulative and differential compound motors are given in Figure 6.15.

In a differentially compound motor, the flux decreases with an increase in the armature current. Therefore, there is a possibility for a differentially compound motor to attain dangerously high speed as the flux due to the series field winding approaches the flux created by the shunt field winding.

A cumulative compound motor has a definite no-load speed, so it does not "run away" like a series motor when the load is removed. It also develops high starting torque when the load is suddenly increased. This makes it suitable for such applications as rolling mills, shears, and punching presses. It is also a preferred motor for applications (such as cranes and elevators) that (a) require high starting torque, (b) are prone to sudden load changes, and (c) present a possibility of going from full load to no load.

EXAMPLE 6.3

To stabilize the shunt motor of Example 6.2, a stabilizing (series field) winding having 4 turns per pole and a resistance of 0.025 Ω is added. Determine (a) the speed of the cumulative compound motor, (b) the power developed, (c) the torque developed, and (d) the percent speed regulation.

● SOLUTION

The mmf of the series field winding at full load is

$$\text{mmf}_s = 4 \times 38.8 = 155.2 \text{ A·t/pole}$$

Thus, the effective shunt field current is

$$I_{fe} = 1.2 + \frac{155.2}{500} = 1.51 \text{ A}$$

From the magnetization curve, Figure 6.12, the back emf at 2400 rpm is 130 V. However, the back emf at full load is

$$E_a = 120 - 38.8(0.1 + 0.025) = 115.15 \text{ V}$$

(a) Since the ratio of back emfs is equal to the ratio of the speeds, the new speed of the cumulative compound motor is

$$N_m = \frac{115.15}{130} \times 2400 \approx 2126 \text{ rpm} \qquad \text{or} \qquad 222.62 \text{ rad/s}$$

(b) The power developed by the cumulative compound motor is

$$P_d = 115.15 \times 38.8 = 4467.82 \text{ W}$$

(c) The torque developed by the cumulative compound motor is

$$T_d = \frac{4467.82}{222.62} = 20.07 \text{ N·m}$$

(d) The speed regulation is

$$SR\% = \frac{2400 - 2126}{2126} \times 100 = 12.89\%$$

■

Exercises

6.8. A 230-V, 1800-rpm, long-shunt, cumulative compound motor has a shunt field resistance of 460 Ω, series field resistance of 0.2 Ω, and armature-circuit resistance of 1.3 Ω. The full-load line current is 10.5 A. The rotational loss is 5% of the power developed. Compute the efficiency of the motor at full load.

6.9. In the motor of Exercise 6.8, the line current rises to 15.5 A when the load is increased. Assume that there is a 20% increase in flux in the motor due to the increase in the line current. Calculate (a) the speed of the motor and (b) its efficiency.

6.8 Methods of Speed Control

The foremost reason that a dc motor is used extensively in the design of a control system is the ease with which its performance can be tailored to meet the demands of the system. Stated differently, a dc motor enables us to change its speed at any desired torque without making any changes in its construction.

The two methods that are commonly used to secure speed control are **armature resistance control** and **field control**.

Armature Resistance Control

In this method, the speed control is achieved by inserting a resistance R_c in the armature circuit of a shunt, series, or compound motor as illustrated in Figure 6.16. In a shunt or a compound motor, the field winding is connected directly across the full-line voltage. The additional resistance in the armature circuit decreases the back emf in the motor for any desired armature current. Since the flux in the motor is constant and the torque depends upon the armature current, the decrease in the back emf forces a drop in the motor speed.

We can express the speed of a dc motor in terms of its armature current as

$$\omega_m = \frac{V_s - I_a R}{K_a \Phi_P} \tag{6.23}$$

where

$$R = R_a + R_s + R_c$$

for a series or a compound motor, and

$$R = R_a + R_c$$

for a shunt motor.

It is obvious from the above equation that any increase in the value of the control resistance R_c decreases the speed of the motor. **The armature resistance control method, therefore, is suitable to operate the motor at a speed lower than its rated speed while delivering the same torque.** The speed-torque characteristics of series, shunt, and compound motors for various values of the control resistor are depicted in Figure 6.17. In fact, the starting resistors, as we have shown

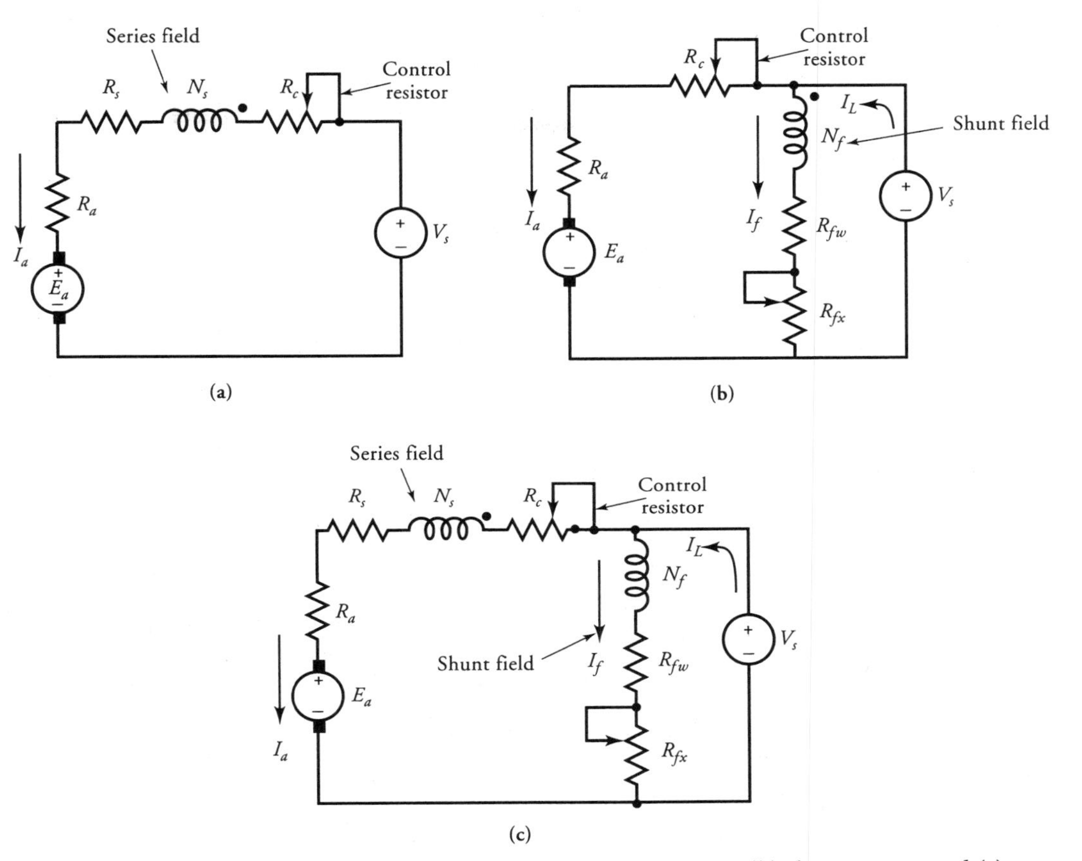

Figure 6.16 Armature resistance control for **(a)** series motor, **(b)** shunt motor, and **(c)** compound motors.

in the equivalent circuits of dc motors, can also be used for speed-control purposes.

The disadvantages of this method of speed control are the following:

(a) A considerable power loss in the control resistance R_c
(b) A loss in the efficiency of the motor
(c) Poor speed regulation for the shunt and the compound motors

The armature resistance control method is essentially based upon the reduction in the applied voltage across the armature terminals of a dc motor. Therefore, it should be possible to control the speed of a dc motor by simply connecting its armature to a variable voltage source. This method of speed control is known as the **Ward–Leonard method** and is discussed in the next section.

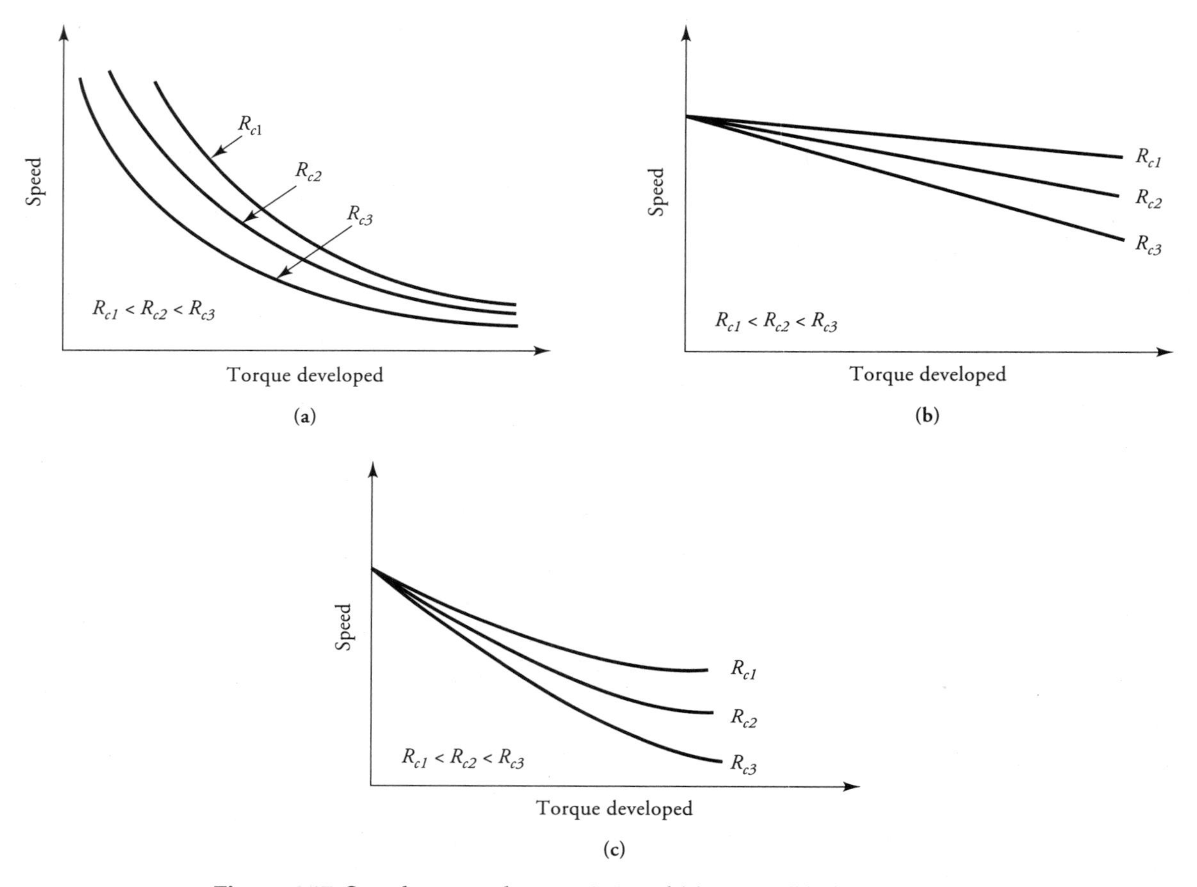

Figure 6.17 Speed-torque characteristics of **(a)** series, **(b)** shunt, and **(c)** compound motors showing the effect of additional resistance in the armature circuit.

EXAMPLE 6.4

A 120-V, 2400-rpm shunt motor has an armature resistance of 0.4 Ω and a shunt field resistance of 160 Ω. The motor operates at its rated speed at full load and takes 24.75 A. The no-load current of the motor is 2 A. If an external resistance of 3.6 Ω is inserted in the armature circuit, calculate the motor speed, the power loss in the external resistance as a percent of total power input, and the efficiency of the motor. Assume that the rotational loss is proportional to the speed.

● SOLUTION

The field current: $I_f = 120/160 = 0.75$ A

From **no-load data**:

$$I_a = 2 - 0.75 = 1.25 \text{ A}$$

$$E_{anL} = 120 - 1.25 \times 0.4 = 119.5 \text{ V}$$

$$P_{dnL} = 119.5 \times 1.25 = 149.38 \text{ W}$$

The power developed at no load accounts for the rotational loss in the motor.

From **full-load data**:

$$I_{afL} = 14.75 - 0.75 = 14 \text{ A}$$

$$E_{afL} = 120 - 14 \times 0.4 = 114.4 \text{ V}$$

$$P_{dfL} = 114.4 \times 14 = 1601.6 \text{ W}$$

$$N_{mfL} = 2400 \text{ rpm}$$

Hence, the no-load speed is

$$N_{mnL} = \frac{119.5}{114.4} \times 2400 \approx 2507 \text{ rpm}$$

The rotational loss at full load is

$$P_{rfL} = \frac{2400}{2507} \times 149.38 = 143 \text{ W}$$

The power output: $P_{ofL} = 1601.6 - 143 = 1458.6$ W

The power input: $P_{infL} = 120 \times 14.75 = 1770$ W

Thus, the efficiency: $\eta = \dfrac{1458.6}{1770} = 0.824 \quad \text{or} \quad 82.4\%$

With **external resistance in the armature circuit**:

$$E_{an} = 120 - 14 \times (0.4 + 3.6) = 64 \text{ V}$$

$$P_{dn} = 64 \times 14 = 896 \text{ W}$$

$$N_{mn} = \frac{64}{114.4} \times 2400 \approx 1343 \text{ rpm}$$

$$P_{rn} = \frac{1343}{2507} \times 149.38 = 80 \text{ W}$$

The power output and the efficiency are

$$P_{on} = 896 - 80 = 816 \text{ W}$$

$$\eta = \frac{816}{1770} = 0.461 \qquad \text{or} \qquad 46.1\%$$

Power loss in the external resistance: $P_c = 14^2 \times 3.6 = 705.6$ W
Percent of the power loss in the control resistance is

$$\frac{705.6}{1770} \times 100 = 39.86\%$$

It is evident from this example that (a) a large percentage of the power supplied to the motor is lost in the control resistance, (b) the efficiency has decreased considerably, and (c) the speed has been reduced to one-half of its rated value. It is left for the reader to verify that there is no change in the torque developed by the motor.

■

Exercises

6.10. A 240-V, 1800-rpm shunt motor has $R_a = 2.5\ \Omega$ and $R_f = 160\ \Omega$. When it operates at full load and its rated speed, it takes 21.5 A from the source. What resistance must be placed in series with the armature in order to reduce its speed to 450 rpm while the torque developed by the motor remains the same?

6.11. A 20-hp, 440-V series motor is 87% efficient when it delivers the rated load at 900 rpm. The armature-circuit resistance is 0.3 Ω, and the series field resistance is 0.2 Ω. If an external resistance of 2.5 Ω is inserted in the armature circuit and the load is reduced by 20%, determine the motor speed. Assume that the motor operates in the linear region.

6.12. A 120-V series motor takes 20 A when it delivers the rated load at 1600 rpm. The armature resistance is 0.5 Ω and the series field resistance is 0.3 Ω. Determine the resistance that must be added to obtain the rated torque (a) at starting and (b) at 1200 rpm.

Field-Control Method

Another approach to control the speed of a dc motor involves the control of the field current, which in turn controls the flux in the motor. The field current in a shunt motor can be controlled by inserting an external resistor in series with the field winding. Because the field current is a very small fraction of the total current intake of a shunt motor, the power dissipated by the external resistor is relatively small. Therefore, the flux-control method is economically better than the armature-resistance-control method.

To control the flux in a series motor, a field diverter resistor can be connected in parallel with the series field winding. If all the coils in a series field winding are connected in series, we can also change the flux in a series motor by connecting the coils in parallel.

The addition of a resistance in series with the shunt field winding or in parallel with the series field winding causes the field current and thereby the flux in the motor to decrease. Since the speed of a motor is inversely proportional to its flux, a decrease in its flux results in an increase in its speed. Thus, **the flux-control method makes a motor operate at a speed higher than its rated speed**.

As the torque developed by a shunt motor is proportional to the product of the armature current and the flux per pole, a decrease in the flux must be accompanied by a corresponding increase in the armature current for the motor to deliver the same torque. This method of speed control is, therefore, not satisfactory for compound motors, because any decrease in the flux produced by the shunt field winding is offset by an increase in the flux produced by the series field winding owing to an increase in the armature current.

EXAMPLE 6.5

If in the shunt motor of Example 6.4 an external resistance of 80 Ω is inserted in series with the shunt field winding instead of the armature circuit, determine (a) the motor speed, (b) the power loss in the external resistance, and (c) the efficiency. Assume that the flux is proportional to the square root of the field-winding current.

● SOLUTION

The new field-winding current is

$$I_{fn} = \frac{120}{160 + 80} = 0.5 \text{ A}$$

Let Φ_P be the flux at full load when the field current is 0.75 A, and Φ_{Pn} be the flux when the field current is 0.5 A. The new flux in the motor is

$$\Phi_{Pn} = \sqrt{\frac{0.5}{0.75}}\,\Phi_P = 0.816\,\Phi_P$$

For the torque developed to be the same, the new armature current must be

$$I_{an} = \frac{\Phi_P}{\Phi_{Pn}} I_a = \frac{1}{0.816} \times 14 = 17.15 \text{ A}$$

Thus, the new back emf in the motor is

$$E_{an} = 120 - 17.15 \times 0.4 = 113.14 \text{ V}$$

(a) Hence, the new speed of the motor is

$$N_{mn} = \frac{\Phi_P E_{an}}{\Phi_{Pn} E_a} N_m = \frac{113.14 \times 2400}{0.816 \times 114.4} \approx 2909 \text{ rpm}$$

(b) The power loss in the external resistance is

$$P_{fx} = 0.5^2 \times 80 = 20 \text{ W}$$

(c) The power input: $P_{in} = 120 \times (17.15 + 0.5) = 2118$ W
The power developed: $P_{dn} = 113.14 \times 17.15 = 1940.35$ W

To determine the efficiency, we need information on the rotational losses. The motor is expected to have a lower core loss due to the reduction in the flux but higher friction and windage loss due to the increase in speed. It is possible that one may offset the other. If we assume that the rotational loss is the same as that at 2400 rpm, then the power output is

$$P_o = 1940.35 - 143 = 1797.35 \text{ W}$$

Hence,
$$\eta = \frac{1797.35}{2118} = 0.849 \quad \text{or} \quad 84.9\%$$

■

Exercises

6.13. A 220-V shunt motor whose magnetization curve is given below runs at a speed of 1200 rpm. The field-winding resistance is 40 Ω. Determine the resistance that must be inserted in series with the field winding to increase the speed to 1500 rpm at no load.

I_f (A):	1	2	3	4	5	6
Φ_P (mWb):	4.5	8	10	11	11.6	12.2

6.14. A 120-V series motor runs at 2000 rpm at full load and takes a current of 12 A. The armature-winding resistance is 0.3 Ω and the series field resistance is 0.2 Ω. A diverter resistance of 0.2 Ω is connected across the series field, and the load on the motor is adjusted. If the motor now takes 20 A, what is the new speed of the motor? Determine the torque developed in each case.

6.15. A 440-V shunt motor has an armature resistance of 0.5 Ω. The field current is 2 A. The motor takes a current of 22 A on full load. Determine the percent reduction in its flux if the motor is to develop the same torque at twice the speed.

6.9 The Ward-Leonard System

A critical examination of the armature–resistance–control method reveals that the insertion of an external resistor in series with the armature circuit is equivalent to applying a voltage lower than the rated value across the armature terminals. It should, therefore, be obvious that we can achieve the same effect by actually applying a reduced voltage across the armature terminals of a dc motor while the voltage across the shunt field winding of a shunt or a compound motor is held constant. This method of speed control is known as the **armature voltage control**. The advantage of this method is that it eliminates the excessive power loss that is inherently associated with the armature resistance control. The main drawback is that it requires two power sources to control the speed of a shunt or a compound motor.

The Ward-Leonard system of speed control is, in fact, based upon the method of armature voltage control. To control the speed of a dc motor, this system requires two generators and an ac motor. The three-phase ac motor acts as a prime mover that drives both generators as illustrated in Figure 6.18. One generator, called the **exciter**, provides a constant voltage that is impressed upon the field windings of the other separately excited generator and the separately excited motor under control as shown. The armature winding of the motor is permanently connected to the armature terminals of the other generator, whose voltage can be varied by varying its field current. The variable armature voltage provides the means by which the motor speed can be controlled.

The motor speed is high when the generator voltage is high. For the generator

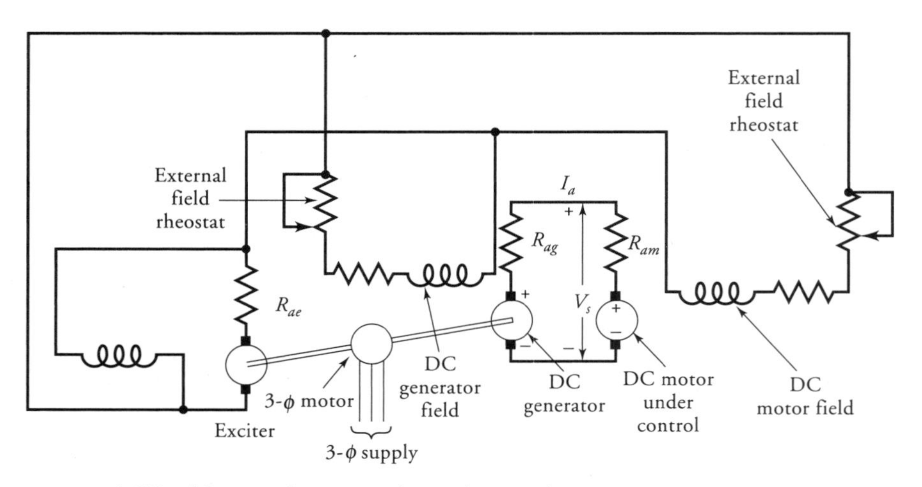

Figure 6.18 A Ward-Leonard system of speed control.

voltage to be high, the total resistance in its field circuit must be low. On the other hand, a high resistance in the field circuit of the generator results in low generator voltage and thereby low motor speed. In other words, by simply controlling the field-winding current of a generator, we can achieve an unlimited speed control of a dc motor.

It must be obvious that we need a set of three machines to control the speed of a dc motor. The system is expensive but is used where an unusually wide and very sensitive speed control is desired. This arrangement offers an excellent stepless speed control and is well suited for such applications as passenger elevators, colliery winders, and electric excavators, to name just a few.

EXAMPLE 6.6

A 120-V shunt motor has an armature resistance of 0.3 Ω and is used as a separately excited motor in a Ward-Leonard system. The field current of the motor is adjusted at 1.2 A. The output voltage of the exciter is 120 V. The dc generator has 200 turns per pole and is running at a speed of 1200 rpm. The magnetization curve for the generator is given in Figure 5.26. The field-winding resistance of the generator is 30 Ω, and its armature resistance is 0.2 Ω. Determine the external resistance in the field-winding circuit of the generator when the motor develops (a) a torque of 30 N·m at 2000 rpm and takes 50 A and (b) the same torque but at a reduced speed of 715 rpm.

● SOLUTION

(a)
$$\omega_m = \frac{2000 \times 2 \times \pi}{60} = 209.44 \text{ rad/s}$$

The power developed by the motor: $P_d = 30 \times 209.44 = 6283.2$ W
The back emf of the motor: $E_{am} = 6283.2/50 = 125.66$ V
The applied voltage across the armature terminals of the motor is

$$V_s = 125.66 + 50 \times 0.3 = 140.66 \text{ V}$$

The induced emf of the generator must be

$$E_{ag} = 140.66 + 50 \times 0.2 = 150.66 \text{ V}$$

The corresponding field current from the magnetization curve is 2 A. The total resistance in the field circuit of the dc generator is

$$R_f = \frac{120}{2} = 60 \ \Omega$$

Thus, the external resistance is

$$R_{ax} = 60 - 30 = 30 \ \Omega$$

(b) Since the torque developed is the same, the armature current of the motor must be the same. However, the new back emf of the motor is

$$E_{amn} = \frac{125.66 \times 715}{2000} = 44.92 \text{ V}$$

The new voltage across the armature terminals of the motor is

$$V_{sn} = 44.92 + 50 \times 0.3 = 59.92 \text{ V}$$

The new induced emf in the generator must be

$$E_{agn} = 59.92 + 50 \times 0.2 = 69.92 \text{ V}$$

From the magnetization curve, the corresponding field current is 0.5 A. Hence the total resistance in the field circuit of the generator must now be

$$R_{fn} = \frac{120}{0.5} = 240 \ \Omega$$

Thus, the external resistance must now be adjusted to a value of

$$R_{axn} = 240 - 30 = 210 \ \Omega$$

■

Exercises

6.16. Repeat Example 6.6 if the demagnetization effect of the armature reaction is equivalent to 80% of the armature current.

6.17. A series motor with $R_a = 0.6 \ \Omega$ and $R_s = 0.4 \ \Omega$ is used in a Ward-Leonard system. The armature terminal voltage of the motor is 280 V when it runs at a speed of 2500 rpm and takes 25 A. The motor drives a fan for which the torque varies as the square of the speed. If the speed of the motor is to be lowered to 2000 rpm, what must be the armature terminal voltage of the motor? Assume that the motor is operating in the linear region. The magnetization curve given in Figure 5.21 applies to the generator. Determine the field current in each case. The armature circuit resistance of the generator is $0.8 \ \Omega$.

6.10 Torque Measurements

The two methods that are commonly used to measure the torque of a dc motor are the **prony brake test** and the **dynamometer test.**

Prony Brake Test

The most common method to measure the torque and the efficiency of a motor is called the prony brake test. The basic arrangement of the test is shown in Figure

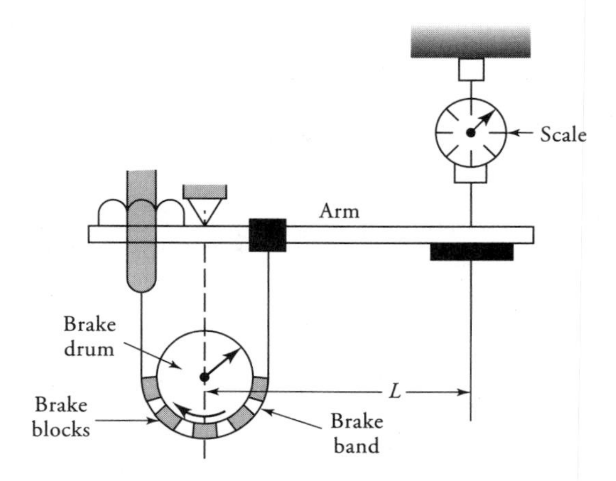

Figure 6.19 The prony brake test.

6.19. In this method, a pulley is mounted on the shaft of the motor that acts as a brake drum. A brake band with or without brake blocks is wrapped around the brake drum as shown. One end of the brake band may be permanently fastened to the torque arm, and the other end can be adjusted by a thumbscrew for tightening against the surface of the brake drum. For high-horsepower motors, cooling of the brake drum may be necessary.

The brake band is slowly tightened until the motor runs at its rated speed. The tendency of the torque arm to move with the drum is resisted by the force of the spring (scale) attached at the far end of the torque arm. The deflection of the spring may be calibrated in terms of either force units or torque units. If the calibration is done to display the force, the torque is simply the product of the force times the effective length of the torque arm. The effective length is simply the distance between the center of the pulley and the place where the spring is attached.

In our discussion we have assumed that the torque arm is perfectly horizontal before the beginning of the test and the dead (tare) weight of the arm is zero. In actual practice, the dead weight of the torque arm must be taken into consideration, as illustrated by the following example.

EXAMPLE 6.7

A 120-V, 0.75-hp motor is tested at 2400 rpm using the prony brake test. The input current is 7 A, and the deflection force on the spring is 4.57 N. The effective length of the torque arm is 50 cm, and its dead weight is 0.03 N. Determine the torque and the efficiency of the motor.

● SOLUTION

Net force on the spring: $F = 4.57 - 0.03 = 4.54$ N
The shaft torque of the motor: $T_s = 4.54 \times 0.5 = 2.27$ N·m
The angular velocity of the shaft is

$$\omega_m = \frac{2\pi \times 2400}{60} = 251.33 \text{ rad/s}$$

The power output: $P_o = 2.27 \times 251.33 = 570.51$ W
The power input: $P_{in} = 120 \times 7 = 840$ W
Hence, the efficiency of the motor is

$$\eta = \frac{570.51}{840} = 0.679 \quad \text{or} \quad 67.9\%$$

■

Another variation of the prony brake test is shown in Figure 6.20. The rope is wound around the pulley and its ends are attached to two spring balances, as shown. For the direction of rotation as shown, the force F_1 is larger than the force F_2. The net pull at the rim of the pulley is $F_1 - F_2$. The torque at the shaft is

$$T_s = (F_1 - F_2)a \tag{6.24}$$

where a is the radius of the pulley.

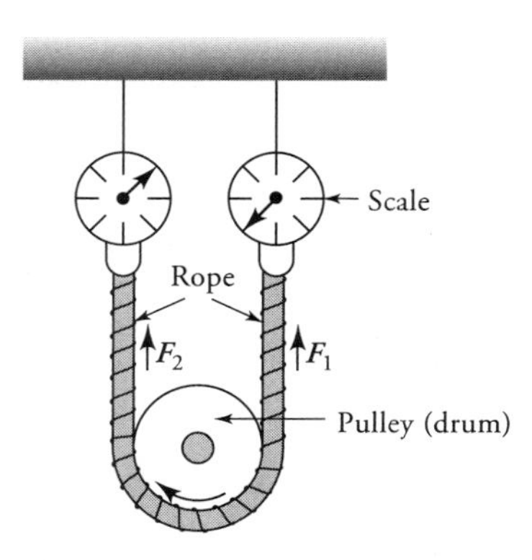

Figure 6.20 The rope test.

Dynamometer Method

The drawbacks of the prony brake test are the vibrations caused by the brake blocks and the necessity of constantly cooling the brake drum. These drawbacks are overcome by a torque-measuring electrical device known as the **dynamometer**. A dynamometer is a dc machine with a separately excited field winding. The only difference between a dc motor and a dynamometer is that the stator of the dynamometer is also free to rotate, whereas it is rigid in a dc motor. The machine is mounted on low-friction ball bearings. On the outside of the stator of a dynamometer is fastened a torque arm, the other side of which is attached to a spring (scale) as shown in Figure 6.21.

The armature of the dynamometer is rotated by coupling it to the shaft of the motor under test. The rotation of the armature coils in a uniform magnetic field results in an induced emf in them (generator action). If the armature circuit is completed by connecting the armature coils to a resistive load, a current will flow in the armature winding. The magnitude of the current depends upon the induced emf and the load resistance. The current-carrying conductors in a uniform magnetic field now experience a force acting on them (motor action). The direction of the force is such that it tends to resist the rotation of the armature. Because the stator, which houses the field windings, is free to rotate, it is pulled around equally by the motor action. The only restraining force acting on the stator is provided by the spring, very much like the prony brake test. By controlling the flux in the motor, we can control the speed of the motor and make it run at any desired speed.

EXAMPLE 6.8

A 5-hp motor rated at 1200 rpm is tested on a dynamometer whose torque arm is 40 cm in length. If the scale is calibrated in terms of the force, what must be the reading on the scale?

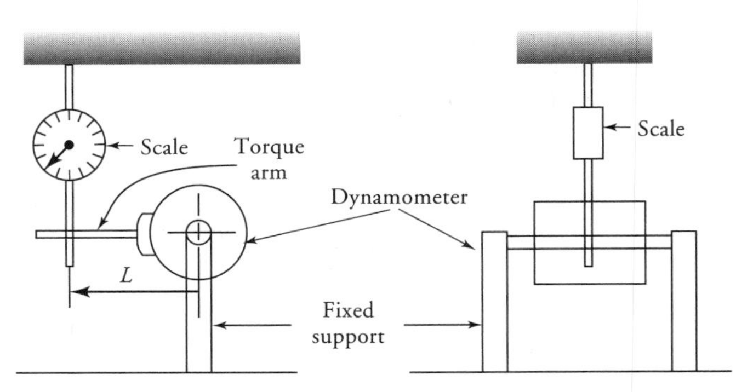

Figure 6.21 A dynamometer arrangement to test a dc motor.

● SOLUTION

The power output: $P_o = 5 \times 746 = 3730$ W
The angular velocity of the shaft is

$$\omega_m = \frac{2\pi \times 1200}{60} = 125.66 \text{ rad/s}$$

The torque available at the shaft is

$$T_s = \frac{3730}{125.66} = 29.68 \text{ N·m}$$

The force reading on the scale is

$$F = \frac{29.68}{0.4} = 74.2 \text{ N}$$

If the scale is calibrated in kilograms, the reading on the scale will be $74.2/9.81 = 7.564$ kg.

■

Exercises

6.18. In a prony brake test of a shunt motor, the ammeter and the voltmeter measuring the input read 35 A and 220 V, respectively. The speed of the motor is 900 rpm. The balance on a 50-cm scale arm reads 12 kg. The dead weight of the arm is 0.8 kg. Determine (a) the power output of the motor and (b) the efficiency.

6.19. In a rope brake test, the readings on the two balances are 12.6 kg and 2.4 kg. The diameter of the pulley is 20 cm. If the motor speed is 900 rpm, determine the horsepower of the motor.

6.20. In a dynamometer test of a shunt motor, the ammeter and voltmeter readings were 120 V and 20 A. The speed of the motor was 1200 rpm. The balance read 3.9 kg. If the length of the torque arm is 40 cm, determine (a) the horsepower rating and (b) the efficiency of the motor.

6.11 Braking or Reversing DC Motors

In certain applications, it may be necessary to either stop the motor quickly or reverse its direction of rotation. The motor may be stopped by using **frictional**

braking. The drawbacks of frictional braking are that the operation is (a) difficult to control, (b) dependent upon the braking surface, and (c) far from being smooth.

Therefore, it is desirable to use other means of stopping and/or reversing the direction of rotation of a dc motor. The three commonly employed methods are **plugging, dynamic braking**, and **regenerative braking**.

Plugging

Stopping and/or reversing the direction of a dc motor merely by reversing the supply connections to the armature terminals is known as **plugging** or **counter-current braking**. The field-winding connections for shunt motors are left undisturbed. This method is employed to control the dc motors used in elevators, rolling mills, printing presses, and machine tools, to name just a few.

Just prior to plugging, the back emf in the motor is opposing the applied source voltage. Because the armature resistance is usually very small, the back emf is almost equal and opposite to the applied voltage. At the instant the motor is plugged, the back emf and the applied voltage are in the same direction. Thus, the total voltage in the armature circuit is almost twice as much as the applied voltage. To protect the motor from a sudden increase in the armature current, an external resistance must be added in series with the armature circuit. The circuit connections, in their simplest forms, for shunt and series motors are given in Figure 6.22.

As the current in the armature winding reverses direction, it produces a force that tends to rotate the armature in a direction opposite to its initial rotation. This causes the motor to slow down, stop, and then pick up speed in the opposite direction. Plugging, therefore, allows us to reverse the direction of rotation of a motor. This technique can also be used to brake the motor by simply disconnecting the power from the motor when it comes to rest. As a further safeguard, mechanical brakes can also be applied when the motor is coming to rest.

At any time during the plugging action, the armature current is

$$I_a = \frac{V_s + E_a}{R + R_a} = \frac{V_s}{R + R_a} + \frac{E_a}{R + R_a}$$

$$= \frac{V_s}{R + R_a} + \frac{K_a \Phi_P \omega_m}{R + R_a} \tag{6.25}$$

Thus, the braking torque is

$$T_b = K_a I_a \Phi_P$$

$$= K_1 \Phi_P + K_2 \Phi_P^2 \omega_m \tag{6.26}$$

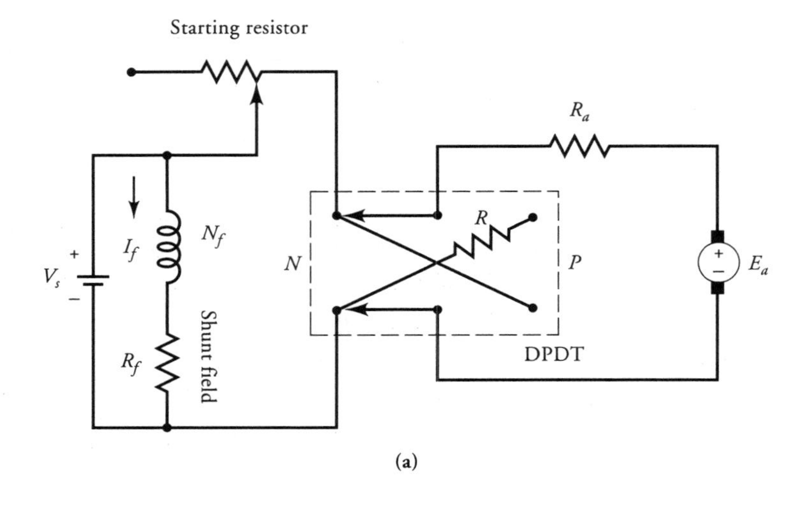

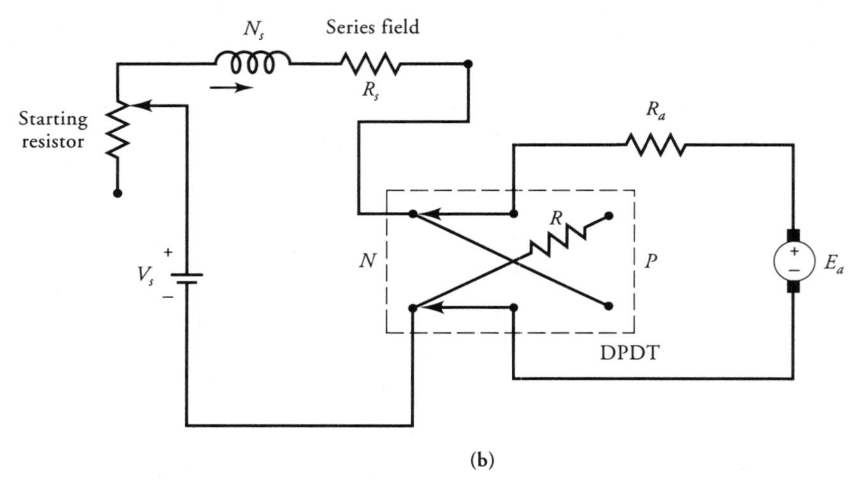

Figure 6.22 Plugging of **(a)** a shunt motor and **(b)** a series motor.

where

$$K_1 = \frac{K_a V_s}{R + R_a} \qquad \text{and} \qquad K = \frac{K_a^2}{R + R_a}$$

For the series motor, the flux also depends upon the armature current, which in turn depends upon the motor speed. Since the flux in a shunt motor is constant, the above equation, for a shunt motor, becomes

$$T_b = K_3 + K_4\omega_m \tag{6.27}$$

where $$K_3 = K_1\Phi_P \quad \text{and} \quad K_4 = K_2\Phi_P^2$$

From the above equation, it is obvious that even when a shunt motor is reaching zero speed, there is some braking torque, $T_b = K_3$. If the supply voltage is not disconnected at the instant the motor reaches zero speed, it will accelerate in the reverse direction.

Dynamic Braking

If the armature winding of a dc motor is suddenly disconnected from the source, the motor will coast to a stop. The time taken by the motor to come to rest depends upon the kinetic energy stored in the rotating system.

Dynamic braking, on the other hand, makes use of the back emf in the motor in order to stop it quickly. If the armature winding, after being disconnected from the source, is connected across a variable resistance R, the back emf will produce a current in the reverse direction. A current in the reverse direction in the armature winding results in a torque that opposes the rotation and forces the motor to come to a halt.

The dynamic braking effect is controlled by varying R. At the time of dynamic braking, R is selected to limit the inrush of armature current to about 150% of its rated value. As the motor speed falls, so does the induced emf and the current through R. Thus, the dynamic braking action is maximum at first and diminishes to zero as the motor comes to a stop. Simple circuits illustrating the principle of dynamic braking for a shunt and a series motor are given in Figure 6.23.

At any time during the dynamic braking process, the armature current is

$$I_a = \frac{E_a}{R + R_a} = \frac{K_a\Phi_P\omega_m}{R + R_a}$$

and the braking torque is

$$T_b = K_a\Phi_P I_a = \frac{K_a^2\Phi_P^2\omega_m}{R + R_a} = K_2\Phi_P^2\omega_m \tag{6.28}$$

Since the flux in a series motor is proportional to the armature current, $\Phi_P = k_f I_a$, the braking torque for a series motor becomes

$$T_{tb} = K_2 k_f I_a^2\omega_m \tag{6.29}$$

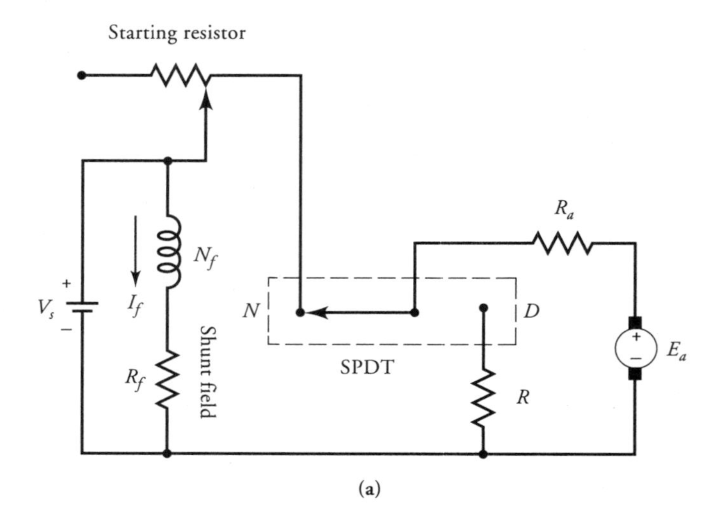

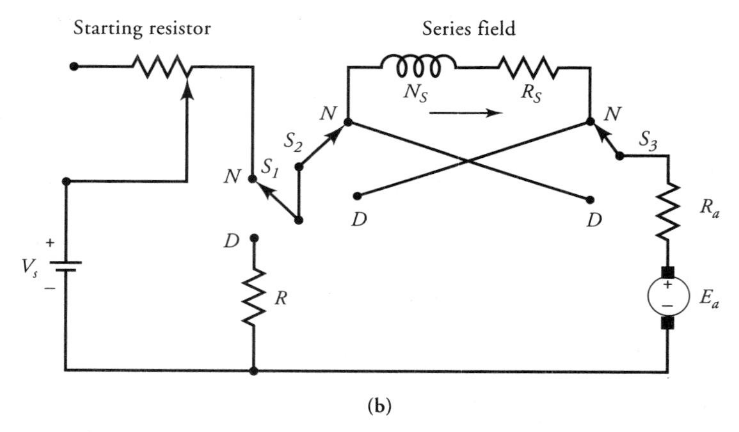

Figure 6.23 Dynamic braking of **(a)** a shunt motor and **(b)** a series motor.

On the other hand, the braking torque for a shunt motor is

$$T_b = K_4\omega_m \tag{6.30}$$

From Eqs. (6.28), (6.29), and (6.30), it is evident that the braking torque vanishes as the motor speed approaches zero.

Regenerative Braking

Regenerative braking is used in applications in which the motor speed is likely to increase from its rated value. Such applications include electric trains, elevators, cranes, and hoists. Under normal operation of a dc motor, say a permanent-magnet (PM) motor in an electric train, the back emf is slightly less than the applied voltage. Note that the back emf in a PM motor is directly proportional to the motor speed. Now assume that the train is going downhill. As the motor speed increases, so does the back emf in the motor. If the back emf becomes higher than the applied voltage, the current in the armature winding reverses its direction and the motor becomes a generator. It sends power back to the source and/or other devices operating from the same source. The reversal of armature current produces a torque in a direction opposite to the motor speed. Consequently, the motor speed falls until the back emf in the motor is less than the applied voltage. The regenerative action not only controls the speed of the motor but also develops power that may be used elsewhere.

EXAMPLE 6.9

A 400-V shunt motor draws 30 A while supplying the rated load at a speed of 100 rad/s. The armature resistance is 1.0 Ω and the field-winding resistance is 200 Ω. Determine the external resistance that must be inserted in series with the armature circuit so that the armature current does not exceed 150% of its rated value when the motor is plugged. Determine the braking torque (a) at the instant of plugging and (b) when the motor is approaching zero speed.

● SOLUTION

The field current: $I_f = 400/200 = 2$ A
The armature current: $I_a = 30 - 2 = 28$ A
The back emf of the motor: $E_a = 400 - 28 \times 1 = 372$ V
The total voltage in the armature circuit at the time of plugging is $400 + 372 = 772$ V. The maximum allowable current in the armature is $1.5 \times 28 = 42$ A. Thus, the external resistance in the armature circuit at the instant of plugging must be

$$R = \frac{772}{42} - 1 = 17.38\ \Omega$$

From the rated operation of the motor, we have

$$K_a\Phi_P = \frac{E_a}{\omega_m} = \frac{372}{100} = 3.72$$

$$K_3 = \frac{K_a\Phi_P V_s}{R + R_a} = \frac{3.72 \times 400}{17.38 + 1} = 80.96$$

$$K_4 = \frac{(K_a\Phi_P)^2}{R + R_a} = \frac{3.72^2}{18.38} = 0.753$$

Thus, the braking torque, from Eq. (6.27) is

$$T_b = 80.96 + 0.753\ \omega_m$$

(a) At the instant of plugging, $\omega_m = 100$ rad/s, and the braking torque is

$$T_b = 80.96 + 0.753 \times 100 = 156.26\ \text{N·m}$$

(b) When the motor attains zero speed, the braking torque is 80.96 N·m. ■

Exercises

6.21. A 400-V, 600-rpm, 20-hp PM motor is operating at full load. The efficiency of the motor is 80%, and the armature resistance is 0.4 Ω. Determine (a) the external resistance that must be added in series with the armature resistance so that the armature current does not exceed twice the full-load current when plugged, (b) the maximum braking torque, and (c) the braking torque and the armature current when the motor speed is approaching zero.

6.22. Repeat Exercise 6.21 if the motor is to be stopped using dynamic braking.

SUMMARY

We devoted this chapter to the study of dc motors. Even though dc motors are not used as widely as ac motors, they do have excellent speed-torque characteristics that cannot be duplicated by ac motors. They also offer more flexible and precise speed control than their ac counterparts.

The principle of operation of a dc motor is no different from that of a dc generator. The induced emf in a dc motor opposes the applied voltage and for this reason is called the back emf. The back emf of the motor adjusts itself depending upon the torque requirements of the load. Because the back emf is zero at the instant of starting a motor, a dc motor should never be started by applying a rated voltage across the armature terminals. We must either use a variable voltage or insert a variable resistance in the armature circuit at the time of starting.

As the load on the motor changes, so does its speed. Therefore, we defined the percent speed regulation of a dc motor as

$$SR\% = \frac{N_{mnL} - N_{mfL}}{N_{mfL}} \times 100 = \frac{\omega_{mnL} - \omega_{mfL}}{\omega_{mfL}} \times 100$$

During our discussion of dc motors, we observed that a series motor is a variable-speed motor while a shunt motor is essentially a constant-speed motor. Since the speed regulation of a compound motor is higher than that of a shunt motor, it is also considered a variable-speed motor.

A shunt motor is used in such applications as a grinders, polishers, wood planers, and washing machines. Its speed can be adjusted by employing (a) field control, (b) armature control, and (c) variable voltage control techniques. An adjustable-speed shunt motor may be found in such applications as lathes, elevators, large blowers, and small printing presses.

A series motor is used in applications such as cranes, turntables, bucket hoists, continuous conveyers, and mine hoists. Its speed can be controlled by using the same techniques mentioned for the shunt motor.

A compound motor is used in compressors, rotary presses, punch presses, elevators, and stamping machines. Field-control and armature-control methods can also be used to control its speed.

The Ward-Leonard method of speed control, although expensive, is employed where an unusually wide and very sensitive speed control is required. It is the system of choice for the regenerative braking process.

We described two methods of testing dc motors, the prony brake test and the dynamometer method. The rope test is a slight variation of the prony brake test. Shunt and compound motors can also be tested by running them at no load. The power input to these machines under no load yields information on the rotational losses.

In order to stop and/or reverse the direction of rotation of a dc motor, we employ such methods as plugging, dynamic braking, and regenerative braking. Each of these techniques is based upon temporarily converting a motor into a generator. The retarding torque of the generator then acts as a brake on the machine.

The plugging technique requires reversal of the supply across the armature terminals. The dynamic braking method involves the connection of a suitable resistance across the armature terminals when the supply is disconnected. When the armature current is fed back to the supply, the braking technique is said to be regenerative. For regenerative braking, the induced emf in the motor must be greater than the line voltage.

Review Questions

6.1. Why is the induced emf in a dc motor called the back emf or the counter emf?

6.2. Explain why a dc motor should not be started by impressing its rated voltage across the armature terminals.

6.3. Explain the armature reaction in a dc motor.

6.4. Define percent speed regulation.

6.5. The no-load and full-load speeds of a dc motor are 2400 rpm and 2000 rpm, respectively. What is the percent speed regulation of the motor?

6.6. What are the two factors responsible for the decrease in the speed when the load on a series motor is increased?

6.7. A 4-pole dc motor has 28 slots with two conductors per slot. The back emf in its lap-wound armature is 240 V at 1800 rpm. What is the flux per pole in the motor?

6.8. A 4-pole dc motor draws 25 A when it is wave wound. Determine the armature current for a lap-wound motor delivering the same torque.

6.9. The speed of a dc motor falls from 1200 rpm at no load to 1000 rpm at full load. What is the percent speed regulation?

6.10. The speed regulation of a dc motor is 2.5%. If the full-load speed is 2400 rpm, what is its no-load speed?

6.11. A 120-V dc machine has an armature circuit resistance of 0.8 Ω. If the full-load current is 15 A, determine the induced emf when the machine acts as (a) a generator and (b) a motor.

6.12. The armature resistance of a 120-V dc motor is 0.25 Ω. When operating at full load, the armature current is 16 A at 1400 rpm. Determine the speed of the motor if the flux is increased by 20% and the armature current is 26 A.

6.13. What is a shunt motor? Why is it referred to as a constant-speed motor?

6.14. Is it possible for a shunt motor to have full-load speed higher than the no-load speed? Give reasons to justify your answer.

6.15. What is a stabilizing winding? How does it affect the operation of a shunt motor?

6.16. A 120-V shunt motor takes 10 A at full load. Total electrical losses are 125 W. Friction and windage loss is 75 W. The core loss is 100 W. Determine the efficiency of the motor. Draw its power-flow diagram. What is the nominal power rating of the motor?

6.17. An electric motor develops 0.75 hp at 1725 rpm. If the pulley has a radius of 10 cm, determine the force (a) in newton-meters and (b) in kilograms (wt) acting at the rim of the pulley.

6.18. Explain the variation in torque when the load is increased upon (a) a shunt motor, (b) a series motor, and (c) a compound motor.

6.19. What is the advantage of a cumulative compound motor over a series motor?

6.20. Why does the torque in a differential compound motor decrease with an increase in the armature current?

6.21. Why does the speed of a differential compound motor increase with an increase in the armature current?

6.22. What happens to the speed-torque characteristic of a cumulative compound motor if the series winding is shunted with a field diverter resistance?

6.23. What methods are used to reverse the direction of rotation of a dc motor?

6.24. Does it make sense to speak of the speed regulation of a series motor? How about a differential compound motor?
6.25. What is an armature resistance–control method? Can it be used to increase the speed of a motor at a given load?
6.26. What are the drawbacks of an armature resistance–control method?
6.27. What is a field-control method? Can it be used to decrease the speed of a motor?
6.28. What are the advantages of a field-control method?
6.29. What is the difference between speed control and speed regulation?
6.30. What are the advantages of the Ward-Leonard method of speed control? What are its disadvantages?
6.31. Describe the Ward-Leonard method of speed control.
6.32. Sketch and explain the speed-torque characteristics of dc motors.
6.33. What is plugging? Why is it necessary to insert a resistance in the armature circuit when plugging a motor?
6.34. When a plugged motor attains zero speed, what is the current in the armature?
6.35. What is dynamic braking? What is the armature current in the motor when the speed approaches zero?
6.36. Explain regenerative braking.
6.37. Cite methods to control the speed of (a) a shunt motor and (b) a series motor.
6.38. Describe (a) the prony brake test and (b) the dynamometer test.
6.39. Force can be measured in either kilograms or newtons. What is the difference between the two measurements?
6.40. What is meant by the dead weight of the brake arm? How can it be determined?

Problems

6.1. A lap-wound, 6-pole dc motor has 180 conductors distributed on the armature's surface. If the armature current is 25 A and the flux per pole is 0.5 Wb, determine the torque developed by the machine. When the motor is operating at 900 rpm, what is the back emf? What is the current in each conductor?

6.2. The armature winding of a 2-pole motor has 1600 conductors. The flux per pole is 50 mWb, and the armature current is 80 A. Determine the torque developed by the motor when (a) the armature reaction is zero, and (b) the armature reaction results in a 12% drop in the flux.

6.3. The torque developed by a dc motor is 20 N·m when the armature current is 25 A. If the flux is reduced by 20% of its original value and the armature current is increased by 40%, what is the developed torque of the motor?

6.4. A 125-V, 10-hp series motor delivers the rated load at 2750 rpm. The armature-circuit resistance is 0.08 Ω, and the series field resistance is 0.07 Ω. If the rotational loss is neglected, determine the torque and the efficiency of the motor. If the motor is derated to deliver 8 hp at 3200 rpm, what is the percent flux reduction in the motor?

6.5. A 230-V, 15-A, dc series motor is rated at 1800 rpm. It has an armature resistance of 2.1 Ω and a series field resistance of 1.25 Ω. Under light load, the motor draws 5 A from the 230-V source. What is the speed of the motor?

6.6. A 240-V, 900-rpm, dc series motor takes 25 A at full load. The armature and field-winding resistances are 0.2 Ω and 0.3 Ω, respectively. The rotational loss is 300 W. Calculate (a) the power developed, (b) the power output, (c) the available torque, (d) the copper loss, and (e) the efficiency. If the pulley diameter is 20 cm, what is the force exerted by the motor at the rim of the pulley?

6.7. A 230-V series motor takes 50 A at a speed of 1200 rpm. The armature resistance is 0.4 Ω and the field-winding resistance is 0.6 Ω. The motor operates in the linear region. When the load is gradually reduced, the current intake is 20 A. What is the percent change in the torque? What is the speed of the motor at the reduced load?

6.8. The magnetization curve of a 200-V series motor is given in Figure 6.7. The other parameters of the motor are $R_a = 0.02\ \Omega$, $R_s = 0.4\ \Omega$, $N_s = 5$ turns/pole, and $P_r = 1500$ W. The field diverter resistance is 0.6 Ω. If the input current is 100 A, find (a) the power developed, (b) the power output, (c) the speed of the motor, (d) the available torque, and (e) the efficiency.

6.9. A fan that requires 8 hp at 700 rpm is coupled directly to a dc series motor. Calculate the efficiency of the motor when the supply voltage is 500 V, assuming that the power required for the fan varies as the cube of the speed. The magnetization curve at 600 rpm is obtained by running the motor as a self-excited generator and is given as follows:

Load current (A):	7	10.5	14	27.5
Terminal voltage (V):	347	393	434	458

The resistance of the armature and the field winding is 3.5 Ω, and the rotational loss is 450 W. The rotational loss may be assumed to be constant for the speeds corresponding to the above range of currents at normal voltage.

6.10. A 6-pole, 3-hp, 120-V, lap-wound shunt motor has 960 conductors in the armature winding. The armature resistance is 0.75 Ω and the flux per pole is 10 mWb. The field-winding current is 1.2 A. The rotational power loss is 200 W. Calculate (a) the armature current, (b) the torque available at the shaft, and (c) the efficiency.

6.11. A 120-V shunt motor takes 2 A at no load and 7 A when running on full load at 1200 rpm. The armature resistance is 0.8 Ω and the shunt field resistance is 240 Ω. Determine (a) the rotational power loss, (b) the no-load speed, (c) the speed regulation, (d) the power output, and (e) the efficiency.

6.12. A 240-V, 20-hp, 2000-rpm shunt motor has an efficiency of 76% at full load. The armature-circuit resistance is 0.2 Ω. The field current is 1.8 A. Determine (a) the full-load line current, (b) the full-load shaft torque, and (c) the starter resistance to limit the starting current to 1.5 times the full-load current. What are the no-load speed and the speed regulation of the motor?

6.13. A 230-V shunt motor draws 40 A at full load. The shunt field current is 2.5 A. The armature-circuit resistance is 0.8 Ω. The rotational loss is 200 W. Determine the power rating and the efficiency of the motor. Sketch its power-flow diagram.

6.14. A 230-V shunt motor draws 5 A at a no-load speed of 2400 rpm. The shunt-field resistance is 115 Ω, and the armature resistance is 0.1 Ω. The motor takes 25 A at full load. Determine (a) the rotational loss, (b) the full-load speed, (c) the speed regulation, (d) the motor efficiency, and (e) the nominal rating of the motor.

6.15. A 4-pole, 120-V shunt motor has 280 lap-wound conductors. It takes 21 A at full load and develops a shaft torque of 10 N·m at 1800 rpm. The field-winding current is 1 A and the armature resistance is 0.5 Ω. Calculate (a) the flux per pole, (b) the rotational loss, and (c) the efficiency.

6.16. The shunt field winding of a 4-pole, 120-V dc compound motor has 200 turns per pole and a resistance of 40 Ω. The series field has 5 turns per pole and a resistance of 0.05 Ω. The armature-circuit resistance is 0.2 Ω. When tested as a shunt motor at full load, it drew 53 A from the 120-V supply. The rotational loss was found to be 10% of the power developed by the shunt motor at a speed of 1200 rpm. The motor is then connected as a cumulative compound motor. For the motor to operate at the same flux density and speed, determine the external resistance that must be added in series with the shunt field winding. Draw the power-flow diagram for the compound motor and compute its efficiency.

6.17. A cumulative compound motor draws 12 A from a 220-V supply and runs at 1800 rpm. The armature-circuit resistance is 0.3 Ω, the series field resistance is 0.2 Ω, and the shunt field resistance is 120 Ω. Determine the developed torque. What must be the percent change in the flux to obtain a speed of 1200 rpm when the armature current is 20 A? What is the torque developed?

6.18. The magnetization curve of a compound motor is obtained at 600 rpm by exciting its shunt field winding only and is given below:

I_f (A):	1	2	3	4	5	6	7	8	9	10
E_a (V):	39	60	82	95	105	112	120	125	128	130

The machine operates as a shunt motor from a 240-V supply and draws 462 A at full load. Its other parameters are $R_f = 20\ \Omega$, $R_s = 0.02\ \Omega$, $R_a = 0.02\ \Omega$, $N_f = 1140$ turns per pole, $N_s = 5$ turns/pole, and $P_r = 2.5$ kW. The demagnetization effect of the armature reaction is equivalent to 1 A of shunt field current. Determine (a) the speed, (b) the shaft torque, (c) the horsepower rating, and (d) the efficiency of the shunt motor.

6.19. The motor in Problem 6.18 is connected as a long-shunt compound motor. For the full-load armature current, determine (a) the speed, (b) the shaft torque, and (c) the efficiency.

6.20. The parameters of a 120-V, long-shunt cumulative compound machine are $R_a = 0.2\ \Omega$, $R_s = 0.1\ \Omega$, $R_f = 60\ \Omega$, and $P_r = 115$ W. If operated as a shunt machine, it draws 52 A and operates a load of 30 N·m. Calculate (a) the speed and (b) the efficiency of the shunt motor. When the motor operates as a compound motor and supplies the same torque, the flux per pole is increased by 10%. Determine (a) the speed and (b) the efficiency of the motor.

6.21. A 240-V, 600-rpm compound motor has $R_a = 0.4\ \Omega$, $R_s = 0.2\ \Omega$, and $R_f = 80\ \Omega$. When operated as a series motor, the motor takes 30 A and runs at a speed of 1450 rpm. When it is connected as a short-shunt, cumulative compound motor, the armature current is the same. The shunt field winding contributes twice as much flux per pole as the series field. Determine (a) the speed of the compound motor and (b) the torque developed by it.

6.22. A 440-V, long-shunt compound motor has $R_a = 0.5\ \Omega$, $R_s = 0.25\ \Omega$, and $I_f = 2$ A. The full-load speed is 1200 rpm, and the load current is 32 A. The motor takes 7 A at no load. If the flux drops to 80% of its full-load value, determine the no-load speed of the motor. What are the efficiency and the horsepower rating of the motor at full load? Assume that the rotational loss is constant.

6.23. A 220-V shunt motor has an armature resistance of 0.25 Ω. The armature takes 5 A at a no-load speed of 2400 rpm. The shunt field current is 1 A. What resistance should be connected in series with the armature to reduce its speed to 800 rpm at its rated load of 31 A? How much power is lost in the control resistance? What is its speed regulation? What is the efficiency of the motor? Assume that the rotational loss is proportional to the speed of the motor.

6.24. A 440-V shunt motor has $R_a = 0.75\ \Omega$ and $I_f = 1$ A. The motor takes 6 A at no load and runs at a speed of 1200 rpm. Calculate the rated speed of the motor when it delivers the rated load and draws 26 A from the supply. Estimate the speed at this load if a resistance of 2.5 Ω is connected in series with the armature.

6.25. A 220-V shunt motor drives a pump whose torque varies as the square of the speed. When the motor runs at 900 rpm, it takes 47 A from the supply.

The shunt field current is 2 A, and the armature resistance is 0.5 Ω. What resistance must be inserted in the armature circuit in order to reduce its speed to 600 rpm?

6.26. A 500-V series motor has an armature resistance of 0.25 Ω and a series field resistance of 0.15 Ω. The rated current is 60 A. What resistance must be inserted in the armature circuit in order to obtain a starting torque equal to 300% of its rated value? What is the armature current at no load? Assume a linear saturation curve.

6.27. A series motor with a total resistance in the armature circuit of 1.5 Ω takes 30 A from a 440-V supply. If the load torque varies as the cube of the speed, calculate the resistance required to reduce the speed by 20%.

6.28. A 120-V shunt motor has a full-load speed of 1800 rpm and takes a current of 51 A. The shunt field current is 1 A, and the armature resistance is 0.2 Ω. How much resistance should be inserted in series with the armature circuit so that the motor develops a torque of 15 N·m at a speed of 1500 rpm?

6.29. A 240-V series motor develops a torque of 120 N·m at a speed of 900 rpm. The armature-winding resistance is 0.2 Ω and the series field–winding resistance is 0.1 Ω. For a constant torque load, find the new speed of the motor when a diverter resistance of 0.1 Ω is placed across the series field winding.

6.30. A 120-V shunt motor has $R_a = 0.25\ \Omega$ and $R_{fw} = 80\ \Omega$. The motor takes 4 A at a no-load speed of 1500 rpm. If the motor is to develop a torque of 20 N·m at a speed of 2000 rpm, determine the external resistance that must be added in series with the field winding. If the rotational loss is proportional to the speed, determine the efficiency and the horsepower rating of the motor at 2000 rpm.

6.31. A 240-V series motor has $R_a = 0.3\ \Omega$ and $R_s = 0.2\ \Omega$. It takes 50 A when running at 900 rpm. Determine the speed at which the motor runs when the series field is shunted by a diverter resistance of 0.2 Ω and the torque developed is increased by 50%.

6.32. When the field coils of a 6-pole series motor are connected in series, it supplies a rated load at 925 rpm and takes 50 A from a 120-V supply. If the field coils are connected in two parallel groups, determine the current intake and the speed of the motor. The load torque varies as the square of the speed. Ignore all losses in the machine.

6.33. A 120-V shunt motor is tested using the prony brake test and the following data are recorded: Torque arm length = 75 cm, initial tare = 1.25 N, balance reading = 9.5 N at 1000 rpm, and line current = 7.5 A. Calculate the torque and the efficiency of the motor.

6.34. The effective reading of the balance in a brake test is 40 kg. The torque arm length is 30 cm. The speed of the motor is 720 rpm. If the motor takes 46 A from a 220-V supply, determine (a) the shaft torque, (b) the efficiency, and (c) the horsepower rating of the motor.

6.35. The following data were obtained when performing a load test of a dc motor using a brake drum: spring balance readings 3 kg and 15 kg, diameter of the drum 50 cm, motor speed 1200 rpm, applied voltage 120 V, and line current 25 A. Determine the efficiency of the motor.

6.36. The load on one side of the spring was 20 kg and on the other side was 5 kg at a speed of 960 rpm. The diameter of the brake drum was 50 cm. The motor drew 36 A from a 120-V supply. Calculate (a) the shaft torque, (b) the efficiency, and (c) the horsepower of the motor.

6.37. A rope test arrangement is shown in Figure P6.37. If the drum speed is 600 rpm, determine the horsepower of the motor.

6.38. A 39-hp, 440-V, PM motor operates at 1000 rpm on full load. The motor efficiency is 86.72%, and the armature resistance is 0.377 Ω. The motor is being controlled by a suitable Ward-Leonard system. The armature resistance of the generator is 0.336 Ω. Determine the induced emf of the generator at full load. What is the rotational loss of the motor? For the following problems, assume that both the motor and the generator are operating in the linear region, and rotational loss is constant.

6.39. Determine the no-load speed of the motor in Problem 6.38 when no adjustments are made in the field circuit of the generator.

6.40. What must be the percent reduction in the induced emf (or in the field

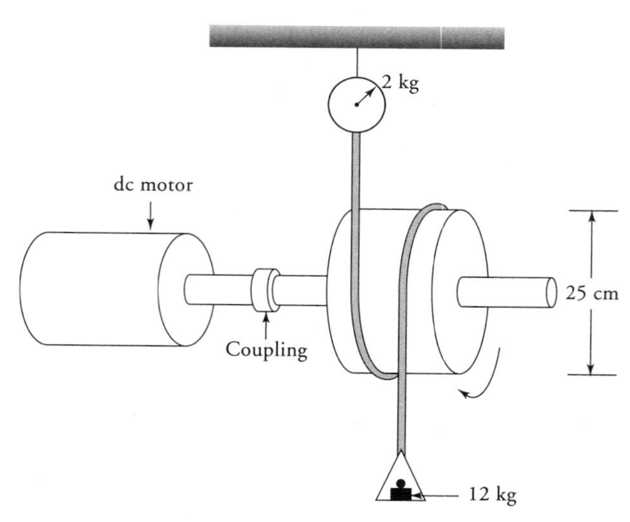

Figure P6.37 Rope test.

current) of the generator in Problem 6.38 to obtain a no-load speed of 1025 rpm?

6.41. What must be the induced emf in the generator if the motor supplies the same torque as in Problem 6.38 but at a speed of 750 rpm? What is the percent change in the field current of the generator? What will the motor speed be under no load?

6.42. A 220-V shunt motor requires 85 A of armature current when it delivers the rated load at 2400 rpm. The total resistance in the armature circuit is 0.5 Ω. Determine (a) the plugging resistor that will limit the inrush of armature current at the time of plugging to 125 A, (b) the maximum braking torque, and (c) the braking torque when the speed approaches zero.

6.43. Repeat Problem 6.42 if the motor is to be stopped using the dynamic braking technique.

CHAPTER 7

Synchronous Generators

Photograph of a synchronous generator and its control. (*Courtesy of Consumers Power*)

7.1 Introduction

Alternating-current (ac) generators are commonly referred to as **synchronous generators** or **alternators**. A synchronous machine, whether it is a generator or a motor, operates at **synchronous speed**, that is, at the speed at which the magnetic field created by the field coils rotates. We have already elaborated this fact in Section 3.5 and obtained an expression for the synchronous speed N_s in revolutions per minute (rpm) as

$$N_s = \frac{120f}{P} \tag{7.1}$$

where f is the frequency in hertz (Hz) and P is the number of poles in the machine. Thus, for a 4-pole synchronous generator to generate power at 60 Hz, its speed of rotation must be 1800 rpm. On the other hand, a 4-pole synchronous motor operating from a 50-Hz source runs at only 1500 rpm. Any attempt to overload the synchronous motor may pull it out of synchronism and force it to stop.

During our discussion of a direct-current (dc) generator we realized that the electromotive force (emf) induced in its armature coils is of the alternating type. Therefore, we can convert a dc generator to an ac generator by (a) replacing its commutator with a set of slip rings and (b) rotating the armature at a constant (synchronous) speed. The idea is novel but is not put into practice for the reasons we will mention shortly.

We also recall that the relative motion of a conductor with respect to the magnetic flux in a machine is responsible for the induced emf in that conductor. In other words, from the induced emf point of view it really does not matter whether the conductors (coils) rotate in a stationary magnetic field or a rotating magnetic field links a stationary conductor (coil). The former arrangement is preferred for dc generators and was discussed in Chapter 5, whereas the latter is more suitable for synchronous generators and is the topic of this chapter. Thus, the stationary member (stator) of a synchronous generator is commissioned as an armature, and the rotating member (rotor) carries the field winding to provide the required flux. There are numerous reasons for such an "inside-out" construction of a synchronous generator, some of which are listed below.

1. Most synchronous generators are built in much larger sizes than their dc counterparts. An increase in power capacity of a generator requires thicker conductors in its armature winding to carry high currents and to minimize copper losses. Deeper slots are therefore needed to house thicker conductors. Because the stator can be made large enough with fewer limitations, it inadvertently becomes the preferred member to house the armature conductors.
2. Since the output of a synchronous generator is of the alternating type, the armature conductors in the stator can be directly connected to the transmission line. This eliminates the need for slip rings for ac power output.

3. Since most of the heat is produced by the armature winding, an outer stationary member can be cooled more efficiently than an inner rotating member.
4. Since the armature winding of a synchronous machine is more involved than the field winding, it is easier to construct it on the stationary member.
5. Since the induced emf in the armature winding is quite high, it is easier to insulate it when it is wound inside the stationary member rather than the rotating member. A rigid frame also enables us to brace the armature winding more securely.
6. The placement of a low-power field winding on the rotor presents no deterrent to the inside-out construction of a synchronous generator. The power to the field winding can be supplied via slip rings. If the field is supplied by permanent magnets, the slip rings can also be dispensed with.

7.2 Construction of a Synchronous Machine

The basic components of a synchronous machine are the stator, which houses the armature conductors, and a rotor, which provides the necessary field as outlined below.

Stator

The stator, also known as the armature, of a synchronous machine is made of thin laminations of highly permeable steel in order to reduce the core losses. The stator laminations are held together by a stator frame. The frame may be of cast iron or fabricated from mild steel plates. The frame is designed not to carry the flux but to provide mechanical support to the synchronous generator. The inside of the stator has a plurality of slots that are intended to accommodate thick armature conductors (coils or windings). The armature conductors are symmetrically arranged to form a balanced polyphase winding. To this end, the number of slots per pole per phase must be an integer. The induced emf per phase in large synchronous generators is in kilovolts (kV) with a power handling capacity in megavolt-amperes (MVA).

The axial length of the stator core is comparatively short for slow-speed, large-diameter generators. These generators have many poles and are left open on both ends for self-cooling. They are installed at locations where hydroelectric power generation is possible.

The axial length of high-speed generators having 2 or 4 poles can be many times its diameter. These generators require forced air circulation for cooling and are totally enclosed. They are used when the rotors are driven by gas or steam turbines.

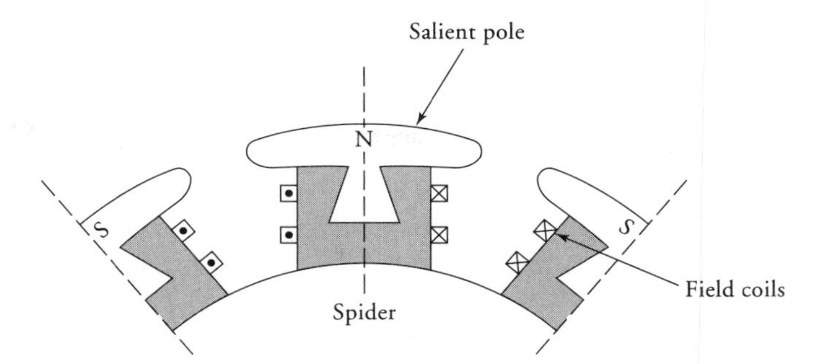

Figure 7.1 A salient pole rotor.

Rotor

Two types of rotors are used in the design of synchronous generators, the **cylindrical rotor** and **a salient-pole rotor**. The rotor is rotated at the synchronous speed by a prime mover such as a steam turbine. The rotor has as many poles as the stator, and the rotor winding carries dc current so as to produce constant flux per pole. The field winding usually receives its power from a 115- or 230-V dc generator. The dc generator may be driven either by the same prime mover driving the synchronous generator or by a separate electric motor.

The salient-pole rotor is used in low- and medium-speed generators because the windage loss is small at these speeds. It consists of an even set of outward projecting laminated poles. Each pole is dovetailed so that it fits into a wedge-shaped recess or is bolted onto a magnetic wheel called the spider. The field winding is placed around each pole, as indicated in Figure 7.1. The poles must alternate in polarity.

The cylindrical rotor is employed in a 2- or 4-pole, high-speed turbo-generator. It is made of a smooth solid forged steel cylinder with a number of slots on its outer periphery. These slots are designed to accommodate the field coils, as shown in Figure 7.2. The cylindrical construction offers the following benefits:

1. It results in a quiet operation at high speed.
2. It provides better balance than the salient-pole rotor.
3. It reduces the windage loss.

7.3 Armature Windings

The stators (armatures) of most synchronous generators are wound with three distinct and independent windings to generate three-phase power. Each winding is said to represent one phase of a three-phase generator. The three windings are

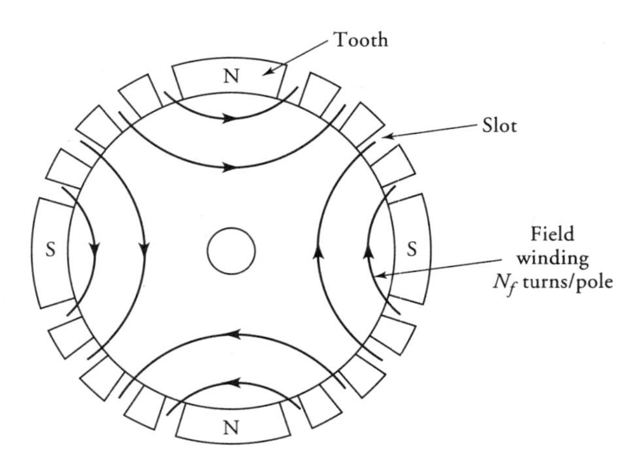

Figure 7.2 A 4-pole cylindrical rotor.

exactly alike in shape and form but are displaced from each other by exactly 120° electrical in order to ensure that the induced emfs in these windings are exactly 120° in time phase. The three-phase windings may be connected to form either a **star** (Y) or a **delta** (Δ) connection. If the windings are internally connected to form a Y connection, the neutral point is also brought out so that it can be properly grounded for safe operation.

The **double-layer** winding is often used to wind the armature of a synchronous generator. As you may recall, a double-layer winding contains as many identical coils as there are slots in the stator. One side of each coil is placed at the bottom half of a slot, and the other side of the same coil fills the top half of another slot. In order to place the coils in this fashion, the coils must be prewound on the winding forms and then inserted into the slots.

The number of coils per phase (or the number of slots per phase for a double-layer winding) must be an integer. Since the coils must be distributed equally among the poles, the number of coils (slots) per pole per phase must also be an integer. In other words, if S is the number of slots in the armature, P is the number of poles, and q is the number of phases, then the number of coils per pole per phase is

$$n = \frac{S}{Pq} \tag{7.2}$$

where n must be an integer. The number of coils per pole per phase, n, is usually referred to as a **phase group** or **phase belt**. When the stator of a three-phase, 4-pole synchronous generator has 24 slots, the number of coils in each phase group is 2. There are 12 phase groups (poles × phases). All coils in a phase group (2 in this case) are connected in series.

Each coil in a phase group can be wound as a full-pitch coil. In other words, each coil in the armature can be made to span 180° electrical. Since the induced emfs in both sides of a full-pitch coil at any time are exactly in phase, theoretical yearning mandates the placement of full-pitch coils from the induced emf point of view. However, a full-pitch coil is rarely used. Instead, the generators are wound with fractional-pitch coils for the following reasons:

1. A properly designed fractional-pitch coil reduces the distorting harmonics and produces a truer sinusoidal waveform.
2. A fractional-pitch coil shortens the end connections of the windings and thereby not only saves copper but also reduces the copper loss in the coil.
3. A shorter coil is easier to manage and reduces the end-turn build-up on both sides of the stator's stack. This slims down the overall length of the generator and minimizes the flux leakages.
4. The elimination of high-frequency harmonics also cuts down the magnetic losses in the generator.

The drawback of a fractional-pitch coil is that the induced emf in it is smaller than in a full-pitch coil. The reason is that the total flux linking the fractional-pitch coil is smaller than that of the full-pitch coil. The ratio of the flux linking the fractional pitch coil to the flux that would link a full-pitch coil is called the **pitch factor**. Later, we will develop an equation to determine the pitch factor.

To illustrate the placement of the phase windings in the slots of a synchronous generator, we make the following assumptions:

(a) All coils are identical.
(b) Each coil is a fractional-pitch coil as long as a phase group contains more than one coil. All the coils in a phase group are connected in series.
(c) Each phase group spans 180° electrical (one full pitch). Thus, the n coils in a phase group must be placed in such a way that the beginning end of the first coil is under the beginning of a pole and the finishing end of the nth coil is under the trailing end of the pole.

The electrical angle from the center of one slot to the center of an adjacent slot is known as the **slot span** or **slot pitch**. The **coil span** or **coil pitch**, the number of slots spanned by each coil, can be expressed in terms of either electrical degrees or the number of slots, as illustrated by the following example.

EXAMPLE 7.1

A three-phase, 4-pole, synchronous generator has 24 slots. Determine the slot pitch, the coil pitch, and the number of coils in a phase group. Sketch the placement of the coils for one of the phases.

● SOLUTION

Since there are 24 slots in the stator, the number of coils for a double-layer winding is also 24. The number of coils per phase is $24/3 = 8$. The number of coils per pole per phase (phase group) is $8/4 = 2$. Thus, $n = 2$. The number of slots (coils) per pole is $24/4 = 6$. Since a pole spans 180° electrical, the slot span (γ) is $180/6 = 30°$ electrical. In other words, the electrical angle from the center of one slot to the center of an adjacent slot is 30°.

The placement of coils for one phase of a 4-pole synchronous generator is shown in Figure 7.3. The slots are numbered for convenience from 1 to 24. The pole span for phase group A_1 includes only half of slot 1 and half of slot 7. The starting end of coil 1, s_1, is placed in slot 1. The starting end of coil 2, s_2, must be placed in slot 2. Since there are only two coils in a phase group, the finishing end of coil 2, f_2, must be placed in slot 7 because the pole span is 6 slots. Another way to determine the pole span is to count the number of teeth. Because the number of teeth spanned by a coil is the same as the number of slots spanned by it, a pole must span 6 teeth. There are exactly 6 teeth between slot 1 and slot 7. Once we have determined where to place the finishing end of coil 2, we immediately know where to place the finishing end of coil 1. It must go in slot 6 because all coils are identical. Note that each coil spans 5 slots or 5 teeth. Thus, the coil pitch is $30° \times 5 = 150°$ electrical.

The phase group A_2 begins where the phase group A_1 ends. Thus, the starting end of coil 1 for phase group A_2 must be placed in slot 7, and so on.

Figure 7.4 shows another way to determine the placement of the coils and thereby the coil pitch. It is referred to as the developed diagram.

■

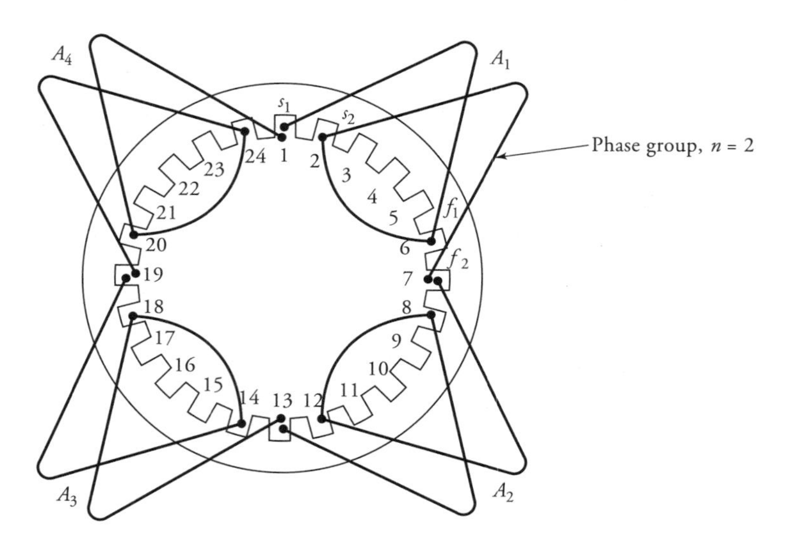

Figure 7.3 Phase A coils for the 4-pole, three-phase, 24-slot stator of Example 7.1.

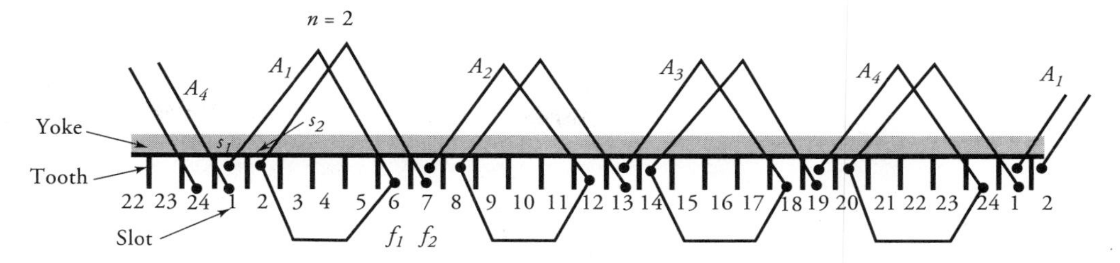

Figure 7.4 A developed diagram indicates the placement of phase *A* coils for the 4-pole, three-phase, 24-slot stator of Example 7.1.

Exercises

7.1. The stator of a three-phase, 60-Hz, 8-pole synchronous generator has 48 slots. Determine (a) the speed of the rotor, (b) the number of coils in a phase group, (c) the coil pitch, and (d) the slot pitch. Sketch the complete winding arrangement of the stator.

7.2. A 54-slot stator of a synchronous generator is wound for three-phase, 50-Hz, 6-pole applications. Determine (a) the number of coils in a phase group, (b) the slots per pole, (c) the slot span, (d) the coil span, and (e) the speed at which the rotor must be rotated.

7.4 Pitch Factor

Owing to the spatial distribution of field windings on each pole of a cylindrical rotor, we can approximate the flux density emanating from the surface of a pole as

$$B = B_m \cos \theta \tag{7.3}$$

where B_m is the maximum flux density per pole, as shown in Figure 7.5. The total flux linking a full-pitch coil is

$$\Phi_p = \int \mathbf{B} \cdot d\mathbf{s}$$

where

$$ds = Lrd\theta_m = \frac{2Lrd\theta}{P}$$

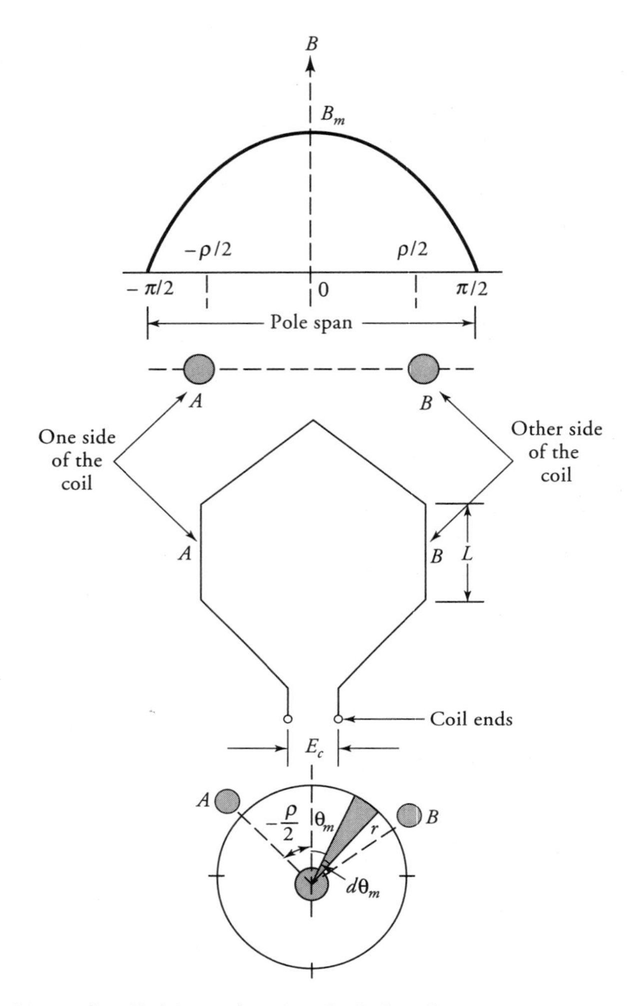

Figure 7.5 Maximum flux linking a fractional-pitch coil.

L is the axial length of the coil (rotor), r is radius of the rotor, and P is the number of poles. For a full-pitch coil, θ varies from $-\pi/2$ to $\pi/2$. Evaluating the integral, we obtain

$$\Phi_P = \frac{4LrB_m}{P} \tag{7.4}$$

Let us now assume that the coil span for a fractional-pitch coil is ρ. The flux linking the coil is maximum when the coil is symmetrically placed along the magnetic axis of the pole, as illustrated in Figure 7.5. Thus, the maximum flux linking the coil is

$$\begin{aligned}\Phi_{cm} &= \int_{-\rho/2}^{\rho/2} B_m \cos\theta\, 2Lr \frac{d\theta}{P} \\ &= \frac{4LrB_m}{P} \sin(\rho/2) \\ &= \Phi_P \sin(\rho/2) = \Phi_P k_p \end{aligned} \tag{7.5}$$

where $k_p = \sin(\rho/2)$ is the **pitch factor** and $k_p \leq 1$. If e_c is the induced emf in a full-pitch coil, then the induced emf in the fractional-pitch coil will be $k_p e_c$. We will use this fact later when we compute the induced emf in each phase group of a synchronous generator.

EXAMPLE 7.2

Calculate the pitch factor for a 48-slot, 4-pole, three-phase winding.

● SOLUTION

The number of slots per pole is $48/4 = 12$. Thus, the slot span is $180/12 = 15°$ electrical.
The number of coils in a phase group is

$$n = \frac{48}{4 \times 3} = 4$$

Since there are 12 slots per pole and 4 coils in a phase group, one side of coil 1 must be placed in slot 1 and the other side in slot 10. Thus, the coil span is 9 slots or $9 \times 15° = 135°$ electrical, as illustrated in Figure 7.6.
The pitch factor is

$$k_p = \sin(135°/2) = 0.924$$

Thus, the induced emf in each fractional-pitch coil is 92.4% of the induced emf in a full-pitch coil.

■

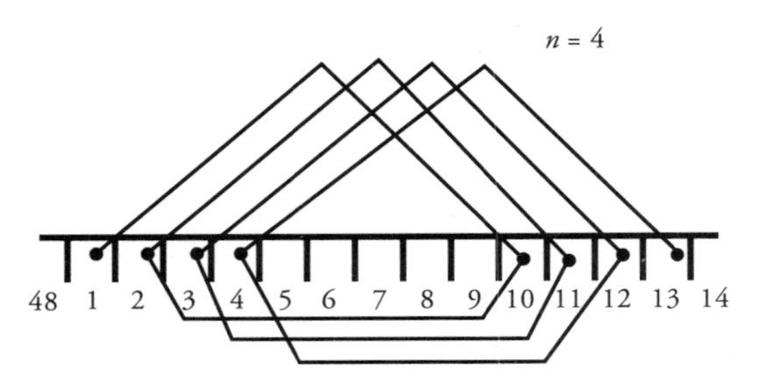

Figure 7.6 Developed diagram for the 4-pole, 48-slot, three-phase stator of Example 7.2.

Exercises

7.3. A 72-slot stator has a 4-pole, three-phase winding. Determine its coil pitch and the pitch factor. How many coils are in each phase group?

7.4. Calculate the pitch factors for the following three-phase windings: (a) 36-slot, 4-pole, (b) 36-slot, 6-pole, and (c) 72-slot, 6-pole. Illustrate the placement of coils by sketches.

7.5 Distribution Factor

In order to make the induced emf approach a sinusoidal function, there are always more than one coil in a phase group. These coils are connected in series as depicted by the winding diagrams. Since the coils are displaced spatially from each other, the induced emfs in these coils are not in phase. If E_c is the induced emf in one coil and n is the number of coils in a phase group, the induced emf in the phase group E_{pg} is

$$E_{pg} = k_d n E_c \tag{7.6}$$

where k_d is called the distribution factor and $k_d \leq 1$.The distribution factor is unity only when all the coils are placed in the same slots. Since that defies the purpose of obtaining a sinusoidal waveform, the distribution factor is always less than unity.

In order to verify Eq. (7.6), let us assume that there are n coils connected in series to form a phase group and the root-mean-square (rms) value of the induced emf in each coil is E_c. Since the coils are distributed, the induced emf in each coil is out of phase with the next by the slot pitch γ, as shown in Figure 7.7. Since the

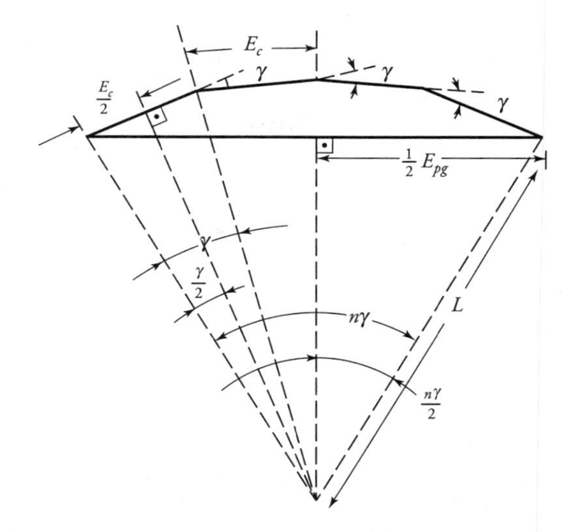

Figure 7.7 Phasor diagram for the induced emf in a phase group.

angle subtended by each phase-voltage phasor is γ, the total angle for the phase group is $n\gamma$. Therefore, we can write

$$\frac{1}{2}E_{pg} = L \sin\left(\frac{n\gamma}{2}\right)$$

and

$$\frac{1}{2}E_c = L \sin\left(\frac{\gamma}{2}\right)$$

From the above equations, we obtain

$$\frac{E_{pg}}{E_c} = \frac{\sin(n\gamma/2)}{\sin(\gamma/2)} = \frac{n \sin(n\gamma/2)}{n \sin(\gamma/2)}$$

Thus, the induced emf per phase group is

$$E_{pg} = n\frac{\sin(n\gamma/2)}{n \sin(\gamma/2)}E_c = nk_dE_c \tag{7.7}$$

where

$$k_d = \frac{\sin(n\gamma/2)}{n \sin(\gamma/2)} \tag{7.8}$$

is the distribution factor.

EXAMPLE 7.3

Calculate the distribution factor for a 108-slot, 12-pole, three-phase winding.

● SOLUTION

The number of coils in a phase group is

$$n = \frac{108}{12 \times 3} = 3$$

Since there are $108/12 = 9$ slots per pole, the slot span, in electrical degrees, is

$$\gamma = \frac{180}{9} = 20°$$

The distribution factor, from Eq. (7.8), is

$$k_d = \frac{\sin(3 \times 20/2)}{3 \sin(20/2)} = 0.96$$

■

Exercises

7.5. Determine the distribution factor for a 72-slot, 4-pole, three-phase winding.
7.6. Compute the distribution factor for a 120-slot, 8-pole, three-phase winding.

7.6 Winding Connections

In order to explain the winding connections, let us scrutinize the phase windings of a 6-pole, three-phase stator (armature) of a synchronous generator with a salient-pole rotor, as depicted in Figure 7.8. Each phase group is indicated by a single coil. For a clockwise rotation and the positive phase sequence, phases *B* and *C* are displaced by 120° and 240° electrical with respect to phase *A* for each of the 6 poles. The polarity of the induced emf in each phase group at one instant of time is marked by a dot (●). The six phase groups for each of the three phases can be connected in three distinct ways. Let us now consider only one of the three phases. The other two phases are connected in exactly the same way.

The magnitude of the induced emf at any time in each phase group of phase *A* is the same. We can establish a **series connection** simply by connecting all six phase groups in series, as indicated in Figure 7.9**a**. The generated voltage per

Figure 7.8 Phase windings of a three-phase, 6-pole armature.

phase (phase voltage) is six times the induced emf in each phase group, but there is only one current path. In general, if E_{pg} is the induced emf in a phase group and P is the number of poles, the phase voltage is PE_{pg}. For a synchronous generator to be connected to a high-voltage transmission line, this is the preferred connection.

The three phase groups under the north poles can be connected in series. Similarly, we can connect the three phase groups under the south poles in series. The two groups can then be connected in parallel, as indicated in Figure 7.9**b**. The phase voltage is one-half of the series connection, but the current-carrying capacity is twice as much.

We can also establish a **parallel connection** by connecting all the phase groups in parallel, as indicated in Figure 7.9**c**. The phase voltage is simply equal to the induced emf in a phase group, but there are six parallel paths.

In general, for a synchronous generator having P poles and $\boldsymbol{a}$ parallel paths, the phase voltage, E_a, is

$$E_a = \frac{P}{a} E_{pg} \tag{7.9}$$

The two terminals, the starting end and the finishing end, of each phase are brought out to form either a Y or a Δ connection. If the generator is to be connected to a high-voltage transmission line, the preferred connection is the Y connection.

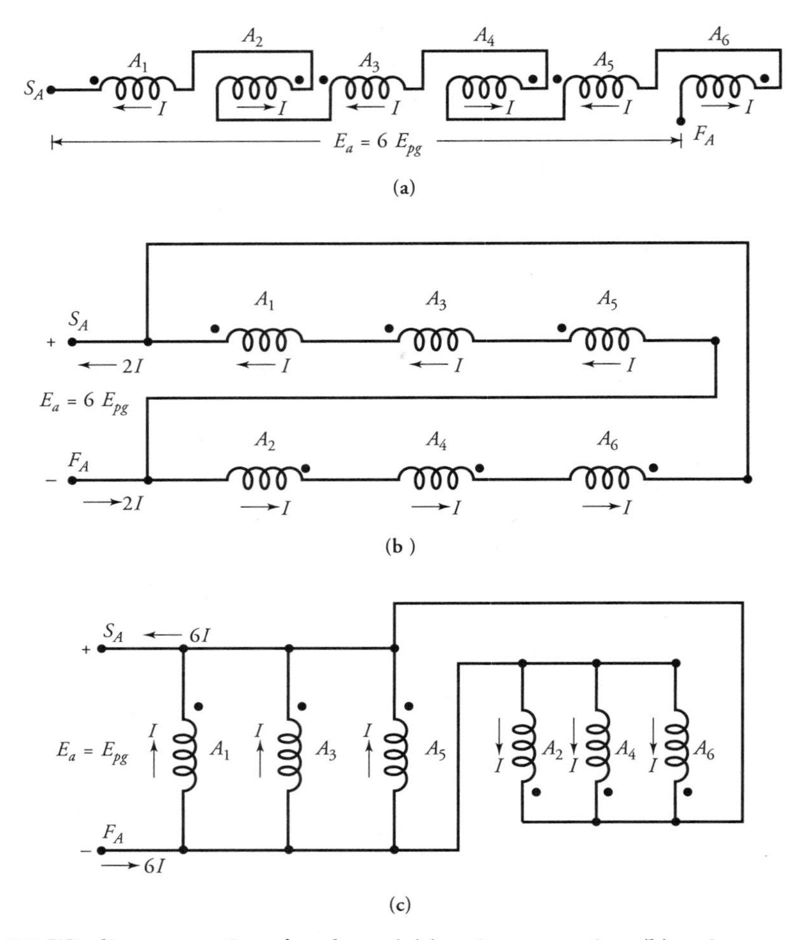

Figure 7.9 Winding connections for phase A **(a)** series connection, **(b)** series-parallel connection, and **(c)** parallel connection.

EXAMPLE 7.4

Show the placement of each phase winding and the winding connections for a 24-slot, 4-pole, three-phase (a) Y-connected and (b) Δ-connected synchronous generator.

● SOLUTION

There are 6 slots per pole and the slot span is 30° electrical. There are two coils in a phase group, and the coil spans 5 slots. Thus, the coil pitch is 150° electrical.

The starting ends of the two coils for phase group A_1 are placed in slots 1 and

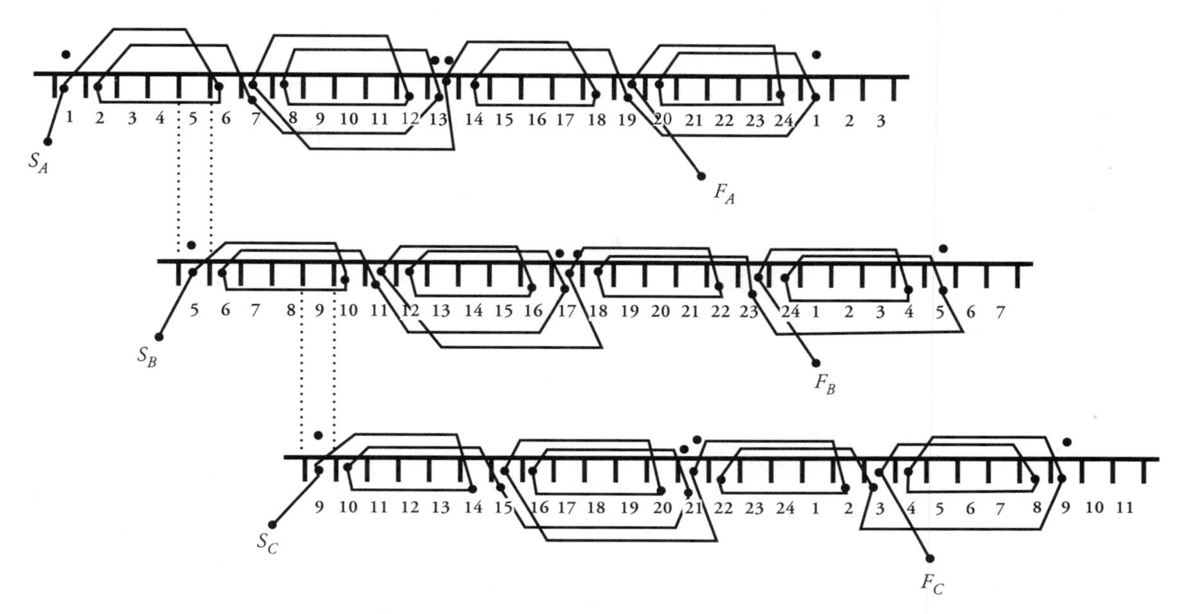

Figure 7.10 Placement of phase windings *A*, *B*, and *C* for a 24-slot, 4-pole, three-phase synchronous generator.

2, while the finishing ends are inserted in slots 6 and 7, as shown in Figure 7.10**a**. The finishing end f_1 is internally connected to starting end s_2 to form a phase group. If the phase group A_1 is under the north pole, the phase group A_3 is also under the north pole. At the same time the phase groups A_2 and A_4 are under the south poles. Thus, if the induced emf in the phase group A_1, that is, the voltage between s_1 and f_2, is positive, the same is true for the phase group A_3. However, the induced emf in phase groups A_2 and A_4 has the opposite polarity. With this understanding, we can connect the four phase groups in series to form a series connection, as illustrated in Figure 7.10**a**.

In order to displace phase group B_1 by 120° electrical from the phase group A_1, the starting end of its first coil must be inserted in slot 5, as indicated in Figure 7.10**b**. In other words, each phase group is displaced 4 slots to the right. The phase groups for phase *B* can be connected in the same fashion as we did for phase *A*.

The starting end of the first coil for phase *C* must be placed in slot 9 in order to achieve a phase difference of 240° electrical with respect to phase *A*. The rest of the connections are shown in Figure 7.10**c**.

Figure 7.11 shows how the phase windings can be connected to a Y and a Δ connection.

■

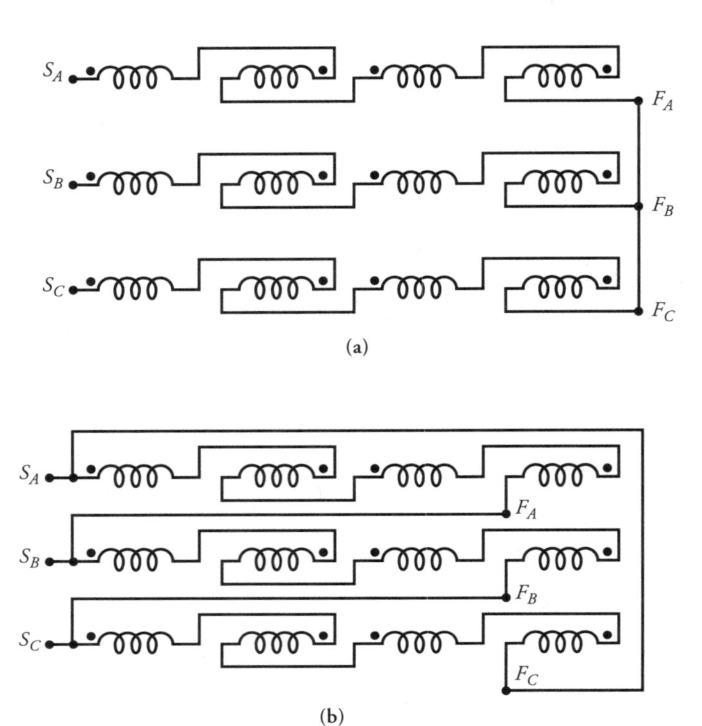

Figure 7.11 **(a)** Y-connected, and **(b)** Δ-connected armature of a three-phase synchronous generator.

Exercises

7.7. Show the placement of each phase winding and the winding connections for a 36-slot, 6-pole, three-phase (a) Y-connected and (b) Δ-connected synchronous generator. Assume series connection.

7.8. Show the placement of each phase winding and the winding connections for a 12-slot, 2-pole, three-phase (a) Y-connected and (b) Δ-connected synchronous generator. Assume series connection.

7.7 Induced EMF Equation

Let us assume that the total flux emanating per pole of a round rotor revolving at an angular velocity of ω_s is Φ_P. The maximum flux linking the fractional pitch coil is $\Phi_P\, k_p$ where $k_p = \sin(\rho/2)$ is the pitch factor and ρ is the coil span in electrical degrees. As the flux revolves, the flux linking the coil at any time t can be expressed as

$$\Phi_c = \Phi_P k_p \cos \omega t \tag{7.10}$$

where $\omega = 2\pi f$ is the angular frequency in rad/s.

For a coil with N_c turns, the induced emf in the coil, from Faraday's law is

$$e_c = N_c k_p \omega \Phi_P \sin \omega t \tag{7.11}$$

The maximum value of the induced emf is

$$E_m = N_c k_p \omega \Phi_P \tag{7.12}$$

and its rms value is

$$E_c = \frac{1}{\sqrt{2}} E_m$$

$$= 4.44 f N_c k_p \Phi_P \tag{7.13}$$

where the factor $\sqrt{2}\pi$ has been approximated as 4.44.

Since a phase group usually has more than one coil connected in series and each coil is displaced by a slot pitch, the induced emf in the phase group, from Eq. (7.7), is

$$E_{pg} = nk_d E_c = 4.44\, nN_c k_p k_d f \Phi_P \tag{7.14}$$

where n is the number of coils in a phase group and k_d is the distribution factor as given by Eq. (7.8). For a given generator, the product $k_p k_d$ is constant and is referred to as the **winding factor**. That is, the winding factor, k_w, is

$$k_w = k_p k_d \tag{7.15}$$

The rms value of the induced emf in each phase group can be expressed in terms of the winding factor as

$$E_{pg} = 4.44\, nN_c k_w f \Phi_P \tag{7.16}$$

For a generator having P poles and ***a*** parallel paths, the induced emf per phase (**phase voltage**) is

$$E_a = \frac{P}{a} 4.44\, nN_c k_w f \Phi_P \tag{7.17}$$

The factor PnN_c/a in the above equation represents the actual number of turns per phase connected in series when there are a parallel paths. By taking into account the winding factor, k_w, we can define the **effective turns per phase** as

$$N_e = \frac{PnN_c k_w}{a} \tag{7.18}$$

Finally, we obtain an expression for the per-phase (no-load) voltage as

$$E_a = 4.44\ fN_e\Phi_P \tag{7.19}$$

Note that Eq. (7.19) is very similar to the one obtained for a transformer. In the case of a transformer, the effective number of turns is the same as the actual number of turns because each transformer winding consists of one coil that embraces the total flux in the magnetic core. The winding factor for a synchronous generator could also have been unity if (a) we used a full-pitch coil and (b) all the coils in a phase group were placed in the same slots.

EXAMPLE 7.5

Each coil of a double-layer wound, 16-pole, 144-slot, three-phase, Y-connected, synchronous generator has 10 turns. The rotor is driven at a speed of 375 rpm. The flux per pole is 25 mWb. Each phase winding is connected in two parallel paths. Determine (a) the frequency of the induced emf, (b) the phase voltage, and (c) the line voltage.

● SOLUTION

Since there are as many coils as there are slots in the armature for a double-layer winding, the number of coils in a phase group is

$$n = \frac{144}{16 \times 3} = 3$$

The number of slots per pole: $S_p = 144/16 = 9$
Thus, the slot-span: $\gamma = 180/9 = 20°$ electrical. Since there are 9 slots per pole and 3 coils in a phase group, each coil must span 7 slots. Hence, the coil pitch is

$20 \times 7 = 140°$ electrical. We can now compute the pitch factor, the distribution factor, and the winding factor as

$$k_p = \sin(140/2) = 0.94$$

$$k_d = \frac{\sin(3 \times 20/2)}{3 \times \sin(20/2)} = 0.96$$

$$k_w = 0.94 \times 0.96 = 0.902$$

The effective turns per phase are

$$N_e = \frac{16 \times 3 \times 10 \times 0.902}{2} = 216.48$$

(a) The frequency of the generated voltage is

$$f = \frac{375 \times 16}{120} = 50 \text{ Hz}$$

(b) The rms value of the generated voltage per phase is

$$E_a = 4.44 \times 50 \times 216.48 \times 0.025 = 1201.46 \text{ V}$$

(c) The rms value of the line voltage is

$$E_L = \sqrt{3} \times 1201.46 \approx 2081 \text{ V}$$

■

Exercises

7.9. A 4-pole, 36-slot, 60-Hz, three-phase, Y-connected, synchronous generator has 10 turns per coil. The flux per pole is 50 mWb. For a series connection, determine (a) the synchronous speed, (b) the per-phase voltage, and (c) the line voltage.

7.10. The no-load (line) voltage of a 6-pole, 72-slot, 1200-rpm, three-phase, Y-connected, synchronous generator is 1732 V. There are 10 conductors per slot. For a series connection, determine (a) the frequency of the induced emf and (b) the flux per pole.

7.8 The Equivalent Circuit

During our discussion of dc generators, we discerned that the terminal voltage of a dc generator is smaller than the generated voltage owing to (a) the voltage drop across its armature winding and (b) the decrease in the armature flux caused by the armature reaction. However, the terminal voltage of an ac generator depends upon the load and may be larger or smaller than the generated voltage. In fact, we aim to show that the terminal voltage **may actually be higher** than the generated voltage when the power factor (pf) is leading. For unity and lagging power factors, the terminal voltage is smaller than the generated voltage.

Armature Resistance Voltage Drop

Let $\tilde{E}_a$ be the per-phase generated voltage of a synchronous generator and $\tilde{I}_a$ the per-phase current supplied by it to the load. If R_a is the per-phase resistance of the armature winding, then $\tilde{I}_a R_a$ is the voltage drop across it. The $\tilde{I}_a R_a$ voltage drop is in phase with the load current $\tilde{I}_a$. Since R_a also causes a power loss in the generator, it is kept as small as possible, especially for large machines.

Armature Leakage-Reactance Voltage Drop

The current $\tilde{I}_a$ in the armature winding produces a flux. A part of the flux, the so-called leakage flux, links the armature winding only and gives rise to a leakage reactance X_a. The leakage reactance causes a voltage drop $j\tilde{I}_a X_a$, which leads $\tilde{I}_a$ by 90°. The phasor diagrams depicting relationships between the per-phase generated voltage $\tilde{E}_a$, the per-phase terminal voltage $\tilde{V}_a$, and the voltage drops $\tilde{I}_a R_a$ and $j\tilde{I}_a X_a$ for three types of loads are shown in Figure 7.12.

Armature Reaction

The flux produced by the armature winding reacts with the flux set up by the poles on the rotor, causing the total flux to change. Such an interaction between the two fluxes is known as the **armature reaction**. To understand the effect of armature reaction on the terminal voltage of a synchronous generator, let us examine a sequence of events when the generator delivers a load at a unity power factor.

(a) If Φ_P is the flux per pole in the generator under no load, then the generated voltage E_a must lag Φ_P by 90°, as shown in Figure 7.13.

(b) Since the power factor is unity, the phase current $\tilde{I}$ is in phase with the terminal phase voltage $\tilde{V}_a$.

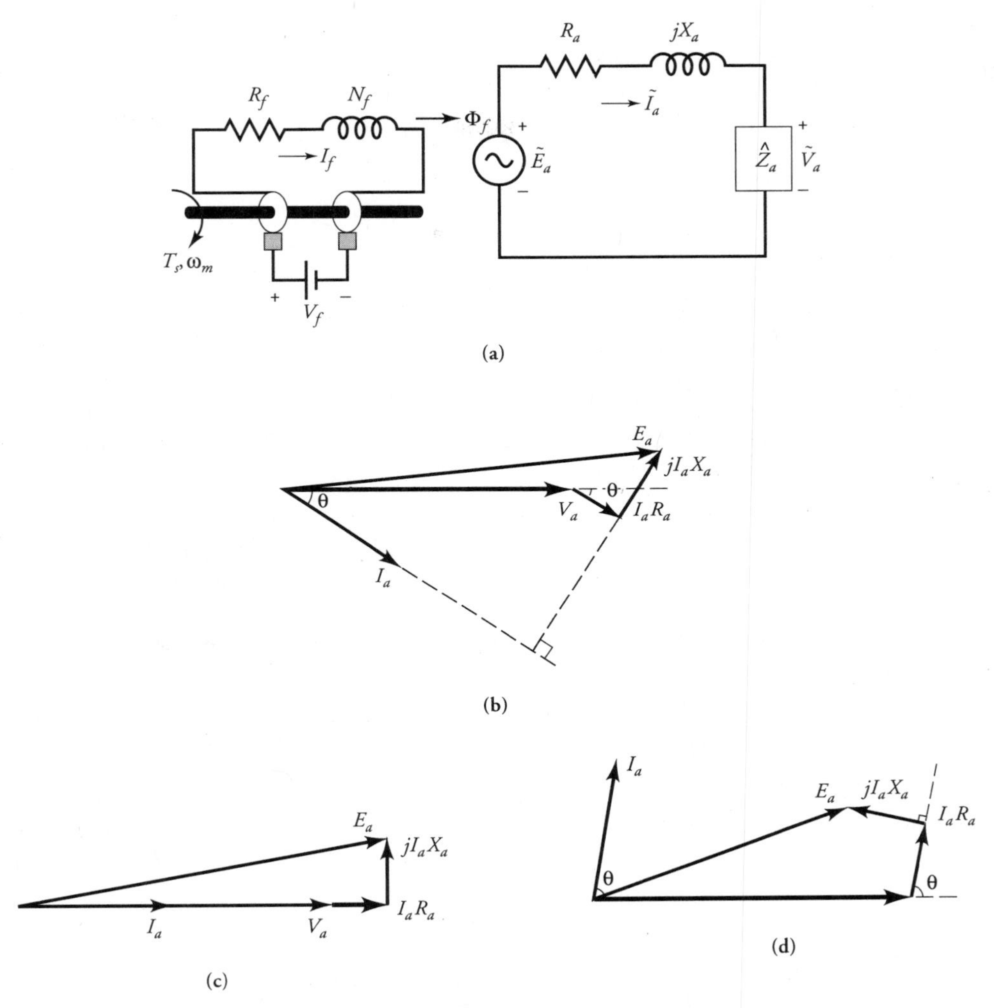

Figure 7.12 **(a)** The per-phase equivalent circuit of a synchronous generator without armature reaction while depicting the revolving field produced by the rotor. The phasor diagrams for a **(b)** lagging pf, **(c)** unity pf, and **(d)** leading pf.

(c) As the phase current $\tilde{I}_a$ passes through the armature winding, its magnetomotive force (mmf) produces a flux Φ_{ar} which is in phase with $\tilde{I}_a$. The effective flux Φ_e per pole in the generator is the algebraic sum of the two fluxes; that is, $\Phi_e = \Phi_P + \Phi_{ar}$, as shown in the figure.

(d) The flux Φ_{ar}, in turn, induces an emf $\tilde{E}_{ar}$ in the armature winding. $\tilde{E}_{ar}$ is called the **armature reaction emf**. The armature reaction emf $\tilde{E}_{ar}$ lags the

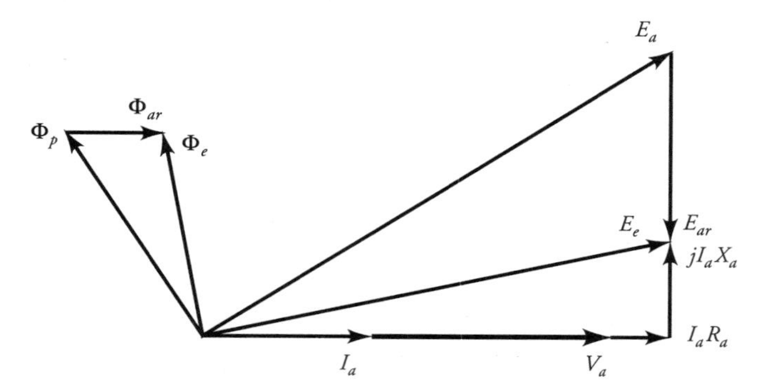

Figure 7.13 Phasor diagram depicting the effect of armature reaction when the power factor is unity.

flux Φ_{ar} by 90°. Hence the effective generated voltage per-phase $\tilde{E}_e$ is the algebraic sum of the no-load voltage $\tilde{E}_a$ and the armature reaction emf $\tilde{E}_{ar}$. That is, $\tilde{E}_e = \tilde{E}_a + \tilde{E}_{ar}$. An equivalent circuit showing the armature reaction emf is given in Figure 7.14.

(e) The per-phase terminal voltage $\tilde{V}_a$ is obtained by subtracting the voltage drops $\tilde{I}_a R_a$ and $j\tilde{I}_a X_a$ from $\tilde{E}_e$. In other words,

$$\tilde{E}_e = \tilde{V}_a + \tilde{I}_a(R_a + jX_a) \tag{7.20}$$

From the phasor diagram, it should be obvious that the armature reaction has reduced the effective flux per pole when the power factor of the load is unity. Also, the terminal voltage is smaller than the generated voltage.

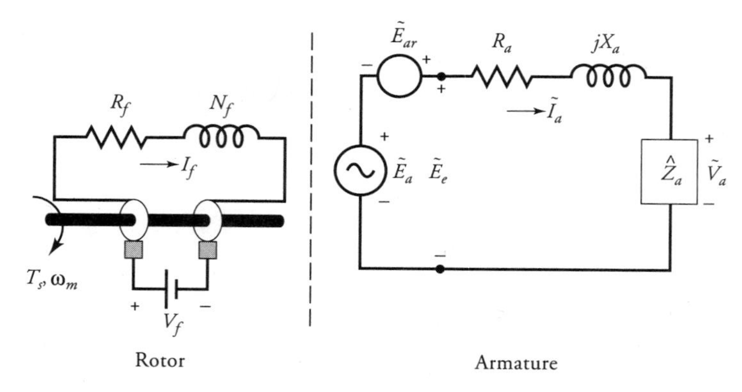

Figure 7.14 A per-phase equivalent circuit showing the induced emf in the armature winding due to the armature reaction.

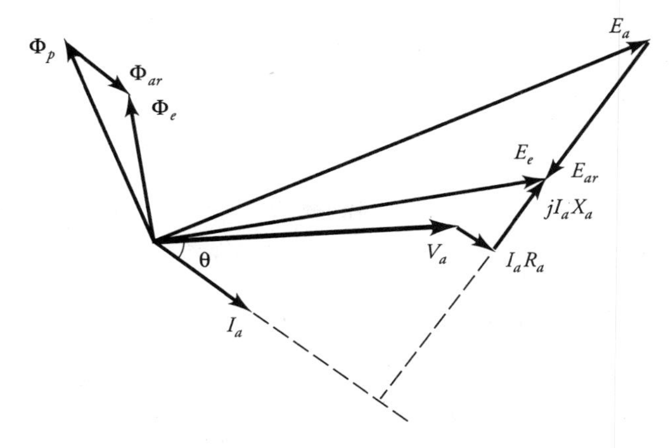

Figure 7.15 The phasor diagram showing the effect of armature reaction when the power factor is lagging.

By following the above sequence of events, we can obtain the phasor diagrams for the lagging (Figure 7.15) and the leading (Figure 7.16) power factors. From these figures it is evident that the resultant flux is (smaller/larger) with armature reaction for the (lagging/leading) power factor than without it. In addition, the terminal voltage $\tilde{V}_a$ is (higher/lower) than the generated voltage $\tilde{E}_a$ when the power factor is (leading/lagging). Since the flux per pole Φ_P is different for each of the three load conditions, the field current I_f must be adjusted each time the load is changed.

Since the armature reaction emf $\tilde{E}_{ar}$ lags the current $\tilde{I}_a$ by 90°, we can also express it as

$$\tilde{E}_{ar} = -j\tilde{I}_a X_m \tag{7.21}$$

where X_m, a constant of proportionality, is known as the **magnetization reactance**.

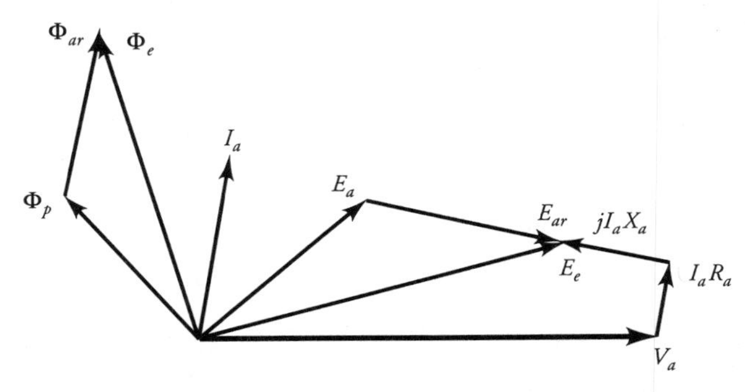

Figure 7.16 The phasor diagram showing the effect of armature reaction when the power factor of the load is leading.

Both the magnetization reactance and the leakage reactance are present at the same time. It is rather difficult to separate one reactance from the other. For this reason, the two reactances are combined together and the sum

$$X_s = X_m + X_a \tag{7.22}$$

is called the **synchronous reactance**. The synchronous reactance is usually very large compared with the resistance of the armature winding. We can now define the **synchronous impedance** on a per-phase basis as

$$\hat{Z}_s = R_a + jX_s \tag{7.23}$$

The Equivalent Circuit and Phasor Diagrams

The exact equivalent circuit of a synchronous generator on a per-phase basis embodying the synchronous reactance is given in Figure 7.17. The per-phase terminal voltage is

$$\tilde{V}_a = \tilde{E}_a - \tilde{I}_a(R_a + jX_s) = \tilde{E}_a - \tilde{I}_a\hat{Z}_s \tag{7.24}$$

and the corresponding phasor diagrams for three types of loads are given in Figure 7.17.

Voltage Regulation

The voltage regulation of a synchronous generator is defined as the ratio of the change in the terminal voltage from no load to full load to the full-load voltage. Since E_a is the rms value of the per-phase terminal voltage and V_a is the terminal voltage at full load, the percent voltage regulation is

$$VR\% = \frac{E_a - V_a}{V_a} \times 100 \tag{7.25}$$

EXAMPLE 7.6

A 9-kVA, 208-V, three-phase, Y-connected, synchronous generator has a winding resistance of 0.1 Ω/phase and a synchronous reactance of 5.6 Ω/phase. Determine its voltage regulation when the power factor of the load is (a) 80% lagging, (b) unity, and (c) 80% leading.

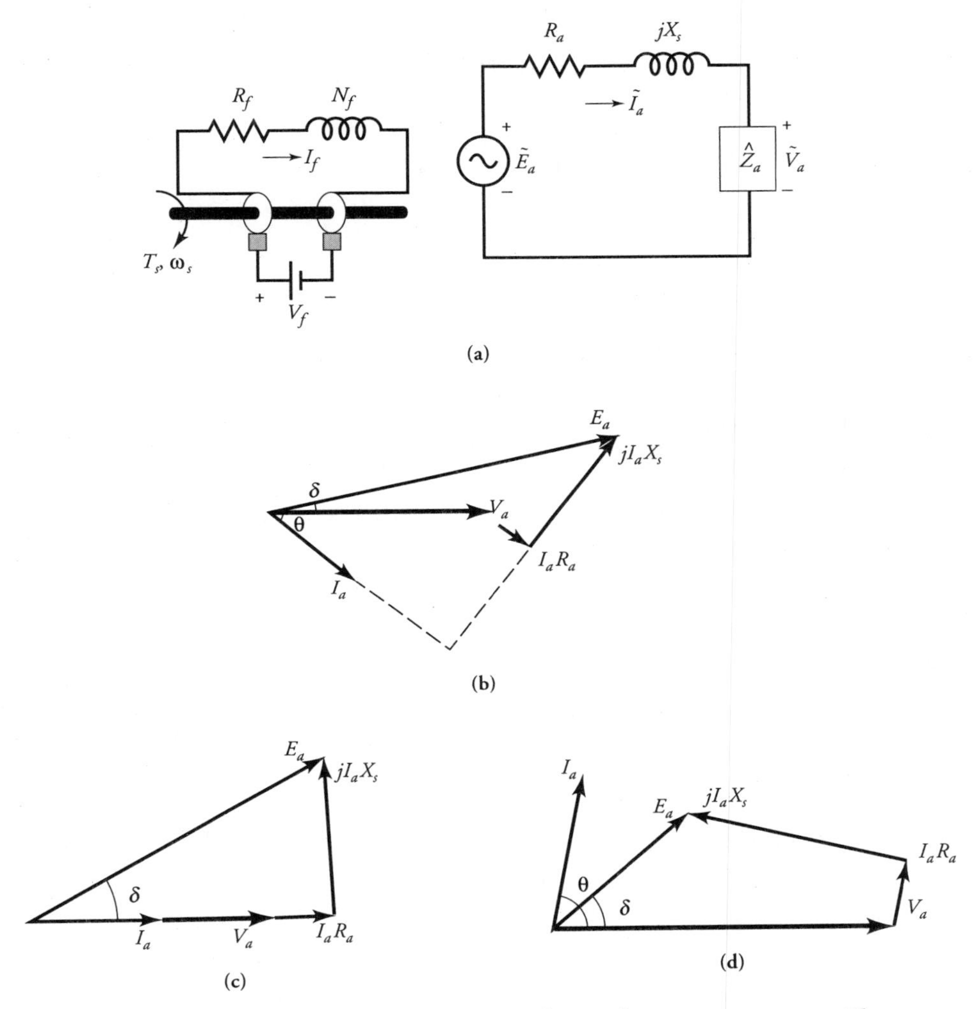

Figure 7.17 **(a)** The per-phase equivalent circuit of a synchronous generator with armature reaction and the corresponding phasor diagrams for **(b)** lagging, **(c)** unity, and **(d)** leading pf loads.

● SOLUTION

Since the voltage regulation is defined only at full load, we assume that the generator is delivering full load at its rated voltage. Thus,

$$V_a = \frac{208}{\sqrt{3}} \approx 120 \text{ V}$$

The rms value of the per-phase current at full load is

$$I_a = \frac{9000}{120 \times 3} = 25 \text{ A}$$

(a) For a lagging power factor of 0.8, $\theta = -36.87°$.

$$\tilde{E}_a = 120 + (0.1 + j5.6) \times 25\underline{/-36.87°}$$
$$= 233.77\underline{/28.21°} \text{ V}$$
$$VR\% = \frac{233.77 - 120}{120} \times 100 = 94.8\%$$

(b) For a unity power factor, $\theta = 0°$

$$\tilde{E}_a = 120 + (0.1 + j5.6) \times 25\underline{/0°}$$
$$= 186.03\underline{/48.81°} \text{ V}$$
$$VR\% = \frac{186.03 - 120}{120} \times 100 = 55.02\%$$

(c) For a leading power factor of 0.8, $\theta = 36.87°$

$$\tilde{E}_a = 120 + (0.1 + j5.6) \times 25\underline{/36.87°}$$
$$= 119.69\underline{/71.49°} \text{ V}$$
$$VR\% = \frac{119.69 - 120}{120} \times 100 = -0.26\%$$

■

EXERCISES

7.11. The synchronous impedance of a 13.2-kVA, 440-V, Δ-connected, three-phase, synchronous generator is $1 + j10$ Ω/phase. Calculate the voltage regulation when the generator supplies the full load at a power factor of (a) 0.866 lagging, (b) unity, and (c) 0.866 leading.

7.12. A three-phase, Y-connected, synchronous generator supplies a rated load of 12 MW at a power factor of 0.8 lagging. The synchronous impedance is $0.2 + j2$ Ω/phase. If the terminal (line) voltage is 11 kV, calculate the no-load line voltage. What is the voltage regulation?

7.13. A 60-kVA, three-phase, 50-Hz, Y-connected, synchronous generator is designed to supply a full load at 480 V. The power factor of the load is 0.9 lagging. The synchronous impedance is $0.15 + j1.3\ \Omega$/phase. Determine the voltage regulation of the generator.

7.9 Power Relationships

The rotor of a synchronous generator is connected to a prime mover which may, in fact, be a dc motor, a steam turbine, a gas turbine, a diesel engine, or the like. If the prime mover exerts a torque T_s at the shaft at an angular velocity of ω_s, the mechanical power supplied to the rotor is $T_s\omega_s$. The dc power input to a wound rotor is $V_f I_f$, where V_f is the dc voltage across the field winding and I_f is the dc current through it. Thus, the total power input is

$$P_{in} = T_s\omega_s + V_f I_f \tag{7.26}$$

The losses in a synchronous generator are the same as in a dc generator and consist of rotational loss (mechanical loss and magnetic loss), the copper loss in the armature winding, the field-excitation loss in the field winding, and the stray-load loss, if any. As with a dc generator, we subtract the rotational loss, the field-winding loss, and the stray-load loss from the input power to obtain the power developed by the armature. By subtracting the copper losses in the armature from the developed power, we obtain the output power of the machine, as illustrated by the power-flow diagram of Figure 7.18.

If V_a is the load voltage, I_a is the load current, and θ is the phase angle between V_a and I_a, the power output of a synchronous generator is

$$P_o = 3V_a I_a \cos\theta \tag{7.27}$$

The copper loss in the armature winding is

$$P_{cu} = 3I_a^2 R_a \tag{7.28}$$

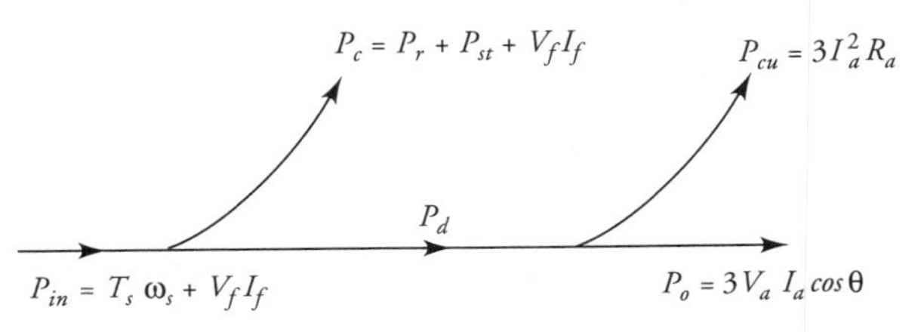

Figure 7.18 Power-flow diagram of a synchronous generator.

If P_r is the rotational loss of a synchronous generator and P_{st} is the stray-load loss, then the power input is

$$P_{in} = 3V_aI_a \cos\theta + 3I_a^2R_a + P_r + P_{st} + V_fI_f \tag{7.29}$$

Since the rotor revolves at a constant speed, the rotational loss is constant. The field-winding loss is constant. Assuming the stray-load loss to be a constant, we can group these losses together and consider them a constant loss. Thus, the constant loss is

$$P_c = P_r + P_{st} + V_fI_f \tag{7.30}$$

Since the copper loss in the armature depends upon the load current, it is considered a variable loss.

The efficiency of the generator is

$$\eta = \frac{3V_aI_a \cos\theta}{3V_aI_a \cos\theta + P_c + 3I_a^2R_a} \tag{7.31}$$

From the above equation we obtain a condition for the maximum efficiency as

$$3I_a^2R_a = P_c \tag{7.32}$$

Approximate Power Relationship

As mentioned earlier, the per-phase armature-winding resistance of a synchronous generator is usually very small and can be neglected in comparison with its synchronous reactance. The approximate equivalent circuit and a corresponding phasor diagram for a lagging load are given in Figure 7.19. Note that the terminal-

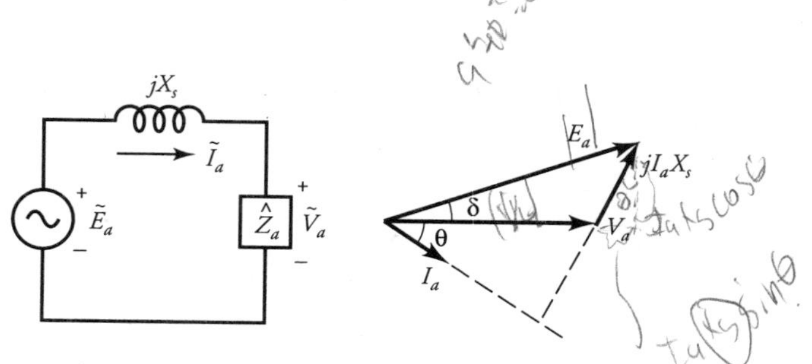

Figure 7.19 An approximate equivalent circuit of a synchronous generator and its phasor diagram for a lagging pf load.

phase voltage $\tilde{V}_a$ has been taken as a reference and the per-phase generated voltage $\tilde{E}_a$ leads $\tilde{V}_a$ by an angle δ. The phase current $\tilde{I}_a$ lags $\tilde{V}_a$ by an angle θ. Thus,

$$\tilde{E}_a = E_a \cos \delta + jE_a \sin \delta \tag{7.33}$$

and

$$\tilde{I}_a = I_a \cos \theta + jI_a \sin \theta \tag{7.34}$$

where E_a and I_a are the rms values of $\tilde{E}_a$ and $\tilde{I}_a$.

The per-phase terminal voltage is

$$\tilde{V}_a = \tilde{E}_a - j\tilde{I}_a X_s$$

or

$$\tilde{I}_a = \frac{\tilde{E}_a - \tilde{V}_a}{jX_s} = \frac{E_a \sin \delta}{X_s} - j\frac{E_a \cos \delta - V_a}{X_s}$$

Thus,

$$I_a \cos \theta = \frac{E_a \sin \delta}{X_s}$$

Hence, the approximate power output of the generator is

$$P_o = 3V_a I_a \cos \theta = \frac{3V_a E_a \sin \delta}{X_s} \tag{7.35}$$

When a synchronous generator operates at a constant speed with a constant field current, X_s and E_a are both constants. V_a is the terminal voltage, which is usually held constant. Thus, the power output of the generator varies as $\sin \delta$ where δ is the angle from $\tilde{V}_a$ to $\tilde{E}_a$ and is called the **power angle**. Equation (7.35) is known as the **power-angle relation**. This relation is also shown in the power-angle curve of Figure 7.20. In the development of Eq. (7.35) we have tacitly assumed a round rotor. For a salient-pole rotor, the curve is somewhat modified and is indicated by the dashed curve. Later in this chapter, we will develop a relationship between the power-angle and the power output for salient-pole rotors.

For a given amount of field current and a certain terminal voltage, the maximum power output P_{om} (or developed P_{dm}) of a synchronous generator is

$$P_{dm} = P_{om} = \frac{3V_a E_a}{X_s} \tag{7.36}$$

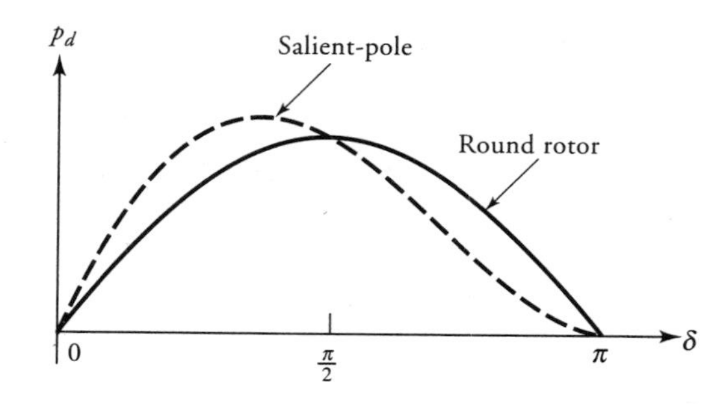

Figure 7.20 Power developed as a function of power-angle for a cylindrical rotor (solid) and a salient-pole rotor (dashed).

The torque developed, from Eq. (7.35), is

$$T_d = \frac{3V_a E_a \sin \delta}{X_s \omega_s} \tag{7.37}$$

Since the torque developed is also proportional to $\sin \delta$, the angle δ is also referred to as the **torque angle**. The torque developed by a synchronous generator opposes the torque applied by the prime mover.

The maximum torque developed by the synchronous generator, from Eq. (7.36), is

$$T_{dm} = \frac{3V_a E_a}{X_s \omega_s} \tag{7.38}$$

EXAMPLE 7.7

A 9-kVA, 208-V, 1200-rpm, three-phase, 60-Hz, Y-connected, synchronous generator has a field-winding resistance of 4.5 Ω. The armature-winding impedance is $0.3 + j5$ Ω/phase. When the generator operates at its full load and 0.8 pf lagging, the field-winding current is 5 A. The rotational loss is 500 W. Determine (a) the voltage regulation, (b) the efficiency of the generator, and (c) the torque applied by the prime mover.

● SOLUTION

The per-phase terminal voltage: $V_a = 208/\sqrt{3} \approx 120$ V
The per-phase apparent power of the generator is $9/3 = 3$ kVA. Hence the rated

current on a per-phase basis is 3000/120 = 25 A. For a lagging power factor of 0.8,

$$\tilde{I}_a = 25\underline{/-36.87^\circ} \text{ A}$$

Hence, the per-phase generated voltage is

$$\begin{aligned}\tilde{E}_a &= 120 + (0.3 + j5) \times 25\underline{/-36.87^\circ} \\ &= 222.534\underline{/25.414^\circ} \text{ V}\end{aligned}$$

(a) The voltage regulation is

$$VR\% = \frac{222.534 - 120}{120} \times 100 = 85.45\%$$

(b) The power output: $P_o = 3 \times 120 \times 25 \times 0.8 = 7200$ W
The copper loss: $P_{cu} = 3 \times 25^2 \times 0.3 = 562.5$ W
The power developed: $P_d = 7200 + 562.5 = 7762.5$ W
We can also compute the power developed as

$$P_d = 3 \times \text{Re}[222.534\underline{/25.414^\circ} \times 25\underline{/36.87^\circ}] = 7762.4 \text{ W}$$

The constant loss: $P_c = 500 + 5^2 \times 4.5 = 612.5$ W
Hence, the total power input: $P_{in} = 7762.5 + 612.5 = 8375$ W
The efficiency of the generator is

$$\eta = \frac{7200}{8375} \approx 0.86 \quad \text{or} \quad 86\%$$

(c) The angular velocity of the prime mover is

$$\omega_s = \frac{2\pi \times 1200}{60} = 40\pi \text{ rad/s}$$

The mechanical power in terms of the electrical power input is

$$P_{inm} = P_d + P_r = 7762.5 + 500 = 8262.5 \text{ W}$$

Hence, the torque applied by the prime mover is

$$T_s = \frac{8262.5}{40\pi} = 65.75 \text{ N·m}$$

■

Exercises

7.14. Repeat Example 7.7 when the synchronous generator supplies the rated load at a unity power factor. The corresponding value of the field current is 3 A.

7.15. Repeat Example 7.7 when the synchronous generator supplies the rated load at 0.8 pf leading. The corresponding value of the field current is 6 A.

7.16. A 40-kVA, 240-V, 50-Hz, 4-pole, three-phase, Y-connected alternator has a synchronous reactance of 0.08 Ω/phase. The armature winding resistance is negligibly small, and the revolving field is established by permanent magnets. The rotational loss is 5% of the power developed. When the generator delivers the rated load at a leading power factor of 0.866, determine (a) the power angle, (b) the efficiency, (c) the voltage regulation, and (d) the torque supplied by the prime mover.

7.10 Synchronous Generator Tests

To obtain the parameters of a synchronous generator, we perform three simple tests as described below.

The Resistance Test

This test is conducted to measure the armature-winding resistance of a synchronous generator when it is at rest and the field winding is open. The resistance is measured between two lines at a time and the average of the three resistance readings is taken to be the measured value of the resistance, R_L, from line to line. If the generator is Y-connected, the per-phase resistance is

$$R_a = 0.5R_L \tag{7.39a}$$

However, for a Δ-connected generator, the per-phase resistance is

$$R_a = 1.5R_L \tag{7.39b}$$

The resistance can be measured by applying a dc voltage across each pair of terminals (lines) and allowing the dc current to be about equal to the rated current of the generator. The dc resistance thus obtained may be multiplied by a factor of 1.05 to 1.25 to account for its rise owing to the fact that the generator carries alternating current.

The Open-Circuit Test

The open-circuit test, or the **no-load test**, is performed by driving the generator at its rated speed while the armature winding is left open. The field current is varied in suitable steps, and the corresponding values of the open-circuit voltage between any two pair of terminals of the armature windings are recorded as depicted in Figure 7.21 for a Y-connected generator. The field current can be raised until the open-circuit voltage is twice the rated value. From the recorded data for the open-circuit voltage we can compute the per-phase (open-circuit) voltage. When the per-phase (open-circuit) voltage is plotted as a function of the field current, the graph is referred to as the **open-circuit saturation characteristic (curve)**, or **OCC** for short.

The OCC follows a straight-line relation as long as the magnetic circuit of the synchronous generator does not saturate. Since, in the linear region, most of the applied mmf is consumed by the air-gap, the straight line is appropriately called the **air-gap line**. As the saturation sets in, the OCC starts deviating from the air-gap line, as depicted in Figure 7.22. The OCC is taken to be the **magnetization curve** of the generator under load conditions.

The Short-Circuit Test

The short-circuit test provides information about the current capabilities of a synchronous generator. It is performed by driving the generator at its rated speed

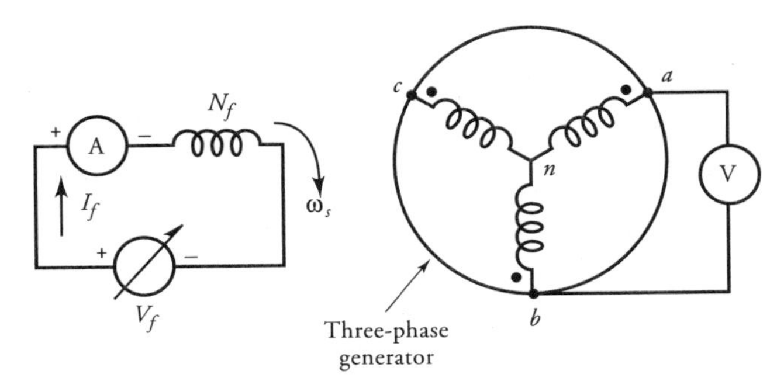

Figure 7.21 Circuit diagram to perform open-circuit test.

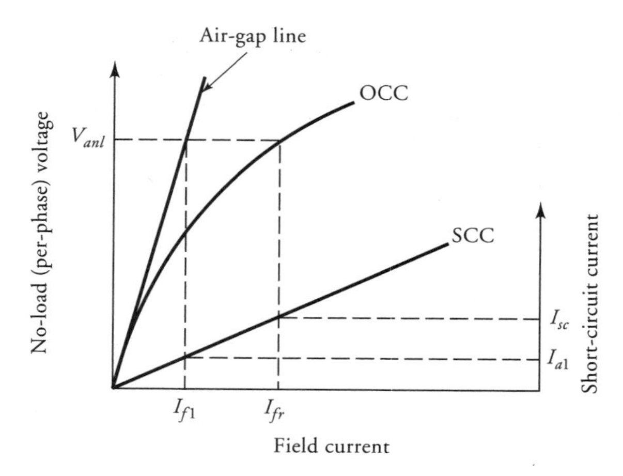

Figure 7.22 Open-circuit and short-circuit characteristics of a synchronous generator.

when the terminals of the armature winding are shorted, as shown in Figure 7.23 for a Y-connected generator. An ammeter is placed in series with one of the three shorted lines. The field current is gradually increased and the corresponding value of the current is recorded. The maximum armature current under short circuit should not exceed twice the rated current of the generator. From the recorded data we can compute the per-phase short-circuit current. When the per-phase short-circuit current is plotted as a function of the field current, the graph is called the **short-circuit characteristic** (**SCC** for short) of a generator. For convenience, the OCC and SCC are plotted on the same graph, as shown in Figure 7.22.

Since the terminal voltage under short-circuit condition is zero, the per-phase generated voltage must be equal to the voltage drop across the synchronous impedance. To calculate the per-phase synchronous impedance from the OCC and SCC of a synchronous generator at its rated voltage, the procedure is as follows:

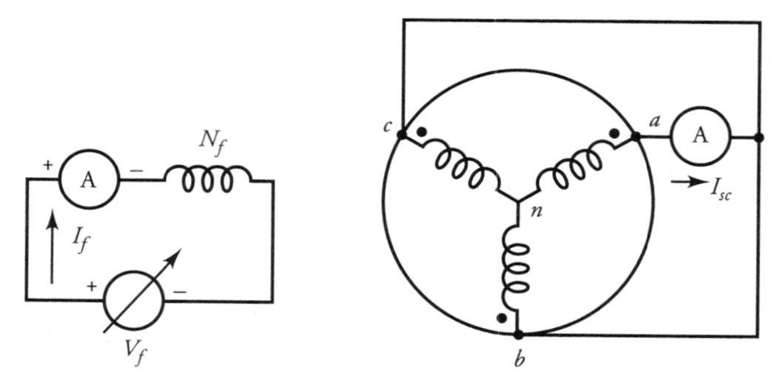

Figure 7.23 Circuit diagram to perform short-circuit test.

1. Find the value of the field current (I_{fr}) that gives the rated per-phase voltage (V_{anL}) from the OCC of the generator.
2. Find the value of the short-circuit current (I_{sc}) from the SCC for the same value of the field current, I_{fr}.
3. The magnitude of the synchronous impedance is equal to the open-circuit voltage divided by the short-circuit current. That is,

$$Z_s = \frac{V_{anL}}{I_{sc}} \tag{7.40}$$

Since the resistance of each phase winding of the armature is already known from the resistance test, the synchronous reactance of the generator is

$$X_s = \sqrt{Z_s^2 - R_a^2} \tag{7.41}$$

From the OCC and SCC we can, in fact, plot the synchronous impedance as a function of the field current. A typical plot is given in Figure 7.24. As long as the flux density is below the knee of the saturation curve (the flux is proportional to the applied mmf), the synchronous impedance is fairly constant and is referred to as the **unsaturated synchronous impedance**. As the generator operates above the knee of its saturation curve, the generated voltage is smaller than what it would have been without saturation. Consequently, the **saturated synchronous impedance** is smaller than its unsaturated value. Both the unsaturated and saturated synchronous impedances can be determined from Figure 7.22 at the rated voltage of the generator. The air-gap line gives the necessary field current I_{f1} at the rated voltage V_{anL} for the unsaturated synchronous impedance and the corresponding short-circuit current I_{a1}. Thus, the magnitude of the unsaturated synchronous impedance is

$$Z_{su} = \frac{V_{anL}}{I_{a1}} \tag{7.42}$$

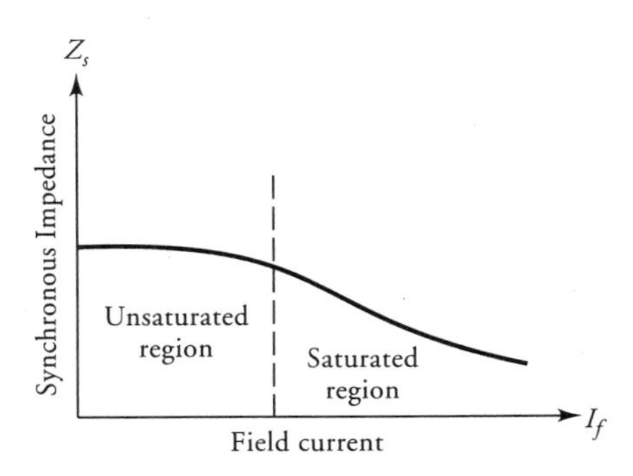

Figure 7.24 Synchronous impedance as a function of the field current.

and the magnitude of the (saturated) synchronous impedance has already been determined in Eq. (7.40).

It must, therefore, be evident that the synchronous impedance of a synchronous generator may vary considerably from light load to full load. For our calculations, we assume that Z_s is constant and its value corresponds to the rated no-load voltage.

EXAMPLE 7.8

A 500-kVA, 2300-V, three-phase, Y-connected, synchronous generator is operated at its rated speed to obtain its rated no-load voltage. When a short circuit is established the phase current is 150 A. The average resistance of each phase is 0.5 Ω. Determine the synchronous reactance per phase. Using the per-unit system, determine the percent regulation when the generator delivers the rated load at its rated voltage and 0.8 pf lagging.

● SOLUTION

The open-circuit phase voltage: $V_{anL} = 2300/\sqrt{3} = 1327.91$ V
The short-circuit phase current: $I_{sc} = 150$ A

Thus
$$Z_s = \frac{1327.91}{150} = 8.85\ \Omega$$

and
$$X_s = \sqrt{8.85^2 - 0.5^2} = 8.84\ \Omega$$

The rated (full-load) current is

$$I_a = \frac{500{,}000}{3 \times 1327.91} = 125.51\ \text{A}$$

Let us use the per-phase voltage and the rated current as the base values. That is,

$$V_b = 1327.91\ \text{V}$$
$$I_b = 125.51\ \text{A}$$
$$Z_b = \frac{1327.91}{125.51} = 10.58\ \Omega$$

The per-unit quantities when the generator operates at its rated load are

$$\tilde{I}_{apu} = 1\angle -36.87^\circ$$

$$\tilde{V}_{apu} = 1\angle 0^\circ$$

$$\hat{Z}_{spu} = \frac{0.5 + j8.84}{10.58} = 0.047 + j0.836$$

Hence,

$$\tilde{E}_{apu} = 1 + (0.047 + j0.836) \times 1\angle -36.87^\circ$$

$$= 1.667\angle 22.58^\circ$$

Thus, the generated voltage per phase is

$$\tilde{E}_a = 1327.91 \times 1.667\angle 22.58^\circ = 2213.36\angle 22.58^\circ \text{ V}$$

In terms of the per-unit values, the percent voltage regulation is

$$VR\% = (1.667 - 1) \times 100 = 66.7\%$$

■

Exercises

7.17. In a three-phase, Y-connected, 750-kVA, 1732-V, synchronous generator a field current of 20 A at the rated speed produces a current of 300 A on short circuit and a voltage of 1732 V on open circuit. The resistance between any two terminals of the generator is 0.8 Ω. Determine the synchronous reactance of the generator. The rotational loss is 20 kW. Using the per-unit system, determine the voltage regulation and the efficiency of the generator when it delivers the rated load at its rated voltage and unity power factor.

7.18. The test data obtained at the rated speed on a three-phase, Y-connected, synchronous generator are given as follows:
Short-circuit test: Field current = 1.2 A
Short-circuit current = 25 A
Open-circuit test: Field current = 1.2 A
Open-circuit voltage = 440 V
The per-phase winding resistance is 1.2 Ω. Determine the synchronous impedance of the generator.

7.19. A 2-MVA, 4400-V, 25-Hz, Y-connected, three-phase, synchronous generator is tested at its rated speed of 500 rpm. The open-circuit voltage and the short-circuit current are 4800 V and 260 A, respectively, when the field current is 15 A. The generator is designed to have a resistive voltage drop of 5% of its rated voltage per phase. Determine the synchronous impedance. Using the per-unit system, find the voltage regulation when the generator delivers the rated load at the rated voltage and unity power factor. How many poles are in the generator?

7.11 The External Characteristic

The external characteristic of a synchronous generator shows the variation of the terminal voltage with the load of an independent generator. An independent generator is a relatively small standby generator (which may be driven by a gasoline engine at a constant speed) that supplies independent electrical loads. Its terminal voltage varies with the load. If the excitation current is held constant, then the generated voltage per phase, $\tilde{E}_a$, will be constant.

From the approximate equivalent circuit (Figure 7.19), the terminal voltage is

$$\tilde{V}_a = \tilde{E}_a - j\tilde{I}_a X_s \tag{7.43}$$

Since $\tilde{E}_a$ and X_s are constant, the terminal voltage $\tilde{V}_a$ depends upon the magnitude of the load-current and its power factor. Although the magnitude of the phase voltage $\tilde{E}_a$ is constant, its phase (the power angle) is free to change. It should, therefore, be obvious that the locus of $\tilde{E}_a$ must be a circle. Let us now explore the changes in $\tilde{V}_a$ as a function of load with unity, lagging, and leading power factors.

Unity Power Factor

For a purely resistive load, $\hat{Z}_L = R_L$, the terminal voltage $\tilde{V}_a$ and the load-current $\tilde{I}_a$ are in phase. The increase in the load-current causes (a) the voltage drop across the synchronous reactance to increase, (b) the power angle to increase, and (c) the terminal voltage to decrease, as illustrated by Figure 7.25. The external characteristic—that is, the terminal voltage as a function of the load for the unity power factor—is shown in Figure 7.26.

Lagging Power Factor

For an inductive load, $\hat{Z}_L = R_L + jX_L$, the load-current lags the terminal voltage by an angle θ. For a given power-factor angle θ, the increase in the load-current results in the decrease in the terminal voltage, as shown in Figure 7.27. However, for a constant magnitude of the load-current, the terminal voltage decreases even further with the decrease in the power factor, as shown in Figure 7.28. The external characteristics for two lagging power factors are given in Figure 7.26.

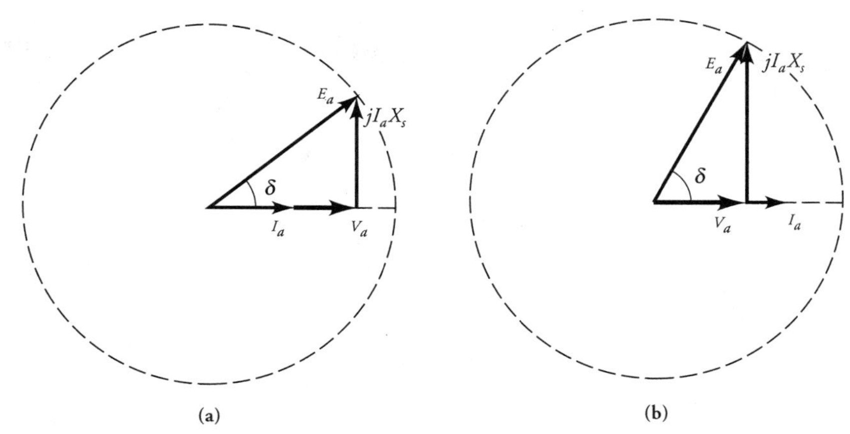

Figure 7.25 Effect of resistive loading on the terminal voltage of an independent synchronous generator for **(a)** small and **(b)** large armature currents.

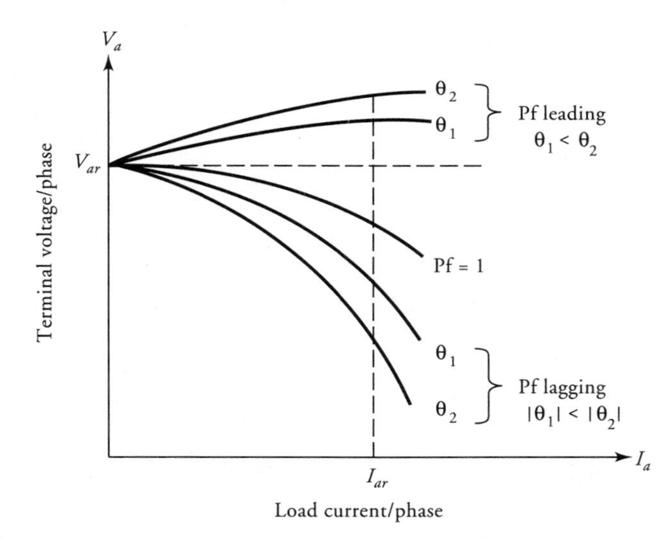

Figure 7.26 External characteristics of a synchronous generator under various load conditions.

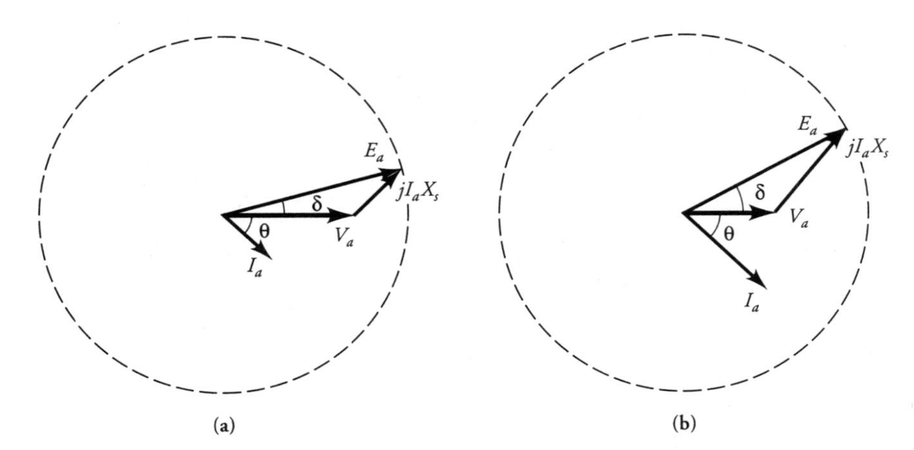

Figure 7.27 Effect of inductive loading on the terminal voltage of an independent synchronous generator for **(a)** small and **(b)** large load currents.

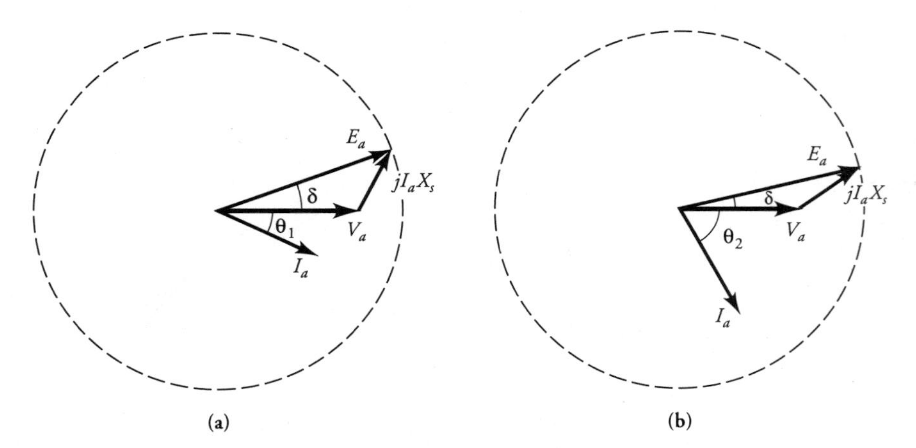

Figure 7.28 Effect of lagging power factor on the terminal voltage of an independent synchronous generator for **(a)** small and **(b)** large power factor angles.

Leading Power Factor

For a capacitive load, $\hat{Z}_L = R_L - jX_c$ the load-current leads the terminal voltage by an angle θ. The phasor diagrams for small and large armature currents for a given power-factor angle θ are sketched in Figure 7.29. It is evident from Figure 7.29 that the terminal voltage increases with the increase in the load-current. Figure 7.30 shows the phasor diagrams for small and large power-factor angles for the same magnitude of the load-current. It is evident that the terminal voltage increases even further with an increase in the power-factor angle.

For a generator operating independently we can adjust the field (excitation) current under each load in such a way that the generator delivers the rated load

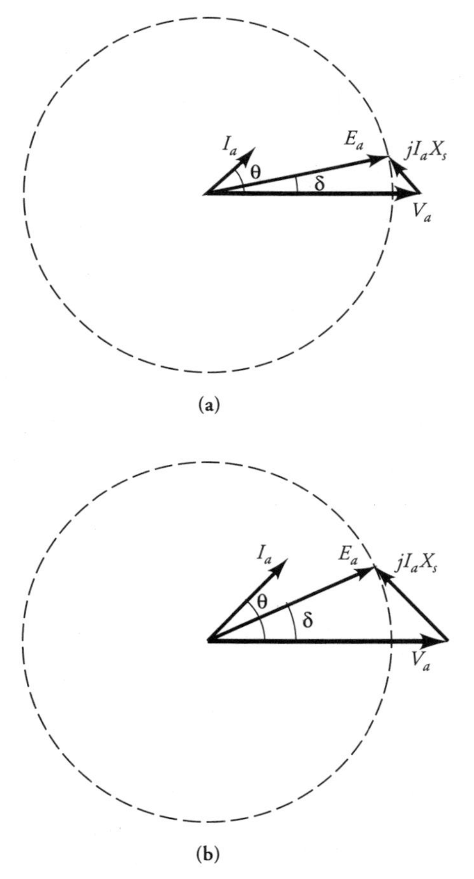

Figure 7.29 Effect of capacitive loading on the terminal voltage of an independent synchronous generator for **(a)** small and **(b)** large load currents.

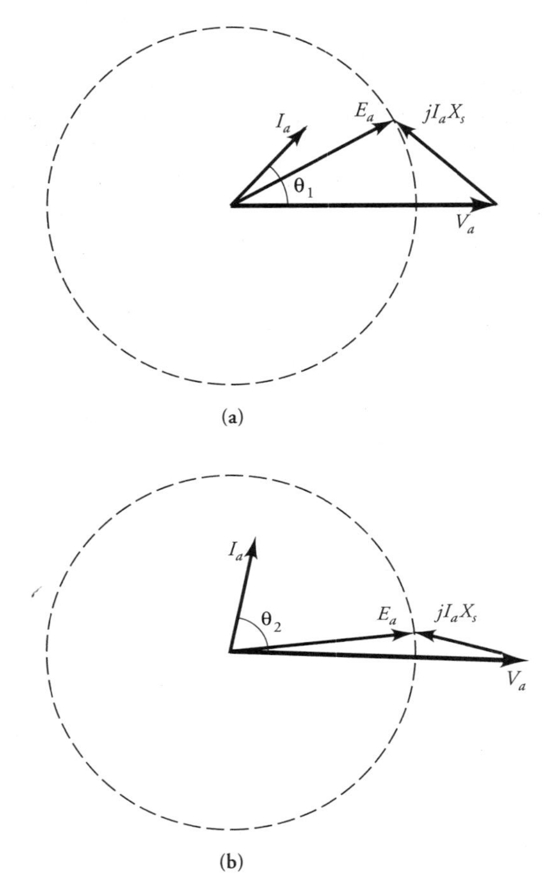

Figure 7.30 Effect of leading power factor on the terminal voltage of an independent synchronous generator for **(a)** small and **(b)** large power factor angles.

at its rated terminal voltage. If we now vary the load-current without making any further adjustments in the excitation current, we observe the following:

1. For a resistive load, the terminal voltage increases with the decrease in the load-current.
2. For an inductive load, the terminal voltage also increases with the decrease in the load-current. However, the increase is larger for an inductive load than a resistive load.
3. For a capacitive load, the terminal voltage decreases with the decrease in the line current.

The external characteristics of an independent generator adjusted to operate at its rated voltage and the load are plotted in Figure 7.31.

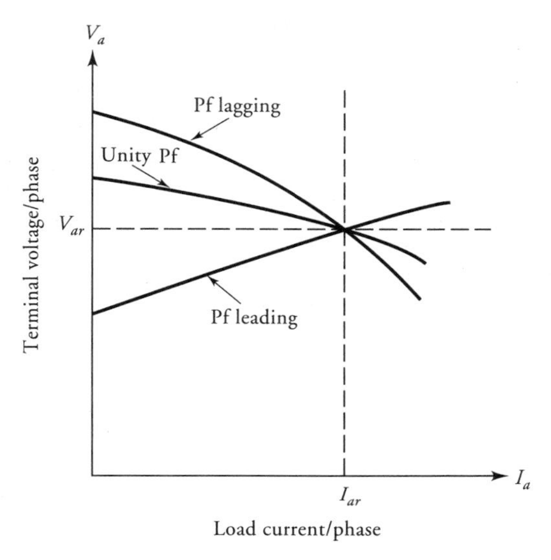

Figure 7.31 External characteristics of an independent synchronous generator adjusted to operate at the rated load under various load conditions.

7.12 Salient-Pole Synchronous Generator

The foregoing analysis of a synchronous generator is satisfactory only when the rotor has a cylindrical construction. A cylindrical rotor presents an almost uniform air-gap, and the variation in the air-gap reluctance around its periphery owing to the slots is negligible. On the other hand, a salient-pole rotor has a larger air-gap in the region between the poles than in the region just above the poles, as is evident from Figure 7.32. In other words, the reluctances of the two regions in a salient-pole generator differ significantly. In order to account for this difference, the synchronous reactance is split into two reactances. The component of the synchronous reactance along the pole-axis (the *d*-axis) is commonly called the **direct-axis synchronous reactance** X_d, and the other component along the axis between the poles (the *q*-axis) is called the **quadrature-axis synchronous reactance** X_q. The armature current I_a is also resolved into two components: **the direct component** $\tilde{I}_d$ and the **quadrature component** $\tilde{I}_q$. The direct component $\tilde{I}_d$ produces the field along the *d*-axis and lags $\tilde{E}_a$ by 90°. The quadrature component $\tilde{I}_q$ produces the field along the *q*-axis and is in phase with $\tilde{E}_a$.

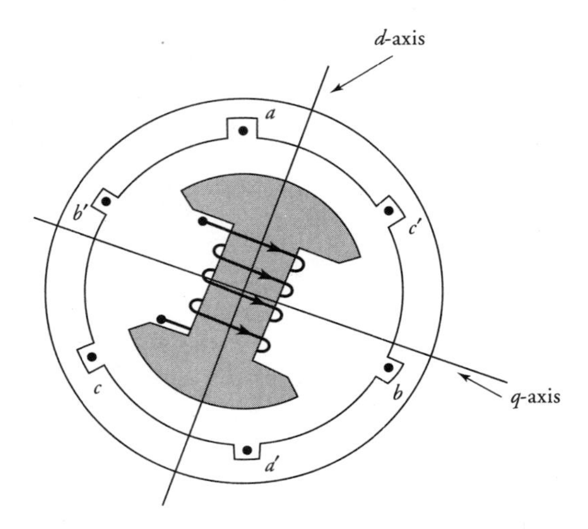

Figure 7.32 A 2-pole, salient-pole synchronous generator.

If $\tilde{E}_a$ is the per-phase generated voltage under no load and $\tilde{E}_d$ and $\tilde{E}_q$ are the induced emfs in the armature winding by the current components $\tilde{I}_d$ and $\tilde{I}_q$, respectively, then the per-phase terminal voltage of the generator is

$$\tilde{V}_a = \tilde{E}_a + \tilde{E}_d + \tilde{E}_q - \tilde{I}_a R_a \tag{7.44}$$

We can, however, express the induced emfs $\tilde{E}_d$ and $\tilde{E}_q$ in terms of X_d and X_q as

$$\tilde{E}_d = -j\tilde{I}_d X_d \tag{7.45}$$

and

$$\tilde{E}_q = -j\tilde{I}_q X_q \tag{7.46}$$

Substituting the above expressions in Eq. (7.44), we obtain

$$\tilde{E}_a = \tilde{V}_a + \tilde{I}_a R_a + j\tilde{I}_d X_d + j\tilde{I}_q X_q \tag{7.47}$$

If we express $j\tilde{I}_d X_d$ as

$$j\tilde{I}_d X_d = j\tilde{I}_d X_q + j\tilde{I}_d(X_d - X_q)$$

then

$$\begin{aligned}\tilde{E}_a &= \tilde{V}_a + \tilde{I}_a R_a + j\tilde{I}_d X_q + j\tilde{I}_q X_q + j\tilde{I}_d(X_d - X_q) \\ &= \tilde{V}_a + \tilde{I}_a R_a + j\tilde{I}_a X_q + j\tilde{I}_d(X_d - X_q) \\ &= \tilde{E}'_a + j\tilde{I}_d(X_d - X_q) \qquad (7.48a)\end{aligned}$$

where
$$\tilde{E}'_a = \tilde{V}_a + \tilde{I}_a R_a + j\tilde{I}_a X_q \qquad (7.48b)$$

and
$$\tilde{I}_a = \tilde{I}_d + \tilde{I}_q \qquad (7.48c)$$

Based upon these equations, we can represent the salient-pole synchronous generator with an equivalent circuit as shown in Figure 7.33. If we consider E'_a as the effective generated emf (excitation voltage), then the equivalent circuit of a salient-pole generator is similar to that of a round-rotor generator.

The phasor diagram for a load with a lagging power factor based upon Eq. (7.48a) is given in Figure 7.34a. Since $\tilde{E}'_a$ is in phase with $\tilde{E}_a$, the phase angle by

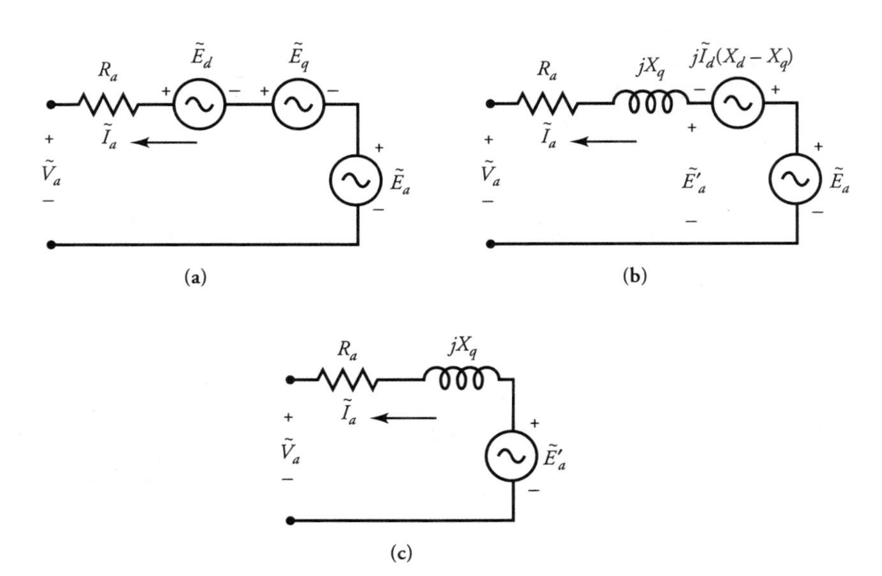

Figure 7.33 Equivalent-circuit representations of a salient-pole synchronous generator.

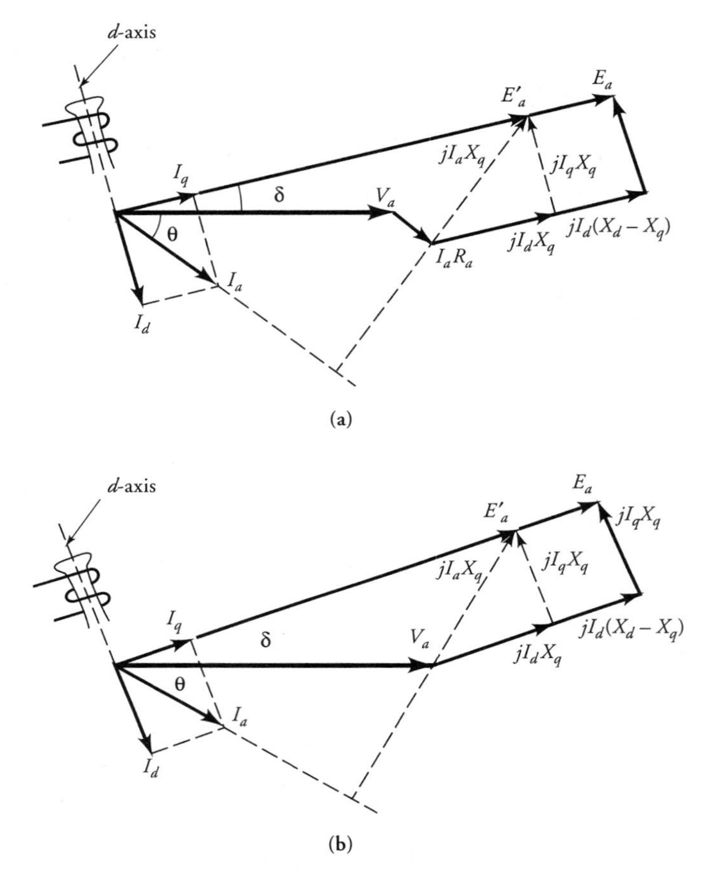

Figure 7.34 **(a)** Phasor diagram of a salient-pole synchronous generator having finite armature-winding resistance and lagging power factor. **(b)** Phasor diagram of a salient-pole synchronous generator with neglible armature-winding resistance and lagging power factor.

which $\tilde{E}_a'$ leads the terminal voltage $\tilde{V}_a$ is the power angle δ. Therefore, we can use Eq. (7.48b) to determine the power angle δ when the terminal voltage, the load-current, and the parameters of the generator are known. The generated voltage per phase $\tilde{E}_a$ can then be determined from Eq. (7.48c). From the phasor diagram we can also obtain the following equation to determine the power angle δ.

$$\tan \delta = \frac{I_a X_q \cos \theta}{V_a + I_a(R_a \cos \theta + X_q \sin \theta)} \tag{7.49}$$

We can now obtain expressions $\tilde{I}_d$ and $\tilde{I}_a$ in terms of I_a as

$$\tilde{I}_d = I_a \sin (\delta + \theta) \angle \delta - 90^\circ \tag{7.50a}$$

$$\tilde{I}_q = I_a \cos (\delta + \theta) \angle \delta \tag{7.50b}$$

where I_a is the rms value of the armature current.

The power output can now be computed as

$$\begin{aligned} P_o &= 3 \operatorname{Re}[\tilde{V}_a \tilde{I}_a^*] = 3 \operatorname{Re}[\tilde{V}_a(\tilde{I}_d^* + \tilde{I}_q^*)] \\ &= 3\, V_a[I_d \sin \delta + I_q \cos \delta] \end{aligned} \tag{7.51}$$

When the armature resistance is so small that it can be neglected, the power output is the same as the power developed. The phasor diagram for a load with a lagging power factor is given in Figure 7.34. From this phasor diagram, we obtain

$$I_q = \frac{V_a \sin \delta}{X_q} \tag{7.52a}$$

and

$$I_d = \frac{E_a - V_a \cos \delta}{X_d} \tag{7.52b}$$

Thus,

$$\tan \delta = \frac{I_a X_q \cos \theta}{V_a + I_a X_q \sin \theta} \tag{7.53}$$

Substituting for I_d and I_q in Eq. (7.51), we obtain an expression for the power developed (output) as

$$P_d = \frac{3V_a E_a \sin \delta}{X_d} + \frac{3(X_d - X_q)}{2X_d X_q} V_a^2 \sin 2\delta \tag{7.54}$$

The first term in the above equation is the same as that obtained for a cylindrical-rotor generator. It represents the power due to the field excitation (generated voltage E_a). The second term highlights the effect of saliency. It represents the reluctance power and is independent of the field excitation. Thus, a salient-pole generator can deliver higher power than a cylindrical-rotor generator at a power angle less than 90° for the same terminal and excitation voltages.

The torque developed by a salient-pole generator is

$$T_d = \frac{3V_a E_a \sin \delta}{X_d \omega_s} + \frac{3(X_d - X_q)}{2X_d X_q \omega_s} V_a^2 \sin 2\delta \tag{7.55}$$

The torque developed by a salient-pole generator as a function of torque angle δ is plotted in Figure 7.35. The variation in the excitation torque (first term) and the reluctance torque (second term) are also indicated.

EXAMPLE 7.9

A 70-MVA, 13.8-kV, 60-Hz, Y-connected, three-phase, salient-pole, synchronous generator has $X_d = 1.83\ \Omega$, and $X_q = 1.21\ \Omega$. It delivers the rated load at 0.8 pf lagging. The armature resistance is negligible. Determine (a) the voltage regulation and (b) the power developed by the generator.

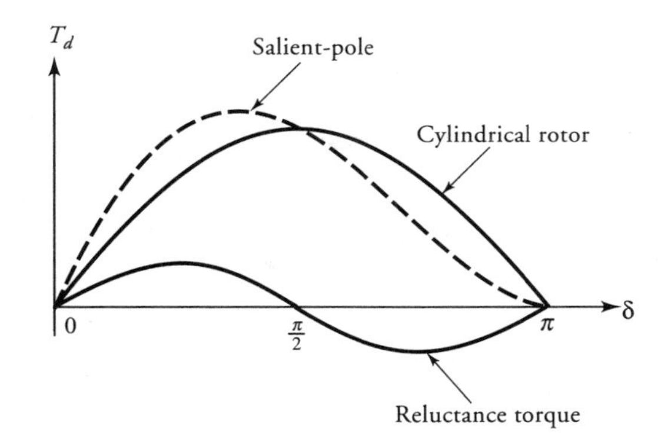

Figure 7.35 Torque developed by a salient-pole synchronous generator as a function of torque angle.

● SOLUTION

On a per-phase basis, the rms values of the terminal voltage and the load-current are

$$V_a = \frac{13{,}800}{\sqrt{3}} = 7967.43 \text{ V}$$

$$I_a = \frac{70 \times 10^6}{\sqrt{3} \times 13{,}800} = 2928.59 \text{ A}$$

The power factor angle: $\theta = -\cos^{-1}(0.8) = -36.87°$

From Eq. (7.53), we have

$$\tan\delta = \frac{2928.59 \times 1.21 \times 0.8}{7967.43 + 2928.59 \times 1.21 \times 0.6}$$

$$= 0.28$$

or

$$\delta = 15.69°$$

If we assume $\tilde{V}_a$ to be the reference phasor (Figure 7.34), then E_a leads $\tilde{V}_a$ by 15.69° and I_a lags $\tilde{V}_a$ by 36.87°. Thus,

$$\tilde{V}_a = 7967.43\underline{/0°} \text{ V}$$

$$\tilde{I}_a = 2928.59\underline{/-36.87°} \text{ A}$$

The d- and q-axis currents are

$$\tilde{I}_d = I_a \sin(52.56°)\underline{/-74.31°} = 2325.20\underline{/-74.31°} \text{ A}$$
$$\tilde{I}_q = I_a \cos(52.56°)\underline{/15.69°} = 1780.47\underline{/15.69°} \text{ A}$$

The generated voltage per phase, from Eq. (7.47), is

$$\tilde{E}_a = 7967.43\underline{/-15.69°} + (2325.20\underline{/-74.31°})(j1.83) + (1780.47\underline{/15.69°})(j1.21)$$
$$\approx 11{,}925.76\underline{/15.69°} \text{ V}$$

Thus, the percent voltage regulation is

$$VR\% = \frac{11{,}925.76 - 7967.43}{7967.43} \times 100 = 49.68\%$$

Since the power developed is the same as the power output, the power developed is

$$P_d = P_o = 3 \times 7967.43 \times 2928.59 \times 0.8$$
$$= 56 \text{ MW}$$

We could also have used Eq. (7.54) to determine the power developed by a salient-pole synchronous generator when its armature-winding resistance is negligible. ■

Exercises

7.20. The d-axis and the q-axis reactances on a per-unit basis of a salient-pole synchronous generator are 0.92 and 0.54, respectively. The armature resistance is negligible. When the generator delivers its rated load at 0.8 pf lagging and rated terminal voltage, determine (a) the voltage regulation and (b) the per-unit power developed by the generator.

7.21. The d-axis and the q-axis reactances on a per-unit basis of a salient-pole synchronous generator are 0.8 and 0.6, respectively. The armature resistance is negligible. When the generator delivers its rated load at unity power factor and rated terminal voltage, determine (a) the voltage regulation and (b) the per-unit power developed by the generator.

7.13 Parallel Operation of Synchronous Generators

The generation of electric power, its transmission, and its distribution must be conducted in an **efficient** and **reliable** way at a reasonable cost with the least number of interruptions. By "efficient" we mean that an alternator must operate not only at its maximum efficiency, but the efficiency must be maximum at or near its full load. As the demand for electric energy can fluctuate from a light load to a heavy load several times during the day, it is almost impossible to operate a single alternator at its maximum efficiency at all times.

The term "reliable" implies that consumers must not be conscious of a loss in electric power at any time. A single alternator cannot ensure such a reliable operation owing to the possibility of its failure or a deliberate shut-off for periodic inspection. Therefore, **a single alternator supplying a variable load cannot be very efficient, cost-effective, and reliable.**

To guarantee reliability and the continuity of electric service, it becomes necessary to generate electric power at a central location where several alternators can be connected in parallel to meet the power demand. When the demand is light, some of the alternators can be **taken off** line while the other alternators are operating at their maximum efficiencies. As the demand increases, another alternator can be **put on** line without causing any service interruption. Not only are all the alternators at one location connected in parallel to a common line known as an **infinite bus** but there may be many power generating stations feeding the same bus. Thus, there is hardly a change in either the voltage or the frequency of the infinite bus. Therefore, the following requirements have to be satisfied prior to connecting an alternator to the infinite bus.

1. The line voltage of the (incoming) alternator must be equal to the constant voltage of the infinite bus.
2. The frequency of the incoming alternator must be exactly equal to that of the infinite bus.
3. The phase sequence of the incoming alternator must be identical to the phase sequence of the infinite bus.

Figure 7.36 shows a wiring diagram for the parallel operation of two alternators. Alternator A is already connected to the infinite bus and is supplying the load. In order to meet the increased load demand, let us now bring alternator B on line.

Step 1: Alternator B is driven at or near its rated speed, and the field current is raised to a level at which its no-load voltage is nearly equal to that of the grid. The no-load voltage is checked by placing a voltmeter between any two lines of the incoming alternator, as shown in the figure when the circuit breaker is in the open position.

Step 2: To verify the phase sequence, three lamps are connected asymmetrically as shown. When the phase sequence of alternator B is the same as that of alternator A (or the grid), lamp L_1 is dark while the other two lamps glow brightly. If the phase sequence is not proper, the three lamps become bright or dark simultaneously.

Step 3: When the phase sequence is proper and the frequency of the incoming generator is exactly equal to that of the grid, lamp L_1 stays dark while the other

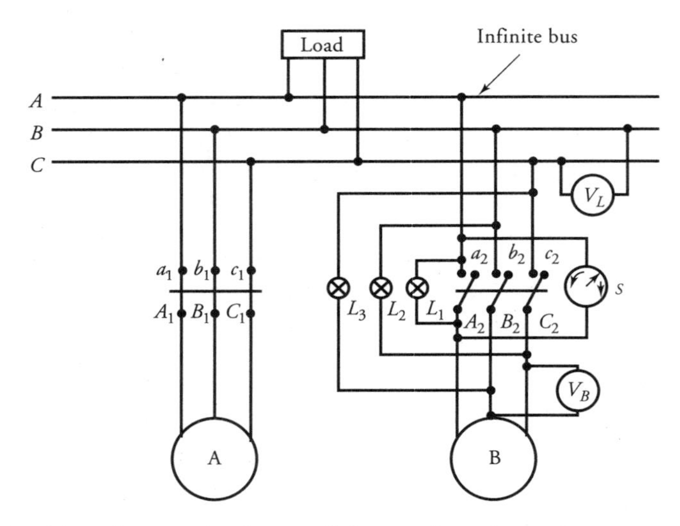

Figure 7.36 A wiring diagram for the parallel operation of two alternators.

lamps stay bright. Any small mismatch in the frequency forces the three lamps to go from dark to bright in a sequential order. In addition to the lamps to check the condition for synchronism, a device called a **synchroscope** (Figure 7.37) is also connected across one of the phases. The synchroscope is shown connected across phase *a* in Figure 7.36. The synchroscope measures the phase angle between the *a* phases of the incoming alternator B and the grid. When the two frequencies are the same and the phase sequence is proper, the phase difference between the two

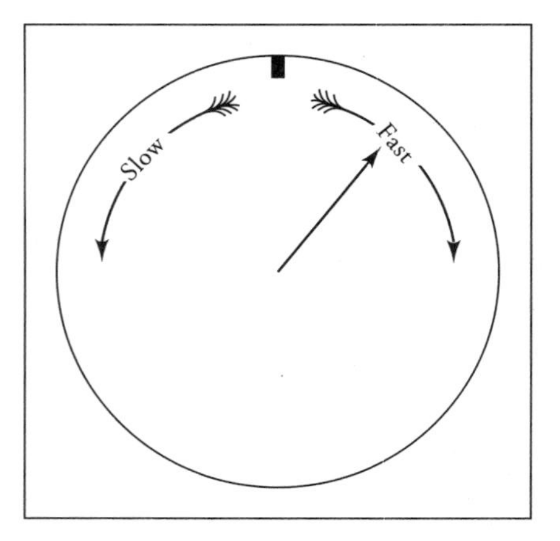

Figure 7.37 A synchroscope.

a phases must be zero. This corresponds to the vertically up position of the synchroscope's pointer. The slow clockwise rotation of the synchroscope's pointer indicates that the phase *a* of alternator B is moving ahead of phase *a* of the grid. In other words, the frequency (and therefore the speed) of alternator B is slightly greater than that of the infinite bus, and vice versa for the counterclockwise rotation. The speed is altered by controlling the mechanical input to the alternator.
Step 4: Alternator B is ready to be put into service by closing the circuit breaker when (a) the line voltage of incoming alternator B is equal to that of the infinite grid, (b) lamp L_1 is dark while the other two lamps are bright, and (c) the synchroscope pointer is pointing vertically up (zero phase-difference position).

Once the circuit breaker is closed, alternator B is on line. At this time, it is neither receiving nor delivering power. This is referred to as the **floating stage** of the alternator (Figure 7.38**a**).

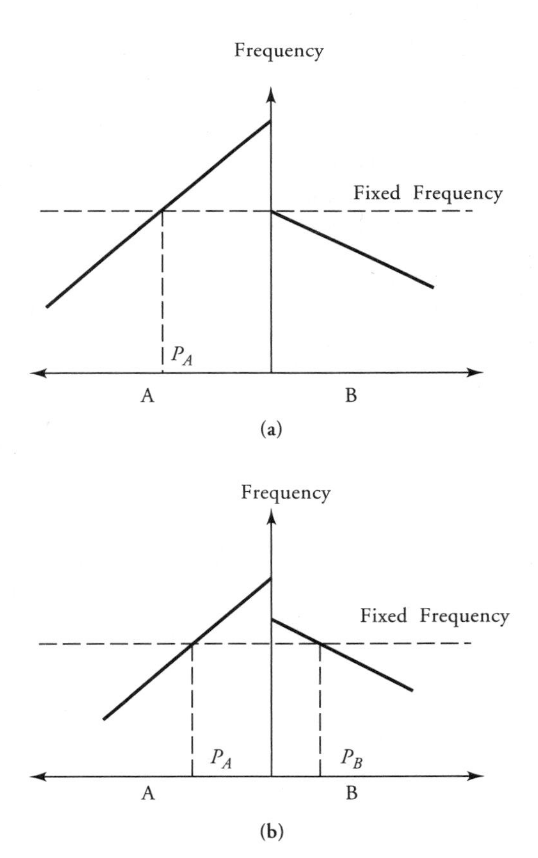

Figure 7.38 **(a)** Alternator A supplies the load while B is floating. **(b)** Alternator B shares the load with A.

Loading the Alternator

Increasing the field current (excitation) of alternator B changes the power factor of the alternator, as explained below. On the other hand, increasing the mechanical power input to alternator B tends to increase its speed and thereby its frequency. Since the frequency of the alternator is fixed by the infinite bus and cannot be changed, the alternator begins supplying the load as indicated in Figure 7.38**b**. Thus, some of the load on alternator A can be transferred to alternator B by simply increasing the mechanical power input to alternator B and simultaneously decreasing the mechanical power input to alternator A.

When both alternators are on line and the load demand is decreasing, it may become feasible to shut off one of the alternators. To do so, the entire load of the alternator that is to be taken off the line must be transferred to the other by following the reverse of the above-mentioned process. The circuit breaker is opened when the alternator attains the floating stage.

The V Curves

Let us assume that a cylindrical rotor synchronous generator is connected to an infinite bus and is delivering power at a lagging power factor. The corresponding phasor diagram is shown in Figure 7.39 when its armature winding resistance can be neglected. The power developed by the synchronous generator is

$$P_d = \frac{3E_a V_a \sin\delta}{X_s}$$

In the absence of the winding resistance, the power output is exactly equal to the power developed. Let us assume that I_f is the necessary field current to generate the per-phase voltage E_a.

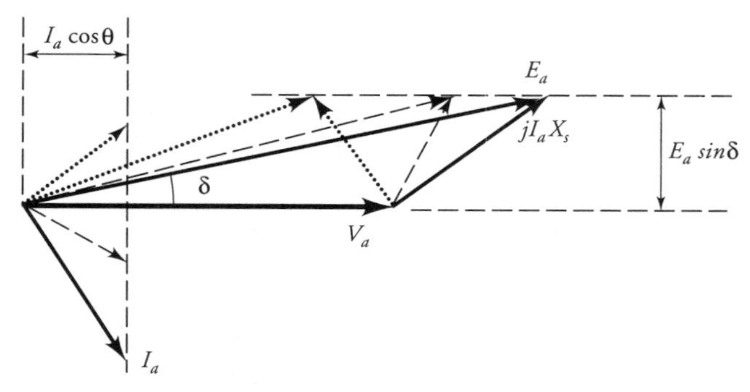

Figure 7.39 Loci of I_a and E_a when the synchronous generator delivers constant power.

Let us now reduce the field current but keep the power output of the generator the same as before. Since V_a is held constant by the infinite bus and X_s is constant, the same power output implies that the product $E_a \sin\delta$ must remain the same. However, a reduction in the field current must reduce the generated voltage E_a. A reduction in E_a for the same power output suggests that the power angle δ must increase. This means that the tip of generated voltage phasor $\tilde{E}_a$ must move along the horizontal line, as shown in Figure 7.39. A reduction in E_a and an increase in δ are now evident from the figure.

The power output is also given as

$$P_o = 3V_aI_a \cos\theta$$

For the power output to be the same when V_a is constant, $I_a \cos\theta$ must also be constant. In other words, the projection of the current phasor onto $\tilde{V}_a$ must be the same even when I_f has been reduced. Thus, the tip of the current phasor must move along the vertical line in Figure 7.39. The phasor diagram for a reduced field current is shown by the dashed lines in Figure 7.39. Note the decrease in the power factor angle or the improvement in the power factor.

If we keep on reducing the field current I_f and therefore E_a, the following sequence of events occurs:

(a) δ keeps on increasing until it becomes 90°.
(b) The power factor angle continues changing from lagging to unity and finally to leading.

From Figure 7.39 it is obvious that as long as the power factor is changing from lagging to unity, the armature current is decreasing. The armature current is minimum when the power factor is unity. As the power factor changes from unity to leading, the armature current starts increasing again. The change in the armature current as a function of the field current is given in Figure 7.40. Because

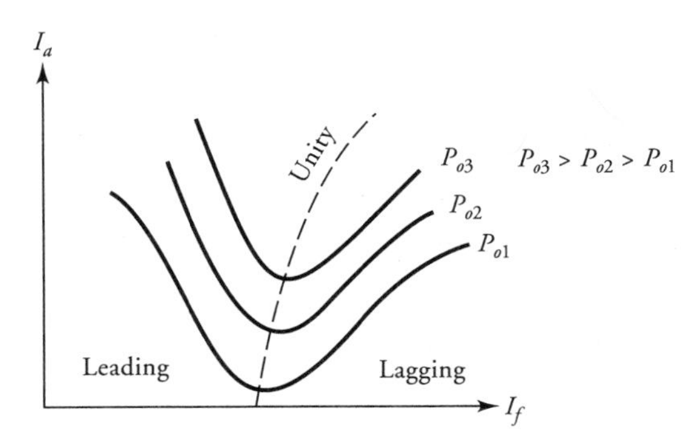

Figure 7.40 V curves for a synchronous generator.

of the shape of the curve, it is called the **V curve**. We can plot similar V curves for different values of power output. As the power output of the generator increases, the field current I_f corresponding to the unity power factor also increases. Do you know why?

Parallel Operation of Two Synchronous Generators

Figure 7.41 shows two synchronous generators with identical speed-torque characteristics connected in parallel. If $\tilde{V}_a$ is the terminal voltage, $\tilde{I}_L$ is the load current, $\hat{Z}_L$ is the load impedance, $\tilde{I}_{a1}$ and $\tilde{I}_{a2}$ are the armature currents, $\tilde{E}_{a1}$ and $\tilde{E}_{a2}$ are the generated voltages, $\hat{Z}_{s1}$ and $\hat{Z}_{s2}$ are the synchronous impedances on a per-phase basis, then

$$\tilde{V}_a = \tilde{E}_{a1} - \tilde{I}_{a1}\hat{Z}_{s1} = \tilde{E}_{a2} - \tilde{I}_{a2}\hat{Z}_{s2} = \tilde{I}_L\hat{Z}_L$$

Also,

$$\tilde{I}_L = \tilde{I}_{a1} + \tilde{I}_{a2}$$

Thus,

$$\tilde{E}_{a1} = \tilde{I}_{a1}\hat{Z}_{s1} + \tilde{I}_L\hat{Z}_L \tag{7.56a}$$

$$= \tilde{I}_{a1}(\hat{Z}_{s1} + \hat{Z}_L) + \tilde{I}_{a2}\hat{Z}_L \tag{7.56b}$$

and

$$\tilde{E}_{a2} = \tilde{I}_{a2}\hat{Z}_{s2} + \tilde{I}_L\hat{Z}_L \tag{7.57a}$$

$$= \tilde{I}_{a2}(\hat{Z}_{s2} + \hat{Z}_L) + \tilde{I}_{a1}\hat{Z}_L \tag{7.57b}$$

Manipulating the above equations, we obtain

$$\tilde{I}_{a1} = \frac{(\tilde{E}_{a1} - \tilde{E}_{a2})\hat{Z}_L + \tilde{E}_{a1}\hat{Z}_{s2}}{\hat{Z}_L(\hat{Z}_{s1} + \hat{Z}_{s2}) + \hat{Z}_{s1}\hat{Z}_{s2}} \tag{7.58}$$

$$\tilde{I}_{a2} = \frac{(\tilde{E}_{a2} - \tilde{E}_{a1})\hat{Z}_L + \tilde{E}_{a2}\hat{Z}_{s1}}{\hat{Z}_L(\hat{Z}_{s1} + \hat{Z}_{s2}) + \hat{Z}_{s1}\hat{Z}_{s2}} \tag{7.59}$$

$$\tilde{I}_L = \frac{\tilde{E}_{a1}\hat{Z}_{s2} + \tilde{E}_{a2}\hat{Z}_{s1}}{\hat{Z}_L(\hat{Z}_{s1} + \hat{Z}_{s2}) + \hat{Z}_{s1}\hat{Z}_{s2}} \tag{7.60}$$

$$\tilde{V}_a = \tilde{I}_L\hat{Z}_L = \frac{\tilde{E}_{a1}\hat{Z}_{s2} + \tilde{E}_{a2}\hat{Z}_{s1}}{\hat{Z}_L(\hat{Z}_{s1} + \hat{Z}_{s2}) + \hat{Z}_{s1}\hat{Z}_{s2}}\hat{Z}_L \tag{7.61}$$

We can use these equations to determine how the load is being shared by the two synchronous generators, as illustrated by the following example.

EXAMPLE 7.10

Two three-phase, Y-connected, synchronous generators have per-phase generated voltages of $120\angle 10°$ V and $120\angle 20°$ V under no load, and reactances of $j5\ \Omega$/phase

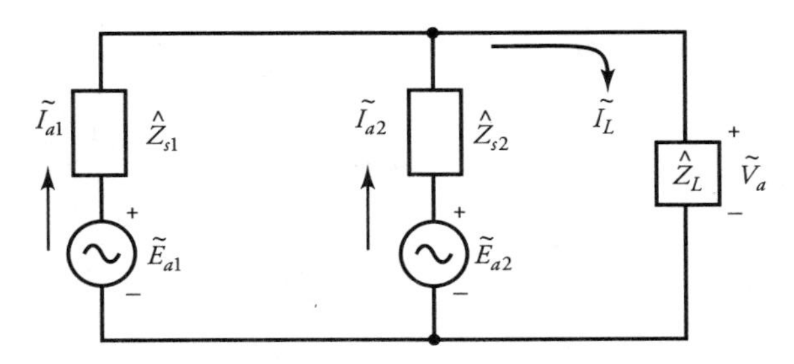

Figure 7.41 Two synchronous generators connected in parallel.

and $j8\ \Omega$/phase, respectively. They are connected in parallel to a load impedance of $4 + j3\ \Omega$/phase. Determine (a) the per-phase terminal voltage, (b) the armature current of each generator, (c) the power supplied by each generator, and (d) the total power output.

● SOLUTION

On a per-phase basis: $\tilde{E}_{a1} = 120\underline{/10^\circ}$ V $\qquad \tilde{E}_{a2} = 120\underline{/20^\circ}$ V

$$\hat{Z}_{s1} = j5\ \Omega \qquad \hat{Z}_{s2} = j8\ \Omega \qquad \hat{Z}_L = 4 + j3 = 5\underline{/36.87^\circ}\ \Omega$$

(a) From Eq. (7.61), the terminal voltage on a per-phase basis is

$$\tilde{V}_a = \frac{(120\underline{/10^\circ})(j8) + (120\underline{/20^\circ})(j5)}{(4 + j3)(j5 + j8) + (j5)(j8)} \times (4 + j3) = 82.17\underline{/-5.93^\circ}\ \text{V}$$

(b) The armature current for each generator is

$$\tilde{I}_{a1} = \frac{\tilde{E}_{a1} - \tilde{V}_a}{\hat{Z}_{s1}} = \frac{120\underline{/10^\circ} - 82.17\underline{/-5.93^\circ}}{j5} = 9.36\underline{/-51.17^\circ}\ \text{A}$$

$$\tilde{I}_{a2} = \frac{\tilde{E}_{a2} - \tilde{V}_a}{\hat{Z}_{s2}} = \frac{120\underline{/20^\circ} - 82.17\underline{/-5.93^\circ}}{j8} = 7.31\underline{/-32.06^\circ}\ \text{A}$$

The load current: $\tilde{I}_L = 9.36\underline{/-51.17^\circ} + 7.31\underline{/-32.06^\circ}$
$= 16.43\underline{/-42.8^\circ}$ A

(c) The power output of each generator is

$$P_{o1} = 3\ \text{Re}[(82.17\underline{/-5.93^\circ})(9.36\underline{/51.17^\circ})] = 1624.15\ \text{W}$$

$$P_{o2} = 3\ \text{Re}[(82.17\underline{/-.93^\circ})(7.31\underline{/32.06^\circ})] = 1617.13\ \text{W}$$

(d) The total power output is

$$P_o = 1624.15 + 1617.13 = 3241.28 \text{ W}$$

■

Exercises

7.22. When the two generators connected in parallel in Example 7.10 are operating under no load, what is the circulating current in the two generators?

7.23. Two Y-connected, three-phase, synchronous generators supply a load of 415.68 kW at a terminal voltage of 3000 V and a lagging power factor of 0.8. The synchronous impedance of generator A is $0.5 + j5\ \Omega$/phase and that of generator B is $0.2 + j10\ \Omega$/phase. The field excitation and the mechanical power input of A are adjusted so that it delivers half the power at a unity power factor. Determine (a) the current, (b) the power factor, (c) the per-phase generated voltage, and (d) the power angle of each generator.

7.24. If the load is removed without making any adjustments to the generators in Exercise 7.23, what is the circulating current at no load?

SUMMARY

In this chapter, we explained the mechanical construction and electrical operation of an ac generator, which we referred to as an alternator or synchronous generator. The rotor, which may be of either the cylindrical type or the salient-pole type, houses the dc field winding and is driven by a prime mover at its synchronous speed. The synchronous speed, in revolutions per minute (rpm), is given as

$$N_s = \frac{120f}{P}$$

Since the stator houses the three-phase windings of the alternator, it is referred to as an armature. The winding is usually of the double-layer type. The number of coils in a phase group is given as

$$n = \frac{S}{Pq}$$

where S is the number of slots (number of coils in a double-layer winding) in the armature and q is the number of phases. In order to reduce copper and end-turn buildup, a fractional-pitch winding is commonly used. The fraction by which the flux linking the coil has been reduced as a result of fractional-pitch winding is called the pitch factor. If ρ is the coil span, the pitch factor is

$$k_p = \sin(\rho/2)$$

Since the coils in a phase group are connected in series and are displaced spatially from each other, the induced emf in one coil is out of phase with that of the next coil by an angle equal to slot span (γ). To account for the reduction in the induced emf in a phase group, we defined the distribution factor as

$$k_d = \frac{\sin(n\gamma/2)}{n \sin(\gamma/2)}$$

If E_c is the rms value of the induced emf in each coil, the induced emf in a phase group is

$$E_{pq} = nk_d E_c$$

However, the rms value of the induced emf in each coil, from Faraday's law of induction, is

$$E_c = 4.44\ N_c f k_p \Phi_P$$

where $k_p\Phi_P$ is the maximum flux that links the coil with N_c number of turns.

Since there are P phase groups for each of the three phases, these groups can be connected in a parallel paths. Hence, the phase voltage is

$$E_a = 4.44\, f N_e \Phi_P$$

where

$$N_e = \frac{PnN_c k_w}{a}$$

is the effective number of turns in each phase and $k_w = k_p k_d$ is the winding factor.

The current in each phase winding gives rise to a leakage reactance X_a and the magnetization reactance X_m due to the armature reaction. The synchronous reactance is simply the sum of the two reactances; that is, $X_s = X_a + X_m$. The winding resistance R_a of each phase winding is the same. Thus, if $\tilde{V}_a$ is the per-phase terminal voltage and $\tilde{I}_a$ is the armature current, then for the round-rotor alternator

$$\tilde{E}_a = \tilde{V}_a + \tilde{I}_a(R_a + jX_s)$$

and

$$\tilde{E}_a = \tilde{V}_a + \tilde{I}_a R_a + j(\tilde{I}_d X_d + \tilde{I}_q X_q)$$

for the salient-pole alternator when

$$\tilde{I}_a = \tilde{I}_d + \tilde{I}_q$$

where $\tilde{I}_q$ is along $\tilde{E}_a$ and $\tilde{I}_d$ lags $\tilde{E}_a$ by 90°.

To minimize the power loss in the armature winding, R_a is so small that for large alternators it can be neglected. In that case, the power output of an alternator is the same as the power developed. That is

$$P_o = P_d = \frac{3V_a E_a \sin\delta}{X_s}$$

for a round-rotor alternator, and

$$P_o = P_d = \frac{3V_a E_a \sin\delta}{X_d} + \frac{3(X_d - X_q)}{2X_d X_q} V_a^2 \sin 2\delta$$

for a salient-pole alternator.

The power developed is maximum when the power angle δ is 90°. However, for a stable operation, the power angle is usually small.

The percent voltage regulation of the alternator was defined as

$$VR\% = \frac{E_a - V_a}{V_a} \times 100$$

At a location where the power is being generated, many alternators are connected in parallel to an infinite bus. The terminal voltage and the frequency of the bus are fixed. The power supplied by an alternator to a load via an infinite bus can be (increased/decreased) by (increasing/decreasing) its mechanical power input. By changing its field excitation we can control the power factor of the alternator. If I_f is the field current corresponding to the unity power factor for a certain load on the alternator, the (leading/lagging) power factor can be obtained by making the field current (smaller/larger) than I_f.

Review Questions

7.1. What is the difference between a dc generator and an ac generator?
7.2. Explain how a dc generator can be converted into an ac generator.
7.3. Can an ac generator be operated at any speed?
7.4. What are the advantages of inside-out construction of an ac generator?
7.5. Why is an ac generator usually referred to as a synchronous generator?
7.6. What is synchronous speed?
7.7. Calculate the synchronous speed of a 60-Hz ac generator when it has (a) 2 poles, (b) 4 poles, (c) 6 poles, (d) 8 poles, and (e) 12 poles.
7.8. The synchronous speed of a 12-pole generator is 500 rpm. What is the frequency of its operation?
7.9. Determine the number of poles of a 60-Hz generator when its rotor speed is (a) 600 rpm, (b) 400 rpm, and (c) 150 rpm.
7.10. Explain the construction of an ac generator.
7.11. Cite the differences between a cylindrical and a salient-pole rotor. Enumerate the advantages and drawbacks of each rotor type.
7.12. What is a full-pitch winding? Is it feasible to wind an ac generator using full-pitch winding? What are its advantages and disadvantages?
7.13. What is a fractional-pitch winding? What are its advantages and shortcomings?
7.14. Explain the following terms: pitch factor, phase belt, distribution factor, winding factor, phase voltage, and line voltage.
7.15. What is a double-layer winding? If the stator (armature) of an ac generator has 144 slots, how many coils are required for the double-layer winding?
7.16. A 48-slot, 8-pole, three-phase generator is wound using the double-layer winding. Determine the pitch factor, the distribution factor, the winding factor, the phase voltage, and the line voltage.
7.17. Repeat Question 7.16 if the generator has 72 slots.
7.18. What is meant by the number of parallel paths? The phase windings in Questions 7.16 and 7.17 are connected in two parallel paths. Calculate the effective turns in each phase.
7.19. A 25-Hz, three-phase alternator is designed to operate at 1500 rpm. There are 4 coils per pole per phase. Determine the number of slots in the armature. How many poles must the alternator have?
7.20. A 6-pole, three-phase alternator has 72 slots. Determine its pitch factor, its distribution factor, and the winding factor.
7.21. A three-phase, Δ-connected alternator supplies a rated load of 13 A at 2300 V. The power delivered to a balanced load is 40 kW. Determine the phase voltage, phase current, power per phase, and power factor.
7.22. Define voltage regulation. What does it mean?
7.23. Two alternators are designed to supply the same rated load at the rated voltage. The voltage regulation of one alternator is 20% while that of the

other is 25%. Which alternator is better suited for a continuous operation at the rated load?

7.24. Under what loading condition can the voltage regulation be negative?

7.25. Does the terminal voltage increase or decrease when the load is removed for a positive voltage regulation?

7.26. Why is the armature-winding resistance of a synchronous generator kept as small as possible?

7.27. Define leakage reactance, synchronous reactance, and synchronous impedance.

7.28. Explain clearly how an open-circuit test is performed.

7.29. Describe clearly the procedure to perform the short-circuit test.

7.30. List all the conditions that must be fulfilled before one alternator can be connected in parallel with another.

7.31. Why must the alternators be connected in parallel?

7.32. What is an infinite bus or grid?

7.33. Why is the d-axis reactance larger than the q-axis reactance in a salient-pole alternator?

7.34. Explain the nature of the external characteristics of a synchronous generator for unity, leading, and lagging power factors.

7.35. Since the power developed is maximum when the power angle is 90°, give reasons why an alternator is never operated at that power angle.

7.36. Draw the phasor diagram for a salient-pole alternator when the power factor of the load is (a) unity and (b) leading.

7.37. What is a V curve?

7.38. Why does the field current increase with the increase in the load when the power factor is unity? Answer using phasor diagrams.

7.39. What is a synchroscope? What is its function? Can you explain its operation?

7.40. What is the importance of the air-gap line?

Problems

7.1. A 96-slot, 16-pole, 60-Hz, three-phase, Y-connected, synchronous generator has 6 turns per coil. The flux per pole is 21 mWb. Determine (a) the synchronous speed, (b) the number of coils in a phase group, (c) the coil pitch, (d) the slot span, (e) the generated voltage per phase, and (f) the line voltage. Sketch the winding arrangement for a phase group. Assume series connection.

7.2. A 4-pole, three-phase, 60-Hz, Y-connected, synchronous generator has 4 slots per pole per phase. There are 6 conductors per slot. The flux per pole is 51 mWb. Calculate the phase and the line voltages. Show the placement of coils for a phase group. Assume series connection.

7.3. A three-phase, Δ-connected, 16-pole, synchronous generator has 144 slots with 10 conductors per slot. The flux per pole is 83.3 mWb, and the speed is 375 rpm. Determine the phase and the line voltages if each winding is connected in two parallel paths.

7.4. A 6-pole, 25-Hz, three-phase, Y-connected, synchronous generator has 36 slots. There are 17 turns per coil, and the flux per pole is 94.8 mWb. Find the line voltage if there are two parallel paths. Sketch the placement of all coils and show the winding connections.

7.5. An 8-pole, 60-Hz, Y-connected, three-phase generator has 48 slots. There are 6 turns per coil. The rotor radius is 40 cm and the pole length is 80 cm. The maximum flux density per pole is 1.2 T. Compute the line voltage for a series connection.

7.6. A three-phase, 12-pole, synchronous generator is required to generate a no-load voltage of 5.6 kV per phase at 50 Hz. The flux per pole is 185 mWb. The stator has 108 slots. Determine the number of turns per coil for a series connection.

7.7. A 20-pole, Y-connected, three-phase, 400-Hz alternator has 3 coils per phase group. Each coil has 2 turns and the flux per pole is 23 mWb. The coils are connected in two parallel groups. Determine (a) the rotor speed, (b) the number of slots in the armature, (c) the winding factor, (d) the per-phase voltage, and (e) the line voltage.

7.8. A three-phase, Y-connected alternator delivers a rated load of 50 A at 230 V with 0.8 pf lagging. When the load is removed, the terminal voltage is found to be $280\underline{/30^\circ}$ V. Calculate (a) the synchronous impedance per phase and (b) the voltage regulation.

7.9. A 110-V, three-phase, Y-connected, 8-pole, 48-slot, 6000-rpm, double-layer wound, synchronous generator has 12 turns per coil. If one side of the coil is in slot 1, the other side is in slot 6. There are 4 parallel paths. When the generator delivers the rated load at a line voltage of 110 V, the voltage regulation is 5%. What is the flux per pole?

7.10. A 10-kVA, 380-V, 60-Hz, 2-pole, three-phase, Y-connected, synchronous generator delivers the rated load at 0.8 pf lagging. The synchronous impedance is $1.2 + j4\ \Omega$/phase. Determine (a) the synchronous speed, (b) the generated voltage per phase, and (c) the efficiency if the fixed loss is 1 kW.

7.11. A 120-kVA, 1-kV, Δ-connected, three-phase, synchronous generator has a winding resistance of 1.5 Ω per phase and a synchronous reactance of 15 Ω per phase. If the fixed loss is 1500 W, determine (a) the voltage regulation and (b) the efficiency when the generator delivers the rated load at 0.707 pf lagging.

7.12. A 1732-V, 120-kVA, Y-connected, three-phase, synchronous generator has a synchronous reactance of 1.2 Ω/phase. It supplies the rated load at 0.9 pf leading. The fixed loss is 5% of the power developed. Determine (a) the generated voltage, (b) the power angle, (c) the voltage regulation, and (d) the efficiency. Draw the phasor and power-flow diagrams.

7.13. Modify the expressions for the pitch factor and the distribution factor to include the effect of harmonics.

7.14. A three-phase, Y-connected, 216-slot, 8-pole, 60-Hz, 360-rpm generator has 5 turns per coil. The flux per pole has a fundamental component of 120 mWb, third harmonic component of 40 mWb, and fifth harmonic component of 10 mWb. Calculate the rms value of the induced emf per phase.

7.15. The no-load voltage of a Y-connected synchronous generator is 3464 V. When the generator delivers the rated load of 432 kW at 0.8 pf lagging, the terminal voltage is 3117.69 V. If the armature winding resistance is negligible, determine the synchronous reactance of the generator. When the load is changed to $80 + j60\ \Omega$/phase, determine the terminal (line) voltage.

7.16. A 72-kVA, 208-V, Y-connected, three-phase synchronous generator delivers the rated load at 0.866 pf lagging. The armature winding resistance is 20 mΩ/phase. The core loss is 800 W. The friction and the windage loss is 350 W. The field winding is connected across a 120-V dc source and the field current is 5.5 A. Calculate the efficiency of the generator.

7.17. A 25-kVA, 480-V, three-phase, 60-Hz, synchronous generator has a synchronous reactance of 8 Ω/phase. The fixed loss is 1.5 kW. When the generator delivers full load at a power factor of 0.8 leading, determine its efficiency. What is the maximum power developed by the machine? Neglect the winding resistance.

7.18. A balanced Δ-connected load is connected to a 3.6-kVA, 208-V, Y-connected, three-phase, synchronous generator via a three-wire transmission line. The impedance per line is $0.5 + j5\ \Omega$. The synchronous impedance of the generator is $0.25 + j4\ \Omega$/phase. The fixed losses are 175 W. When the generator delivers the rated load at the rated voltage and 0.8 pf lagging, determine (a) its efficiency, (b) the voltage regulation, and (c) the load impedance per phase.

7.19. A balanced Y-connected load is connected to a 7.2-kVA, 208-V, Δ-connected, three-phase, synchronous generator via a three-wire transmission line. The impedance per line is $0.4 + j2.4\ \Omega$. The synchronous impedance per phase is $0.3 + j6\ \Omega$. The fixed loss is 140 W. When the generator delivers full load at its rated voltage and 0.8 pf leading, determine (a) the voltage regulation, (b) the efficiency, and (c) the load impedance per phase.

7.20. A 1732-V, 300-kVA, Y-connected, synchronous generator has a synchronous impedance of $0.5 + j4\ \Omega$/phase. Determine the power factor of the load that yields zero voltage regulation.

7.21. A 230-kVA, 1100-V, Δ-connected, three-phase, synchronous generator has an average resistance of 0.3 Ω between any two of its terminals. With a particular field excitation and at its rated speed, the no-load voltage was 1100 V and the short-circuit current in each line was 121.24 A. Determine the synchronous impedance of the generator. If the rotational loss is 12 kW and the generator delivers the rated load at 0.866 pf lagging, determine its voltage regulation and efficiency using the per-unit system.

7.22. A 6-MVA, 6.6-kV, Y-connected, 2-pole, 50-Hz turbo-generator has an average resistance of 0.45 Ω between any two lines. With a certain field current and at its rated speed, the open-circuit voltage was 8 kV and the short-circuit current was 800 A. Find the voltage regulation of the generator on full load at unity power factor.

7.23. A 4.8-kV, three-phase, Y-connected, synchronous generator is rated at 50 A. The winding resistance is 0.8 Ω/phase. When driven at its rated speed with a certain field current, the open-circuit voltage was 480 V and the short-circuit current was 50 A. What is its voltage regulation at 0.8 pf leading?

7.24. A 1000-kVA, 2400-V, 60-Hz, Y-connected, three-phase alternator has an armature resistance of 0.5 Ω between any two of its lines. A field current of 25 A at the rated speed produces a short-circuit current of 240 A and an open-circuit voltage of 1800 V. Calculate the percent voltage regulation at full load and 0.707 pf lagging.

7.25. A 10.8-kVA, 208-V, Y-connected, three-phase, synchronous generator supplies the rated load at 0.8 pf lagging. The synchronous impedance is $0.5 + j5$ Ω/phase. The field-winding resistance is 20 Ω. Its per-phase OCC at the rated speed is given as

$$E_a = \frac{2400 I_f}{7.5 + 6.5 I_f}$$

If the rotational loss is 1.2 kW, determine the voltage regulation and the efficiency of the generator. If the field voltage is 120 V (dc), what must be the external resistance in the field-winding circuit?

7.26. Repeat Problem 7.25 if the power factor of the load is 0.8 leading.

7.27. A 15-kW, 120-V, 60-Hz, Δ-connected, three-phase alternator has a synchronous impedance of $0.1 + j3$ Ω/phase. The field-winding resistance is 25 and the field voltage is 120 V (dc). The per-phase OCC curve at its rated speed is given as follows:

I_f (A):	0	0.5	1.0	1.5	2.0	2.5	3.0	3.5	4.0	4.5	5.0
E_a (V):	0	90	140	185	209	223	230	236	238	248	250

If the rotational loss is 800 W, determine the voltage regulation and the efficiency of the generator when it supplies the rated load at 0.9 pf lagging. What must be the external resistance in the field-winding circuit?

7.28. Repeat Problem 7.27 if the power factor is 0.9 leading.

7.29. A 500-V, 150-kVA, Δ-connected, three-phase alternator has an effective resistance of 0.2 Ω between its two lines. A field current of 12 A produces a short-circuit current of 173.2 A and an open-circuit voltage of 450 V. Calculate the full-load voltage regulation when the power factor is unity.

7.30. Modify Eq. (7.35) to include the effect of armature resistance.

7.31. A 300-kVA, 500-V, Δ-connected, three-phase, synchronous generator draws a field current of 2 A to maintain the rated current under short-circuit condition. For the same field current, the open-circuit voltage is 572 V. Determine the synchronous reactance of the generator if its winding resistance is negligible. What is its voltage regulation when the generator delivers the rated load at 0.707 pf lagging and rated terminal voltage?

7.32. A 5-MVA, 6.6-kV, 60-Hz, three-phase, Y-connected generator has the OCC given as follows:

I_f (A):	10	15	20	25	30
E_a (kV):	4.5	6.6	7.5	8.25	8.95

A field current of 20 A is needed to circulate a full-load current on short circuit. The armature-winding resistance is 1.2 Ω/phase, the field-winding resistance is 500 Ω, and the rotational loss is 250 kW. Determine the power angle, the voltage regulation, and the efficiency when the generator delivers the rated load at unity power factor and rated terminal voltage.

7.33. A 4160-V, 3.5-MVA, 60-Hz, three-phase, Y-connected, salient-pole generator has a *d*-axis reactance of 2.75 Ω/phase and a *q*-axis reactance of 1.8 Ω/phase. The armature-winding resistance is negligible. If the generator delivers the rated load at 0.8 pf lagging, determine the voltage regulation and the power developed by the generator.

7.34. The per-unit *d*-axis and *q*-axis reactances of a salient-pole generator are 0.75 and 0.5, respectively. The armature-winding resistance is negligible. Determine the voltage regulation if the generator delivers the rated load at 0.866 pf lagging.

7.35. Repeat Problem 7.34 if the per-unit armature-winding resistance is 0.02 Ω/phase.

7.36. A three-phase, 3.3-kV, Y-connected, salient-pole generator delivers 900 kW at unity power factor. The *d*-axis and *q*-axis synchronous reactances are 1.2 Ω/phase and 0.8 Ω/phase, respectively. The armature-winding resistance is negligible. Determine the percent voltage regulation.

7.37. Repeat Problem 7.36 if the armature-winding resistance is 0.2 Ω/phase.

7.38. Two identical three-phase, Y-connected, synchronous generators are connected in parallel to equally share a load of 900 kW at 11 kV and 0.8 pf lagging. The synchronous impedance of each generator is $0.5 + j10$ Ω/phase. The field current of one generator is adjusted so that its armature current is 25 A at a lagging power factor. Determine (a) the armature current of the other generator, (b) the power factor of each generator, (c) the per-phase generated voltage, and (d) the power angle of each generator. What is the circulating current under no load?

7.39. Two three-phase, Y-connected, synchronous generators have induced voltages of $480\angle 0^\circ$ V and $480\angle 15^\circ$ V. The synchronous impedance of each generator is $0.2 + j8$ Ω/phase. When the generators are connected in parallel to a load impedance of $30 + j40$ Ω/phase, determine (a) the terminal voltage, (b) the armature currents, and (c) the power delivered by each generator. Compute the circulating current under no load.

7.40. Two three-phase, Y-connected, synchronous generators operate in parallel to supply a lighting load of 480 kW and a motor load of 240 kW at 0.8 pf lagging. The mechanical input and the field excitation of one generator are adjusted so that it supplies 240 kW at 0.9 pf lagging. Find the load and the power factor of the other generator.

CHAPTER

Synchronous Motors

A synchronous motor with a gear drive. (*Courtesy of Bodine Electric Company*)

8.1 Introduction

A synchronous motor, as the name suggests, runs under steady-state conditions at a fixed speed called the **synchronous speed**. The synchronous speed, as discussed in the preceding chapter, depends only upon (a) the frequency of the applied voltage and (b) the number of poles in the machine. In other words, the speed of a synchronous motor is independent of the load as long as the load is within the capability of the motor. If the load torque exceeds the maximum torque that can be developed by the motor, the motor simply comes to rest and the average torque developed by it is zero. For this reason, a synchronous motor is not inherently self-starting. Therefore, it must be brought up almost to its synchronous speed by some auxiliary means before it can be synchronized to the supply.

Because of its constant speed-torque characteristic, a small synchronous motor is used as a timing device. A large synchronous motor may be used not only to drive a certain load but also to improve the overall power factor (pf) of an industrial plant because it can be operated at a leading power factor. However, when a synchronous motor is operated at no load just to improve the power factor, it is usually referred to as a **synchronous condenser**.

A synchronous motor can be either a single-phase or a polyphase motor. Only three-phase synchronous motors are discussed in this chapter, but the development is valid for any polyphase synchronous motor.

8.2 Construction and Operation of a Synchronous Motor

The armature of a synchronous motor is exactly the same as that of a synchronous generator. It has a large number of slots that are designed to house the three identical, double-layer phase windings. The phase windings are spatially displaced by 120° electrical from one another and are excited by a balanced three-phase source. As outlined in Section 3.5, the phase windings upon excitation produce a uniform magnetic field that rotates along the periphery of the air-gap at the synchronous speed. If Φ_m is the maximum value of the flux produced by the maximum current I_m in each phase, the strength of the uniform revolving magnetic field [from Chapter 3, Eq. (3.37)], is

$$\Phi_r = 1.5\Phi_m \tag{8.1}$$

The synchronous speed in revolutions per minute (rpm) at which the flux revolves around the periphery of the air-gap [Chapter 3, Eq. (3.39)] is

$$N_s = \frac{120f}{P} \tag{8.2}$$

where f is the frequency of the three-phase power source and P is the number of poles in the motor.

The rotor of the synchronous motor has a field winding that produces the constant flux in the motor in exactly the same fashion as it does in a synchronous generator. Once the field winding is excited by a direct-current (dc) source, it produces alternate poles on the surface of the rotor. Thus far, it must be obvious that there is no difference between a synchronous motor and a synchronous generator.

Let us now assume that the rotor is at rest (standstill condition) and the field winding is excited to produce alternate poles on its periphery. The revolving field created by the armature can be visualized as if two magnets, a north pole and a south pole, are rotating at a constant (synchronous) speed just above the poles of the rotor. When the south pole of the revolving field is just above the north pole of the rotor, the force of attraction between them tends to move the rotor in the direction of the revolving field. Owing to the heavy mass of the rotor, it takes time before it can start moving, but by then the revolving field has reversed its polarity. Now the force of repulsion between the two like polarity poles tends to move the rotor in the opposite direction. As the rotor tries to rotate in the opposite direction, the revolving field has reversed its polarity once again. Thus, each pole on the rotor is acted upon by a rapidly reversing force of equal magnitude in both directions. The average torque thus developed by the rotor is zero. Therefore, **a synchronous motor is not self-starting**. Hence, to start a synchronous motor we must either provide some means for it to develop starting torque by itself or drive the rotor at nearly its synchronous speed by another prime mover and then synchronize it by exciting the field winding.

To make a synchronous motor self-starting, an additional winding known as the **damper winding** (the **induction winding** or the **amortisseur winding**) is provided in the pole faces of the motor. The damper winding, also called a **squirrel-cage** winding, is a short-circuit winding. For small machines, a squirrel-cage winding requires the placement of the rotor laminations in a mold and then forcing of the molten conducting material (often aluminum) into the slots. The mold has cavities on either side of the rotor, which are filled by the molten conducting material at the same time. The conducting material from one end of the slot to the other forms a **conducting bar**. The conducting bars are shorted by the **end rings** as shown in Figure 8.1. The entire one-piece construction looks like a squirrel cage and hence the name. For large machines, the squirrel-cage winding may be formed by driving metal bars into the slots one at a time and then shorting them with annular conducting strips on both ends.

The damper winding in a synchronous motor may also be of the wound-rotor type. The wound-rotor winding is used when we want to (a) control the speed of the motor and (b) develop high starting torque. A three-phase winding with as many poles as there are in the armature is placed in the rotor slots. One end of the three-phase windings is internally connected to form a common node while the other ends are connected to the slip rings. External resistance can then be

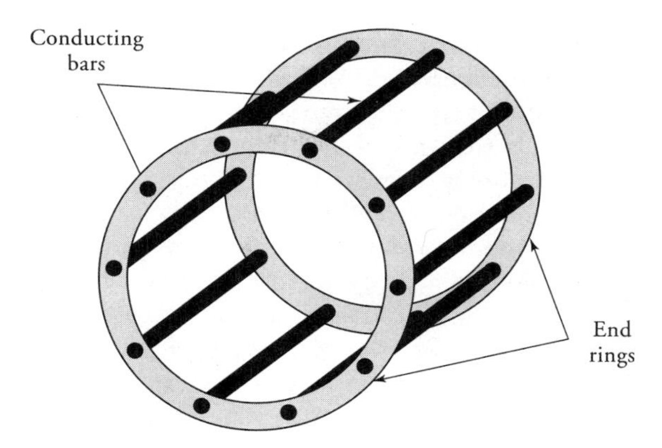

Figure 8.1 A squirrel-cage damper winding.

included in series with each phase winding, as shown in Figure 8.2, in order to increase the starting torque (see Chapter 9 for details).

In any case, the damper winding forms a closed loop. The uniform revolving field induces an electromotive force (emf) in the damper winding which, in turn, results in an induced current in it. As explained in Chapter 3, the induced current exerts a torque on the damper-winding conductors and forces them to rotate in the direction of the revolving field. Under no load, the rotor speed is almost (but not exactly) equal to the synchronous speed of the motor. This, in fact, is the principle of operation of an induction motor, as explained in Chapter 3. A synchronous motor is, therefore, brought to its no-load speed as an induction motor.

During the rotor acceleration period, the field winding must be shorted through an appropriate bank of resistors. The field winding should never be left

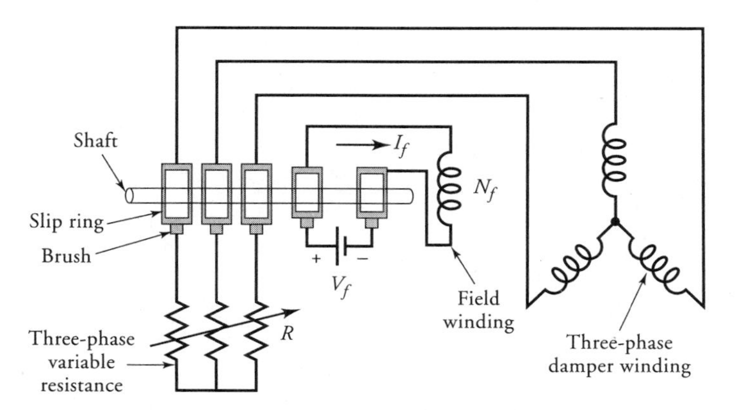

Figure 8.2 Three-phase damper winding on the rotor of a synchronous motor.

open because it can develop high voltage, just like the secondary winding of a step-up transformer. Once the rotor attains a steady speed, the short circuit is removed and the field winding is energized by connecting it to a dc source. The field poles thus formed on the rotor's periphery pull the rotor in step with the revolving field. In other words, the strong field poles are locked in step with the revolving poles of opposite polarity created by the armature winding (armature poles). The motor is then said to be **synchronized**. The damper winding becomes ineffective when the rotor rotates at synchronous speed.

At no load, the magnetic axes of the armature poles and the rotor poles are nearly aligned, as shown in Figure 8.3**a**. The magnetic lines of force are perpendicular to the rotor surface. Thus, they exert no torque on the rotor. The rotor poles start slipping behind the armature poles as the load on the motor is increased, as depicted in Figure 8.3**b**. The magnetic lines of force then have a component parallel to the armature surface and exert a force on the rotor. The angle, more appropriately the power angle, between the two magnetic axes keeps increasing with increasing load on the motor. The motor reaches an unstable region at about 60° electrical of angular displacement between the magnetic axes. Any further increase in the load may pull the motor out of synchronism.

The damper winding not only enables a synchronous motor to develop the starting torque but also serves another useful purpose. It tends to minimize motor **hunting**. Hunting comprises the successive overshoots and undershoots in the motor speed due to sudden changes in the load. When the load on a synchronous motor is changed suddenly, it takes time for the motor to adjust its power angle

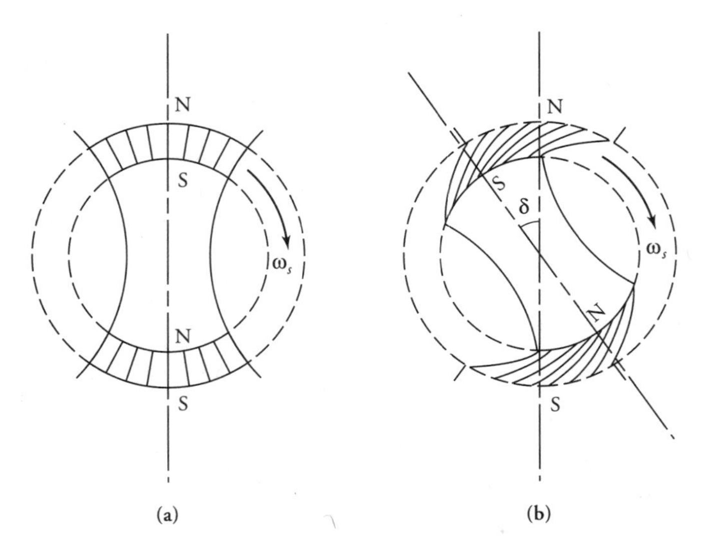

Figure 8.3 Magnetic lines of force **(a)** at no load and **(b)** at some load between the revolving field and the rotor of a synchronous motor.

owing to its inertia. During these power-angle adjustments, the speed of the motor fluctuates above and below its synchronous speed. These changes in speed induce current in the damper winding, thereby creating a torque that opposes the change. For example, when the load on the motor is suddenly increased, the rotor tends to slow down owing to an increase in the applied torque. As soon as it slows down, the current induced in the damper-winding conductors exerts an accelerating force on the rotor in the direction of its rotation. On the other hand, if the load is suddenly reduced, the inertia of the motor tends to increase the rotor speed. Again, a current is induced in the damper-winding conductors. However, the induced current is in the opposite direction now. Thus, it creates a torque in the opposite direction and forces the rotor to slow down. It is possible for hunting to become intolerably severe if the motor is not equipped with a damper winding.

8.3 Equivalent Circuit of a Synchronous Motor

As the rotor turns at the synchronous speed, so does the constant flux produced by the direct current in the field winding. The rotor flux induces an emf in the armature winding just like the generator action we discussed in Chapter 7. On the other hand, the current in each phase winding gives rise to (a) a leakage flux that links the phase winding only and (b) the armature reaction. Following the procedure discussed in the preceding chapter, the effects of leakage flux and the armature reaction can be assimilated in the equivalent circuit as leakage and magnetization reactances. Because both effects occur simultaneously, the two reactances can be replaced by either a single synchronous reactance X_s for a round-rotor synchronous motor or the d- and q-axis synchronous reactances X_d and X_q for the salient-pole motor.

Round-Rotor Synchronous Motor

We can now draw the equivalent circuit of a round-rotor synchronous motor on a per-phase basis, as shown in Figure 8.4, where R_a and X_s are the per-phase

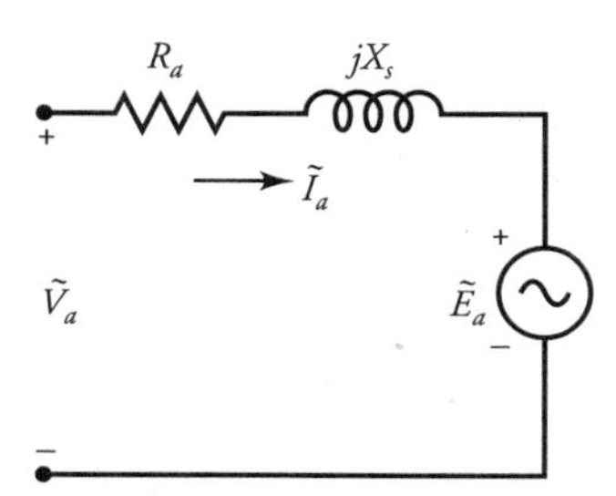

Figure 8.4 The per-phase equivalent circuit of a round-rotor synchronous motor.

winding resistance and synchronous reactance, respectively. From the equivalent circuit, we can write

$$\tilde{V}_a = \tilde{E}_a + \tilde{I}_a R_a + j\tilde{I}_a X_s \tag{8.3}$$

or

$$\tilde{I}_a = \frac{\tilde{V}_a - \tilde{E}_a}{R_a + jX_s} \tag{8.4}$$

One of the most important aspects of a synchronous motor is that it can be operated at either lagging, unity, or leading power factor simply by controlling its field current. We will have more to say about this in a later section. At this moment, our aim is to draw its phasor diagram when it operates at any power factor. These phasor diagrams are given in Figure 8.5, where the per-phase applied voltage has been taken as a reference. From these phasor diagrams, it is apparent that the per-phase excitation voltage lags the applied voltage. In other words, the power angle δ is a negative quantity.

Just like the alternator, the synchronous motor is also a doubly fed machine.

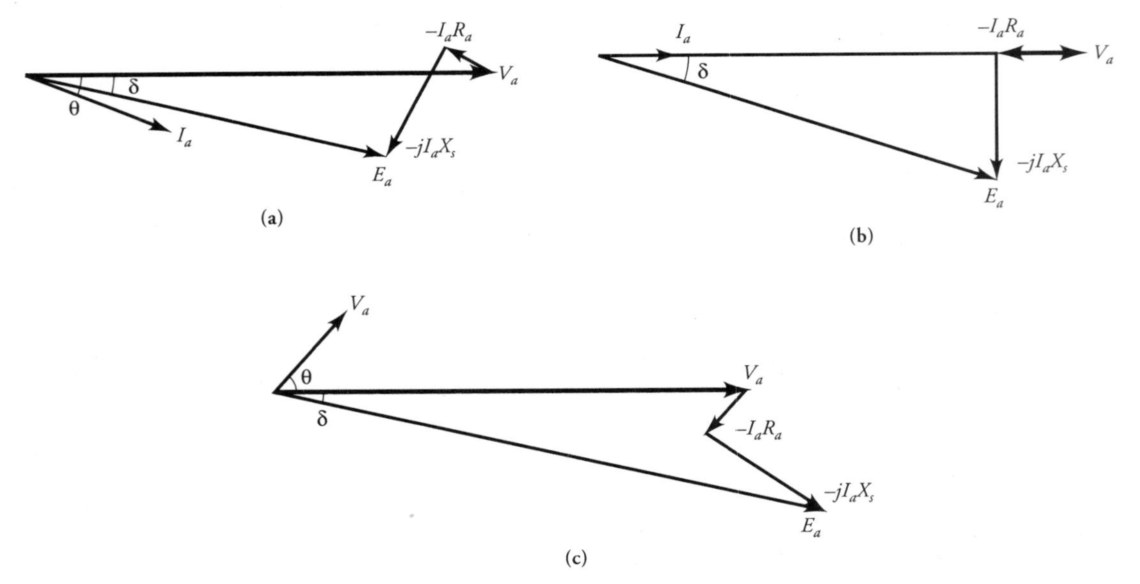

Figure 8.5 Phasor diagrams for **(a)** lagging pf, **(b)** unity pf, and **(c)** leading pf of a round-rotor synchronous motor.

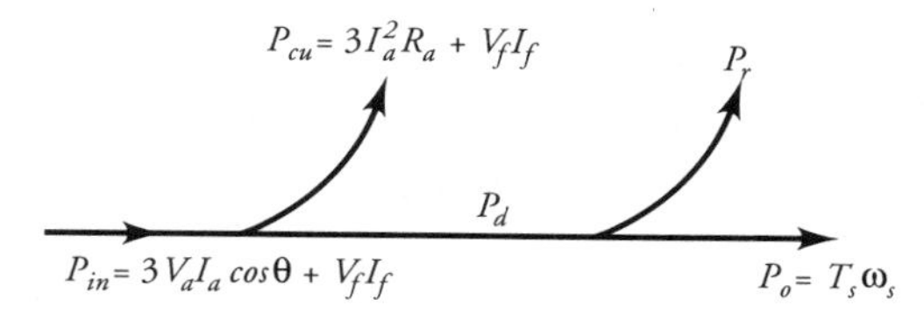

Figure 8.6 Power-flow diagram of a synchronous motor.

However, both power inputs are electrical in nature. Thus, the average power input to the machine is

$$P_{in} = 3V_aI_a \cos\theta + V_fI_f \tag{8.5}$$

where θ is the power-factor angle between the applied voltage and the armature-winding current, and V_fI_f represents the dc power supplied to the field winding, which simply materializes as a power loss. This power loss can be avoided by installing permanent magnets on the surface of the rotor.

The copper (electrical) loss in a synchronous motor takes place in the armature winding. If the power loss in the field winding is known, it can also be included as a part of the copper loss. Thus, the total copper loss in a synchronous motor is

$$P_{cu} = 3I_a^2R_a + V_fI_f \tag{8.6}$$

By subtracting the copper loss from the power input, we obtain the power developed by a round-rotor synchronous motor as

$$P_d = 3V_aI_a \cos\theta - 3I_a^2R_a \tag{8.7}$$

If ω_s is the synchronous angular velocity (rad/s) of the motor, the torque developed by the motor is

$$T_d = \frac{P_d}{\omega_s} \tag{8.8}$$

We subtract the rotational losses and the stray-load losses, if there are any, from the power developed to obtain the power output, P_o. Since the power output is mechanical, it is an accepted practice to express it in terms of horsepower. Figure 8.6 shows a typical power-flow diagram for a synchronous motor.

EXAMPLE 8.1

A 10-hp, 230-V, 60-Hz, three-phase, Y-connected, synchronous motor delivers full load at a power factor of 0.707 leading. The synchronous reactance of the motor is $j5\ \Omega$/phase. The rotational loss is 230 W and the field-winding loss is 70 W.

Calculate the generated voltage and the efficiency of the motor. Neglect the armature-winding resistance.

● SOLUTION

The power output: $P_o = 10 \times 746 = 7460$ W
The power developed: $P_d = P_o + P_r = 7460 + 230 = 7690$ W
Since there is no copper loss in the armature winding, the power supplied by the 230-V alternating-current (ac) source is 7690 W. Hence, the armature current is

$$I_a = \frac{7690}{\sqrt{3} \times 230 \times 0.707} \approx 27.3 \text{ A}$$

Total power input: $P_{in} = 7690 + 70 = 7760$ W

Hence, $$\eta = \frac{7460}{7760} = 0.961 \quad \text{or} \quad 96.1\%$$

If we assume the per-phase applied voltage as a reference, then

$$\tilde{V}_a = 132.79\angle 0^\circ \text{ V}$$

and $$\tilde{I}_a = 27.3\angle 45^\circ \text{ A}$$

The generated voltage per phase is

$$\tilde{E}_a = 132.79\angle 0^\circ - (27.3\angle 45^\circ)(j5)$$
$$= 248.78\angle -22.8^\circ \text{ V}$$

■

Salient-Pole Synchronous Motor

The equivalent circuit of a round-rotor synchronous motor is exactly the same as that of a round-rotor synchronous generator except for the direction of the phase current $\tilde{I}_a$. By analogy, we can obtain the equivalent circuit of a salient-pole synchronous motor, as shown in Figure 8.7. The polarities of the induced emfs in the *d*- and *q*-axis have been reversed to account for the change in the direction of the phase current. From Figure 8.7**a**, we can express the excitation voltage per phase as

$$\tilde{E}_a = \tilde{V}_a - \tilde{I}_a R_a - j\tilde{I}_d X_d - j\tilde{I}_q X_q$$
$$= \tilde{V}_a - \tilde{I}_a R_a - j\tilde{I}_a X_q - j\tilde{I}_d (X_d - X_q) \quad (8.9)$$

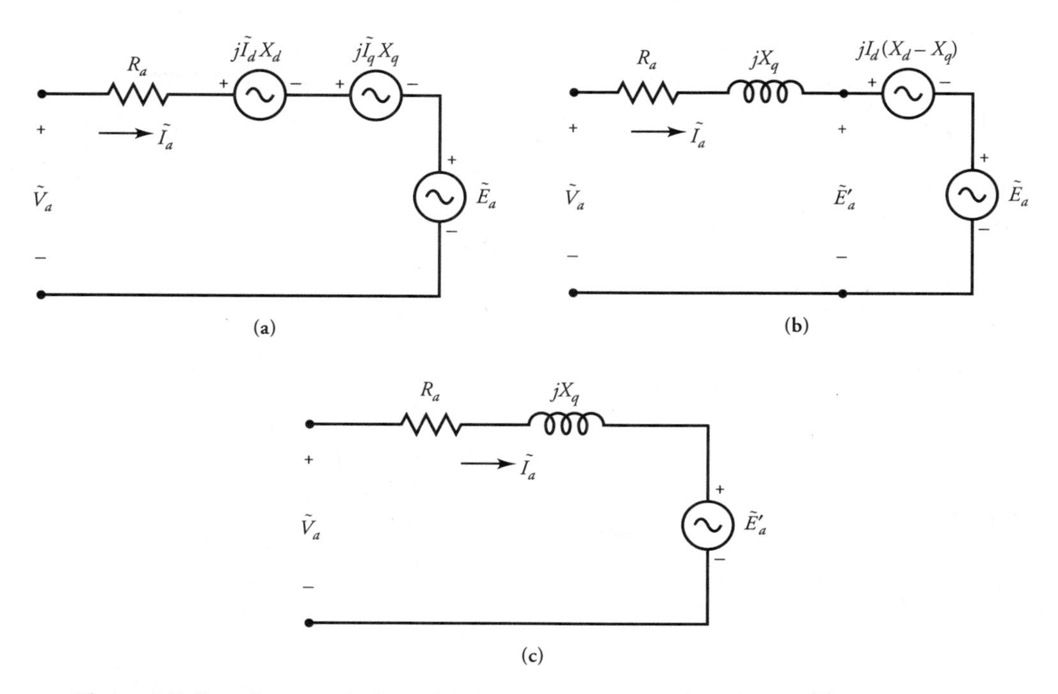

Figure 8.7 Per-phase equivalent circuit representations of a salient-pole synchronous motor.

If we define $\tilde{E}'_a$ as

$$\tilde{E}'_a = \tilde{V}_a - \tilde{I}_a R_a - j\tilde{I}_a X_q \tag{8.10}$$

then Eq. (8.9) can be expressed as

$$\tilde{E}_a = \tilde{E}'_a - j\tilde{I}_d(X_d - X_q) \tag{8.11}$$

The equivalent circuits of Figure 8.7**b** and **c** are based upon Eqs. (8.11) and (8.10), respectively. Also, note that the phase angle of $\tilde{E}'_a$ is the same as that of $\tilde{E}_a$, which, in fact, is the power angle. Thus, Eq. (8.10) can be used to compute (a) the power angle δ if it is not known and (b) the power developed by the motor. That is,

$$P_d = 3\text{Re}[\tilde{E}'_a \tilde{I}^*_a] \tag{8.12}$$

The power developed can also be computed from Eq. (8.7).

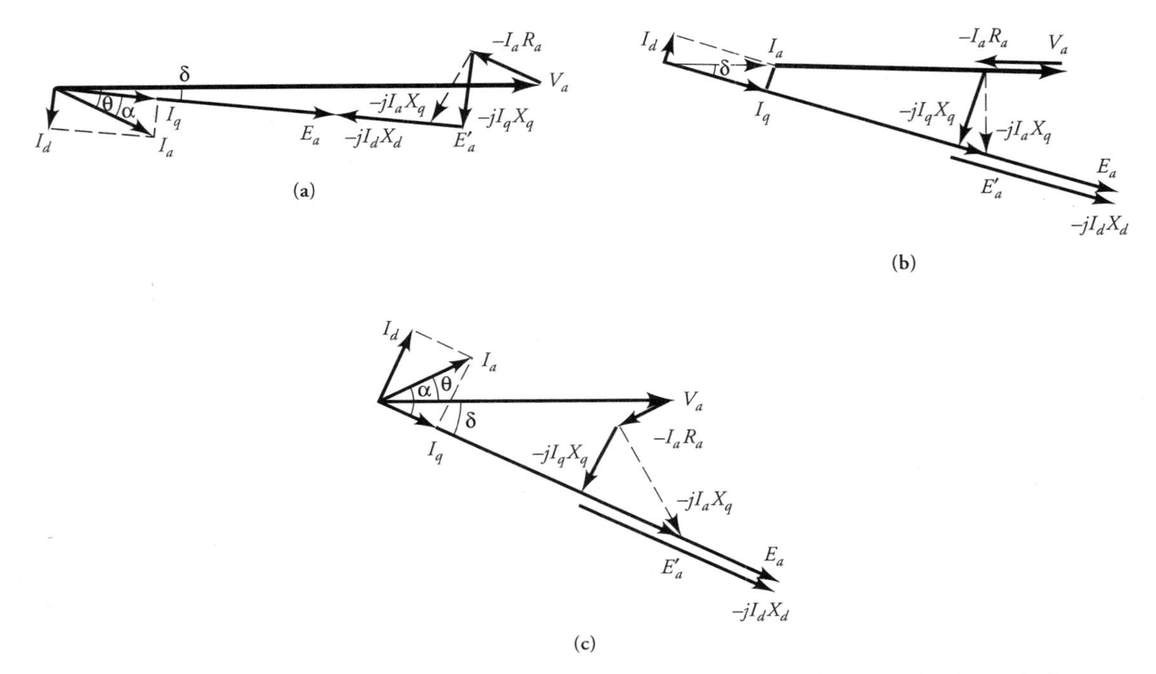

Figure 8.8 Phasor diagrams for a **(a)** lagging pf, **(b)** unity pf, and **(c)** leading pf of a salient-pole synchronous motor.

If we write Eq. (8.11) as

$$\tilde{E}'_a = \tilde{E}_a + j\tilde{I}_d(X_d - X_q) \tag{8.13}$$

then $\tilde{E}'_a$ can be referred to as the **effective excitation voltage**. It represents the induced emf in the armature winding of a synchronous motor when the effect of saliency is taken into consideration. As expected, when $X_d = X_q$ (round rotor), $\tilde{E}'_a = \tilde{E}_a$. In fact, the excitation voltage $\tilde{E}_a$ need not be computed unless it is necessary to identify the induced emf per phase in the armature winding due to the rotor flux per pole. The same can be said for $\tilde{I}_d$ and $\tilde{I}_q$. In any case, we can sketch the phasor diagrams (Figure 8.8) for a salient-pole synchronous motor depicting both $\tilde{E}'_a$ and $\tilde{E}_a$ when the power factor is either lagging, unity, or leading.

EXAMPLE 8.2

A 480-V, three-phase, Y-connected, salient-pole, synchronous motor is operating at its full load and draws a current of 50 A at a unity power factor. The d- and q-axis reactances are 3.5 Ω/phase and 2.5 Ω/phase, respectively. The armature-winding resistance is 0.5 Ω/phase. Determine (a) the excitation voltage of the motor and (b) the power developed by it.

● SOLUTION

Assuming the per-phase applied voltage as a reference, we have

$$V_a = \frac{480}{\sqrt{3}} = 277.128 \text{ V}$$

$$\tilde{I}_a = 50\angle 0^\circ \text{ A}$$

Thus,

$$\tilde{E}'_a = \tilde{V}_a - \tilde{I}_a R_a - j\tilde{I}_a X_q$$

$$= 277.128 - 50 \times 0.5 - j50 \times 2.5 = 281.414\angle -26.37^\circ$$

Hence, the power angle is

$$\delta = -26.37^\circ$$

From the phasor diagram (Figure 8.8), the d-axis current is

$$\tilde{I}_d = 50 \times \sin(26.37^\circ)\angle 90^\circ - 26.37^\circ$$

$$= 22.209\angle 63.63^\circ \text{ A}$$

Thus, the excitation voltage on a per-phase basis is

$$\tilde{E}_a = \tilde{E}'_a - j\tilde{I}_d(X_d - X_q)$$

$$= 281.414\angle -26.37^\circ - j(22.209\angle 63.63^\circ)(3.5 - 2.5)$$

$$= 303.623\angle -26.37^\circ \text{ V}$$

Finally, the rms value of the excitation voltage from line to line is

$$E_L = \sqrt{3} \times 303.623 = 525.89 \text{ V}$$

The power developed by the motor is

$$P_d = 3\text{Re}[\tilde{E}'_a \tilde{I}^*_a] = 3\text{Re}\left[(281.414\angle -26.37^\circ)(50)\right]$$

$$= 37.82 \text{ kW}$$

Check:

$$P_d = 3[\text{Re}(\tilde{V}_a \tilde{I}_a^*) - I_a^2 R_a]$$
$$= 3[\text{Re}(277.128 \times 50) - 50^2 \times 0.5]$$
$$= 37.82 \text{ kW}$$

■

Exercises

8.1. A 2-hp, 120-V, 4-pole, three-phase, Y-connected, synchronous motor has a synchronous impedance of $0.2 + j6\ \Omega$/phase. The friction and windage loss is 20 W, the core loss is 35 W, and the field-winding loss is 30 W. Determine the power angle and the efficiency of the motor when it delivers the rated power at a unity power factor.

8.2. A 5-hp, 208-V, 60-Hz, 6-pole, three-phase, Y-connected, salient-pole, synchronous motor delivers the rated power at 0.8 pf lagging. If $X_d = 2.5\ \Omega$/phase, $X_q = 1.7\ \Omega$/phase, and $P_r = 260$ W, determine (a) the excitation voltage, (b) the power angle, (c) the power developed, (d) the torque developed, and (e) the efficiency of the motor if the rotational loss is 5% of the power developed.

8.4 Power Expressions

In this section, our aim is to obtain expressions for the power developed in terms of the applied voltage, the excitation voltage, and the power angle by both the round-rotor and salient-pole synchronous motors.

Round-Rotor Synchronous Motor

From the equivalent circuit of a round-rotor synchronous motor (Figure 8.4), we obtain an expression for the phase current as

$$\tilde{I}_a = \frac{\tilde{V}_a - \tilde{E}_a}{R_a + jX_s}$$

The power developed is

$$P_d = 3\text{Re}[\tilde{E}_a \tilde{I}_a^*]$$
$$= 3\text{Re}\left[\frac{\tilde{E}_a \tilde{V}_a^* - E_a^2}{R_a - jX_s}\right]$$
$$= 3\text{Re}\left[\frac{\tilde{E}_a \tilde{V}_a^*(R_a + jX_s)}{Z_s^2} - \frac{E_a^2 R_a}{Z_s^2} - j\frac{E_a^2 X_s}{Z_s^2}\right] \tag{8.14}$$

where $$\hat{Z}_s = R_s + jX_s \tag{8.15}$$

is the synchronous impedance. Note that $Z_s^2 = \hat{Z}_s\hat{Z}_s^*$.

If the applied voltage is assumed as a reference—that is, $\tilde{V}_a = V_a\angle 0°$—then

$$\tilde{E}_a = E_a\angle -\delta$$

where **δ is now the magnitude of the power angle** by which the excitation voltage $\tilde{E}_a$ lags the applied voltage $\tilde{V}_a$.

Equation (8.14) can thus be written as

$$P_d = \frac{3E_aV_a}{Z_s^2}[R_a \cos\delta + X_s \sin\delta] - \frac{3E_a^2R_a}{Z_s^2} \tag{8.16}$$

Equation (8.16) yields an exact expression for the power developed by a round-rotor synchronous motor. When the armature-winding resistance is so small that it can be neglected, the above equation can be approximated as

$$P_d = \frac{3E_aV_a \sin\delta}{X_s} \tag{8.17}$$

The torque developed by a round-rotor synchronous motor is

$$T_d = \frac{3E_aV_a \sin\delta}{X_s\omega_s} \tag{8.18}$$

The power (or torque) developed by a round-rotor synchronous motor depends upon the power angle and the excitation voltage when the applied voltage is held constant. When the field excitation is reduced to zero, the power (or torque) developed is also zero. In other words, if the field winding is accidently open-circuited, the motor stalls. From the above equations, it is also evident that the power (or torque) developed by a round-rotor synchronous motor is maximum when δ is −90°.

EXAMPLE 8.3

A 440-V, three-phase, Δ-connected, synchronous motor has a synchronous reactance of 36 Ω/phase. Its armature-winding resistance is negligible. When the motor runs at a speed of 188.5 rad/s, it consumes 9 kW and the excitation voltage is 560 V. Determine (a) the power factor, (b) the power angle, (c) the line-to-line excitation voltage for a positive phase sequence, and (d) the torque developed by the motor.

● SOLUTION

Taking the per-phase supply voltage as a reference, we write

$$\tilde{V}_a = \frac{440}{\sqrt{3}} \approx 254\angle 0° \text{ V}$$

Using Δ-Y transformation, the equivalent per-phase reactance of a Y-connected, synchronous motor is $X_s = 36/3 = 12\ \Omega$, and the per-phase excitation voltage is $E_a = 560/\sqrt{3} = 323.32$ V. Thus,

$$\sin|\delta| = \frac{P_d X_s}{3V_a E_a} = \frac{9000 \times 12}{3 \times 254 \times 323.32} = 0.438 \qquad => \qquad \delta = -26°$$

Hence, $\tilde{E}_a = 323.32\angle -26°$ V.

The phase current (or the line current) in an equivalent Y-connected, synchronous motor is

$$\tilde{I}_a = \frac{\tilde{V}_a - \tilde{E}_a}{jX_s} = \frac{254 - 323.32\angle -26°}{j12} = 12.2\angle 14.48° \text{ A}$$

(a) The power factor: pf = cos (14.48°) = 0.97 (lead)
(b) The power angle: δ = −26° electrical
(c) For a positive phase sequence, the line-to-line excitation voltage is

$$\tilde{E}_L = \sqrt{3}\tilde{E}_a\angle 30° = 560\angle 4° \text{ V}$$

and the line-to-line applied voltage is

$$\tilde{V}_L = 440\angle 30° \text{ V}$$

(d) Since the power developed is given, the torque developed is

$$T_d = \frac{P_d}{\omega_s} = \frac{9000}{188.5} = 47.75 \text{ N·m}$$

■

Salient-Pole Synchronous Motor

By comparing the equivalent circuit of a round-rotor synchronous motor (Figure 8.4) with that of a salient-pole synchronous motor (Figure 8.7**c**), the approximate expression for the power developed by a salient-pole synchronous motor can be obtained, by modifying Eq. (8.17), as

$$P_d = \frac{3V_a E_a' \sin \delta}{X_q} \tag{8.19}$$

From the phasor diagrams (Figure 8.8) we can write

$$E_a' = E_a \pm I_d(X_d - x_q) \tag{8.20}$$

where the plus (+) sign is for a lagging power factor and the minus sign is for the unity or leading power factor. Eliminating E_a' in Eq. (8.19), we obtain

$$\begin{aligned} P_d &= \frac{3V_a E_a \sin \delta}{X_q} \pm 3V_a I_d \sin \delta \left[\frac{X_d - X_q}{X_q}\right] \\ &= \frac{3V_a E_a \sin \delta}{X_d} + 3V_a \sin \delta \left[\frac{X_d - X_q}{X_d X_q}\right][E_a \pm I_d X_d] \end{aligned} \tag{8.21}$$

However, from the phasor diagrams, we have

$$E_a \pm I_d X_d = V_a \cos \delta \tag{8.22}$$

Substituting Eq. (8.22) in Eq. (8.21), we obtain an approximate expression for the power developed by a salient-pole synchronous motor as

$$P_d = \frac{3V_a E_a \sin \delta}{X_d} + 3V_a^2 \sin 2\delta \left[\frac{X_d - X_q}{2X_d X_q}\right] \tag{8.23}$$

From the above equation, it is obvious that a salient-pole motor can still develop power (or torque) even when the field excitation is reduced to zero. The reason, of course, is the saliency of the rotor structure. In other words, even when the field winding is open-circuited, the motor keeps on operating at its synchronous speed as long as the load on the motor is less than or equal to the power developed by it owing to its saliency. On the other hand, when $X_d = X_q = X_s$, Eq. (8.23) reduces to Eq. (8.17) for a round-rotor synchronous motor.

If we write Eq. (8.23) as

$$P_d = A \sin \delta + B \sin 2\delta \tag{8.24a}$$

where

$$A = \frac{3V_a E_a}{X_d} \qquad \text{and} \qquad B = 3V_a^2 \left[\frac{X_d - X_q}{2X_d X_q}\right]$$

then the power developed owing to the field excitation is

$$P_{df} = A \sin \delta \tag{8.24b}$$

and the power developed due to the saliency of the rotor is

$$P_{ds} = B \sin 2\delta \tag{8.24c}$$

When the applied voltage and the field excitation are held constant, the condition for the power developed to be maximum can be obtained by setting $dP_d/d\delta$ to zero. This done, the power developed is maximum when $\delta \rightarrow \delta_m$ where

$$\delta_m = -\cos^{-1}\left[\frac{-A + \sqrt{A^2 + 32B^2}}{8B}\right] \tag{8.25}$$

EXAMPLE 8.4

A 208-V, 60-Hz, three-phase, Y-connected, salient-pole, synchronous motor operates at full load and draws a current of 40 A at 0.8 pf lagging. The *d*- and *q*-axis reactances are 2.7 Ω/phase and 1.7 Ω/phase, respectively. The armature-winding resistance is negligible, and the rotational loss is 5% of the power developed by the motor. Determine (a) the excitation voltage, (b) the power developed due to the field excitation, (c) the power developed due to saliency of the motor, (d) the total power developed, (e) the efficiency of the motor, and (f) the maximum power developed by the motor.

● SOLUTION

$$\tilde{V}_a = 120\angle 0^\circ \text{ V} \qquad \text{and} \qquad \tilde{I}_a = 40\angle -36.87^\circ \text{ A}$$

$$\begin{aligned}\tilde{E}'_a &= \tilde{V}_a - j\tilde{I}_a X_q \\ &= 120 - j1.7 \times 40\angle -36.87^\circ \\ &= 96.083\angle -34.48^\circ \text{ V}\end{aligned}$$

Thus, the absolute value of the power angle δ is 34.48°.

$$\alpha = |\theta - \delta| = 2.39°$$

$$\begin{aligned}\tilde{I}_d &= |I_a| \sin\alpha \underline{/-34.48° - 90°} \\ &= 40 \times \sin(2.39°)\underline{/-124.34°} \\ &= 1.668\underline{/-124.34°} \text{ A}\end{aligned}$$

(a) The per-phase excitation voltage is

$$\begin{aligned}\tilde{E}_a &= \tilde{E}'_a - j\tilde{I}_d(X_d - X_q) \\ &= 96.083\underline{/-34.48°} - j1.668\underline{/-124.34°} \times (2.7 - 1.7) \\ &= 94.418\underline{/-34.48°} \text{ V}\end{aligned}$$

(b) The power developed due to the field excitation, P_{df}, is

$$P_{df} = \frac{3 \times 120 \times 94.418 \times \sin(34.48°)}{2.7} = 7126.9 \text{ W}$$

(c) The power developed due to saliency of the motor, P_{ds}, is

$$P_{ds} = \frac{(2.7 - 1.7)\sin(2 \times 34.48°)}{2 \times 2.7 \times 1.7} \times 3 \times 120^2 = 4392.1 \text{ W}$$

(d) The total power developed is

$$P_d = P_{df} + P_{ds} = 11{,}519 \text{ W}$$

Check:

$$\begin{aligned}P_d &= 3\text{Re}[\tilde{E}'_a \tilde{I}^*_a] \\ &= 3\text{Re}\left[96.083\underline{/-34.48°} \times 40\underline{/36.87°}\right] \approx 11{,}520 \text{ W}\end{aligned}$$

(e) The rotation loss: $P_r = 0.05 \times 11{,}519 \approx 576$ W

The power input:

$$\begin{aligned}P_{in} &= 3\ \text{Re}[\tilde{V}_a \tilde{I}^*_a] \\ &= 3\ \text{Re}\left[120 \times 40\underline{/36.87°}\right] \\ &= 11{,}520 \text{ W}\end{aligned}$$

As expected, the power input is equal to the power developed when the armature-winding resistance is neglected.

The power output: $P_o = 11{,}520 - 576 = 10{,}944$ W

Hence, the efficiency of the motor is

$$\eta = \frac{10{,}944}{11{,}520} = 0.95 \qquad \text{or} \qquad 95\%$$

(f) From Eq. (8.24),

$$A = \frac{3 \times 120 \times 94.418}{2.7} = 12{,}589.07$$

$$B = \frac{3(2.7 - 1.7)}{2 \times 2.7 \times 1.7} \times 120^2 = 4705.88$$

$$P_d = 12{,}589.07 \sin \delta + 4{,}705.88 \sin 2\delta$$

From Eq. (8.25), $\delta_m = -63.4°$
Finally, the maximum power developed is

$$P_{dm} = 12{,}589.07 \sin (63.4°) + 4{,}705.88 \sin (2 \times 63.4°)$$
$$\approx 15{,}025 \text{ W}$$

■

Exercises

8.3. Verify Eq. (8.25).

8.4. A 20-hp, 480-V, 60-Hz, 12-pole, three-phase, Y-connected, salient-pole, synchronous motor delivers the rated load at a unity power factor. The *d*- and *q*-axis synchronous reactances are 1.5 Ω/phase and 0.9 Ω/phase, respectively. The rotational power loss is 800 W. The armature-winding resistance is negligible. Determine (a) the excitation voltage, (b) the power angle, (c) the power developed due to the field excitation, (d) the power developed due to the saliency of the motor, (e) the total power developed, (f) the torque developed, (g) the maximum power developed, and (h) the efficiency of the motor.

8.5. If the motor in Exercise 8.4 is connected to a 480-V, three-phase source via a three-wire transmission line with a reactance of 2.5 Ω/line, determine the power angle and the power developed by the motor.

8.5 Exact Condition for Maximum Power

The power developed by a round-rotor synchronous motor depends upon the terminal voltage, its synchronous impedance, the excitation voltage, and the power angle. The synchronous impedance is constant as long as the motor operates in the linear region. The terminal voltage is constant when the motor is connected to an infinite bus. If the field excitation is held constant, the excitation voltage is also constant. Under these constraints, the change in the power developed must be accompanied by a change in the power angle. Thus, the power developed is maximum when $dP_d/d\delta = 0$. Differentiating Eq. (8.16) with respect to δ and setting it to zero, we obtain

$$-R_a \sin \delta_m + X_s \cos \delta_m = 0$$

or

$$\tan \delta_m = \frac{X_s}{R_a} \tag{8.26}$$

which clearly indicates that the power angle $|\delta_m| \rightarrow -90°$ when $R_a \rightarrow 0$. This is in harmony with Eq. (8.18).

Substituting Eq. (8.26) in Eq. (8.16) and after some simplifications, we obtain an expression for the **per-phase maximum power developed**, P_{dma}, as

$$P_{dma} = \frac{V_a E_a}{Z_s} - \frac{E_a^2 R_a}{Z_s^2} \tag{8.27a}$$

When $R_a \rightarrow 0$, the above expression simplifies to

$$P_{dma} = \frac{V_a E_a}{X_s} \tag{8.27b}$$

Note that Eq. (8.27) yields the maximum power for a given value of E_a. However, E_a can be varied by varying the field excitation. Therefore, P_{dma} (increases/decreases) as E_a (increases/decreases) with an (increase/decrease) in the field excitation. By differentiating Eq. (8.27a) with respect to E_a (dP_{dma}/dE_a) and setting it to zero, we obtain an equation for E_a as

$$E_a = \frac{V_a Z_s}{2R_a} \tag{8.28}$$

This is the value of the excitation voltage that gives the maximum power developed by a round-rotor synchronous motor. We hasten to add that **this is not the maximum value of the excitation voltage.**

Substituting Eq. (8.28) in Eq. (8.27), we obtain the per-phase maximum power developed as

$$(P_{dma})_{\max} = \frac{V_a^2}{4R_a} \tag{8.29}$$

Equation (8.27a) can also be written as

$$E_a^2 - \frac{Z_s}{R_a} V_a E_a + \frac{Z_s^2}{R_a} P_{dma} = 0$$

Solving for E_a, we obtain

$$E_a = \frac{Z_s}{2R_a}\left[V_a \pm \sqrt{V_a^2 - 4R_a P_{dma}}\right] \tag{8.30}$$

Corresponding to the per-phase maximum power developed, P_{dma}, there are two values of the excitation voltage E_a as given by the above equation. These values are said to represent the excitation limits for any load on the motor.

We can also obtain the criterion for the per-phase maximum power developed as a function of E_a for fixed values of V_a, Z_s, and δ by differentiating Eq. (8.16) with respect to E_a and setting it to zero. By doing so, we obtain

$$E_a = \frac{V_a}{2R_a}\left[R_a \cos\delta + X_s \sin\delta\right] \tag{8.31}$$

Substitution of Eq. (8.31) in Eq. (8.16) yields the maximum power developed as

$$P_{dma} = \frac{V_a^2}{4R_a}\left[\frac{R_a}{Z_s}\cos\delta + \frac{X_s}{Z_s}\sin\delta\right]^2 \tag{8.32}$$

However, from the impedance triangle (Figure 8.9),

$$R_a = Z_s \cos\phi$$

and

$$X_s = Z_s \sin\phi$$

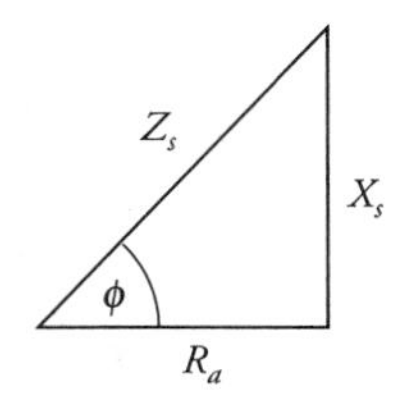

Figure 8.9 Impedance triangle for the synchronous impedance of a round-rotor synchronous motor.

Substituting for R_a and X_s in Eq. (8.32), we obtain the maximum power developed as

$$P_{dma} = \frac{V_a^2}{4R_a} \cos^2 (\phi - \delta) \tag{8.33}$$

From the above equation it is evident that P_{dma} is maximum when $\delta = \phi$. When $\delta = \phi$, Eqs. (8.31) and (8.33) reduce to Eqs. (8.28) and (8.29), respectively.

EXAMPLE 8.5

A 120-V, 60-Hz, three-phase, Y-connected, round-rotor, synchronous motor has an armature-winding resistance of $0.5 + j3$ Ω/phase. The motor takes a current of 10 A at 0.8 pf leading when operating at a certain field current. The load is gradually increased until the motor develops the maximum torque while the field current is held constant. Determine the new line current, the power factor, and the torque developed by the motor.

● SOLUTION

Initial operation:

$$\hat{Z}_s = 0.5 + j3 = 3.04\underline{/80.54°}\ \Omega$$

$$\tilde{V}_a = 69.28\underline{/0°}\ \text{V}$$

$$\tilde{I}_{a1} = 10\underline{/36.87°}\ \text{A}$$

$$\tilde{E}_{a1} = \tilde{V}_a - \tilde{I}_{a1}\hat{Z}_s = 87.55\underline{/-17.96°}\ \text{V}$$

Operation at maximum power:

The power angle associated with the maximum torque is

$$\delta_m = \tan^{-1}\left[\frac{3}{0.5}\right] = 80.54° \text{ (lag)}$$

$$\tilde{E}_{a2} = 87.55\underline{/-80.54°} \text{ V}$$

The line current is

$$\tilde{I}_{a2} = \frac{69.28 - 87.55\underline{/-80.54°}}{0.5 + j3} = 33.64\underline{/-22.98°} \text{ A}$$

The per-phase maximum power developed by the motor is

$$P_{dma} = 3\text{Re}\left[(87.55\underline{/-80.54°})(33.64\underline{/22.98°})\right]$$
$$\approx 1580 \text{ W/phase}$$

The total power developed by the motor is

$$P_{dm} = 3P_{dma} = 4740 \text{ W}$$

We can verify the maximum power developed by using Eq. (8.27a) as

$$P_{dma} = \frac{69.28 \times 87.55}{3.04} - \frac{87.55^2 \times 0.5}{3.04^2} \approx 1580 \text{ W/phase}$$

■

Exercises

8.6. A 440-V, 60-Hz, 4-pole, three-phase, Δ-connected, synchronous motor has a synchronous impedance of $1.2 + j8\ \Omega$/phase. The field current is adjusted so that the excitation voltage is 560 V. When the motor develops the maximum power, determine (a) the line current, (b) the power factor, (c) the torque angle, (d) the maximum power developed, and (e) the maximum torque developed.

8.7. A 208-V, 50-Hz, 12-pole, three-phase, Y-connected, synchronous motor has a synchronous impedance of $0.75 + j7.5\ \Omega$/phase. The field excitation is so adjusted that the motor takes 25 A on full load at unity power factor. The rotational loss is 1.2 kW. If the field excitation is held constant, calculate (a) the line current, (b) the power factor, (c) the maximum power developed, (d) the maximum power output, (e) the maximum shaft torque, and (f) the efficiency.

8.6 Effect of Excitation

No-Load Condition

In order to discern the effect of excitation on the behavior of a synchronous motor, we assume that the motor is ideal in the sense that (a) it has no armature-winding resistance, and (b) there are no rotational losses. We also assume that the motor is connected to an infinite bus so that its terminal voltage is constant and there is no change in its frequency. The operation of an ideal motor at no load requires no armature current. For this to happen, the applied voltage must be equal and opposite to the excitation voltage. Since the motor rotates at a constant speed and the frequency of the applied source is also constant, the excitation voltage can be changed only by changing the excitation (field-winding) current. When the excitation current is adjusted to obtain the excitation voltage equal in magnitude to the applied voltage, it is referred to as **normal (100%) excitation**. Since there is no power developed by the motor, the power angle δ must be zero, as illustrated in Figure 8.10**a**.

If the excitation current is increased beyond what is needed for normal excitation, the magnitude of the excitation voltage increases and the motor is said to be **overexcited**. The difference between the applied voltage and the excitation voltage is responsible for the armature current in the motor. We can express the armature current as

$$\tilde{I}_a = j\frac{\tilde{E}_a - \tilde{V}_a}{X_s} \tag{8.34}$$

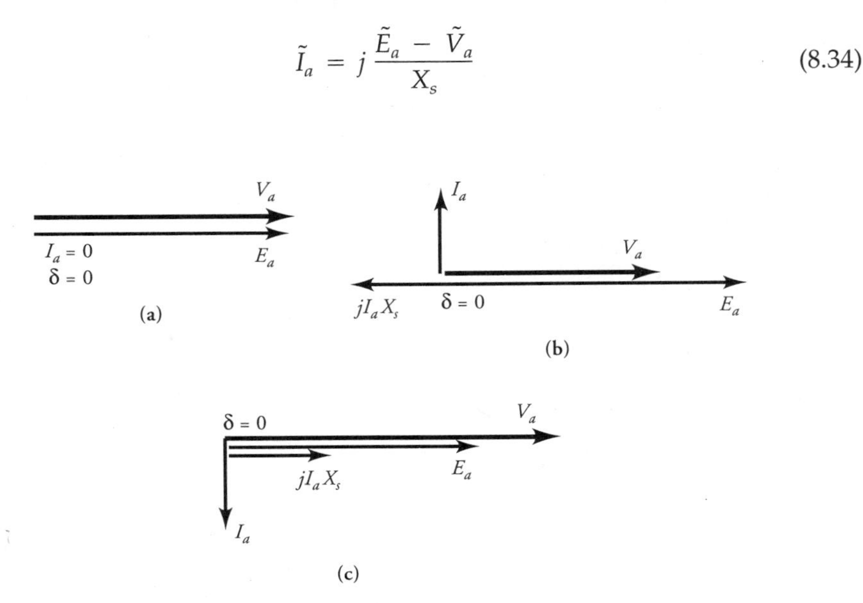

Figure 8.10 Operation of a synchronous motor at no load for **(a)** normal excitation, **(b)** overexcitation, and **(c)** underexcitation.

Since $\tilde{E}_a > \tilde{V}_a$, the armature current leads the phasor $\tilde{E}_a - \tilde{V}_a$ by 90°. The magnitude of armature current depends upon the level of overexcitation. Since the power output (developed) is zero at no load, the power angle δ must still be zero, as shown in Figure 8.10**b**. However, the complex power input to the motor is

$$\hat{S}_m = -j3V_aI_a \tag{8.35}$$

where V_a and I_a are the effective values of the applied (terminal) voltage and the armature current. The complex power is purely reactive, and the presence of the minus sign highlights the fact that the motor behaves as a capacitor when it is overexcited. When an overexcited synchronous motor operates at no load, it is customary to refer to it as a **synchronous condenser**. It is used in this fashion to improve the overall power factor of a plant. We will shed more light on its function as a synchronous capacitor in the next section.

When the field current is below its normal value, the motor is said to be **underexcited**. In this case, the excitation voltage is smaller than the terminal voltage, and the current in the armature winding is

$$\tilde{I}_a = -j\frac{\tilde{V}_a - \tilde{E}_a}{X_s} \tag{8.36}$$

Since $\tilde{V}_a$ is greater than $\tilde{E}_a$ and the power angle δ must still be zero, the presence of $-j$ in the above equation indicates that the current is lagging the applied voltage by 90°, as depicted in Figure 8.10**c**. The complex power input to the motor

$$\hat{S}_m = j3V_aI_a \tag{8.37}$$

is purely inductive. Thus, an underexcited synchronous motor behaves like an inductor. If used in this way, a synchronous motor may be called a **synchronous inductor**.

From the above discussion, it must be obvious that a change in the level of excitation affects the reactive power only when the motor operates at no load. There is no change in the average (real) power either supplied to or developed by the motor. The change in the average power can be accomplished only by varying the load on the motor.

Load Condition

Let us now examine the condition when an ideal synchronous motor is supplying some load. The power developed by a round-rotor synchronous motor is

$$P_d = \frac{3V_aE_a \sin \delta}{X_s} \tag{8.38}$$

Since V_a and X_s are constants and P_d is held constant, $E_a \sin \delta$ must also be constant. As E_a changes with a change in the field excitation, the power angle δ must also change in such a way that $E_a \sin \delta$ is constant. In other words, the locus of the tip of $\tilde{E}_a$ must trace a line parallel to $\tilde{V}_a$, as shown in Figure 8.11.

A change in the excitation voltage is also accompanied by a change in the armature current because

$$\tilde{I}_a = \frac{\tilde{V}_a - \tilde{E}_a}{jX_s} \tag{8.39}$$

However, the power developed by an ideal motor is also equal to the power input to the motor. That is

$$P_{in} = 3V_a I_a \cos \theta \tag{8.40}$$

Since the power output of the motor is constant, the power input to the motor must also be constant. In other words, $I_a \cos \theta$ must be constant. As the magnitude of the armature current changes with the change in $\tilde{E}_a$, the power-factor angle θ adjusts itself in such a way that the product $I_a \cos \theta$ remains constant. This is possible when the locus of the tip of $\tilde{I}_a$ traces a line perpendicular to $\tilde{V}_a$, as illustrated in Figure 8.11.

V Curves

We can make the following observations when the level of excitation is changed gradually from underexcitation to overexcitation for a constant power output:

1. The armature current is minimum when it is in phase with the terminal voltage; that is, the power factor is unity.

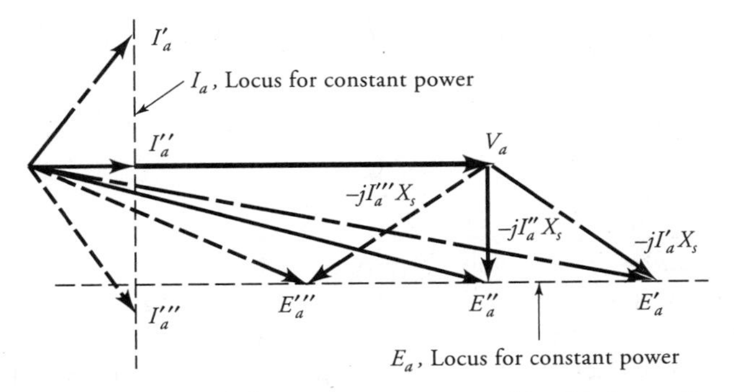

Figure 8.11 Constant power loci of armature current and excitation voltage.

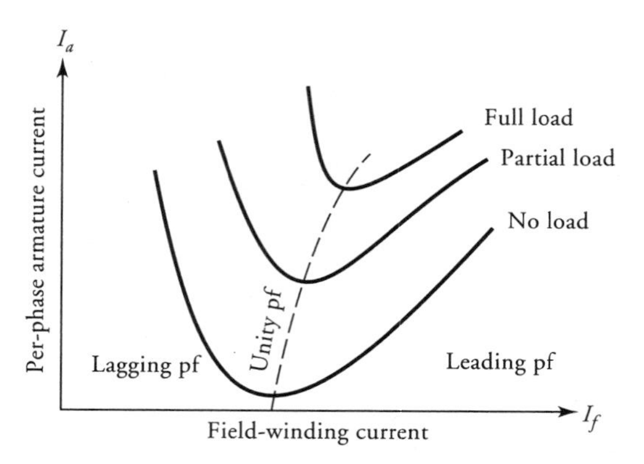

Figure 8.12 V curves for a synchronous motor.

2. The power factor is lagging when the motor is underexcited. In this case, the motor behaves like an inductive load.
3. The power factor is leading when the motor is overexcited. The operation of an overexcited synchronous motor is analogous to that of a capacitive load. Once again, the motor can be used not only to deliver a load torque but also to improve the power factor of the three-phase power supply.

When the magnitude of the armature current is plotted as a function of the excitation current for various load conditions, we obtain a set of nested curves, the so-called V curves, as shown in Figure 8.12. These curves are very similar to those obtained for synchronous generators in the preceding chapter. The only difference is that the power factor is lagging when the motor is underexcited and leading when it is overexcited.

We can also plot the variations in the power factor as a function of the excitation current, as depicted in Figure 8.13. Note the sharpness in the curve at no load compared with that at full load. Do you know why?

EXAMPLE 8.6

A 208-V, Y-connected, three-phase, synchronous motor has a synchronous reactance of 4 Ω/phase and negligible armature-winding resistance. At a certain load the motor takes 7.2 kW at 0.8 pf lagging. If the power developed by the motor remains the same while the excitation voltage is increased by 50% by raising the field excitation, determine (a) the new armature current and (b) the power factor.

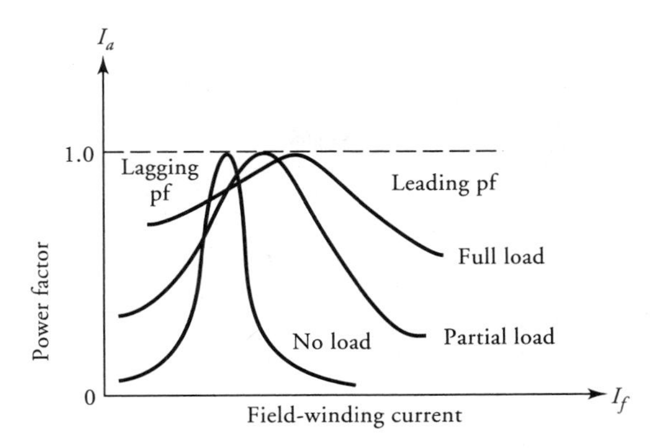

Figure 8.13 Power-factor characteristic of a synchronous motor.

● SOLUTION

$$\tilde{V}_a = 120\angle 0^\circ \text{ V}$$

$$I_a = \frac{7200}{3 \times 120 \times 0.8} = 25 \text{ A}$$

Thus,

$$\tilde{I}_a = 25\angle -36.87^\circ \text{ A}$$

$$\tilde{E}_a = 120\angle 0^\circ - j4 \times 25\angle -36.87^\circ$$

$$= 100\angle -53.13^\circ \text{ V}$$

The new excitation voltage: $E_{an} = 100 \times 1.5 = 150$ V

If δ_n is the new torque angle, then

$$\sin \delta_n = \frac{7200 \times 4}{3 \times 120 \times 150} = 0.533 \quad => \quad \delta_n = -32.23^\circ$$

Hence, the new armature current is

$$\tilde{I}_{an} = \frac{120 - 150\angle -32.23^\circ}{j4} = 20.074\angle 4.92^\circ \text{ A}$$

and the new power factor is

$$\text{pf} = \cos(4.92) = 0.996 \text{ (leading)}$$

■

Exercises

8.8. A 208-V, three-phase, Y-connected, synchronous motor takes 7.2 kW at 0.8 pf leading. The synchronous reactance of the motor is 4 Ω/phase and the armature-winding resistance is negligible. If the field excitation is adjusted to decrease the excitation voltage by 50% while the load on the motor remains the same, determine (a) the new torque angle, (b) the armature current, and (c) the power factor.

8.9. Repeat Exercise 8.8 when the field excitation is adjusted to obtain a unity power factor. What is the change in the excitation voltage?

8.10. Plot the V-curve for the motor in Exercise 8.8 by varying the power factor from 0.5 lagging to 0.5 leading. Assume $E_a = 100I_f$, where I_f is the field-winding current.

8.7 Power Factor Correction

In an electrical system in which fluorescent lamps are used for lighting and induction motors are used as workhorses, the overall power factor of the system is low. Since the total power supplied by an infinite bus to the load is $3V_aI_a \cos \theta$, a decrease in the power factor $\cos \theta$ or an increase in the power-factor angle θ is accompanied by an increase in the current I_a. For the same load, the increase in the current owing to the decrease in the power factor results not only in a greater voltage drop over the transmission line but also in greater power losses. For this reason, a utility company usually charges a higher rate to a customer for maintaining a power factor lower than 50%.

Since an overexcited synchronous motor takes power with a leading power factor, it behaves like a capacitor. Advantage is taken of this fact in order to improve the overall power factor of an electrical system. To do so, a high-efficiency synchronous motor is placed in parallel with the rest of the load. A synchronous motor may be installed to replace an induction motor and also to improve the power factor. Simply by changing the excitation current, the power factor of a synchronous motor can be varied from unity (0°) to zero (**leading** 90°).

When a synchronous motor is intentionally built without any shaft extension just to improve the power factor, it is commonly called the **synchronous condenser (capacitor)**. Such a motor requires very little real (average) power for its own rotational loss but supplies the needed reactive power with a leading power factor.

Although it may seem very logical to improve the overall power factor of an electrical system to unity to minimize energy consumption, it is customarily not done because such a correction imposes high reactive power requirements on the synchronous motor. It may not be economically viable to install such a synchronous motor. Therefore, in most of the cases the reactive power supplied by the

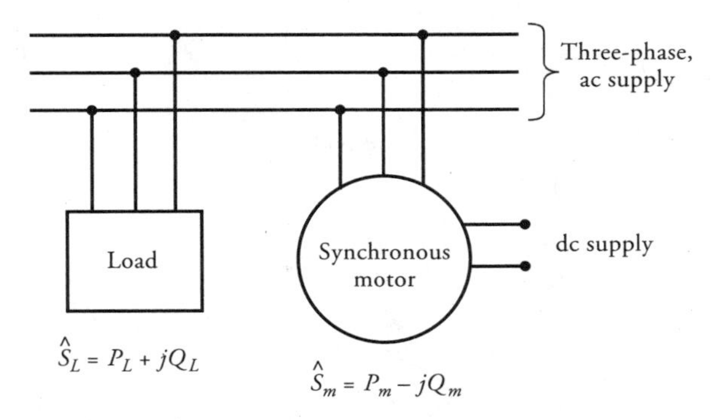

Figure 8.14 Power factor correction using a synchronous motor.

synchronous motor is smaller than the reactive power demand of the rest of the electrical system.

Let $\hat{S}_L = P_L + jQ_L$ be the complex power required by an electrical system as shown in Figure 8.14, and $\hat{S}_m = P_m - jQ_m$ be the complex power intake of the synchronous motor. The overall power requirements are

$$\hat{S}_t = P_L + P_m + j(Q_L - Q_m) \tag{8.41a}$$

$$= P_t + jQ_t \tag{8.41b}$$

In light of our earlier discussion, $Q_m \leq Q_L$. The overall power factor is

$$\text{pf} = \frac{P_t}{S_t} \tag{8.42}$$

where $S_t = |\hat{S}_t|$. As long as $Q_m \leq Q_L$, the overall power factor, although improved by the addition of a synchronous motor, is still lagging, as depicted by the power triangles in Figure 8.15.

EXAMPLE 8.7

A manufacturing plant uses 100 kVA at 0.6 pf lagging under normal operation. A synchronous motor is added to the system to improve the overall power factor. The power required by the synchronous motor is 10 kW. Determine the overall power factor when the synchronous motor operates at 0.5 pf leading. What must be the power factor of the motor to improve the overall power factor to 0.9 lagging?

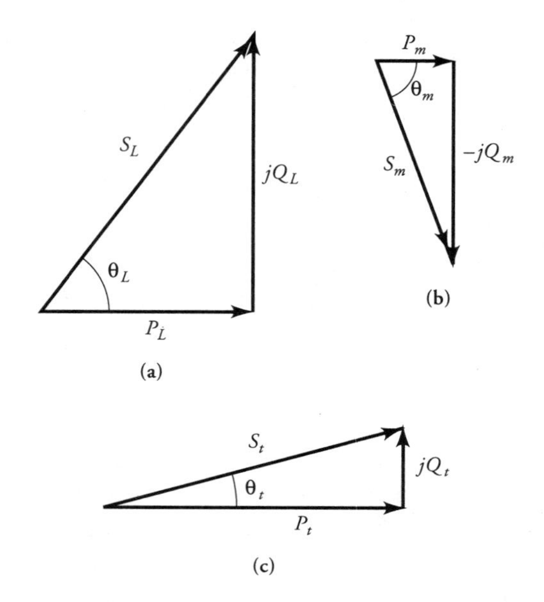

Figure 8.15 Power triangles for **(a)** load, **(b)** synchronous motor, and **(c)** improved load on the three-phase, power source.

● SOLUTION

For the load:

$$\theta_L = \cos^{-1}(0.6) = 53.13^\circ \text{ (lag)}$$

$$\hat{S}_L = 100\underline{/53.13^\circ} = 60 + j80 \text{ kVA}$$

For the synchronous motor:

$$\theta_m = \cos^{-1}(0.5) = 60^\circ \text{ (lead)}$$

$$S_m = \frac{10}{0.5} = 20 \text{ kVA}$$

Thus,

$$\hat{S}_m = 20\underline{/-60^\circ} = 10 - j17.32 \text{ kVA}$$

The overall power requirements are

$$\begin{aligned}\hat{S}_t &= 60 + 10 + j(80 - 17.32)\\ &= 70 + j62.68\\ &= 93.96\underline{/41.84^\circ} \text{ kVA}\end{aligned}$$

and the power factor is

$$\text{pf} = \cos(41.84°) = 0.74 \text{ lagging}$$

For the power factor to be 0.9 lagging, $\theta_t = 25.84°$ (lag). The real (average) power requirement is still 70 kW, and the corresponding apparent power must be $70/0.9 = 77.778$ kVA. Thus,

$$\hat{S}_t = 77.778\underline{/25.84°} = 70 + j33.9 \text{ kVA}$$

For no change in the power requirements of the load, the new power demand on the motor is

$$\begin{aligned}\hat{S}_m &= (70 + j33.9) - (60 + j80) \\ &= 10 - j46.1 \\ &= 47.172\underline{/-77.76°} \text{ kVA}\end{aligned}$$

Hence, the power factor of the motor is

$$\text{pf} = \cos(77.76) = 0.21 \text{ (lead)}$$

■

Exercises

8.11. The power consumption of a factory is 2000 kVA at 0.45 pf lagging. A synchronous motor is added to raise the power factor to 0.8 lagging. If the power intake of the motor is 100 kW, determine the power factor and the kVA rating of the motor.

8.12. A factory load of 360 kW at a power factor of 0.6 lagging includes an induction motor delivering 50 hp at an efficiency of 80% and 0.866 pf lagging. When the induction motor is replaced by a synchronous motor of the same horsepower and efficiency, the power factor becomes 0.8 lagging. Determine the kVA rating and the power factor of the motor.

SUMMARY

A synchronous motor is electrically the same as a synchronous generator. However, it is not self-starting. When the motor is equipped with a damper winding, it develops a starting torque owing to the induced current in its damper winding.

The damper winding may be either a three-phase winding or a squirrel-cage winding. During the start-up period, the field winding is shorted. When the motor attains a speed nearly equal to its synchronous speed, the field winding is excited and the motor pulls into synchronism. Once the motor attains its synchronous speed, the damper winding becomes ineffective. The motor keeps on rotating at its synchronous speed as long as the load torque is less than the pull-out torque. The pull-out torque is the maximum torque the motor can handle without pulling out of synchronism.

The equations for the power (or torque) developed by a synchronous motor are the same as those for a synchronous generator. The only difference is that the motor rotates in the same direction as the torque it develops. On the other hand, the generator is forced to rotate by the prime mover in a direction opposite to the torque developed.

For any given load, a synchronous motor can be made to operate at either a lagging power factor, unity power factor, or a leading power factor. The field current (and thereby the excitation voltage) that results in the armature current in phase with the terminal voltage (unity power factor) at any load is known as the **normal excitation**. A decrease in the field current gives rise to **underexcitation** and a lagging power factor. Likewise, an increase in the field current from its normal value results in **overexcitation** and a leading power factor.

An overexcited synchronous motor with a leading power factor is not only used as a motor to supply the needed mechanical power but also acts as a power factor correction device. In fact, there are some synchronous motors called **synchronous condensers** that are designed just to control the power factor.

Review Questions

8.1. What is the difference between a synchronous motor and a synchronous generator?
8.2. What is a damper winding? How does it help to minimize hunting?
8.3. If a synchronous generator is equipped with a damper winding, what effect does it have on the operation of the generator?
8.4. What is the effect of the change in the excitation current on the motor speed?
8.5. How does the change in the excitation current affect the power factor of a synchronous motor? Explain, succinctly, what happens when the field excitation is (a) increased and (b) decreased.
8.6. Explain, briefly, the operation of a synchronous motor.
8.7. A synchronous motor is operating at a unity power factor. What must be done to make the motor operate at a leading power factor?
8.8. A synchronous machine is operating from an infinite bus. The excitation

voltage leads the bus voltage by 15° and the phase current lags the terminal voltage by 36.87°. If the armature-winding resistance is negligible, is the machine acting as a motor or a generator? What are the effects on the power angle, the power factor, and the power developed when the field current is increased?

8.9. The armature current increases with an increase in the field current of a synchronous motor. What is the initial power-factor angle? What is the effect on the torque angle?

8.10. A synchronous motor is operating at its full load. The armature current decreases with an increase in the field current. Does the armature current lead or lag the terminal voltage?

8.11. Why does the power angle lag in a synchronous motor?

8.12. When a synchronous motor is loaded, what happens to the magnetic axis of the rotor with respect to the magnetic axis of the revolving field?

8.13. Is it possible for a single-phase, synchronous motor to establish a revolving field?

8.14. Is it possible for a two-phase, synchronous motor to set up a revolving field?

8.15. Is there any difference between a synchronous motor and a synchronous condenser?

8.16. Why is it necessary to improve the overall power factor of a manufacturing plant?

8.17. Why does the field current increase with the increase in the load on a synchronous motor in order to keep it operating at unity power factor?

8.18. If the excitation current is held constant and the load on the synchronous motor is increased, what happens to the power factor of the motor?

8.19. A 100-hp, 90% efficient, synchronous motor delivers the full load at 0.707 pf leading. What is the power input to the motor? What is its kVA requirement? Sketch its power triangle.

8.20. A machine uses 500 kW at 0.8 pf lagging. What sort of a load does the machine represent? What is the kVA requirement of this machine? Draw its power triangle.

8.21. A 200-hp, 80% efficient, synchronous motor operates at full load and takes 600 kVA from the supply. What is the power factor of the motor? What must be done to reduce its kVA requirements?

8.22. The angle between the magnetic axes of the rotor and the revolving field of a 12-pole, 50-Hz, three-phase, synchronous motor is 2.5° mechanical. What is the power angle of the motor?

8.23. Why does the armature current increase with the increase in the load on a synchronous motor even when there is no change in its field excitation?

8.24. A synchronous motor is operating at a unity power factor (normal excitation) under a certain load. If the load is increased, explain why the current changes faster than the power factor.

Problems

8.1. A 220-V, three-phase, Y-connected, synchronous motor has a synchronous impedance of $0.25 + j2.5\ \Omega$/phase. The motor delivers the rated load of 80 A at 0.707 pf leading. Determine (a) the excitation voltage, (b) the power angle, and (c) the power developed by the motor.

8.2. The synchronous impedance of a 22-kV, 50-A, three-phase, Y-connected synchronous motor is $0.5 + j12\ \Omega$/phase. The motor operates at full load and 0.866 pf lagging. The field-winding resistance is 25 Ω and the field current at full load is 40 A. The rotational loss is 20 kW. Determine (a) the power angle, (b) the excitation voltage, and (c) the efficiency of the motor.

8.3. A 450-kVA, 2080-V, 50-Hz, 6-pole, three-phase, Y-connected, synchronous motor has a synchronous impedance of $1 + j15\ \Omega$/phase. Determine (a) the excitation voltage, (b) the power angle, (c) the power developed when the motor is fully loaded at 0.866 pf leading, and (d) the torque developed.

8.4. A 2200-V, 60-Hz, three-phase, Y-connected, 8-pole, synchronous motor has a synchronous impedance of $0.8 + j8\ \Omega$/phase. At no load the field excitation is adjusted so that the excitation voltage is equal in magnitude to the applied voltage. When the motor is loaded, the power angle is 22° electrical (lagging). Compute the torque developed by the motor.

8.5. The synchronous reactance of a 1000-V, Δ-connected, three-phase, synchronous motor is 9 Ω/phase. When the motor operates at a certain load, the power input is 138 kW and the excitation voltage is 1200 V (line to line). If the armature-winding resistance is negligible, determine (a) the line current and (b) the power factor.

8.6. The synchronous impedance of a 2200-V, three-phase, Y-connected synchronous motor is $0.5 + j10\ \Omega$/phase. At a certain load, the power input to the motor is 762 kW and the excitation voltage is 2600 V. If the rotational loss is 50 kW, determine (a) the power angle, (b) the power factor, and (c) the efficiency of the motor.

8.7. A 100-hp, 2300-V, 50-Hz, three-phase, Δ-connected, synchronous motor delivers full load at 0.8 pf leading. The synchronous reactance is 12 Ω/phase and the armature-winding resistance is negligible. The rotational loss is 3 kW and the field-excitation loss is 1.5 kW. Determine (a) the armature current, (b) the power angle, and (c) the motor efficiency.

8.8. A 10-hp, 208-V, 60-Hz, 6-pole, three-phase, Y-connected, synchronous motor has a synchronous impedance of $0.25 + j8.5\ \Omega$/phase. The friction and windage loss is 450 W, the core loss is 250 W, and the field-excitation loss is 100 W. Determine (a) the armature current, (b) the torque angle, (c) the shaft torque, and (d) the efficiency when the motor operates at full load with a unity power factor.

8.9. A 460-V, three-phase, Y-connected, synchronous motor consumes 60 kW for a certain load at 1200 rpm. The excitation voltage is 580 V. The synchronous reactance is 5 Ω/phase and the armature-winding resistance is negligible. Calculate the line current and the torque developed by the motor.

8.10. A 208-V, Δ-connected, synchronous motor takes 30 A from the infinite bus (supply) at full load and 0.866 pf leading. The synchronous impedance of the motor is $0.5 + j6$ Ω/phase. Determine the power developed by the motor. What is the maximum power developed by the motor when the field excitation is held constant?

8.11. A three-phase synchronous motor is connected to a three-phase synchronous generator. Obtain an expression for the power developed by the synchronous motor in terms of the per-phase no-load voltages of the generator E_{ga} and the motor E_{ma}, the synchronous reactances of the generator X_{sg} and the motor X_{sm}, and the power angles of the generator δ_g and the motor δ_m.

8.12. A 500-V, three-phase, Y-connected, salient-pole, synchronous motor operates at full load and draws a current of 30 A at 0.8 pf leading. The *d*- and *q*-axis synchronous reactances are 1.85 Ω/phase and 1.2 Ω/phase, respectively. The armature-winding resistance is 0.25 Ω/phase. The rotational loss is 5% of the power developed at full load. Determine (a) the excitation voltage, (b) the torque angle, (c) the total power developed, and (d) the efficiency of the motor.

8.13. The *d*- and *q*-axis synchronous reactances of a salient-pole motor are 1.0 per unit and 0.7 per unit. The armature-winding resistance is negligible. The torque angle is 30°. The motor operates at its rated terminal voltage, and the excitation voltage is 1.2 per unit. What is the ratio of the power developed due to saliency to the total power developed by the motor? If the power factor is unity, what is the per-unit current intake of the motor?

8.14. A 460-V, 60-Hz, 12-pole, three-phase, Y-connected, synchronous motor operates at full load with 0.8 pf lagging. The full-load current is 45 A. The armature-winding resistance is 0.2 Ω/phase, the *d*-axis synchronous reactance is 15 Ω/phase, and the *q*-axis synchronous reactance is 8 Ω/phase. The rotational power loss is 1.5 kW. Determine (a) the torque angle, (b) the shaft torque, and (c) the efficiency of the motor.

8.15. Repeat Problem 8.14 when the armature-winding resistance is negligible. Also compute the maximum power developed by the motor.

8.16. A 208-V, three-phase, Y-connected, synchronous motor has a synchronous impedance of $0.4 + j4$ Ω/phase. The motor takes 30 A on full load at a unity power factor. If the field current is held constant and the load is gradually increased until the motor develops the maximum torque, determine (a) the new power factor, (b) the torque angle, and (c) the power developed by the motor. Also determine the excitation limits for the excitation voltage.

8.17. A 2080-V, 8-pole, 50-Hz, three-phase, Y-connected, synchronous motor has a synchronous impedance of $0.4 + j8\ \Omega$/phase. The field current is adjusted to obtain an excitation voltage of 1732 V on full load. What is the maximum torque the motor can deliver before it pulls out of synchronism? Also determine (a) the torque angle, (b) the armature current, (c) the power factor, and (d) the excitation limits.

8.18. A 208-V, three-phase, Y-connected, synchronous motor has a synchronous impedance of $0.25 + j5\ \Omega$/phase. The motor operates at its lower excitation limit and delivers a maximum power of 15 kW. Determine (a) the torque angle, (b) the excitation voltage, (c) the line current, (d) the power factor, and (e) the efficiency.

8.19. A 2300-V, Y-connected, three-phase, synchronous motor has a synchronous impedance of $0.2 + j10\ \Omega$/phase. The motor takes 40 A at 0.8 pf lagging on full load. The field current is adjusted in such a way that the motor develops maximum power without any significant change in its power angle. Determine (a) the new excitation voltage, (b) the power angle, (c) the line current, (d) the power factor, and (e) the maximum power developed.

8.20. Repeat Problem 8.19 if the field excitation is held constant and the load on the motor is increased until it develops the maximum power.

8.21. The synchronous reactance of a 2200-V, three-phase, Y-connected, synchronous motor is $8\ \Omega$/phase and the armature-winding resistance is negligible. At full load, the motor takes 90 kW at 0.8 pf leading. If the load torque is held constant but the field excitation is increased by 30%, determine (a) the new torque angle, (b) the armature current, and (c) the power factor.

8.22. Obtain the V curve for the synchronous motor of Problem 8.21 at its full load. Assume that the excitation voltage per phase is 500 times the field current. Vary the power factor from 0.5 lagging to 0.5 leading.

8.23. A 460-V, Y-connected, three-phase, synchronous motor takes 30 kW at full load when the power factor is unity. The synchronous reactance is $4\ \Omega$/phase and the winding resistance is negligible. For a constant power output at 0.8 pf leading, what must be the change in the excitation voltage? What are the new power, angle, armature current, and power factor?

8.24. A 4160-V, three-phase, Y-connected source supplies power to two inductive loads connected in parallel via a three-wire transmission line. Load A is Y-connected and has an impedance of $30 + j40\ \Omega$/phase. Load B is Δ-connected and has an impedance of $150 + j360\ \Omega$/phase. For each load, determine (a) the current, power, and power factor, (b) the total power supplied by the source, and (c) the power factor of the combined load. A Y-connected synchronous condenser that takes 30 kW at 0.2 pf leading is connected in parallel with the source. What is the current intake of the motor? What is the overall power factor?

8.25. A company is expending 500 kVA at 0.45 pf lagging on a regular basis. A synchronous motor is placed in parallel with the supply in order to improve the power factor. If the power intake of the motor is 30 kW at 0.1 pf leading, determine the new power factor.

8.26. A 2.4-MVA load operates at an average power factor of 0.6 lagging. What must be the kVA rating and the power factor of a synchronous condenser in order to obtain an overall power factor of (a) 0.85 lagging and (b) unity? The winding resistance of the synchronous condenser is negligible, and the rotational loss is 24 kW. What conclusions can be drawn by comparing the kVA ratings of the synchronous motor?

8.27. The average power requirement of a plant is 1500 kVA at a power factor of 0.707 lagging. A 200-hp, 90% efficient, synchronous motor is added to operate a new assembly line. When the motor operates at its rated load and takes 300 kVA, what is the overall power factor of the plant?

8.28. A load takes 3 MVA at 0.5 pf lagging. A 400-hp, 80% efficient, synchronous motor is installed. When the motor operates at its rated load, the overall power factor improves to 0.866 lagging. Determine the kVA rating and the power factor of the motor.

CHAPTER 9

Polyphase Induction Motors

Sectional view of a three-phase, high efficiency, induction motor. (*Courtesy, MagneTek*)

9.1 Introduction

The direct-current (dc) and synchronous motors we have discussed thus far have one thing in common: both are the doubly fed type. These motors have direct current in their field windings and alternating current (ac) in their armature windings. Since the electrical power is delivered directly to the armature of a dc motor via a commutator, it can also be referred to as a **conduction motor**.

We now consider a motor in which the rotor receives its power not by conduction but by induction and is therefore called an **induction motor**. A winding that receives its power exclusively by induction constitutes a transformer. Therefore, an induction motor is a transformer with a rotating secondary winding. From the above discussion, the following must be evident:

1. An induction motor is a singly fed motor. Therefore, it does not require a commutator, slip-rings, or brushes. In fact, there are no moving contacts between the stator and the rotor. This results in a motor that is rugged, reliable, and almost maintenance free. Thus, an induction motor has a relatively high efficiency.
2. The absence of brushes eliminates the electrical loss due to the brush voltage drop and the mechanical loss due to friction between the brushes and commutator or the slip-rings.
3. An induction motor carries alternating current in both the stator and the rotor windings.
4. An induction motor is a rotating transformer in which the secondary winding receives energy by induction while it rotates.

There are two basic types of induction motors: single-phase induction motors and polyphase induction motors. Single-phase induction motors are favored for domestic applications. A large number of these motors are built in the fractional-horsepower range. We will discuss single-phase induction motors in Chapter 10. On the other hand, polyphase induction motors cover the entire spectrum of horsepower ratings and are preferably installed at locations where a polyphase power source is easily accessible.

Owing to the widespread generation and transmission of three-phase power, most polyphase induction motors are of the three-phase type. In this chapter, we confine our discussion exclusively to three-phase induction motors. The theoretical development, however, can be easily extended to an n-phase induction motor where $n \geq 2$.

9.2 Construction

The essential components of an induction motor are a **stator** and a **rotor**.

Stator

The outer (stationary) member of an induction motor is called the stator and is formed by stacking thin-slotted, highly permeable steel laminations inside a steel or cast-iron frame. The frame provides mechanical support to the motor. Although the frame is made of a magnetic material, it is not designed to carry magnetic flux.

Identical coils are wound (or placed) into the slots and then connected to form a balanced three-phase winding. For further insight on the placement of the coils and their internal connections, refer to Chapter 7.

Rotor

The rotor is also composed of thin-slotted, highly permeable steel laminations that are pressed together onto a shaft. There are two types of rotors: a **squirrel-cage rotor** and a **wound rotor**.

The **squirrel-cage rotor** is commonly used when the load requires little starting torque. For small motors, such a winding is molded by forcing a molten conducting material (quite often, aluminum) into the slots in a die-casting process. Circular rings called the **end-rings** are also formed on both sides of the stack. These end-rings short-circuit the bars on both ends of the rotor, as explained in Chapter 8, where we referred to the squirrel-cage winding as a damper winding. For large motors, the squirrel-cage winding is formed by inserting heavy conducting bars (usually of copper, aluminum, or their alloys) into the slots and then welding or bolting them to the end-rings.

Each pair of poles has as many rotor phases as there are bars because each bar behaves independently of the other. It is a common practice to **skew** the rotor laminations to reduce cogging and electrical noise in the motor. We will have more to say about it in a later section.

It becomes necessary to use a **wound rotor** when the load requires a high starting torque. A wound rotor must have as many poles and phases as the stator. In fact, the placement of coils in a wound rotor is no different from that in the stator. The three-phase windings on the rotor are internally connected to form an **internal neutral** connection. The other three ends are connected to the slip-rings, as explained in Chapter 8. With the brushes riding on the slip-rings, we can add external resistances in the rotor circuit. In this way the total resistance in the rotor circuit can be controlled. By controlling the resistance in the rotor circuit we are, in fact, controlling the torque developed by the motor. We will show that the speed at which an induction motor develops the maximum torque (called the **breakdown speed**) depends upon the rotor resistance. As the rotor resistance increases, the breakdown speed decreases. Therefore, it is possible to obtain maximum torque at starting (zero speed) by inserting just the right amount of resistance in the rotor circuit. However, a wound-rotor induction motor is more expensive and less efficient than a squirrel-cage induction motor of the same rating.

For these reasons, a wound-rotor induction motor is used only when a squirrel-cage induction motor cannot deliver the high starting torque demanded by the load.

9.3 Principle of Operation

When the stator winding of a three-phase induction motor is connected to a three-phase power source, it produces a magnetic field that (a) is constant in magnitude and (b) revolves around the periphery of the rotor at the synchronous speed. The details of how the revolving field is produced and the torque is developed are given in Chapter 3. A brief review is presented here.

If f is the frequency of the current in the stator winding and P is the number of poles, the synchronous speed of the revolving field is

$$N_s = \frac{120f}{P} \tag{9.1a}$$

in revolutions per minute (rpm), or

$$\omega_s = \frac{4\pi f}{P} \tag{9.1b}$$

in radians per second.

The revolving field induces electromotive force (emf) in the rotor winding. Since the rotor winding forms a closed loop, the induced emf in each coil gives rise to an induced current in that coil. When a current-carrying coil is immersed in a magnetic field, it experiences a force (or torque) that tends to rotate it. The torque thus developed is called the **starting torque**. If the load torque is less than the starting torque, the rotor starts rotating. The force developed and thereby the rotation of the rotor are in the same direction as the revolving field. This is in accordance with Faraday's law of induction. Under no load, the rotor soon achieves a speed nearly equal to the synchronous speed. However, the rotor can never rotate at the synchronous speed because the rotor coils would appear stationary with respect to the revolving field and there would be no induced emf in them. In the absence of an induced emf in the rotor coils, there would be no current in the rotor conductors and consequently no force would be experienced by them. In the absence of a force, the rotor would tend to slow down. As soon as the rotor slows down, the induction process takes over again. In summary, the rotor receives its power by induction only when there is a relative motion between the rotor speed and the revolving field. Since the rotor rotates at a speed lower than the synchronous speed of the revolving field, an induction motor is also called an **asynchronous motor**.

Let N_m (or ω_m) be the rotor speed at a certain load. With respect to the motor, the revolving field is moving ahead at a relative speed of

$$N_r = N_s - N_m \tag{9.2a}$$

or

$$\omega_r = \omega_s - \omega_m \tag{9.2b}$$

The relative speed is also called the **slip speed**. This is the speed with which the rotor is slipping behind a point on a fictitious revolving pole in order to produce torque. However, it is a common practice to express slip speed in terms of the slip (s), which is a ratio of the slip speed to the synchronous speed. That is,

$$s = \frac{N_r}{N_s} = \frac{\omega_r}{\omega_s}$$

or

$$s = \frac{N_s - N_m}{N_s} = \frac{\omega_s - \omega_m}{\omega_s} \tag{9.3}$$

Although the above equation yields the slip on a per-unit basis, it is customary to express it as a percentage of synchronous speed (**percent slip**).

In terms of the synchronous speed and the per-unit slip, we can express the rotor speed as

$$N_m = (1 - s)N_s \tag{9.4a}$$

or

$$\omega_m = (1 - s)\omega_s \tag{9.4b}$$

When the rotor is stationary, the per-unit slip is 1 and the rotor appears exactly like a short-circuited secondary winding of a transformer. The frequency of the induced emf in the rotor winding is the same as that of the revolving field. However, when the rotor rotates, it is the relative speed of the rotor N_r (or ω_r) that is responsible for the induced emf in its windings. Thus, the frequency of the induced emf is

$$\begin{aligned} f_r &= \frac{PN_r}{120} \\ &= \frac{P(N_s - N_m)}{120} = \frac{PN_s}{120}\left[\frac{N_s - N_m}{N_s}\right] \\ &= sf \end{aligned} \tag{9.5}$$

The above equation highlights the fact that the rotor frequency depends upon the slip of the motor. At standstill, the slip is 1 and the rotor frequency is the same as that of the revolving field. However, the rotor frequency decreases with the decrease in the slip. As the slip approaches zero, so does the rotor frequency. An induction motor usually operates at low slip. Hence the frequency of the induced emf in the rotor is low. For this reason, the core loss in the rotor magnetic circuit is most often ignored.

EXAMPLE 9.1

A 208-V, 60-Hz, 4-pole, three-phase induction motor has a full-load speed of 1755 rpm. Calculate (a) its synchronous speed, (b) the slip, and (c) the rotor frequency.

● SOLUTION

(a) The synchronous speed of the induction motor is

$$N_s = \frac{120 \times 60}{4} = 1800 \text{ rpm}$$

(b) At full load, the slip is

$$s = \frac{1800 - 1755}{1800} = 0.025 \quad \text{or} \quad 2.5\%$$

(c) The rotor frequency at full load is

$$f_r = 0.025 \times 60 = 1.5 \text{ Hz}$$

■

Exercises

9.1. The rotor speed of a 440-V, 50-Hz, 8-pole, three-phase induction motor is 720 rpm. Determine (a) the synchronous speed, (b) the slip, and (c) the rotor frequency.

9.2. If the rotor frequency of a 6-pole, 50-Hz, three-phase induction motor is 3 Hz, determine (a) the slip and (b) the rotor speed.

9.3. The magnetic field produced by a three-phase induction motor revolves at a speed of 900 rpm. If the frequency of the applied voltage is 60 Hz, determine the number of poles in the motor. When the rotor turns at a speed of 800 rpm, what is the percent slip of the motor?

9.4 Development of an Equivalent Circuit

When a balanced three-phase induction motor is excited by a balanced three-phase source, the currents in the phase windings must be equal in magnitude and 120° electrical apart in phase. The same must be true for the currents in the rotor windings as the energy is transferred across the air-gap from the stator to the rotor by induction. However, the frequency of the induced emf in the rotor is proportional to its slip [Eq. (9.5)]. Since the stator and the rotor windings are coupled inductively, an induction motor resembles a three-phase transformer with a rotating secondary winding. The similarity becomes even more striking when the rotor is at rest (**blocked-rotor condition**, $s = 1$). Thus, a three-phase induction motor can be represented on a per-phase basis by an equivalent circuit at any slip s as depicted in Figure 9.1. In this figure,

$\tilde{V}_1$ = applied voltage on a per-phase basis
R_1 = per-phase stator winding resistance
L_1 = per-phase stator winding leakage inductance
$X_1 = 2\pi f L_1$ = per-phase stator winding leakage reactance
R_r = per-phase rotor winding resistance
L_b = per-phase rotor winding leakage inductance
$X_b = 2\pi f L_b$ = per-phase rotor winding leakage reactance under blocked-rotor condition ($s = 1$)
$X_r = 2\pi s f L_b = sX_b$ = per-phase rotor winding leakage reactance at slip s.
X_m = per-phase magnetization reactance
R_c = per-phase equivalent core-loss resistance
N_1 = actual turns per phase of the stator winding
N_2 = actual turns per phase of the rotor winding
k_{w1} = winding factor for the stator winding
k_{w2} = winding factor for the rotor winding
Φ_m = amplitude of the per-phase flux
$\tilde{E}_1 = 4.44 f N_1 k_{w1} \Phi_m$ = per-phase induced emf in the stator winding

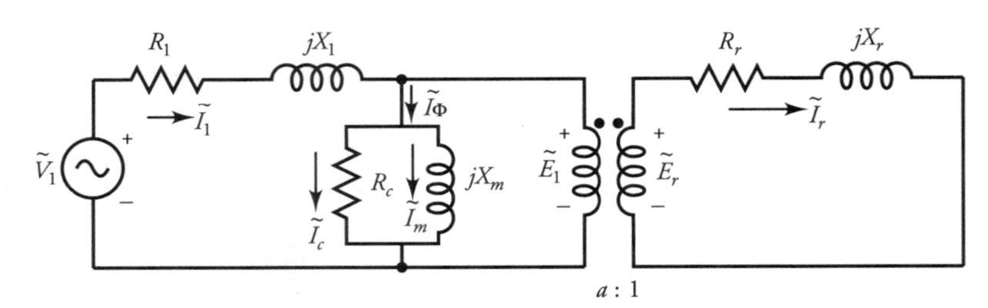

Figure 9.1 Per-phase equivalent circuit of a balanced three-phase induction motor.

$\tilde{E}_b = 4.44 f N_2 k_{w2} \Phi_m$ = per-phase induced emf in the rotor winding under blocked-rotor condition ($s = 1$)
$\tilde{E}_r = s\tilde{E}_b$ = per-phase induced emf in the rotor winding at slip s
$\tilde{I}_r$ = per-phase rotor winding current
$\tilde{I}_1$ = per-phase current supplied by the source
$\tilde{I}_\phi = \tilde{I}_c + \tilde{I}_m$ = per-phase excitation current
$\tilde{I}_c$ = per-phase core-loss current
$\tilde{I}_m$ = per-phase magnetization current

From the per-phase equivalent circuit (Figure 9.1), it is evident that the current in the rotor circuit is

$$\tilde{I}_r = \frac{\tilde{E}_r}{R_r + jX_r} = \frac{s\tilde{E}_b}{R_r + jsX_b} = \frac{\tilde{E}_b}{(R_r/s) + jX_b} \tag{9.6}$$

Based upon the above equation, we can develop another circuit of an induction motor as given in Figure 9.2. In this circuit, the hypothetical resistance R_r/s in the rotor circuit is called the **effective resistance**. The effective resistance is the same as the actual rotor resistance when the rotor is at rest (standstill or blocked-rotor condition). On the other hand, when the slip approaches zero under no-load condition, the effective resistance is very high ($R_r/s \to \infty$).

By defining the ratio of transformation, the a-ratio, as

$$a = \frac{N_1 k_{w1}}{N_2 k_{w2}} \tag{9.7}$$

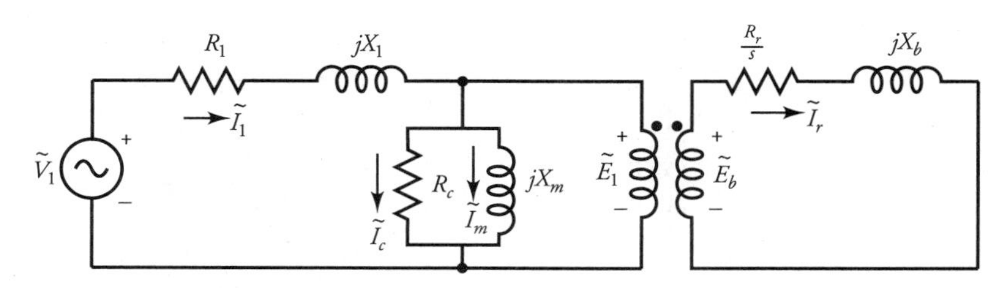

Figure 9.2 Modified equivalent circuit of a balanced three-phase motor on a per-phase basis.

we can represent the induction motor by its per-phase equivalent circuit as referred to the stator. Such an equivalent circuit is shown in Figure 9.3, where

$$R_2 = a^2 R_r \tag{9.8a}$$

$$X_2 = a^2 X_b \tag{9.8b}$$

and

$$\tilde{I}_2 = \frac{\tilde{I}_r}{a} \tag{9.8c}$$

For this equivalent circuit

$$\tilde{E}_1 = \frac{\tilde{I}_2 R_2}{s} + j\tilde{I}_2 X_2$$

$$\tilde{I}_\phi = \tilde{I}_c + \tilde{I}_m$$

where

$$\tilde{I}_c = \frac{\tilde{E}_1}{R_c} \quad \text{and} \quad \tilde{I}_m = \frac{\tilde{E}_1}{jX_m}$$

The per-phase stator winding current and the applied voltage are

$$\tilde{I}_1 = \tilde{I}_\phi + \tilde{I}_2$$

and

$$\tilde{V}_1 = \tilde{E}_2 + \tilde{I}_1(R_1 + jX_1)$$

The equivalent circuit of the rotor in Figure 9.3 is in terms of the hypothetical resistance R_2/s. In this circuit, $I_2^2 R_2/s$ represents the per-phase power delivered to

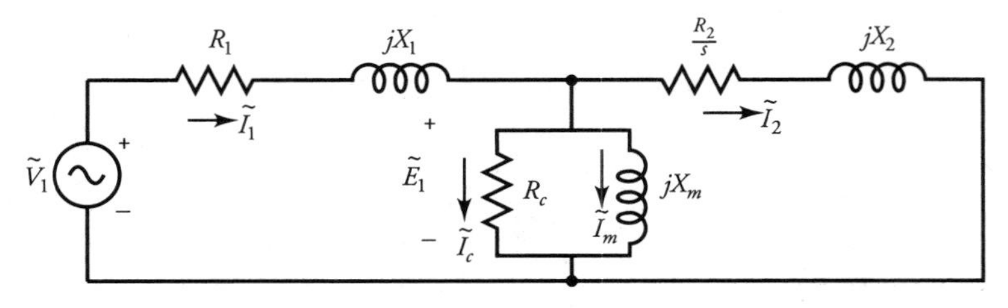

Figure 9.3 Per-phase equivalent circuit of a balanced three-phase induction motor as referred to the stator side.

the rotor. However, the per-phase copper loss in the rotor must be $I_2^2R_2$. Thus, the per-phase power developed by the motor is

$$I_2^2\frac{R_2}{s} - I_2^2R_2 = I_2^2R_2\left[\frac{1 - s}{s}\right]$$

or

$$\frac{R_2}{s} = R_2 + R_2\left[\frac{1 - s}{s}\right] \tag{9.9}$$

The above equation establishes the fact that the hypothetical resistance R_2/s can be divided into two components: the actual resistance of the rotor R_2 and in series with an additional resistance R_2 $[(1 - s)/s]$. The additional resistance is called the **load resistance** or the **dynamic resistance**. The load resistance depends upon the speed of the motor and is said to represent the load on the motor because the mechanical power developed by the motor is proportional to it. In other words, the load resistance is the electrical equivalent of a mechanical load on the motor.

An equivalent circuit of an induction motor in terms of the load resistance is given in Figure 9.4. This circuit is proclaimed as the **exact equivalent circuit of a balanced three-phase induction motor on a per-phase basis**.

Power Relations

Since the load resistance varies with the slip and the slip adjusts itself to the mechanical load on the motor, the power delivered to the load resistance is equivalent to the power developed by the motor. Thus, the performance of the motor at any slip can be determined from its equivalent circuit, as given in Figure 9.4.

For a balanced three-phase induction motor

$$P_{in} = 3V_1I_1 \cos \theta \tag{9.10}$$

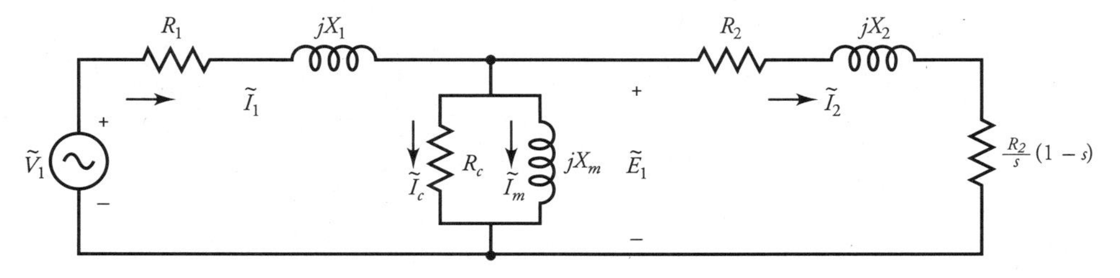

Figure 9.4 The equivalent circuit of Figure 9.3 modified to show the rotor and the load resistances.

where θ is the phase difference between the applied voltage $\tilde{V}_1$ and the stator winding current $\tilde{I}_1$. Since the power input is electrical in nature, we must account for the electrical losses first. The immediate electrical loss that must be taken into consideration is the stator copper loss.

The total stator copper loss is

$$P_{sc\ell} = 3I_1^2R_1 \tag{9.11}$$

If the core loss is modeled by an equivalent core-loss resistance, as shown in the figure, we must also take into account the total core loss (magnetic loss) as

$$P_m = 3I_c^2R_c \tag{9.12}$$

The net power that is crossing the air-gap and is transported to the rotor by electromagnetic induction is called the **air-gap power**. In this case, the air-gap power is

$$P_{ag} = P_{in} - P_{sc\ell} - P_m \tag{9.13a}$$

The air-gap power must also equal the power delivered to the hypothetical resistance R_2/s. That is,

$$P_{ag} = \frac{3I_2^2R_2}{s} \tag{9.13b}$$

The electrical power loss in the rotor circuit is

$$P_{rc\ell} = 3I_2^2R_2 = sP_{ag} \tag{9.14}$$

Hence, the power developed by the motor is

$$\begin{aligned} P_d &= P_{ag} - P_{rc\ell} \\ &= \frac{3I_2^2(1 - s)R_2}{s} = (1 - s)P_{ag} = SP_{ag} \end{aligned} \tag{9.15}$$

where

$$S = 1 - s = \frac{N_m}{N_s} = \frac{\omega_m}{\omega_s}$$

is the per-unit (normalized) speed of the motor.

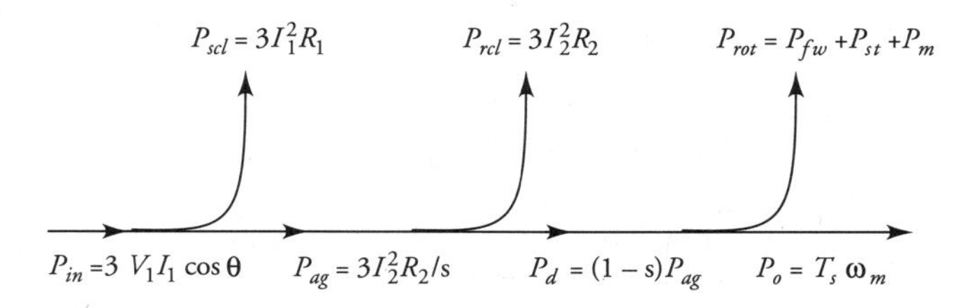

Figure 9.5 Power-flow diagram when the core loss is treated as a part of the rotational loss.

The electromagnetic torque developed by the motor is

$$T_d = \frac{P_d}{\omega_m} = \frac{P_{ag}}{\omega_s} = 3I_2^2\,\frac{R_2}{s\omega_s} \tag{9.16}$$

By subtracting the rotational loss from the power developed, we obtain the power output of the motor as

$$P_o = P_d - P_{rot} \tag{9.17}$$

Since the core loss has already been accounted for, the rotational loss includes the friction and windage loss P_{fw} and the stray-load loss P_{st}. When the core loss P_m is also considered a part of the rotational loss, the core-loss resistance R_c in Figure 9.4 must be omitted. The power-flow diagram when the core loss is a part of the rotational loss is given in Figure 9.5.

Speed-Torque Curve

Equation (9.16) reveals that the torque developed by an induction motor is directly proportional to the square of the current in the rotor circuit and the equivalent hypothetical resistance of the rotor. However, the two quantities, the rotor current and the hypothetical rotor resistance, are inversely related to each other. For instance, if the rotor resistance is increased, we expect the torque developed by the motor to increase linearly. But any increase in the rotor resistance is accompanied by a decrease in the rotor current for the same induced emf in the rotor. A decrease in the rotor current causes a reduction in the torque developed. Whether the overall torque developed increases or decreases depends upon which parameter plays a dominant role.

Let us try to examine the situation from standstill to a no-load condition. At standstill, the rotor slip is unity and the effective rotor resistance is R_2. The magnitude of the rotor current is

$$I_2 = \frac{E_b}{\sqrt{R_2^2 + X_2^2}} \tag{9.18}$$

Note that the rotor winding resistance R_2 is usually very small compared with its leakage reactance X_2. That is, $R_2 << X_2$.
The **starting torque** developed by the motor is

$$T_{ds} = 3I_2^2 \frac{R_2}{\omega_s} \tag{9.19}$$

As the rotor starts rotating, an increase in its speed is accompanied by a decrease in its slip. As s decreases, R_2/s increases. As long as R_2/s is smaller than X_2, the reduction in the rotor current is minimal. Thus, in this speed range, the rotor current may be approximated as

$$I_2 \approx \frac{E_b}{X_2} \tag{9.20}$$

Since the rotor current is almost constant, the torque developed by the motor increases with the increase in the effective resistance R_2/s. Thus, the torque developed by the motor keeps increasing with the decrease in the slip as long as the rotor resistance has little influence on the rotor current.

When the slip falls below a certain value called the **breakdown slip** s_b, the hypothetical resistance becomes the dominating factor. In this range, $R_2/s >> X_2$ and the rotor current can be approximated as

$$I_2 = \frac{sE_b}{R_2} \tag{9.21}$$

The torque developed by the motor is now proportional to the slip s. As the slip decreases, so does the torque developed. At no load, the slip is almost zero, the hypothetical rotor resistance is nearly infinite, the rotor current is approximately zero, and the torque developed is virtually zero. With this understanding, we are able to sketch the speed-torque curve of an induction motor. Such a curve is depicted in Figure 9.6.

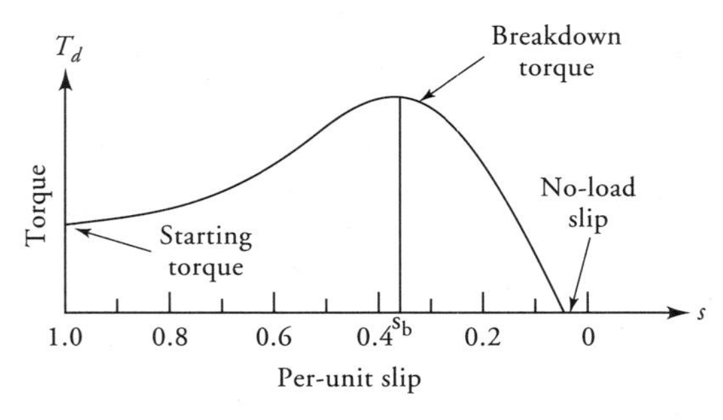

Figure 9.6 Typical speed-torque characteristic of a three-phase induction motor.

EXAMPLE 9.2

A 6-pole, 230-V, 60-Hz, Y-connected, three-phase induction motor has the following parameters on a per-phase basis: $R_1 = 0.5\ \Omega$, $R_2 = 0.25\ \Omega$, $X_1 = 0.75\ \Omega$, $X_2 = 0.5\ \Omega$, $X_m = 100\ \Omega$, and $R_c = 500\ \Omega$. The friction and windage loss is 150 W. Determine the efficiency of the motor at its rated slip of 2.5%.

● SOLUTION

The synchronous speed of the motor is

$$N_s = \frac{(120 \times 60)}{6} = 1200 \text{ rpm} \qquad \text{or} \qquad \omega_s = 125.66 \text{ rad/s}$$

The per-phase applied voltage is

$$V_1 = \frac{230}{\sqrt{3}} = 132.791 \text{ V}$$

The effective rotor impedance as referred to the stator is

$$\hat{Z}_2 = \frac{R_2}{s} + jX_2 = \frac{0.25}{0.025} + j0.5 = 10 + j0.5\ \Omega$$

The stator winding impedance is

$$\hat{Z}_1 = R_1 + jX_1 = 0.5 + j0.75\ \Omega$$

Since R_c, jX_m, and $\hat{Z}_2$ are in parallel, we can compute the equivalent impedance $\hat{Z}_e$ as

$$\frac{1}{\hat{Z}_e} = \frac{1}{500} + \frac{1}{j100} + \frac{1}{10 + j0.5}$$

$$= 0.102 - j0.015 \text{ S}$$

or

$$\hat{Z}_e = 9.619 + j1.417\ \Omega$$

Hence, the total input impedance is

$$\hat{Z}_{in} = \hat{Z}_1 + \hat{Z}_e = 10.119 + j2.167\ \Omega$$

The stator current: $\tilde{I}_1 = \dfrac{\tilde{V}_1}{\hat{Z}_{in}} = 12.832\underline{/-12.09°}$ A

The power factor: pf = cos (12.09°) = 0.978 lagging

Power input: $P_{in} = 3V_1I_1 \cos\theta = 4998.463$ W

Stator copper loss: $P_{sc\ell} = 3I_1^2R_1 = 246.978$ W

$$\tilde{E}_1 = \tilde{V}_1 - \tilde{I}_1\hat{Z}_1 = 124.763\underline{/-3.71°}\ \text{V}$$

Core-loss current: $\tilde{I}_c = \dfrac{\tilde{E}_1}{R_c} = 0.25\underline{/-3.71°}$ A

Magnetization current: $\tilde{I}_m = \dfrac{\tilde{E}_1}{jX_m} = 1.248\underline{/-93.71°}$ A

Excitation current: $\tilde{I}_\phi = \tilde{I}_c + \tilde{I}_m = 1.272\underline{/-82.40°}$ A

Hence, the rotor current: $\tilde{I}_2 = \tilde{I}_1 - \tilde{I}_\phi = 12.461\underline{/-6.57°}$ A

Core loss: $P_m = 3I_c^2R_c = 93.395$ W

Air-gap power: $P_{ag} = P_{in} - P_{sc\ell} - P_m = 4658.091$ W

Rotor copper loss: $P_{rc\ell} = 3I_2^2R_2 = 116.452$ W

Power developed: $P_d = P_{ag} - P_{rc\ell} = 4541.638$ W

Power output: $P_o = P_d - 150 = 4391.638$ W

Efficiency: $\eta = \dfrac{P_o}{P_{in}} = 0.879$ or 87.9%

Shaft torque: $T_s = \dfrac{P_o}{\omega_m} = \dfrac{P_o}{(1 - s)\omega_s}$

$$= \frac{4391.638}{(1 - 0.025) \times 125.66} = 35.845 \text{ N·m}$$

■

Exercises

9.4. A 10-hp, 4-pole, 440-V, 60-Hz, Y-connected, three-phase induction motor runs at 1725 rpm on full load. The stator copper loss is 212 W, and the rotational loss is 340 W. Determine (a) the power developed, (b) the air-gap power, (c) the rotor copper loss, (d) the total power input, and (e) the efficiency of the motor. What is the shaft torque?

9.5. A 2-hp, 120-V, 60-Hz, 4-pole, Y-connected, three-phase induction motor operates at 1650 rpm on full load. The rotor impedance at standstill is $0.02 + j0.06$ Ω/phase. Determine the rotor current if the rotational loss is 160 W. What is the magnitude of the induced emf in the rotor?

9.6. A 208-V, 50-Hz, 12-pole, Y-connected, three-phase induction motor has a stator impedance of $0.1 + j0.3$ Ω/phase and a rotor impedance of $0.06 + j0.8$ Ω/phase at standstill. The core-loss resistance is 150 Ω, and the magnetization reactance is 750 Ω. The friction and windage loss is 2 kW. When the motor operates at its full-load slip of 5%, determine (a) the power input, (b) the stator copper loss, (c) the rotor copper loss, (d) the air-gap power, (e) the power developed, (f) the power output, (g) the efficiency, (h) the shaft torque, and (i) the horsepower rating of the motor.

9.5 An Approximate Equivalent Circuit

A well-designed three-phase induction motor usually meets most of the following guidelines:

1. The stator winding resistance is kept small in order to reduce the stator copper loss.
2. The stator winding leakage reactance is minimized by reducing the mean-turn length of each coil.

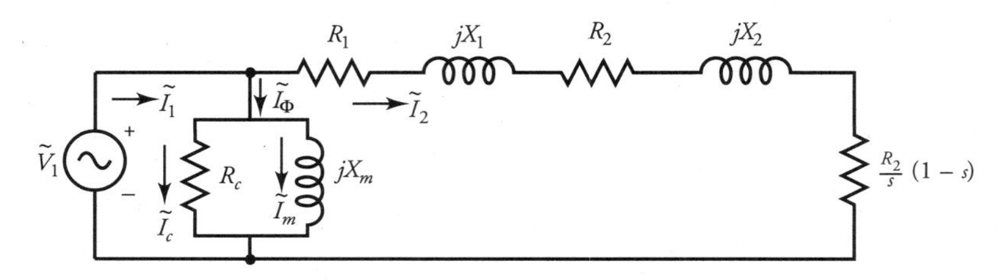

Figure 9.7 An approximate equivalent circuit on a per-phase basis of a balanced three-phase induction motor.

3. Thin laminations of low-loss steel are used to cut down the core loss. Thus, the equivalent core-loss resistance is usually high.
4. The permeability of steel selected for laminations is high, and the operating flux density in the motor is kept below the knee of the magnetization curve. Thus, the magnetization reactance is usually high.

An induction motor conforming to the above stipulations can be represented by an approximate equivalent circuit, as shown in Figure 9.7. In this case, we have placed the parallel branch (the excitation circuit) across the power source. We admit that the analysis of an induction motor using the approximate equivalent circuit is somewhat inaccurate, but the inaccuracy is negligible for a well-designed motor. On the other hand, the approximate equivalent circuit not only simplifies the analysis but also aids in comprehending various characteristics of the motor. For instance, we use the approximate equivalent circuit to determine the speed at which (a) the torque developed is maximum, (b) the power developed is maximum, and (c) the motor efficiency is maximum. Prior to proceeding further, let us examine the error introduced when an induction motor is analyzed using an approximate equivalent circuit.

EXAMPLE 9.3

Using the data of Example 9.2 and the approximate equivalent circuit, determine the efficiency of the motor at its rated slip.

SOLUTION

Core-loss current: $\tilde{I}_c = \dfrac{132.791}{500} = 0.266 \text{ A}$

Magnetization current: $\tilde{I}_m = \dfrac{132.791}{j100} = -j1.328 \text{ A}$

The equivalent impedance of the series circuit is

$$\hat{Z}_e = R_1 + \frac{R_2}{s} + j(X_1 + X_2) = 10.5 + j1.25\ \Omega$$

Hence, the rotor current is

$$\tilde{I}_2 = \frac{132.791}{10 + j1.25} = 12.558\underline{/-6.79^\circ}\ \text{A}$$

The per-phase current supplied by the source is

$$\tilde{I}_1 = \tilde{I}_2 + \tilde{I}_c + \tilde{I}_m = 13.042\underline{/-12.45^\circ}\ \text{A}$$

Power input: $P_{in} = 3\text{Re}[132.791 \times 13.042\underline{/12.45^\circ}]$

$= 5073.525$ W

Stator copper loss: $P_{sc\ell} = 3I_2^2R_1 = 236.558$ W

Rotor copper loss: $P_{rc\ell} = 3I_2^2R_2 = 118.279$ W

Core loss: $P_m = 3I_c^2R_c = 105.801$ W

Power output: $P_o = P_{in} - P_{sc\ell} - P_{rc\ell} - P_m - 150$

$= 4462.887$ W

Finally, the efficiency: $\eta = \dfrac{4462.887}{5073.525} = 0.8796$ or 87.96%

■

Exercises

9.7. Redo Exercise 9.6 using the approximate equivalent circuit.

9.8. The equivalent circuit parameters of a 208-V, 60-Hz, 6-pole, Y-connected, three-phase induction motor in ohms/phase are $R_1 = 0.21$, $R_2 = 0.33$, $X_1 = 0.6$, $X_2 = 0.6$, $R_c = 210$, and $X_m = 450$. When the motor runs at a slip of 5% on full load, determine the torque developed by the motor using the approximate equivalent circuit. What is the starting torque developed by the motor?

9.6 Maximum Power Criterion

From the equivalent circuit as given in Figure 9.7, the rotor current is

$$\tilde{I}_2 = \frac{\tilde{V}_1}{R_e + jX_e + R_2(1 - s)/s} \tag{9.22}$$

where $$R_e = R_1 + R_2$$

and $$X_e = X_1 + X_2$$

The power developed by the three-phase induction motor, from Eq. (9.15), is

$$P_d = \frac{3V_1^2 R_2(1 - s)/s}{R_e^2 + X_e^2 + [R_2(1 - s)/s]^2 + 2R_e R_2(1 - s)/s} \tag{9.23}$$

From the above equation it is evident that the power developed by a three-phase induction motor is a function of slip. Therefore, we can determine the slip s_p at which the power developed by the motor is maximum by differentiating the above equation and setting the derivative equal to zero. After differentiating and canceling most of the terms, we obtain

$$R_e^2 + X_e^2 = \left[\frac{R_2}{s_p}(1 - s_p)\right]^2$$

or (9.24)

$$Z_e = \frac{R_2}{s_p}(1 - s_p)$$

where Z_e is the magnitude of the equivalent impedance of the stator and the rotor windings at rest. That is

$$Z_e = |R_e + jX_e| \tag{9.25}$$

Equation (9.24) states that the power developed by a three-phase induction motor is maximum when the load (dynamic) resistance is equal to the magnitude of the standstill impedance of the motor. This, of course, is the well-known result we obtained from the **maximum power transfer theorem** during the study of electrical circuit theory.

From Eq. (9.24) we obtain the slip at which the induction motor develops maximum power as

$$s_p = \frac{R_2}{R_2 + Z_e} \tag{9.26}$$

Substituting for the slip in Eq. (9.23), we obtain an expression for the maximum power developed by a three-phase induction motor as

$$P_{dm} = \frac{3}{2}\left[\frac{V_1^2}{R_e + Z_e}\right] \tag{9.27}$$

The net power output, however, is less than the power developed by an amount equal to the rotational loss of the motor.

EXAMPLE 9.4

A 120-V, 60-Hz, 6-pole, Δ-connected, three-phase induction motor has a stator impedance of $0.1 + j0.15\ \Omega$/phase and an equivalent rotor impedance of $0.2 + j0.25\ \Omega$/phase at standstill. Find the maximum power developed by the motor and the slip at which it occurs. What is the corresponding value of the torque developed by the motor?

● SOLUTION

$$\begin{aligned}\hat{Z}_e &= R_e + jX_e = R_1 + R_2 + j(X_1 + X_2)\\ &= 0.1 + 0.2 + j(0.15 + 0.25) = 0.5\underline{/53.13^\circ}\ \Omega\end{aligned}$$

From Eq. (9.26), the slip at which the motor develops maximum power is

$$s_p = \frac{0.2}{0.2 + 0.5} = 0.286 \quad \text{or} \quad 28.6\%$$

The maximum power developed by the motor is

$$P_{dm} = \frac{3 \times 120^2}{2(0.3 + 0.5)} = 27{,}000\ \text{W} \quad \text{or} \quad 27\ \text{kW}$$

The synchronous speed of the motor is

$$N_s = \frac{120 \times 60}{6} = 1200 \text{ rpm} \qquad \text{or} \qquad 125.664 \text{ rad/s}$$

The motor speed: $\omega_m = (1 - s_p)\omega_s = (1 - 0.286)125.664 = 89.724$ rad/s

Thus, the torque developed by the motor is

$$T_d = \frac{P_{dm}}{\omega_m} = \frac{27{,}000}{89.724} = 300.923 \text{ N·m}$$

■

Exercises

9.9. A 440-V, 50-Hz, 8-pole, three-phase, Δ-connected induction motor has a stator impedance of $0.05 + j0.25$ Ω/phase and an equivalent rotor impedance of $0.15 + j0.35$ Ω/phase at standstill. Determine (a) the slip at which the motor develops maximum power, (b) the maximum power developed by the motor, and (c) the corresponding value of the torque developed.

9.10. A 208-V, 60-Hz, 4-pole, three-phase, Y-connected induction motor has the following constants in ohms/phase: $R_1 = 0.08$, $X_1 = 0.5$, $R_2 = 0.1$, $X_2 = 0.6$. Determine the slip at which the motor develops maximum power. What is the maximum power developed by the motor? What is the corresponding torque developed by the motor?

9.7 Maximum Torque Criterion

The torque developed by a three-phase induction motor, from Eq. (9.23), is

$$T_d = \frac{\dfrac{3V_1^2R_2}{s}}{\left[R_e^2 + X_e^2 + \left[\dfrac{R_2(1 - s)}{s}\right]^2 + \dfrac{2R_eR_2(1 - s)}{s}\right]\omega_s} \tag{9.28}$$

where $R_e = R_1 + R_2$ and $X_e = X_1 + X_2$.

Differentiating the above equation with respect to s and setting it equal to

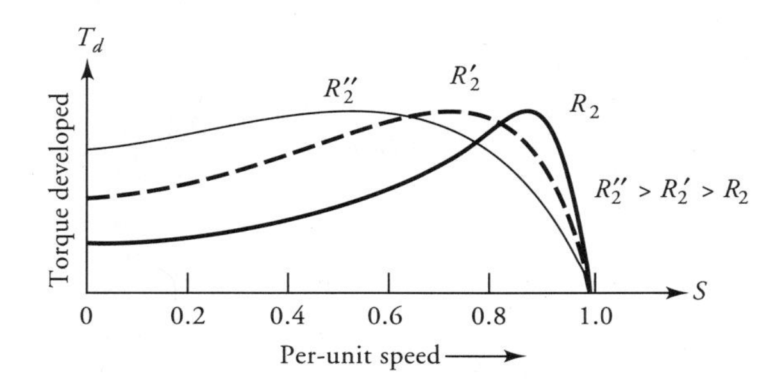

Figure 9.8 Effect of rotor resistance on the breakdown slip.

zero, we obtain an expression for the breakdown slip s_b at which the motor develops the maximum (breakdown) torque as

$$s_b = \frac{R_2}{\sqrt{R_1^2 + (X_1 + X_2)^2}} \tag{9.29}$$

Note that the breakdown slip is directly proportional to the rotor resistance. Since the rotor resistance can be easily adjusted in a wound-rotor induction motor by means of an external resistor, we can obtain the maximum torque at any desired speed, including the zero speed (starting). Substituting the above expression for the breakdown slip in Eq. (9.28), we obtain an expression for the maximum torque developed by the motor as

$$T_{dm} = \frac{3V_1^2}{2\omega_s}\left[\frac{1}{R_1 + \sqrt{R_1^2 + (X_1 + X_2)^2}}\right] \tag{9.30}$$

Note that the maximum torque developed by the motor is independent of the rotor resistance. In other words, the motor develops the same maximum torque regardless of its rotor resistance. The rotor resistance influences only the breakdown slip (or breakdown speed) at which the torque is maximum, as illustrated in Figure 9.8.

EXAMPLE 9.5

Using the data of Example 9.4, determine (a) the breakdown slip, (b) the breakdown torque, and (c) the corresponding power developed by the motor.

● SOLUTION

The motor parameters given in Example 9.4 are

$$R_1 = 0.1\ \Omega,\ X_1 = 0.15\ \Omega,\ R_2 = 0.2\ \Omega,\ X_2 = 0.25\ \Omega,\ V_1 = 120\ \text{V}$$

The breakdown slip from Eq. (9.29) is

$$s_b = \frac{0.2}{\sqrt{0.1^2 + (0.15 + 0.25)^2}} = 0.485$$

From Eq. (9.30), the maximum (breakdown) torque developed by the motor is

$$T_{dm} = \frac{3 \times 120^2}{2 \times 125.664}\left[\frac{1}{0.1 + \sqrt{0.1^2 + (0.15 + 0.25)^2}}\right]$$
$$= 335.513 \text{ N·m}$$

The power developed by the motor at the breakdown slip is

$$P_d = T_{dm}(1 - s_b)\omega_s$$
$$= 335.513(1 - 0.485)125.664 = 21.71 \text{ kW}$$

■

When we compare the expressions for the breakdown slip, Eq. (9.29), with the slip at which the motor develops maximum power, Eq. (9.26), we find that the denominator in Eq. (9.26) is larger than that in Eq. (9.29). In other words, the motor develops maximum power at a slip lower than that at which it develops maximum torque ($s_p < s_b$).

Further Approximations

When the stator impedance is so small that it can be neglected in comparison with the rotor impedance at standstill, we obtain a very useful expression for the breakdown slip, from Eq. (9.29), as

$$s_b = \frac{R_2}{X_2} \tag{9.31}$$

This equation states that the breakdown slip is simply a ratio of rotor resistance to rotor reactance. When the rotor resistance is made equal to the rotor reactance, the breakdown slip is unity. In this case, the motor develops the maximum torque at starting.

The approximate expression for the breakdown torque, Eq. (9.30), becomes

$$T_{dm} = \frac{3V_1^2}{2\omega_s}\left[\frac{1}{X_2}\right] = \frac{3V_1^2}{2\omega_s}\left[\frac{s_b}{R_2}\right] \tag{9.32}$$

In order to use Eq. (9.32), the induction motor must operate at a very low slip. In fact, most three-phase induction motors operate below a slip of 5%, and the approximate relationship can be used to estimate the maximum torque developed by the motor.

The rotor current at any speed when the stator impedance is neglected is

$$\tilde{I}_2 = \frac{\tilde{V}_1}{R_2/s + jX_2} \tag{9.33}$$

The torque developed by the motor at any slip s is

$$\begin{aligned} T_d &= \frac{3I_2^2R_2}{s\omega_s} \\ &= \left[\frac{3V_1^2}{(R_2/s)^2 + X_2^2}\right]\frac{R_2}{s\omega_s} \end{aligned} \tag{9.34}$$

The ratio of the torque developed at any slip s to the breakdown torque is

$$\frac{T_d}{T_{dm}} = \frac{2ss_b}{s^2 + s_b^2} \tag{9.35}$$

This is a very useful, albeit approximate, relation that can be used to determine the torque developed at any slip s once the breakdown torque T_{dm} and the breakdown slip s_b are known.

EXAMPLE 9.6

A 208-V, 60-Hz, 8-pole, Y-connected, three-phase induction motor has negligible stator impedance and a rotor impedance of $0.02 + j0.08\ \Omega$/phase at standstill. Determine the breakdown slip and the breakdown torque. What is the starting torque developed by the motor? If the starting torque of the motor has to be 80% of the maximum torque, determine the external resistance that must be added in series with the rotor.

● SOLUTION

$$s_b = \frac{0.02}{0.08} = 0.25$$

$$N_s = \frac{120 \times 60}{8} = 900 \text{ rpm} \qquad \text{or} \qquad \omega_s = 94.248 \text{ rad/s}$$

The motor speed: $N_m = (1 - s_b)N_s = (1 - 0.25)900 = 675$ rpm

$$V_1 = 120 \text{ V}$$

From Eq. (9.32),

$$T_{dm} = \frac{3 \times 120^2}{2 \times 94.248}\left[\frac{0.25}{0.02}\right] = 2864.782 \text{ N·m}$$

From Eq. (9.35), the starting torque ($s = 1$) in terms of the maximum torque is

$$T_s = \frac{2 \times 1 \times 0.25}{1 + 0.25^2} T_{dm} = 0.47\, T_{dm}$$

Since the starting torque is only 47% of the maximum torque, we must add some resistance to the rotor circuit. As the rotor resistance is increased, the breakdown slip also increases. If s_{bn} is the new breakdown slip, then

$$\frac{T_s}{T_{dm}} = \frac{2ss_{bn}}{s^2 + s_{bn}^2}$$

Since $s = 1$ at starting and $T_s/T_{dm} = 0.8$, the above equation yields

$$s_{bn}^2 - 2.5s_{bn} + 1 = 0$$

The two roots of the above quadratic equation are 2 and 0.5. Hence, the new breakdown slip must be 0.5. The corresponding value of the rotor resistance, from Eq. (9.31), is

$$R_{2n} = s_{bn}X_2 = 0.5 \times 0.08 = 0.04\ \Omega$$

Thus, the external resistance that must be added in each phase in series with the rotor resistance is 0.02 Ω.

■

Exercises

9.11. Compute the breakdown slip and the breakdown torque of the motor of Exercise 9.9. What is the corresponding power developed by the motor?

9.12. Using the data of Exercise 9.10, calculate (a) the breakdown speed, (b) the breakdown torque, and (c) the power developed by the motor at its breakdown slip.

9.13. The rotor impedance at standstill of a three-phase, Y-connected, 208-V,

60-Hz, 8-pole, induction motor is $0.1 + j0.5\ \Omega$/phase. Determine the breakdown slip, the breakdown torque, and the power developed by the motor. What is the starting torque of this motor? Determine the resistance that must be inserted in series with the rotor circuit so that the starting torque is 50% of the maximum torque.

9.8 Maximum Efficiency Criterion

When the core loss is considered a part of the rotational loss and the magnetization current is negligible, the power input is

$$P_{in} = 3V_1 I_2 \cos\theta \tag{9.36}$$

where θ is the power-factor angle between the applied voltage $\tilde{V}_1$ and the rotor current $\tilde{I}_2$.

The power output is

$$P_o = 3V_1 I_2 \cos\theta - 3I_2^2(R_1 + R_2) - P_{rot} \tag{9.37}$$

The motor efficiency is

$$\eta = \frac{3V_1 I_2 \cos\theta - 3I_2^2(R_1 + R_2) - P_{rot}}{3V_1 I_2 \cos\theta} \tag{9.38}$$

Differentiating η with respect to I_2 and setting the derivative equal to zero, we obtain

$$3I_2^2(R_1 + R_2) = P_{rot} \tag{9.39}$$

as the criterion for the maximum efficiency of an induction motor. It simply states that the efficiency of an induction motor is maximum when the sum of the stator and the rotor copper losses is equal to the rotational loss. Note that this conclusion is not quite true when the magnetization current is taken into account.

9.9 Some Important Conclusions

Prior to proceeding further, let us pause and make the following observations regarding the torque developed, the rotor current, and the motor efficiency.

The Torque Developed

When the motor is operating at or near its rated slip, which is usually less than 10%, the hypothetical rotor resistance is considerably greater than the rotor leakage reactance. That is, $R_2/s >> X_2$. We can, therefore, approximate Eq. (9.34) as

$$T_d \approx \frac{3V_1^2 s}{\omega_s R_2} \tag{9.40}$$

1. The torque developed by the motor is proportional to the slip when the applied voltage and rotor resistance are held constant. In this linear range, the ratio of torques developed is equal to the ratio of the slips.
2. The torque developed is inversely proportional to the rotor resistance at a given slip when the applied voltage is kept the same. In other words, the torque developed at any slip can be adjusted by varying the rotor resistance. This can be easily accomplished in a wound-rotor motor.
3. At a definite value of slip and rotor resistance, the torque developed by the motor is directly proportional to the square of the applied voltage.
4. For a constant-torque operation under fixed applied voltage, the motor slip is directly proportional to the rotor resistance.

The Rotor Current

From Eq. (9.33) it is obvious that the rotor current is directly proportional to the applied voltage as long as the rotor resistance and the slip are held the same.

When $R_2/s >> X_2$, the rotor current can be approximated as

$$I_2 = \frac{V_1 s}{R_2} \tag{9.41}$$

(a) The rotor current varies linearly with the slip when the motor operates at low slip.
(b) The rotor current varies inversely with the rotor resistance.

The Motor Efficiency

For an ideal motor we can assume that (a) the stator copper loss is negligible and (b) the rotational loss is zero. In this case, the air-gap power is equal to the power input. That is, $P_{in} = P_{ag}$. However, the power developed is $P_d = (1 - s) P_{ag} = SP_{ag}$, where S is the per-unit speed. Since the rotational loss is zero, the power output is equal to the power developed. Hence, the motor efficiency under the ideal conditions is

$$\eta = 1 - s = S \tag{9.42}$$

The above equation places a maximum limit on the efficiency of a three-phase induction motor. This equation highlights the fact that if a motor is operating at 60% of its synchronous speed, the maximum efficiency under the ideal conditions (theoretically possible) is 60%. Thus, the higher the speed of operation, the higher the efficiency. A motor operating at 5% slip can theoretically have an efficiency of 95%.

EXAMPLE 9.7

A 230-V, 60-Hz, 4-pole, Δ-connected, three-phase induction motor operates at a full-load speed of 1710 rpm. The power developed at this speed is 2 hp and the rotor current is 4.5 A. If the supply voltage fluctuates ±10%, determine (a) the torque range and (b) the current range.

● SOLUTION

$$N_s = \frac{120 \times 60}{4} = 1800 \text{ rpm}$$

$$s = \frac{1800 - 1710}{1800} = 0.05$$

$$\omega_m = \frac{2 \times \pi \times 1710}{60} = 179.07 \text{ rad/s}$$

The torque developed at the rated voltage of 230 V is

$$T_d = \frac{2 \times 746}{179.07} = 8.33 \text{ N·m}$$

When the supply voltage is down by 10%, the torque developed by the motor is

$$T_{dL} = 8.33 \left[\frac{0.9 \times 230}{230}\right]^2 = 6.75 \text{ N·m}$$

The corresponding rotor current is

$$I_{2L} = 4.5 \left[\frac{0.9 \times 230}{230}\right] = 4.05 \text{ A}$$

Similarly, when the supply voltage is up by 10%, the torque developed by the motor is

$$T_{dH} = 8.33[1.1]^2 = 10.08 \text{ N·m}$$

and the rotor current is

$$I_{2H} = 4.5 \times 1.1 = 4.95 \text{ A}$$

Hence, the torque varies from 6.75 N·m to 10.08 N·m, and the rotor current fluctuates between 4.05 A and 4.95 A.

■

Exercises

9.14. A 440-V, 50-Hz, 4-pole, Y-connected, three-phase induction motor runs at a slip of 5% on full load and develops a power of 10 hp. The slip falls to 2% when the motor is lightly loaded. Determine the torque developed and the power output of the motor when it runs at (a) full load and (b) light load.

9.15. A 660-V, 60-Hz, Y-connected, 6-pole, three-phase induction motor runs at 1125 rpm on full load and develops a power of 10 hp. The unregulated supply voltage fluctuates from 600 to 720 V. What is the torque range of the motor if the speed is held constant? What is the speed range for the motor to develop the same torque?

9.10 Equivalent Circuit Parameters

The equivalent circuit parameters and the performance of a three-phase induction motor can be determined by performing four tests. These are (a) the stator-resistance test, (b) the blocked-rotor test, (c) the no-load test, and (d) the load test.

The Stator-Resistance Test

This test is performed to determine the resistance of each phase winding of the stator. Let R be the dc value of the resistance between any two terminals of the motor; then the per-phase resistance is

$$R_1 = 0.5 \text{ R} \qquad \text{for Y-connection}$$

$$R_1 = 1.5 \text{ R} \qquad \text{for } \Delta\text{-connection}$$

The measured value of the resistance may be multiplied by a factor ranging from 1.05 to 1.25 in order to convert it from its dc value to its ac value. This is done to account for the **skin effect**. The multiplying factor may be debatable at power frequencies of 50 or 60 Hz, but it does become significant for a motor operating at a frequency of 400 Hz.

The Blocked-Rotor Test

This test, also called the locked-rotor test, is very similar to the short-circuit test of a transformer. In this case, the rotor is held stationary by applying external torque to the shaft. The stator field winding is connected to a variable three-phase supply. The voltage is carefully increased from zero to a level at which the motor draws the rated current. At this time, the readings of the line current, the applied line voltage, and the power input are taken by using the two-wattmeter method, as illustrated in Figure 9.9.

Since the rotor-circuit impedance is relatively small under blocked-rotor condition ($s = 1$), the applied voltage is considerably lower than the rated voltage of the motor. Thus, the excitation current is quite small and can be neglected. Under this assumption, the approximate equivalent circuit of the motor is given in Figure 9.10 on a per-phase basis. The total series impedance is

$$\hat{Z}_e = R_1 + R_2 + j(X_1 + X_2) = R_e + jX_e \tag{9.43}$$

Let V_{br}, I_{br}, and P_{br} be the applied voltage, the rated current, and the power input on a per-phase basis under the blocked-rotor condition; then

$$R_e = \frac{P_{br}}{I_{br}^2} \tag{9.44}$$

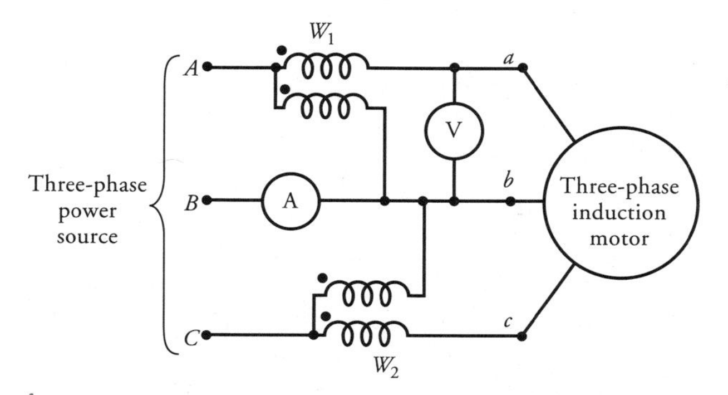

Figure 9.9 Typical connections to perform test on a three-phase induction motor.

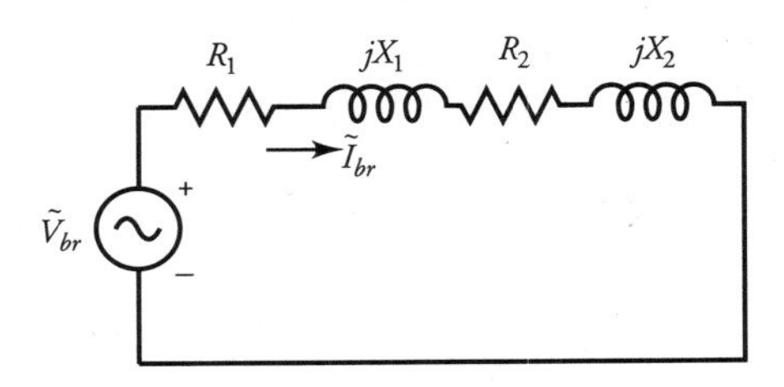

Figure 9.10 An approximate per-phase equivalent circuit of an induction motor under blocked-rotor condition.

Since R_1 is already known from the stator-resistance test, the equivalent rotor resistance is

$$R_2 = R_e - R_1 \tag{9.45}$$

However,

$$Z_e = \frac{V_{br}}{I_{br}} \tag{9.46}$$

Therefore,

$$X_e = \sqrt{Z_e^2 - R_e^2} \tag{9.47}$$

It is rather difficult to isolate the leakage reactances X_1 and X_2. For all practical purposes, these reactances are usually assumed to be equal. That is

$$X_1 = X_2 = 0.5X_e \tag{9.48}$$

The No-Load Test

In this case the rated voltage is impressed upon the stator windings and the motor operates freely without any load. This test, therefore, is similar to the open-circuit test on the transformer except that friction and windage loss is associated with an induction motor. Since the slip is nearly zero, the impedance of the rotor circuit is almost infinite. The per-phase approximate equivalent circuit of the motor with the rotor circuit open is shown in Figure 9.11.

Let W_{oc}, I_{oc}, and V_{oc} be the power input, the input current, and the rated applied voltage on a per-phase basis under no-load condition. In order to repre-

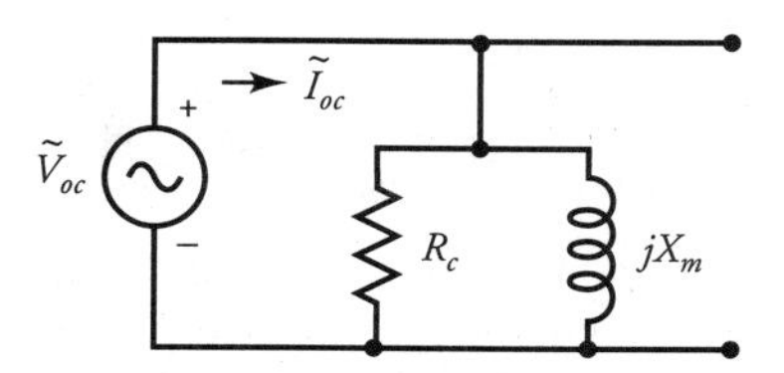

Figure 9.11 An approximate per-phase equivalent circuit of an induction motor under no-load condition.

sent the core loss by an equivalent resistance R_c, we must subtract the friction and windage loss from the power input.

The friction and windage loss can be measured by coupling the motor under test to another motor with a calibrated output and running it at the no-load speed of the induction motor. Let P_{fw} be the friction and windage loss on a per-phase basis. Then, the power loss in R_c is

$$P_{oc} = W_{oc} - P_{fw} \tag{9.49}$$

Hence, the core-loss resistance is

$$R_c = \frac{V_{oc}^2}{P_{oc}} \tag{9.50}$$

The power factor under no load is

$$\cos\theta_{oc} = \frac{W_{oc}}{V_{oc}I_{oc}} \tag{9.51}$$

The magnetization reactance is

$$X_m = \frac{V_{oc}}{I_{oc}\sin\theta_{oc}} \tag{9.52a}$$

The magnetization reactance can also be computed as follows:

$$S_{oc} = V_{oc}I_{oc}$$
$$Q_{oc} = \sqrt{S_{oc}^2 - W_{oc}^2}$$

and

$$X_m = \frac{V_{oc}^2}{Q_{oc}} \tag{9.52b}$$

When using the two-wattmeter method to measure the power under no load, the reading on one wattmeter may actually be negative because the power factor of the motor under no load may be considerably less than 0.5. If this is the case, the total power input is simply the difference of the two wattmeter readings.

The Load Test

To experimentally determine the speed-torque characteristics and the efficiency of an induction motor, couple the shaft to a dynamometer and connect the three-phase stator windings to a balanced three-phase power source. If need be, the direction of rotation may be reversed by interchanging any two supply terminals. Starting from the no-load condition, the load is slowly increased and the corresponding readings for the motor speed, the shaft torque, the power input, the applied voltage, and the line current are recorded. From these data, the motor performance as a function of motor speed (or slip) can be computed. Using analog-to-digital converters, the data can be stored on a magnetic disk for further manipulations.

EXAMPLE 9.8

The test data on a 208-V, 60-Hz, 4-pole, Y-connected, three-phase induction motor rated at 1710 rpm are as follows:
The stator resistance (dc) between any two terminals = 2.4 Ω

	No-Load Test	Blocked-Rotor Test
Power input	450 W	59.4 W
Line current	1.562 A	2.77 A
Line voltage	208 V	27 V

Friction and windage loss = 18 W
Compute the equivalent circuit parameters of the motor.

● SOLUTION

For a Y-connected motor, the per-phase resistance of the stator winding is

$$R_1 = \frac{2.4}{2} = 1.2\ \Omega$$

No-load test: $W_{oc} = \dfrac{450}{3} = 150 \text{ W}$

$$P_{fw} = \frac{18}{3} = 6 \text{ W}$$

$$P_{oc} = 150 - 6 = 144 \text{ W}$$

$$V_{oc} = \frac{208}{\sqrt{3}} = 120 \text{ V}$$

$$I_{oc} = 1.562 \text{ A}$$

The core-loss resistance is

$$R_c = \frac{120^2}{144} = 100 \ \Omega$$

The apparent power input (per phase) is

$$S_{oc} = V_{oc}I_{oc} = 120 \times 1.562 = 187.44 \text{ VA}$$

Thus, the power factor is

$$\cos \theta_{oc} = \frac{W_{oc}}{S_{oc}} = \frac{150}{187.44} = 0.8$$

$$\sin \theta_{oc} = \sqrt{1 - 0.8^2} = 0.6$$

The magnetization current: $I_m = I_{oc} \sin \theta_{oc} = 1.562 \times 0.6 = 0.937 \text{ A}$

$$X_m = \frac{V_{oc}}{I_m} = \frac{120}{0.937} \approx 128 \ \Omega$$

Blocked-rotor test: $V_{br} = \dfrac{27}{\sqrt{3}} = 15.588 \text{ V}$

$$P_{br} = \frac{59.4}{3} = 19.8 \text{ W}$$

$$I_{br} = 2.77 \text{ A}$$

$$R_e = \frac{19.8}{2.77^2} = 2.58 \ \Omega$$

$$R_2 = R_e - R_1 = 2.58 - 1.2 = 1.38 \ \Omega$$

$$Z_e = \frac{15.588}{2.77} = 5.627\ \Omega$$

$$X_e = \sqrt{5.627^2 - 2.58^2} = 5\ \Omega$$

Thus,

$$X_1 = X_2 = 2.5\ \Omega$$

■

Exercises

9.16. Using the equivalent circuit parameters of Example 9.8, determine the shaft torque and the efficiency of the motor at its rated speed.

9.17. The following data were obtained on a 230-V, 60-Hz, 4-pole, Y-connected, three-phase induction motor:
No-load test: power input = 130 W, line current = 0.45 A at the rated voltage.
Blocked-rotor test: power input = 65 W, line current = 1.2 A at a reduced voltage of 47 V.
The friction and windage loss is 15 W, and the stator winding resistance between any two lines is 4.1 Ω. Calculate the equivalent circuit parameters of the motor.

9.11 Starting of Induction Motors

At the time of starting, the rotor speed is zero and the per-unit slip is unity. Therefore, the starting current, from Figure 9.7, is

$$\tilde{I}_{2s} = \frac{\tilde{V}_1}{R_e + jX_e} \tag{9.53}$$

where $R_e = R_1 + R_2$ and $X_e = X_1 + X_2$. The corresponding value of the starting torque is

$$T_{ds} = \frac{3V_1^2 R_2}{\omega_s[R_e^2 + X_e^2]} \tag{9.54}$$

Since the effective rotor resistance, R_2, is very small at the time of starting compared with its value at rated slip, R_2/s, the starting current may be as much as 400% to 800% of the full-load current. On the other hand, the starting torque may only be 200% to 350% of the full-load torque. Such a high starting current is usually unacceptable because it results in an excessive line voltage drop which, in turn, may affect the operation of other machines operating on the same power source.

Since the starting current is directly proportional to the applied voltage, Eq. (9.53) suggests that the starting current can be reduced by impressing a low voltage across motor terminals at the time of starting. However, it is evident from Eq. (9.54) that a decrease in the applied voltage results in a decline in the starting torque. Therefore, we can employ the low-voltage starting only for those applications that do not require high starting torques. For instance, a fan load requires almost no starting torque except for the loss due to friction. The induction motor driving a fan load can be started using low-voltage starting.

The starting current can also be decreased by increasing the rotor resistance. As mentioned earlier, an increase in the rotor resistance also results in an increase in the starting torque which, of course, is desired for those loads requiring high starting torques. However, a high rotor resistance (a) reduces the torque developed at full load, (b) produces high rotor copper loss, and (c) causes a reduction in motor efficiency at full load. These drawbacks, however, do not represent a problem for wound-rotor motors. For these motors, we can easily incorporate high external resistance in series with the rotor windings at the time of starting and remove it when the motor operates at full load.

For rotors using squirrel-cage winding (die-cast rotors), the change in resistance from a high value at starting to a low value at full load is accomplished by using quite a few different designs, as shown in Figure 9.12. In each design, the underlying principle is to achieve a high rotor resistance at starting and a low rotor resistance at the rated speed. At starting, the frequency of the rotor is the same as the frequency of the applied source. At full load, however, the rotor frequency is very low (usually less than 10 Hz). Thus, the skin effect is more pronounced at starting than at full load. Hence, the rotor resistance is higher at

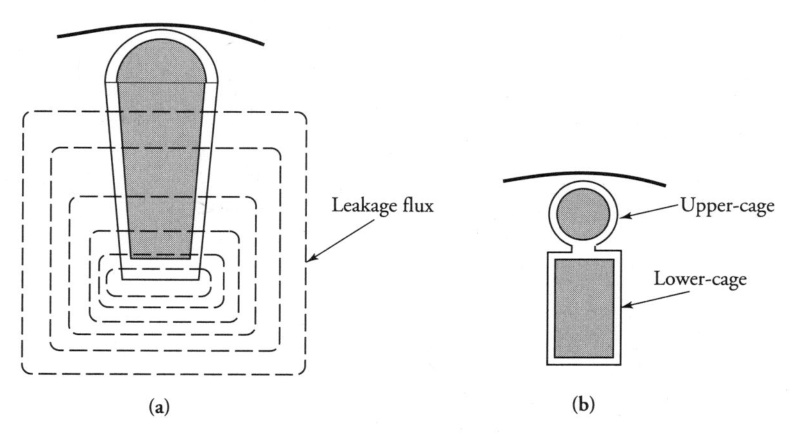

Figure 9.12 **(a)** Deep-bar and **(b)** double-cage rotors.

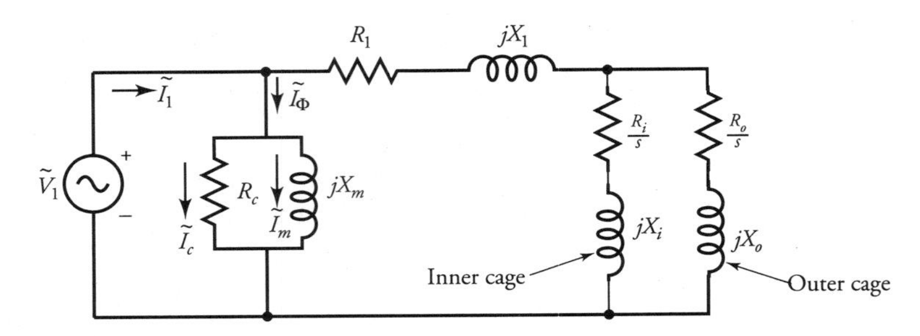

Figure 9.13 Per-phase equivalent circuit of a double-cage induction motor.

starting than at full load owing to the skin effect alone. Also, as the currents are induced in the rotor bars, they produce a secondary magnetic field. Part of the secondary magnetic field links only the rotor conductor and manifests itself as the leakage flux. The leakage flux increases as we move radially away from the air-gap and toward the shaft, and it is significant at starting. Thus, in a multicage rotor at starting, the inner cage presents a high leakage reactance compared with the outer cage. Owing to the high leakage reactance of the inner cage, the rotor current tends to confine itself in the outer cage. If the cross-sectional area of the outer cage is smaller than that of the inner cage, it presents a comparatively high resistance at starting. When the motor operates at full load, the rotor frequency is low. Thereby, the leakage flux is low. In this case, the current tends to distribute equally among all the cages. As a result, the rotor resistance is low.

The approximate equivalent circuit of a **double-cage** induction motor is given in Figure 9.13. The subscripts o and i correspond to the outer and the inner cage of a double-cage rotor shown in Figure 9.12**b**. The change in the speed-torque curve for a single-cage to a double-cage rotor is depicted in Figure 9.14.

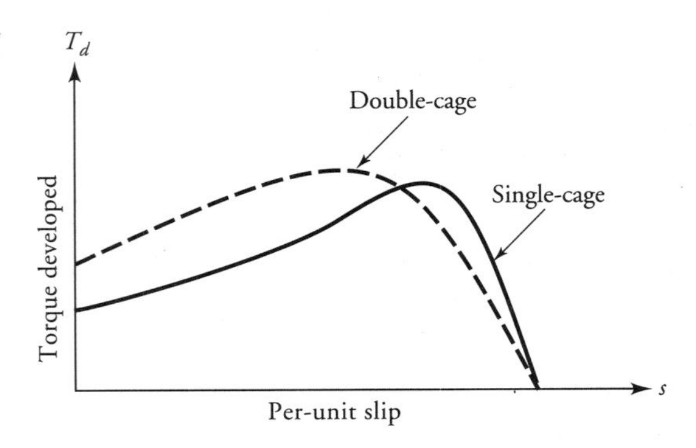

Figure 9.14 Speed-torque characteristics for single-cage and double-cage rotors.

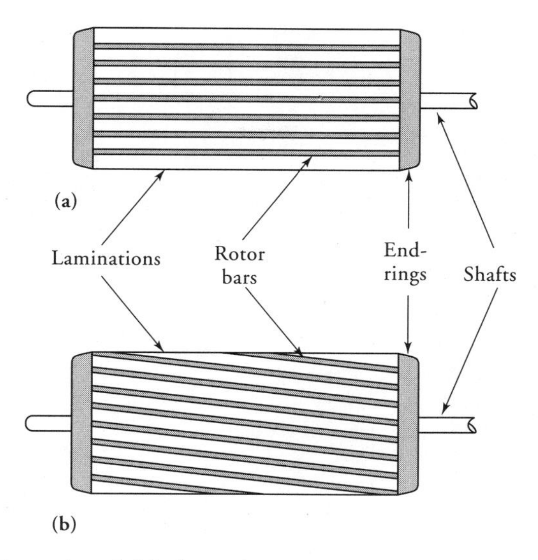

Figure 9.15 **(a)** Unskewed and **(b)** skewed rotor bars.

Another technique that is commonly used to increase the rotor resistance and lessen the effects of harmonics in an induction motor is called **skewing**. In this case, the rotor bars are skewed with respect to the rotor shaft, as illustrated in Figure 9.15. Skew is usually given in terms of bars. The minimum skew must be one bar to avoid cogging. Skews of more than one bar are commonly used.

EXAMPLE 9.9

The impedances of the inner cage and the outer cage rotors of a double-cage, three-phase, 4-pole induction motor are $0.2 + j0.8\ \Omega$/phase and $0.6 + j0.2\ \Omega$/phase, respectively. Determine the ratio of the torques developed by the two cages (a) at standstill and (b) at 2% slip.

● SOLUTION

Since the stator winding impedance, the core-loss resistance, and the magnetization reactance are not given, the approximate circuit neglecting these impedances is given in Figure 9.16. The rotor currents in the inner and the outer cages are

$$\tilde{I}_i = \frac{\tilde{V}_1}{0.2/s + j0.8}$$

and

$$\tilde{I}_o = \frac{\tilde{V}_1}{0.6/s + j0.2}$$

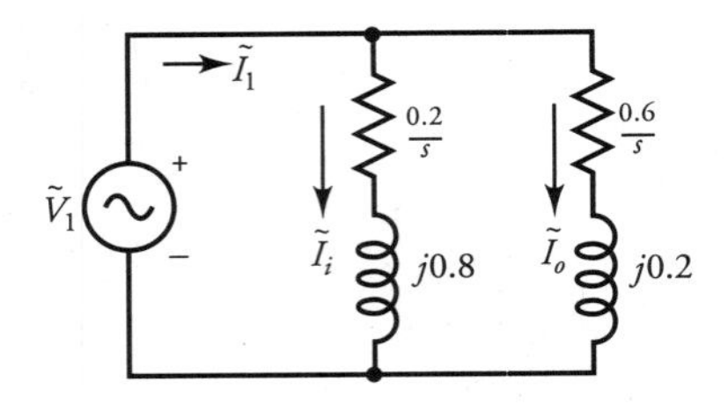

Figure 9.16 Equivalent circuit for Example 9.9.

The torque developed by a three-phase induction motor is

$$T_d = \frac{3I_2^2R_2}{s\omega_s}$$

Thus, the ratio of the torques developed by the outer and inner cages is

$$\frac{T_o}{T_i} = \frac{I_o^2R_o}{I_i^2R_i} = \frac{(0.2/s)^2 + 0.8^2}{(0.6/s)^2 + 0.2^2} \times \frac{0.6}{0.2} \tag{9.55}$$

Substituting $s = 1$ in Eq. (9.55), we obtain T_o in terms of T_i as

$$T_o = 5.1\ T_i$$

Thus, the torque developed by the outer cage is five times the torque developed by the inner cage.

However, substituting $s = 0.02$ in Eq. (9.55), we get

$$T_o = 0.34T_i$$

or

$$T_i = 2.98T_o$$

The inner cage is most effective at a slip of 2% because it develops nearly three times as much torque as the outer cage.

■

EXERCISES

9.18. The stator winding resistance of the double-cage induction motor of Example 9.9 is $2.1 + j8.3\ \Omega$/phase. Find the starting torque developed by the motor if the phase voltage is 110 V, 60 Hz.

9.19. A double-cage, three-phase, 6-pole, Y-connected induction motor has an inner cage impedance of $0.1 + j0.6\ \Omega$/phase and an outer cage impedance of $0.4 + j0.1\ \Omega$/phase. Determine the ratio of the torque developed by the two cages at (a) standstill and (b) 5% slip. What is the slip at which the torque developed by the two cages is the same?

9.20. If the motor in Exercise 9.19 is connected across a 230-V power source and has a stator impedance of $1.5 + j2.5\ \Omega$/phase, obtain the torque developed by the motor at standstill.

9.12 Rotor Impedance Transformation

Thus far we have tacitly assumed for both the squirrel-cage and wound rotors that the rotor circuit impedance can be transformed to the stator side in terms of an a-ratio. The a-ratio was defined on a per-phase basis as the ratio of the effective turns in the stator winding to the effective turns in the rotor winding. That is

$$a = \frac{k_{w1}N_1}{k_{w2}N_2} = \frac{E_1}{E_b} \tag{9.56}$$

For a wound rotor having the same number of poles and phases as that of the stator winding, the total turns per phase N_2 and the winding factor k_{w2} can be calculated the same way as that for the stator winding. However, the problem is somewhat more perplexing for the squirrel-cage (die-cast) rotor. Let us suppose that there are P poles in the stator and Q bars on the rotor. Let us assume that one of the bars is under the middle of the north pole of the stator at any given time. Another bar also exists which is in the middle of the adjacent south pole. The induced emf in both bars is maximum but of opposite polarity. Thus, these two bars carry the maximum current and can be visualized as if they form a single turn. Hence, the total number of turns on the rotor is $Q/2$. The emfs are also induced in other bars. If the flux is distributed sinusoidally, the induced emfs and thereby the induced currents also follow the same pattern. However, the root-mean-square (rms) value of the induced emf in each turn is the same. Since each turn is offset by one slot on the rotor, the induced emf in each bar is offset by that angle. Therefore, we can assume that each turn is equivalent to a phase group, and there are $Q/2$ phase groups in all. Since there are Q bars and P poles, the number of bars per pole is Q/P. As each bar identifies a different phase group, the number of bars per pole is then equivalent to the number of phases m_2 on the rotor. That is

$$m_2 = \frac{Q}{P} \tag{9.57}$$

This realization highlights the fact that the number of bars per pole per phase is 1. Stated differently, the number of turns per pole per phase is 1/2. Now we can determine the total number of turns per phase by multiplying the number of turns per pole per phase by the number of poles. That is,

$$N_2 = \frac{P}{2} \tag{9.58}$$

Since the two bars that are displaced by 180° electrical form a turn, the winding factor is unity because (a) the pitch factor is unity as each turn is a full-pitch turn and (b) the distribution factor is unity as there is only one turn in each phase group.

Since we are trying to transform the rotor circuit elements to the stator side, let m_1 be the number of phases on the stator side, E_2 the induced emf, and I_2 the equivalent rotor current. For the equivalent representation to be valid, the apparent power associated with the rotor circuit on the rotor side must be the same for the equivalent rotor circuit as referred to the stator side. Thus,

$$m_1 E_2 I_2 = m_2 E_b I_b \tag{9.59}$$

Since the induced emf on the stator side is E_1, E_2 must be equal to E_1. Hence,

$$I_2 = \frac{m_2 k_{w2} N_2}{m_1 k_{w1} N_1} I_b \tag{9.60}$$

Also the rotor copper loss prior to and after the transformation must be equal. That is,

$$m_1 I_2^2 R_2 = m_2 I_b^2 R_b \tag{9.61}$$

$$R_2 = \frac{m_1}{m_2}\left[\frac{k_{w1} N_1}{k_{w2} N_2}\right]^2 R_b \tag{9.62}$$

Finally, the magnetic energy stored in the rotor leakage inductance before and after the transformation must also be the same. Therefore,

$$\frac{1}{2} m_1 I_2^2 \frac{X_2}{2\pi f} = \frac{1}{2} m_2 I_b^2 \frac{X_b}{2\pi f}$$

$$X_2 = \frac{m_1}{m_2} \left[\frac{k_{w1} N_1}{k_{w2} N_2} \right]^2 X_b \tag{9.63}$$

Equations (9.62) and (9.63) outline how the actual rotor parameters for a squirrel-cage rotor can be transformed into equivalent rotor parameters on the stator side. R_2 and X_2 are the rotor resistance and leakage reactance that we have used in the equivalent circuit of an induction motor. From the above equations, it is evident that the a-ratio is

$$a = \sqrt{\frac{m_1}{m_2}} \left[\frac{k_{w1} N_1}{k_{w2} N_2} \right] \tag{9.64}$$

Note that for a wound rotor $m_1 = m_2$.

EXAMPLE 9.10

A 4-pole, 36-slot, double-layer wound, three-phase induction motor has 10 turns per coil in the stator winding and a squirrel-cage rotor with 48 bars. The resistance and reactance of each bar are 20 μΩ and 20 mΩ, respectively. Determine the equivalent rotor impedance as referred to the stator on a per-phase basis.

● SOLUTION

The number of coils for a double-layer wound stator is 36. The coils per pole per phase are $n = 3$ [36/(4 × 3)]. The pole span is 36/4 = 9 slots. The slot span is 180/9 = 20° electrical. The coil pitch is 7 slots, 140° electrical, and can be determined from the developed diagram. The pitch factor is $k_{p1} = \sin(140/2) = 0.94$. The distribution factor is

$$k_{d1} = \frac{\sin\left[3 \times \frac{20}{2}\right]}{3 \sin\left(\frac{20}{2}\right)} = 0.96$$

The winding factor for the stator is $k_{w1} = 0.94 \times 0.96 = 0.9$. Total turns per phase are $10 \times 36/3 = 120$. It is assumed that all coils in a phase group are connected in series.

For the rotor, $k_{w2} = 1$. The number of phases $m_2 = Q/P = 12$. Turns per phase are $N_2 = P/2 = 2$. Thus, the a-ratio is

$$a = \sqrt{\frac{3}{12}}\left[\frac{0.9 \times 120}{1 \times 2}\right] = 27$$

Hence, the rotor parameters as referred to the stator are

$$R_2 = 27^2 \times 20 \times 10^{-6} = 14.58 \text{ m}\Omega$$

and

$$X_2 = 27^2 \times 2 \times 10^{-3} = 1.458 \ \Omega$$

■

EXERCISES

9.21. In a wound-rotor induction motor the stator winding has twice as many turns as the rotor winding. The winding distribution factor for the stator is 0.85 and that for the rotor is 0.8. The actual impedance of the rotor on the rotor side at standstill is $0.18 + j0.25\ \Omega$/phase. What is the rotor impedance as referred to the stator side?

9.22. A 6-pole, 36-slot, double-layer wound, three-phase induction motor has 20 turns per coil and a squirrel-cage rotor with 48 bars. Each bar has a resistance of 15 $\mu\Omega$ and a reactance of 1.2 mΩ. Determine the equivalent circuit parameters of the rotor as referred to the stator side.

9.13 Speed Control of Induction Motors

It was pointed out in the preceding sections that the speed of an induction motor for a stable operation must be higher than the speed at which it develops maximum torque. In other words, the slip at full load must be less than the breakdown slip. For an induction motor having a low rotor resistance, the breakdown slip is usually less than 10%. For such a motor, the speed regulation may be within 5%. For all practical purposes, we can refer to a low-resistance induction motor as a **constant-speed motor**. Therefore, we must devise some methods in order to vary its operating speed.

We already know that the synchronous speed is directly proportional to the frequency of the applied power source and inversely proportional to the number of poles; the motor speed at any slip is

$$N_m = \frac{120f}{P}(1 - s) \tag{9.65}$$

From this equation it is evident that the operating speed of an induction motor can be controlled by changing the frequency of the applied voltage source and/or the number of poles. Speed can also be controlled by either changing the applied voltage, the armature resistance, or introducing an external emf in the rotor circuit. Some of these methods are discussed below.

Frequency Control

The operating speed of an induction motor can be increased or decreased by increasing or decreasing the frequency of the applied voltage source. This method enables us to obtain a wide variation in the operating speed of an induction motor. The only requirement is that we must have a variable-frequency supply. To maintain constant flux density and thereby the maximum torque developed, the applied voltage must be varied in direct proportion to the frequency. This is due to the fact that the induced emf in the stator winding is directly proportional to the frequency. The speed-torque characteristics of an induction motor at four different frequencies are given in Figure 9.17. Also shown in the figure is a typical load curve. At each frequency the motor operates at a speed at which the load line intersects the speed-torque characteristic for that frequency.

Changing Stator Poles

This method is quite suitable for an induction motor with a squirrel-cage rotor. In this case, the stator can be wound with two or more entirely independent

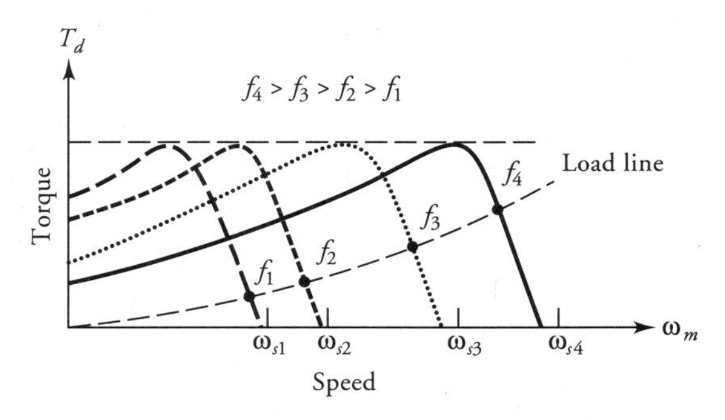

Figure 9.17 Speed-torque characteristics for various frequencies and adjusted supply voltages.

windings. Each winding corresponds to a different number of poles and therefore different synchronous speed. At any time, only one winding is in operation. All other windings are disconnected. For example, an induction motor wound for 4 and 6 poles at a frequency of 60 Hz can operate either at a synchronous speed of 1800 rpm (4-pole operation) or at 1200 rpm (6-pole operation). This method of speed control, although somewhat limited, is very simple, provides good speed regulation, and ensures high efficiency at either speed setting. Use has been made of this method in the design of traction motors, elevator motors, and small motor driving machine tools.

An induction motor is usually wound such that the current in each phase winding produces alternate poles. Thus, the four coils for each phase of a 4-pole induction motor produce two north poles and two south poles, with a south pole located between the two north poles and vice versa. However, if the phase coils are reconnected to produce either four north poles or four south poles, the winding is said to constitute a **consequent-pole winding**. Between any two like poles an unlike pole is induced by the continuity of the magnetic field lines. Thus, a 4-pole motor when reconnected as a consequent-pole motor behaves like an 8-pole motor. Therefore, by simply reconnecting the phase windings on an induction motor, a speed in the ratio of 2:1 can be accomplished by a single winding.

Rotor Resistance Control

We have already discussed the effect of changes in the rotor resistance on the speed-torque characteristic of an induction motor. This method of speed control is suitable only for wound-rotor induction motors. The operating speed of the motor can be decreased by adding external resistance in the rotor circuit. However, an increase in the rotor resistance causes (a) an increase in the rotor copper loss, (b) an increase in the operating temperature of the motor, and (c) a reduction in the motor efficiency. Because of these drawbacks, this method of speed control can be used only for short periods.

Stator Voltage Control

Since the torque developed by the motor is proportional to the square of the applied voltage, the (reduction/augmentation) in operating speed of an induction motor can be achieved by (reducing/augmenting) the applied voltage. The speed-torque characteristics for two values of the applied voltage are depicted in Figure 9.18. The method is very convenient to use but is very limited in its scope because to achieve an appreciable change in speed a relatively large change in the applied voltage is required.

Injecting an EMF in the Rotor Circuit

The speed of a wound-rotor induction motor can also be changed by injecting an emf in the rotor circuit, as shown in Figure 9.19. For proper operation, the fre-

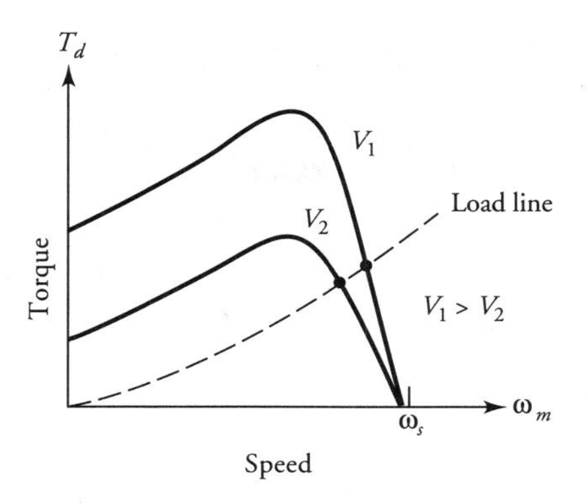

Figure 9.18 Speed-torque characteristics as a function of supply voltage.

quency of the injected emf must be equal to the rotor frequency. However, there is no restriction on the phase of the injected emf. If the injected emf is in phase with the induced emf in the rotor, the rotor current increases. In this case, the rotor circuit manifests itself as if it has a low resistance. On the other hand, if the injected emf is in phase opposition to the induced emf in the rotor circuit, the rotor current decreases. The decrease in the rotor current is analogous to the increase in the rotor resistance. Thus, changing the phase of the injected voltage is equivalent to changing the rotor resistance. The change in the rotor resistance is accompanied by the change in the operating speed of the motor. Further control in the speed can also be achieved by varying the magnitude of the injected emf.

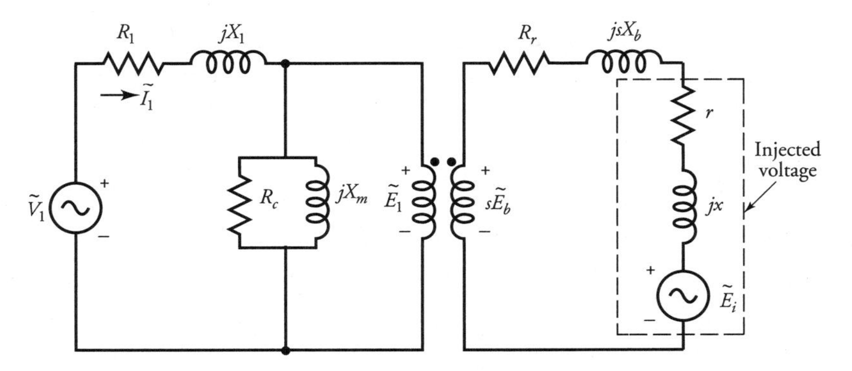

Figure 9.19 An equivalent circuit of an induction motor with an external source in the rotor circuit.

9.14 Types of Induction Motors

The National Electrical Manufacturers Association (NEMA) has categorized squirrel-cage induction motors into six different types by assigning them the letter designations A through F. Each letter designation is intended to satisfy the requirements of a certain application.

Class A Motors

A class A motor is considered a standard motor and is suitable for constant-speed applications. The motor can be started by applying the rated voltage. It develops a starting torque of 125% to 175% of full-load torque. The starting current at the rated voltage is 5 to 7 times the rated current. The full-load slip is usually less than 5% because the rotor resistance is relatively low. The speed regulation is 2% to 4%. The rotor bars are placed close to the surface of the rotor laminations in order to reduce the leakage reactance. These motors drive low-inertia loads and possess high accelerations. They are employed in such applications as fans, blowers, centrifugal type pumps, and machine tools.

Class B Motors

A class B motor is considered a general-purpose motor and can be started by applying the rated voltage. The rotor resistance for a class B motor is somewhat higher than for a class A motor. The rotor conductors are placed deeper in the slots than for the class A motor. Therefore, the rotor reactance of a class B motor is higher than that of a class A motor. The increase in the rotor reactance reduces the starting torque, whereas an increase in the rotor resistance increases the starting torque. Thus, the starting torque range for a class B motor is almost the same as that of the class A motor. Owing to the increase in reactance, the starting current is about 4.5 to 5.5 times the full-load current. The low starting current and almost the same starting torque make class B motors appropriate for class A applications as well. Therefore, class B motors can be substituted in all applications using class A motors. The speed regulation for class B motors is 3% to 5%.

Class C Motors

A class C motor usually has a double-cage rotor and is designed for full-voltage starting. The high-resistance rotor limits the starting current to 3.5 to 5 times the full-load current. The starting torque is 200% to 275% of the full-load torque. The speed regulation is 4% to 5%. Class C motors are used in applications that require high starting torques, such as compression pumps, crushers, boring mills, conveyor equipment, textile machinery, and wood-working equipment.

Class D Motors

A class D motor is a high-resistance motor capable of developing a starting torque of 250% to 300% of the rated torque. The high rotor resistance is created by using high-resistance alloys for the rotor bars and by reducing the cross-sectional area of the bar. Depending upon the design, the starting current may be 3 to 8 times the rated current. The efficiency of a class D motor is lower than that of those discussed above. The speed regulation may be as high as 10%. These motors are used in such applications as bulldozers, shearing machines, punch presses, stamping machines, laundry equipment, and hoists.

Class E Motors

Class E motors in general have low starting torque and operate at low slip at rated load. The starting current is relatively low for motors below 7.5 horsepower. These motors may be started at rated voltage. However, for motors above 7.5 horsepower the starting current may be high enough to require a low-voltage starting circuit.

Class F Motors

A class F motor is usually a double-cage motor. It is a low-torque motor and requires the lowest amount of starting current of all motors. The starting torque is usually 1.25 times the rated torque, whereas the starting current is 2 to 4 times the rated current. The speed regulation is over 5%. These motors can be started by applying the rated voltage. They are designed to replace class B motors and are built in sizes above 25 horsepower.

SUMMARY

In this chapter we shed some light on a three-phase induction motor, which essentially consists of a stator and a rotor. The stator is wound using double-layer winding just like the stator of a synchronous machine. There are two types of rotors: a squirrel-cage rotor and a wound rotor. A wound rotor, although expensive, is wound for the same number of poles as the stator. It provides means to add external resistance in series with the rotor circuit. A squirrel-cage rotor uses bars in the slots that are shorted at either end by the end-rings. For a low-horse-power motor, the bars and the end-rings are formed in a die-casting process.

When the three-phase stator winding is connected to a balanced three-phase source, it sets up a revolving field that rotates around the periphery of the rotor at the synchronous speed given by the following equation:

$$N_s = \frac{120f}{P}$$

where f is the frequency of the applied voltage source and P is the number of poles in the stator.

The uniform revolving field induces emf in the rotor conductors. Since the rotor winding forms a closed circuit, the induced emf gives rise to a current in the rotor conductors. The interaction of the current in the rotor conductors with the magnetic field in the motor creates a torque in accordance with the Lorentz force equation. Therefore, the rotor starts rotating and attains a speed slightly less than the synchronous speed. For this reason, an induction motor is also called an asynchronous motor. The difference between the synchronous speed and the rotor speed is called the slip speed. The per-unit slip is then defined as

$$s = \frac{N_s - N_m}{N_s}$$

By using the transformer analogy we developed an equivalent circuit of an induction motor as referred to the stator side. The rotor circuit parameters were transformed to the stator side by using the a-ratio given as

$$a = \sqrt{\frac{m_1}{m_2}} \left[\frac{k_{w1}N_1}{k_{w2}N_2}\right]$$

where m_1 and m_2 are the number of phases, k_{w1} and k_{w2} are the winding factors, and N_1 and N_2 are the number of turns in each phase of the stator and the rotor windings. For squirrel-cage rotor, $k_{w2} = 1$, $m_2 = Q/P$, and $N_2 = P/2$, where Q is the number of bars in the rotor. For a wound-rotor induction motor $m_1 = m_2$.

We defined $R_2(1 - s)/s$ as the dynamic (or effective) resistance because the power developed is proportional to it. Note that R_2 is the rotor resistance. From the per-phase equivalent circuit, we can compute the power input as

$$P_{in} = 3V_1I_1 \cos \theta$$

where V_1 and I_1 are the per-phase applied voltage and the input current. θ is the power factor angle between the two.

The stator copper loss can be computed as

$$P_{sc\ell} = 3I_1^2R_1$$

If I_c is the core-loss current through R_c, the core loss is

$$P_m = 3I_c^2R_c$$

The air-gap power is

$$P_{ag} = P_{in} - P_{sc\ell} - P_m = \frac{3I_2^2R_2}{s}$$

The rotor copper loss is

$$P_{rc\ell} = 3I_2^2R_2$$

Hence, the power developed is

$$P_d = P_{ag} - P_{rc\ell} = (1 - s)P_{ag} = \frac{3I_2^2R_2(1 - s)}{s}$$

The torque developed is

$$T_d = \frac{P_d}{\omega_m} = \frac{P_{ag}}{\omega_s} = \frac{3I_2^2R_2}{(s\omega_s)}$$

Using the approximate equivalent circuit, we found out that the efficiency of an induction motor is maximum when

$$3I_2^2[R_1 + R_2] = P_{rot}$$

The motor develops maximum torque at a slip known as the breakdown slip such that

$$s_b = \frac{R_2}{\sqrt{R_1^2 + X_e^2}}$$

where $X_e = X_1 + X_2$. The expression for the maximum torque developed is

$$T_{dm} = \frac{3V_1^2}{2\omega_s}\left[\frac{1}{R_1 + \sqrt{R_1^2 + X_e^2}}\right]$$

When the stator winding impedance is negligible, the approximate expressions for the breakdown slip and the breakdown torque are

$$s_b = \frac{R_2}{X_2}$$

$$T_{dm} = \frac{3V_1^2}{2\omega_s}\frac{1}{X_2}$$

If T_d is the torque developed at a slip s, then

$$\frac{T_d}{T_{dm}} = \frac{2ss_b}{s^2 + s_b^2}$$

The power developed by an induction motor is maximum when

$$s_p = \frac{R_2}{R_2 + Z_e}$$

where $Z_e = |R_e + jX_e|$ and $R_e = R_1 + R_2$.
The maximum power developed by the motor is

$$P_{dm} = \frac{3V_1^2}{2[R_e + Z_e]}$$

The motor circuit parameters can be determined by performing the blocked-rotor test, the no-load test, and the stator-resistance test. From the blocked-rotor test

$$R_e = \frac{P_{br}}{I_{br}^2}$$

where P_{br} and I_{br} are the per-phase power input and the current. The test is conducted when the rotor is held stationary and the motor draws the rated current from a carefully applied low voltage, V_{br}. The magnitude of the stator and the rotor winding impedance is

$$Z_e = \frac{V_{br}}{I_{br}}$$

Thus,

$$X_e = \sqrt{Z_e^2 - R_e^2}$$

The individual values of the leakage reactances are

$$X_1 = X_2 = 0.5X_e$$

and the rotor resistance is $R_2 = R_e - R_1$.
The no-load test is conducted at the rated voltage when the rotor is free to rotate without load. If W_{oc}, I_{oc}, and V_{oc} are the power input, the current, and the applied voltage on a per-phase basis, then

$$P_{oc} = W_{oc} - P_{fw}$$

where P_{fw} is the per-phase friction and windage loss.
The core-loss resistance is

$$R_c = \frac{V_{oc}^2}{P_{oc}}$$

The magnetization reactance is

$$X_m = \frac{V_{oc}^2}{Q_{oc}}$$

where

$$Q_{oc} = \sqrt{S_{oc}^2 - W_{oc}^2} \quad \text{and} \quad S_{oc} = V_{oc}I_{oc}$$

We have also examined the effect of changes in the rotor resistance on the speed-torque characteristic of an induction motor. An increase in the rotor resistance increases the starting torque, reduces the starting current, and enables the operation of the motor at a somewhat lower speed. In a wound motor, the rotor resistance is increased by adding external resistance to the rotor circuit via slip-rings. In a squirrel-cage induction motor, the change in the rotor resistance is realized by using a multicage rotor.

We also examined various schemes that enable us to control the speed of an induction motor. Some of the methods we have discussed are frequency control, changing stator poles, rotor resistance control, stator voltage control, and injecting an emf in the rotor circuit.

Review Questions

9.1. Explain the principle of operation of an induction motor.
9.2. Describe the construction of a squirrel-cage induction motor.
9.3. Explain the construction of a wound-rotor induction motor.
9.4. What are the advantages and drawbacks of a wound-rotor induction motor?
9.5. What are the advantages and drawbacks of a squirrel-cage induction motor?
9.6. At what speed does the revolving field rotate in an induction motor? How can it be determined?
9.7. Explain why an induction motor cannot operate at the synchronous speed.
9.8. Explain slip speed, per-unit speed, per-unit slip, and percent slip.
9.9. What is the rotor frequency when the rotor is (a) locked and (b) rotates at 5% slip?
9.10. Define starting torque, breakdown torque, breakdown slip, rated torque, torque developed, and shaft torque.
9.11. How can you minimize the rotor reactance?
9.12. Describe the no-load test, blocked-rotor test, and stator-resistance test.
9.13. Define stator copper loss, rotor copper loss, air-gap power, dynamic resistance, and effective rotor resistance.
9.14. What losses are measured by (a) the no-load test and (b) the blocked-rotor test?
9.15. What are the various techniques used to control the speed of induction motors?
9.16. What are the various classes of squirrel-cage induction motors?
9.17. List several applications for each class of squirrel-cage induction motor.
9.18. How can the starting current be controlled in a wound motor?
9.19. Cite possible reasons why a three-phase induction motor fails to start.
9.20. Why is an induction motor called an asynchronous motor?
9.21. How can the direction of a three-phase induction motor be reversed?
9.22. What happens to the speed of an induction motor if the load is increased?
9.23. What is a consequent-pole winding?
9.24. What happens when a 6-pole induction motor is reconnected as a consequent-pole motor?
9.25. What is the effect of increase in rotor reactance on the starting current? the maximum torque?
9.26. Explain the nature of the speed-torque characteristic of an induction motor.
9.27. What happens to the speed-torque characteristic of an induction motor if the rotor resistance is increased?
9.28. How does the increase in the rotor resistance affect the breakdown slip? the starting torque? the breakdown torque?

9.29. Is it always possible to start an induction motor by applying the rated voltage?

9.30. Is it possible for an induction motor to operate as an induction generator? If yes, how can it be done?

Problems

9.1. The frequency of the induced emf in the secondary winding of an 8-pole, three-phase induction motor is 10 Hz. At what speed does the magnetomotive force (mmf) of the secondary revolve with respect to the secondary winding?

9.2. A 2-pole, 230-V, 50-Hz, three-phase induction motor operates at a speed of 2800 rpm. Determine (a) the per-unit slip and (b) the frequency of the induced emf in the rotor.

9.3. A 12-pole, 440-V, 400-Hz, three-phase induction motor is designed to operate at a slip of 5% on full load. Determine (a) the rated speed, (b) the rotor frequency, and (c) the speed of the rotor revolving field relative to the rotor.

9.4. A three-phase induction motor operates at a slip of 3% and has a rotor copper loss of 300 W. The rotational loss is 1500 W. Determine (a) the air-gap power and (b) the power output. If the rotor impedance is $0.2 + j0.8\ \Omega$/phase, what is the magnitude of the induced emf per phase in the rotor?

9.5. A 10-hp, 6-pole, 440-V, 60-Hz, Δ-connected, three-phase induction motor is designed to operate at 3% slip on full load. The rotational loss is 4% of the power output. When the motor operates at full load, determine (a) the rotor copper loss, (b) the air-gap power, (c) the power developed, and (d) the shaft torque.

9.6. A 4-hp, 230-V, 60-Hz, 6-pole, three-phase, Y-connected, induction motor operates at 1050 rpm on full load. The rotational loss is 300 W. Determine the per-phase rotor resistance if the rotor current is not to exceed 100 A.

9.7. The per-phase equivalent circuit parameters of a 208-V, 4-pole, 60-Hz, three-phase, Y-connected, induction motor are $R_1 = 0.4\ \Omega$, $X_1 = 0.8\ \Omega$, $R_2 = 0.3\ \Omega$, $X_2 = 0.9\ \Omega$, and $X_m = 40\ \Omega$. The core loss is 45 W, and the friction and windage loss is 160 W. When the motor operates at a slip of 5%, determine (a) the input current, (b) the power input, (c) the air-gap power, (d) the power developed, (e) the power output, (f) the shaft torque, and (g) the efficiency of the motor. Draw its power-flow diagram.

9.8. Calculate the starting torque developed by the motor of Problem 9.7.

9.9. A 4-pole, 230-V, 60-Hz, Y-connected, three-phase induction motor has the following parameters on a per-phase basis: $R_1 = 10.12\ \Omega$, $X_1 = 38.61\ \Omega$,

$R_2 = 21.97\ \Omega$, $X_2 = 11.56\ \Omega$, and $X_m = 432.48\ \Omega$. The core loss is 10.72 W, and the friction and windage loss is 5.9 W. When the motor operates at its rated speed of 1550 rpm, determine (a) the stator current, (b) the magnetization current, (c) the rotor current, (d) the power input, (e) the stator copper loss, (f) the rotor copper loss, (g) the power output, (h) the shaft torque, and (i) the efficiency.

9.10. Plot the speed versus torque, power input versus speed, and power developed versus speed characteristics of the induction motor of Problem 9.9. What is the starting torque? Also determine the power input and the power developed at the time of starting.

9.11. A 230-V, 60-Hz, 6-pole, Y-connected, three-phase induction motor has the following parameters in ohms/phase as referred to the stator: $R_1 = 12.5$, $X_1 = 21.3$, $R_2 = 28.6$, $X_2 = 13.6$, $R_c = 4200$, and $X_m = 1800$. The friction and windage loss is 3% of the power developed. If the motor speed is 1125 rpm, determine (a) the power input, (b) the stator copper loss, (c) the rotor copper loss, (d) the air-gap power, (e) the power developed, (f) the shaft torque, and (g) its efficiency.

9.12. A 440-V, 60-Hz, 4-pole, Δ-connected, three-phase induction motor has the following parameters in ohms/phase: $R_1 = 0.3$, $X_1 = 0.9$, $R_2 = 0.6$, $X_2 = 0.9$, $R_c = 150$, $X_m = 60$. If the rotational loss is 4% of the power developed, determine the efficiency of the motor when it runs at 4% slip.

9.13. Using the data of Problem 9.7, calculate (a) the slip at which the motor develops maximum power, (b) the maximum power developed by the motor, and (c) the corresponding torque developed.

9.14. Calculate the slip at which the motor of Problem 9.9 develops maximum power. What is the maximum power developed by the motor? What is the torque developed by the motor at that slip?

9.15. Using the data of Problem 9.11, calculate (a) the slip at which the motor develops maximum power, (b) the maximum power developed, and (c) the associated torque developed by the motor.

9.16. Calculate the maximum power developed by the motor of Problem 9.12. What is the torque developed by it at that slip?

9.17. Using the data of Problem 9.7, calculate (a) the slip at which the motor develops maximum torque, (b) the maximum torque developed by the motor, and (c) the corresponding power developed.

9.18. Calculate the slip at which the motor of Problem 9.9 develops maximum torque. What is the maximum torque developed by the motor? What is the power developed by the motor at that slip?

9.19. Using the data of Problem 9.11, calculate (a) the slip at which the motor develops maximum torque, (b) the maximum torque developed, and (c) the associated power developed by the motor.

9.20. Calculate the maximum torque developed by the motor of Problem 9.12. What is the power developed by it at that slip?

9.21. A 6-pole, 60-Hz, Y-connected, three-phase induction motor develops a maximum torque of 250 N·m at a speed of 720 rpm. The rotor resistance is 0.4 Ω/phase, and the stator winding impedance is negligible. Determine the torque developed by the motor when it operates at a speed of 1125 rpm.

9.22. A 6-pole, 50-Hz, Δ-connected, three-phase induction motor has a rotor impedance of $0.05 + j0.5$ Ω/phase. The stator winding impedance is negligible. Determine the additional resistance required in the rotor circuit so that it develops maximum torque at starting.

9.23. An 8-pole, 230-V, 60-Hz, Δ-connected, three-phase induction motor has a rotor impedance of $0.025 + j0.1$ Ω/phase. The stator winding impedance is negligible. Determine (a) the speed at which the motor develops the maximum torque, (b) the maximum torque of the motor, and (c) the starting torque as a percentage of maximum torque. What additional resistance must be inserted in the rotor circuit to make the starting torque equal to 75% of the maximum torque?

9.24. A 120-V, 60-Hz, 6-pole, three-phase induction motor operates at a speed of 1050 rpm on full load and develops 5 hp. Under reduced load the speed increases to 1125 rpm. Determine the torque and the power developed by the motor at reduced load.

9.25. An 8-pole, 50-Hz, 208-V, Δ-connected, three-phase induction motor develops 20 hp at full-load slip of 5%. Determine the torque and the power developed at the same slip when a reduced voltage of 120 V is applied. What must be the new slip for the motor to develop the same torque?

9.26. The starting current of a 208-V, three-phase, Δ-connected, induction motor is 120 A when the rated voltage is applied to the stator windings. Determine the starting current when the applied voltage is reduced to 120 V. If the starting current is not to exceed 50 A, what must be the applied voltage?

9.27. The following test data were obtained on a 460-V, 60-Hz, Δ-connected, three-phase induction motor:
No-load test: power input = 380 W, line current = 1.15 A at rated voltage.
Blocked-rotor test: power input = 14.7 W, line current = 2.1 A at the line voltage of 21 V.
The friction and windage loss is 21 W, and the winding resistance between any two lines is 1.2 Ω. Determine (a) the equivalent circuit parameters of the motor and (b) its efficiency at a slip of 5%.

9.28. The following test data apply to a 208-V, 4-pole, Y-connected, three-phase induction motor: Running without load the line current and the power input are 2 A and 360 W at the rated voltage. With blocked rotor the current is 20 A and power input is 600 W when the applied voltage is 30 V. The friction and windage loss is 36 W. The resistance between any two lines is 0.2 Ω. Obtain the equivalent circuit parameters of the motor.

9.29. A 440-V, 4-pole, Y-connected, three-phase induction motor gave the following readings when tested:
No load: 440 V, 6.2 A, power factor = 0.1 lagging
Blocked-rotor test: 100 V, 12.5 A, 750 W
The winding resistance between any two lines = 1.2 Ω. The friction and windage loss is 30 W. Determine the equivalent circuit parameters of the motor.

9.30. The following are the test results on a 440-V, 50-Hz, slip-ring, three-phase, Y-connected, induction motor:
No load: 440 V, 7.5 A, 1350 W (including 650-W friction and windage loss)
Blocked-rotor test: 100 V, 32 A, 1800 W. The stator and rotor copper losses are equal under blocked-rotor condition.
Determine the equivalent circuit parameters of the motor.

9.31. The equivalent impedances of the inner and outer cages of a 4-pole, 60-Hz, Y-connected, three-phase induction motor are $0.5 + j2$ Ω/phase and $2 + j0.5$ Ω/phase at standstill. Calculate the ratio of torques developed by the two cages (a) at starting and (b) at 4% slip.

9.32. If the stator winding impedance of the motor in Problem 9.31 is $1 + j3$ Ω/phase and the applied voltage is 208 V, determine the torque developed (a) at starting and (b) at 4% slip.

9.33. The inner and outer cages of a rotor have standstill impedances of $0.5 + j4$ Ω/phase and $2.6 + j1.2$ Ω/phase. Find the ratio of the torques developed by the two cages (a) at starting and (b) at a slip of 0.05.

9.34. The outer and inner cages of a double-cage induction motor have standstill impedances of $2.6 + j2$ Ω/phase and $1.8 + j3$ Ω/phase. If the full-load slip is 5%, determine the ratio of the starting torque to the full-load torque.

CHAPTER 10

Single-Phase Motors

"Cut-away" view of a single-phase, PSC, induction motor. (*Courtesy of MagneTek*)

10.1 Introduction

In the preceding chapter we shed light on the operation of a three-phase induction motor. It employs an ingenious scheme of placing three identical phase windings 120° electrical in space phase with respect to each other. When these windings are excited by a balanced three-phase power source, they set up a uniform magnetic field that revolves around the rotor periphery at synchronous speed. The flux-cutting action induces an electromotive force (emf) and thereby a current in the rotor conductors. The interaction of the rotor current and the revolving magnetic field causes the rotor to rotate at a speed somewhat lower than the synchronous speed of the motor.

What happens when one of the three phases of the source is not connected to the motor? Is the rotor able to start from its standstill position and take up some mechanical load? In fact, we have already answered this question in Chapter 3. The motor behaves more like a two-phase motor and produces a revolving field that induces an emf and the current in the rotor conductors. The motor develops a starting torque and forces the rotor to rotate.

Let us go one step further. What happens if only one of the three phases is excited? At this time you may be surprised to know that if the rotor is already rotating, it continues to rotate in that direction. However, if the motor is at rest and then one of the three phases of the motor is excited by a single-phase source, the rotor buzzes but does not rotate. Hence, a motor operating on a single phase is not capable of developing the starting torque. We will explain why this is so in the next section.

A motor that operates on a single-phase source is called a **single-phase induction motor**. A single-phase induction motor requires only one single-phase winding to keep the motor running. However, such a motor is not self-starting. Therefore, we must provide some external means to start a single-phase induction motor. Most single-phase induction motors are built in the fractional-horsepower range and are used in heating, cooling, and ventilating systems.

A properly designed direct-current (dc) series motor can be made to operate on both dc and alternating-current (ac) sources. For this reason, it is appropriately called a **universal motor**. These motors operate at relatively high speeds and are an integral part of such units as vacuum cleaners, food blenders, and portable electric tools such as saws and routers.

In this chapter we deal with many different types of single-phase induction motors. The operation of the universal motor is also explained. We have reserved another chapter for the study of specialty motors. We begin our study by explaining how a single-phase motor runs.

10.2 Single-Phase Induction Motor

In Chapter 3 we made it clear that in order to make the motor self-starting there must at least be two phase windings placed in space quadrature and excited by a two-phase source. This fact plays a major role in the design of all single-phase induction motors, as discussed later.

A cross-sectional view of a 2-pole, single-phase induction motor with a squirrel-cage rotor is given in Figure 10.1. For clarity, we have placed the rotor conductors on the outer periphery of the rotor. In an actual motor, the rotor conductors are embedded in the rotor slots.

Let us suppose that the supply voltage is increasing in the positive direction and causes a current in the stator (field) windings, as indicated in Figure 10.1, while the rotor is at rest. The current in each winding produces a magnetic field that is also increasing in the upward direction. Since two rotor conductors that are 180° electrical apart form a closed loop, the rotor conductors can then be paired as shown. Let us examine one of the closed loops, say the loop formed by conductors 2 and 2′. The flux passing through this loop induces an emf and thereby a current in this loop. The direction of the current in the loop is such that it produces a magnetic flux which tends to oppose the increase in the magnetic flux set up in the windings. For this to happen, the current must flow out of conductor 2 and into conductor 2′, as shown by the dot and the cross, respectively. In the same way, we can determine the currents in the other conductors. Each current-carrying conductor must experience a force in accordance with the Lorentz force equation. The direction of the force acting on each conductor is also indicated in Figure 10.1. The forces experienced by conductors 1, 2, 3, and 4 in unison with

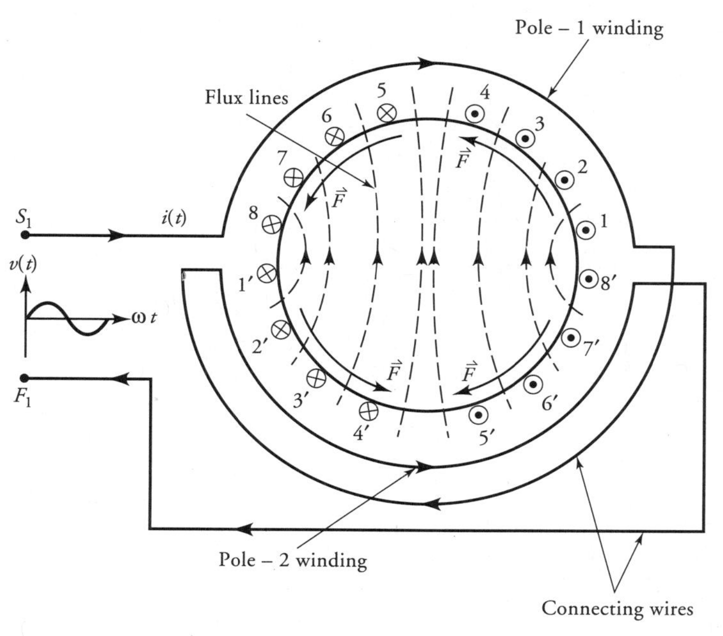

Figure 10.1 A 2-pole, single-phase induction motor.

conductors 1′, 2′, 3′, and 4′ tend to rotate the rotor in the counterclockwise direction. However, the rotation is opposed by the forces acting on the remaining conductors. The symmetric placement of the rotor conductors ensures that the motor develops equal torque in both directions and the net torque developed by it is zero. Hence the rotor remains in its standstill position.

As mentioned earlier, if the rotor is made to rotate in any direction while the single-phase winding is excited, the motor develops torque in that direction. Two theories have been put forth to explain this experimental fact: the **double revolving-field theory** and the **cross-field theory**. In this book, we confine our discussion to the double revolving-field theory.

Double Revolving-Field Theory

According to this theory, a magnetic field that pulsates in time but is stationary in space can be resolved into two revolving magnetic fields that are equal in magnitude but revolve in opposite directions. Let us consider the standstill condition of the rotor again. The magnetic field produced by the motor pulsates up and down with time, and at any instant its magnitude may be given as

$$B = B_m \cos \omega t \tag{10.1}$$

where B_m is the maximum flux density in the motor.

The flux density B can be resolved into two components B_1 and B_2 such that the magnitude of B_1 is equal to the magnitude of B_2. Thus, $B_1 = B_2 = 0.5\ B$. If we assume that B_1 rotates in the clockwise direction, the direction of rotation of B_2 is counterclockwise, as illustrated in Figure 10.2.

We now have two revolving fields of constant but equal magnitude rotating synchronously in opposite directions. An emf is induced in the rotor circuit owing to each revolving field. The polarity of the induced emf in the rotor due to one revolving field is in opposition to the other. Thus, the rotor currents induced by the two revolving fields circulate in opposite directions. However, at standstill, the slip in either direction is the same ($s = 1$) and so is the rotor impedance. Therefore, the starting currents in the rotor conductors are equal and opposite. In other words, the starting torque developed by each revolving field is the same. Since the direction of the starting torque developed by one revolving field opposes the other, the net torque developed by the motor is zero. This is the same conclusion we arrived at before. However, we have gained some insight into the operation of a single-phase induction motor according to the double revolving-field theory. We can look upon a single-phase induction motor as if it consists of two motors with a common stator winding but with rotors revolving in opposite directions. At standstill the two rotors develop equal torques in opposite directions, and the net torque developed is zero. With that understanding we can develop an equivalent circuit of a single-phase induction motor at standstill, as shown in Figure 10.3. Note that the magnetization reactance and the rotor impedance at

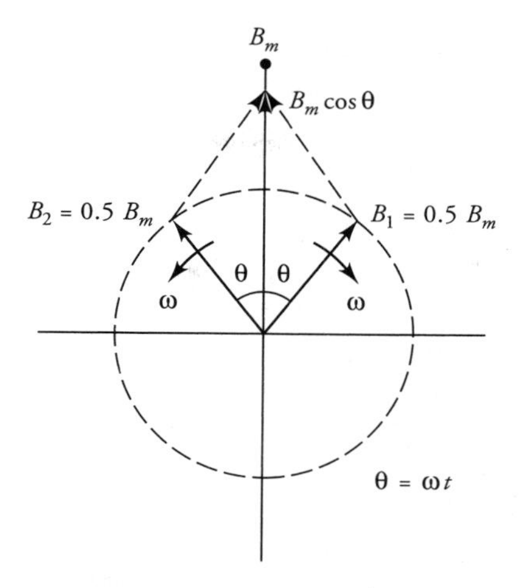

Figure 10.2 Resolution of a pulsating vector into two equal and oppositely revolving vectors.

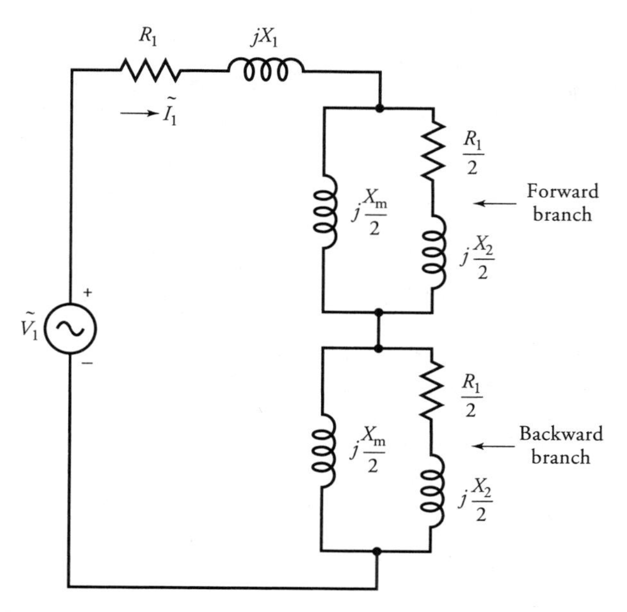

Figure 10.3 An equivalent circuit of a single-phase induction motor at rest.

standstill have been resolved into two sections to highlight the effect of two coupled rotors. For the sake of simplicity we treat core loss as a part of the rotational loss. The core-loss resistance is, therefore, omitted from the equivalent circuit. If the core-loss resistance is known, it can be placed across the power source as we did for the approximate equivalent circuit of a transformer or a three-phase induction motor.

One section of the rotor circuit is usually referred to as the **forward branch**, and the other is called the **backward branch**. When the motor rotates, say in the clockwise direction, the forward branch represents the effect of the revolving field in that direction. In this case, the backward branch corresponds to the rotor circuit associated with the counterclockwise revolving field. At standstill, both branches have the same impedance. The rotor circuit currents are also the same, and the same is true for the torques developed. Thus, when the rotor is at rest, the net torque developed by it is zero. We usually speak of torque developed by a branch. What we really mean is the torque developed by the rotor resistance in that particular branch.

Let us now assume that the rotor is rotating in the clockwise direction with a speed N_m. The magnetic field revolving in the clockwise direction has a synchronous speed of N_s ($N_s = 120f/P$). The synchronous speed of the revolving field in the counterclockwise direction is then $-N_s$. The per-unit slip in the forward (counterclockwise) direction is

$$s = \frac{N_s - N_m}{N_s} = 1 - \frac{N_m}{N_s} \tag{10.2}$$

The per-unit slip in the backward (counterclockwise) direction is

$$s_b = \frac{-N_s - N_m}{-N_s} = 1 + \frac{N_m}{N_s} = 2 - s \tag{10.3}$$

Note that at standstill, $N_m = 0$ and $s = s_b = 1$.

We can now incorporate the effect of slips in the forward and the backward rotor branches as we did for the three-phase induction motor. The modified equivalent circuit is given in Figure 10.4.

In our discussion of three-phase induction motors we found that the torque developed is proportional to the effective resistance in the rotor branch. At standstill, $s = s_b = 1$, the effective resistance in both branches of Figure 10.4 is the same. Thus, the torque developed by the two rotors is equal in magnitude but opposite in direction. That explains why there is no starting torque in a single-phase induction motor. On the other hand, let us now assume that the rotor is rotating with a slip s such that $s < 1$. The effective rotor resistance, R_2/s, in the forward

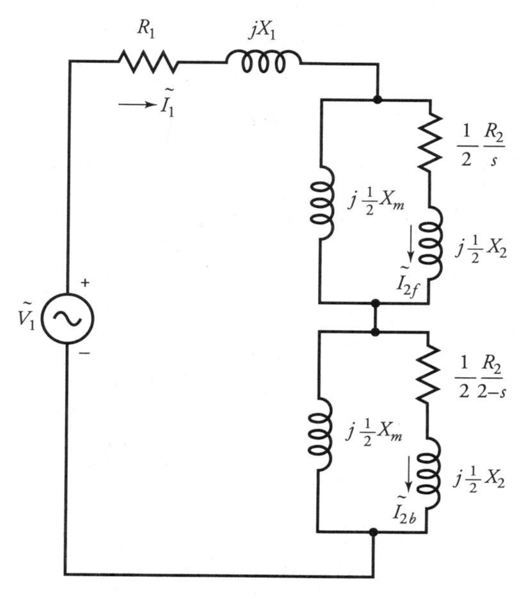

Figure 10.4 An equivalent circuit of a single-phase induction motor at any slip s.

branch is greater than that in the backward branch, $R_2/(2 - s)$. Thus, the torque developed by the forward branch is higher than that developed by the backward branch. The resultant torque is in the forward direction, and it tends to maintain the rotation in that direction. Thus, once a single-phase induction motor is made to rotate in any direction by applying an external torque, it continues rotating in that direction as long as the load torque is less than the maximum net torque developed by it.

EXAMPLE 10.1

A 115-V, 60-Hz, 4-pole, single-phase induction motor is rotating in the clockwise direction at a speed of 1710 rpm. Determine its per-unit slip (a) in the direction of rotation and (b) in the opposite direction. If the rotor resistance at standstill is 12.5 Ω, determine the effective rotor resistance in each branch.

● SOLUTION

$$N_s = \frac{120 \times 60}{4} = 1800 \text{ rpm}$$

(a) Slip in the forward direction is

$$s = \frac{1800 - 1710}{1800} = 0.05 \quad \text{or} \quad 5\%$$

(b) Slip in the backward direction is

$$s_b = 2 - 0.05 = 1.95 \quad \text{or} \quad 195\%$$

The effective rotor resistances are

Forward branch: $$\frac{0.5R_2}{s} = \frac{0.5 \times 12.5}{0.05} = 125\ \Omega$$

Backward branch: $$\frac{0.5R_2}{s_b} = \frac{0.5 \times 12.5}{1.95} = 3.205\ \Omega$$

■

Exercises

10.1. A 230-V, 50-Hz, 2-pole, single-phase induction motor is designed to operate at a slip of 3%. Determine the slip in the other direction. What is the speed of the motor in the direction of rotation? If the rotor resistance at standstill is 2.1 Ω, what is the effective resistance at the rated slip in either direction?

10.2. The rotor resistance of a 208-V, 50-Hz, 6-pole, single-phase induction motor at standstill is 1.6 Ω. When the motor rotates at a slip of 5%, determine (a) its speed, (b) the effective rotor resistance in the forward branch, and (c) the effective rotor resistance in the backward branch.

10.3 Analysis of a Single-Phase Induction Motor

From the equivalent circuit of a single-phase induction motor (Figure 10.4), we obtain

$$\hat{Z}_f = R_f + jX_f = 0.5\,\frac{jX_m[R_2/s + jX_2]}{R_2/s + j(X_2 + X_m)} \tag{10.4}$$

as the effective impedance of the forward branch and

$$\hat{Z}_b = R_b + jX_b = 0.5\,\frac{jX_m[R_2/(2 - s) + jX_2]}{R_2/(2 - s) + j(X_2 + X_m)} \tag{10.5}$$

as the effective impedance of the backward branch. The simplified equivalent circuit in terms of $\hat{Z}_f$ and $\hat{Z}_b$ is given in Figure 10.5.

If $\hat{Z}_1 = R_1 + jX_1$ is the impedance of the stator winding, the input impedance is

$$\hat{Z}_{in} = \hat{Z}_1 + \hat{Z}_f + \hat{Z}_b \tag{10.6}$$

The stator winding current is

$$\tilde{I}_1 = \frac{\tilde{V}_1}{\hat{Z}_{in}} \tag{10.7}$$

The power input is

$$P_{in} = \text{Re}[\tilde{V}_1 \tilde{I}_1^*] = V_1 I_1 \cos\theta \tag{10.8}$$

where θ is the power-factor angle by which the current $\tilde{I}_1$ lags the applied voltage $\tilde{V}_1$. The stator copper loss is

$$P_{sc\ell} = I_1^2 R_1 \tag{10.9}$$

When we subtract the stator copper loss from the total power input, we are left with the air-gap power. However, the air-gap power is distributed between

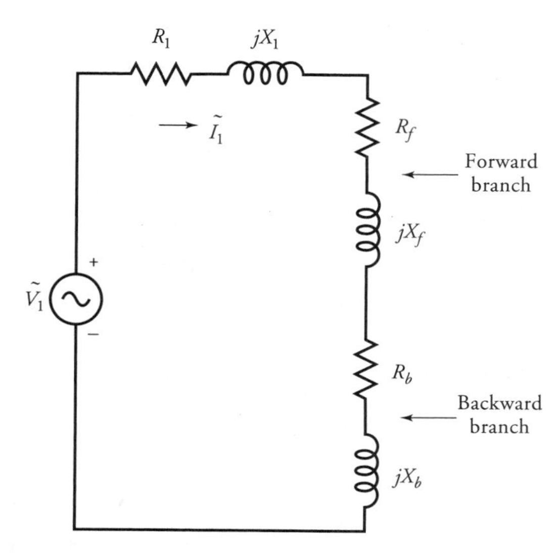

Figure 10.5 Simplified equivalent circuit of a single-phase induction motor.

the two air-gap powers: one due to the forward revolving field and the other due to the backward revolving field. In order to determine the air-gap power associated with each revolving field, we have to determine the rotor currents in both branches. If $\tilde{I}_{2f}$ is the rotor current in the forward branch, then

$$\tilde{I}_{2f} = \tilde{I}_1 \frac{jX_m}{\frac{R_2}{s} + j(X_2 + X_m)} \tag{10.10}$$

Similarly, the rotor current in the backward branch $\tilde{I}_{2b}$ is

$$\tilde{I}_{2b} = \tilde{I}_1 \frac{jX_m}{R_2/(2 - s) + j(X_2 + X_m)} \tag{10.11}$$

Hence, the air-gap powers due to the forward and backward revolving fields are

$$P_{agf} = I_{2f}^2 R_2 \frac{0.5}{s} \tag{10.12a}$$

$$P_{agb} = I_{2b}^2 R_2 \frac{0.5}{2 - s} \tag{10.13a}$$

Since R_f and R_b are the equivalent resistances in the forward and backward branches of the rotor circuit, the power transferred to the rotor must also be consumed by these resistances. In other words, we can also compute the air-gap powers as

$$P_{agf} = I_1^2 R_f \tag{10.12b}$$

for the forward branch and

$$P_{agb} = I_1^2 R_b \tag{10.13b}$$

for the backward branch.

The net air-gap power is

$$P_{ag} = P_{agf} - P_{agb} \tag{10.14}$$

The mechanical power developed by the motor is

$$P_d = (1 - s)P_{ag} = T_d \omega_m = T_d(1 - s)\omega_s \tag{10.15}$$

Hence, the torque developed by the single-phase motor is

$$T_d = \frac{P_{ag}}{\omega_s} \tag{10.16}$$

The power available at the shaft is

$$P_o = P_d - P_{rot} \tag{10.17}$$

where P_{rot} is the rotational loss of the motor. In this case, the rotational loss consists of the friction and windage loss, the core loss, and the stray-load loss.
The load (shaft) torque of the motor is

$$T_s = \frac{P_o}{\omega_m} \tag{10.18}$$

Finally, the motor efficiency is the ratio of the power available at the shaft P_o to the total power input P_{in}.

We could also have computed the torque developed by the forward and the backward revolving fields as

$$T_{df} = \frac{P_{agf}}{\omega_s} \tag{10.19}$$

$$T_{db} = \frac{P_{agb}}{\omega_s} \tag{10.20}$$

The net torque developed by the motor is

$$T_d = T_{df} - T_{db} \tag{10.21}$$

The two expressions of torque developed, Eq. (10.16) and Eq. (10.21), are identical. We can use either expression to compute T_d.

The torques developed, T_{df} and T_{db}, are plotted in Figure 10.6. These curves are also extended into the region of negative speed. This is usually done to show

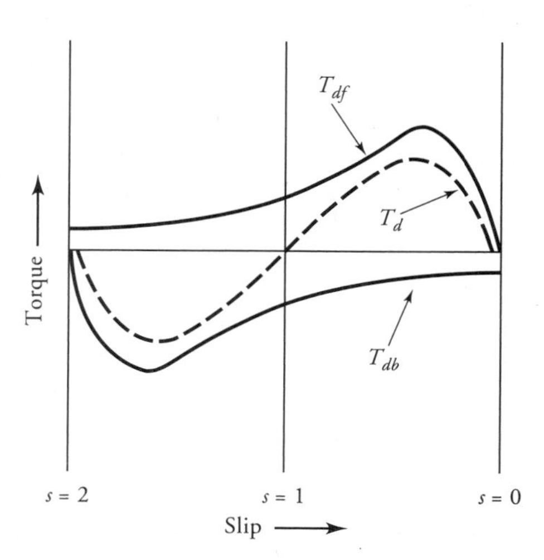

Figure 10.6 Speed-torque characteristic of a single-phase induction motor.

the torque that must be overcome when the motor is driven in the backward direction by a prime mover.

The following example illustrates how to use the above equations to determine the performance of a single-phase induction motor.

EXAMPLE 10.2

A 120-V, 60-Hz, 1/3-hp, 4-pole, single-phase induction motor has the following circuit parameters: $R_1 = 2.5\ \Omega$, $X_1 = 1.25\ \Omega$, $R_2 = 3.75\ \Omega$, $X_2 = 1.25\ \Omega$, and $X_m = 65\ \Omega$. The motor runs at a speed of 1710 rpm and has a core loss of 25 W. The friction and windage loss is 2 W. Determine the shaft torque and the efficiency of the motor.

● SOLUTION

$$N_s = \frac{120 \times 60}{4} = 1800 \text{ rpm}$$

$$s = \frac{1800 - 1710}{1800} = 0.05$$

$$\hat{Z}_f = \frac{j65[(3.75/0.05) + j1.25]0.5}{(3.75/0.05) + j(1.25 + 65)} = 15.822 + j18.524\ \Omega$$

$$\hat{Z}_b = \frac{j65[(3.75/1.95) + j1.25]0.5}{(3.75/1.95) + j(1.25 + 65)} = 0.925 + j0.64\ \Omega$$

$$\begin{aligned}\hat{Z}_{in} &= 2.5 + j1.25 + 15.822 + j18.524 + 0.925 + j0.64 \\ &= 19.246 + j20.414\ \Omega\end{aligned}$$

$$\tilde{I}_1 = \frac{120}{19.246 + j20.414} = 4.277\underline{/-46.69^\circ} \text{ A}$$

$$P_{in} = \text{Re}[120 \times 4.277\underline{/46.69^\circ}] = 352.081 \text{ W}$$

$$\tilde{I}_{2f} = \frac{j65}{(3.75/0.05) + j(1.25 + 65)} 4.277\underline{/-46.69^\circ} = 2.778\underline{/1.86^\circ} \text{ A}$$

$$\tilde{I}_{2b} = \frac{j65}{(3.75/1.95) + j(1.25 + 65)} 4.277\angle -46.69^\circ = 4.195\angle -45.02^\circ \text{ A}$$

We can compute the air-gap powers using the rotor current as

$$P_{agf} = 2.778^2 \times 0.5 \times \frac{3.75}{0.05} = 289.429 \text{ W}$$

$$P_{agb} = 4.195^2 \times 0.5 \times \frac{3.75}{1.95} = 16.918 \text{ W}$$

The air-gap powers can also be determined from Eqs. (10.12b) and (10.13b) as

$$P_{agf} = 4.277^2 \times 15.822 = 289.428 \text{ W}$$

and

$$P_{agb} = 4.277^2 \times 0.925 = 16.92 \text{ W}$$

Thus, the net air-gap power is

$$P_{ag} = 289.429 - 16.918 = 272.511 \text{ W}$$

The gross power developed is

$$P_d = (1 - 0.05) \times 272.511 = 258.886 \text{ W}$$

The net power output is

$$P_o = 258.886 - 25 - 2 = 231.886 \text{ W}$$

Efficiency: $\eta = \frac{231.886}{352.081} = 0.6586$ or 65.86%

$$\omega_m = \frac{2\pi \times 1710}{60} = 179.071 \text{ rad/s}$$

$$T_s = \frac{231.886}{179.071} = 1.295 \text{ N·m}$$

■

Exercises

10.3. A 6-pole, 120-V, 60-Hz, single-phase induction motor has $R_1 = 2.4\ \Omega$, $X_1 = 3.6\ \Omega$, $R_2 = 1.6\ \Omega$, $X_2 = 3.6\ \Omega$, and $X_m = 75\ \Omega$. If the motor slip at full load is 5%, determine (a) the motor speed, (b) the effective rotor resistance in the forward branch, (c) the effective rotor resistance in the backward branch, (d) the forward impedance $\hat{Z}_f$, (e) the backward impedance $\hat{Z}_b$, (f) the shaft torque, and (g) the motor efficiency. Assume that the rotational loss is 5% of the power developed.

10.4. A 4-pole, 110-V, 50-Hz, single-phase induction motor has $R_1 = 2\ \Omega$, $X_1 = 2.8\ \Omega$, $R_2 = 3.8\ \Omega$, $X_2 = 2.8\ \Omega$, and $X_m = 60\ \Omega$. The rotational loss is 20 W. Determine the shaft torque and the motor efficiency when the slip is 4%.

10.4 Types of Single-Phase Induction Motors

Each single-phase induction motor derives its name from the method used to make it self-starting. Some of the motors discussed in this section are **split-phase motor, capacitor-start motor, capacitor-start capacitor-run** motor, and **permanent split-capacitor motor**. Another induction motor discussed later in this chapter is called the **shaded-pole motor**.

For an induction motor to be self-starting, it must have at least two phase windings in space quadrature and must be excited by a two-phase source, as detailed in Chapter 3. The currents in the two phase windings are 90° electrical out of phase with each other. The placement of the two phase windings in space quadrature in a single-phase motor is no problem. However, the artificial creation of a second phase requires some basic understanding of resistive, inductive, and capacitive networks. Let us now examine how the second phase is created in each induction motor.

Split-Phase Motor

This is one of the most widely used induction motors for mechanical applications in the fractional horsepower range. The motor employs two separate windings that are placed in space quadrature and are connected in parallel to a single-phase source. One winding, known as the **main winding**, has a low resistance and high inductance. This winding carries current and establishes the needed flux at the rated speed. The second winding, called the **auxiliary winding**, has a high resistance and low inductance. This winding is disconnected from the supply when the motor attains a speed of nearly 75% of its synchronous speed. A centrifugal switch is commonly used to disconnect the auxiliary winding from the source at a predetermined speed. The disconnection is necessary to avoid the excessive power loss in the auxiliary winding at full load.

At the time of starting, the two windings draw currents from the supply. The main-winding current lags the applied voltage by almost 90° owing to its high inductance (large number of turns) and low resistance (large wire size). The auxiliary-winding current is essentially in phase with the applied voltage owing to its high resistance (small wire size) and low inductance (few number of turns).

As you may suspect, the main-winding current does not lag exactly by 90°, nor is auxiliary-winding current precisely in phase with the applied voltage. In addition, the two phase-winding currents may also not be equal in magnitude. In a well-designed split-phase motor, the phase difference between the two currents may be as high as 60°. It is from this **phase-splitting** action that the **split-phase motor** derives its name.

Since the two phase-windings are wound in space quadrature and carry out-of-phase currents, they set up an unbalanced revolving field. It is this revolving field, albeit unbalanced, that enables the motor to start.

The starting torque developed by a split-phase motor is typically 150% to 200% of the full-load torque. The starting current is about 6 to 8 times the full-load current. The schematic representation of a split-phase motor is given in Figure 10.7 along with its typical speed-torque characteristic. Note the drop in torque at the time the auxiliary winding is disconnected from the supply.

Capacitor-Start Motor

In a capacitor-start motor a capacitor is included in series with the auxiliary winding. If the capacitor value is properly chosen, it is possible to design a capacitor-start motor such that the main-winding current lags the auxiliary-winding current

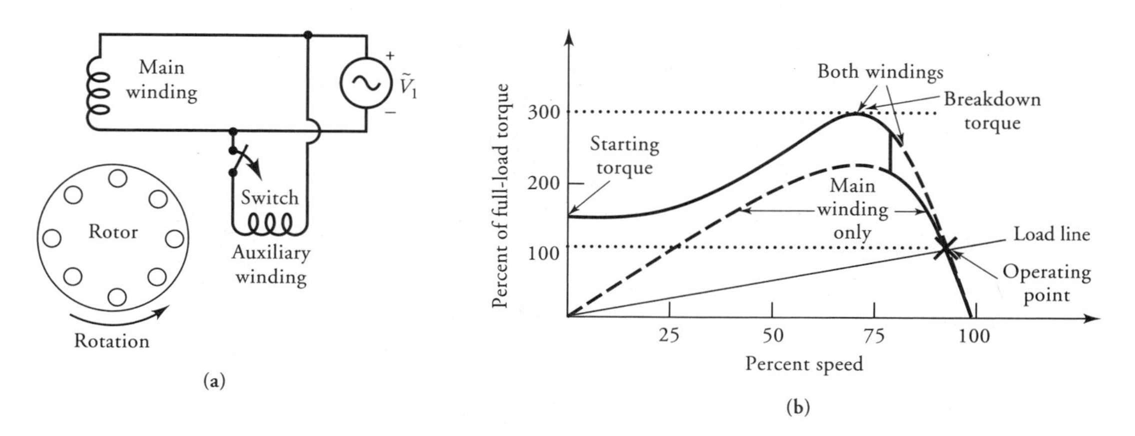

Figure 10.7 **(a)** Schematic representation and **(b)** speed-torque characteristic of a split-phase motor.

by exactly 90°. Therefore, the starting torque developed by a capacitor motor can be as good as that of any polyphase motor.

Once again, the auxiliary winding and the capacitor are disconnected at about 75% of the synchronous speed. Therefore, at the rated speed the capacitor-start motor operates only on the main winding like a split-phase induction motor. The need for an external capacitor makes the capacitor-start motor somewhat more expensive than a split-phase motor. However, a capacitor-start motor is used when the starting torque requirements are 4 to 5 times the rated torque. Such a high starting torque is not within the realm of a split-phase motor. Since the capacitor is used only during starting, its duty cycle is very intermittent. Thus, an inexpensive and relatively small ac electrolytic-type capacitor can be used for all capacitor-start motors. A schematic representation of a capacitor-start motor and its speed-torque characteristic are given in Figure 10.8.

Capacitor-Start Capacitor-Run Motor

Although the split-phase and capacitor-start motors are designed to satisfy the rated load requirements, they have low power factor at the rated speed. The lower the power factor, the higher the power input for the same power output. Thus, the efficiency of a single-phase motor is lower than that of a polyphase induction motor of the same size. For example, the efficiency of a capacitor-start or a split-phase single-phase motor is usually 50% to 60% in the fractional horsepower range. On the other hand, for the same application, a three-phase induction motor may have an efficiency of 70% to 80%.

The efficiency of a single-phase induction motor can be improved by employing another capacitor when the motor runs at the rated speed. This led to the

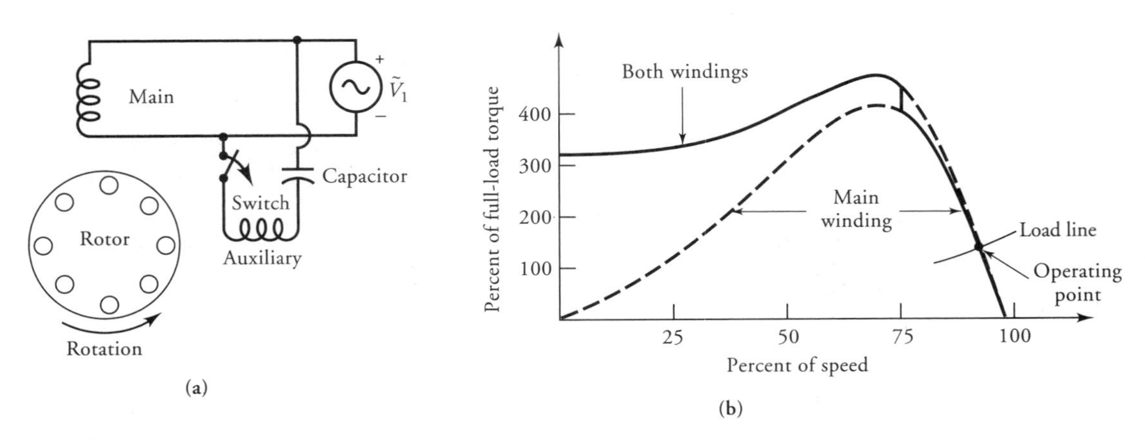

Figure 10.8 **(a)** Schematic representation and **(b)** speed-torque characteristic of a cap-start motor.

development of a capacitor-start capacitor-run (CSCR) motor. Since this motor requires two capacitors, it is also known as the **two-value capacitor motor**. One capacitor is selected on the basis of starting torque requirements (the start capacitor), whereas the other capacitor is picked for the running performance (the run capacitor). The auxiliary winding stays in circuit at all times, but the centrifugal switch helps in switching from the start capacitor to the run capacitor at about 75% of the synchronous speed. The start capacitor is of the ac electrolytic type, whereas the run capacitor is of an ac oil type rated for continuous operation. Since both windings are active at the rated speed, the run capacitor can be selected to make the winding currents truly in quadrature with each other. Thus, a CSCR motor acts like a two-phase motor both at the time of starting and at its rated speed. Although the CSCR motor is more expensive because it uses two different capacitors, it has relatively high efficiency at full load compared with a split-phase or capacitor-start motor. A schematic representation of a CSCR motor and its speed-torque characteristic are given in Figure 10.9.

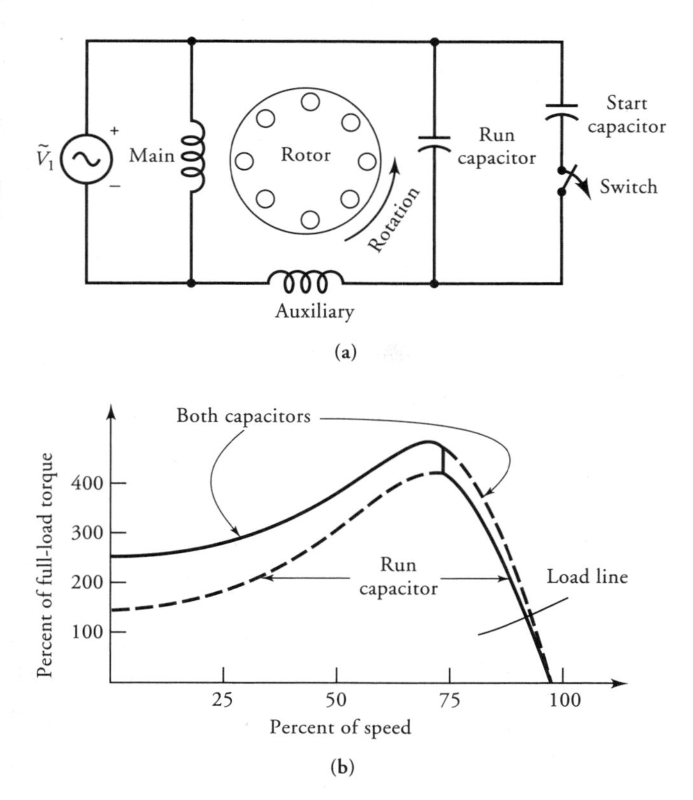

Figure 10.9 **(a)** Schematic representation and **(b)** speed-torque characteristic of a cap-start cap-run motor.

Permanent Split-Capacitor Motor

A less expensive version of a CSCR motor is called a **permanent split-capacitor (PSC) motor**. A PSC motor uses the same capacitor for both starting and full load. Since the auxiliary winding and the capacitor stay in the circuit as long as the motor operates, there is no need for a centrifugal switch. For this reason, the motor length is smaller than for the other types discussed above. The capacitor is usually selected to obtain high efficiency at the rated load. Since the capacitor is not properly matched to develop optimal starting torque, the starting torque of a PSC motor is lower than that of a CSCR motor. PSC motors are, therefore, suitable for blower applications with minimal starting torque requirements. These motors are also good candidates for applications that require frequent starts. Other types of motors discussed above tend to overheat when started frequently, and this may badly affect the reliability of the entire system. With fewer rotating parts, a PSC

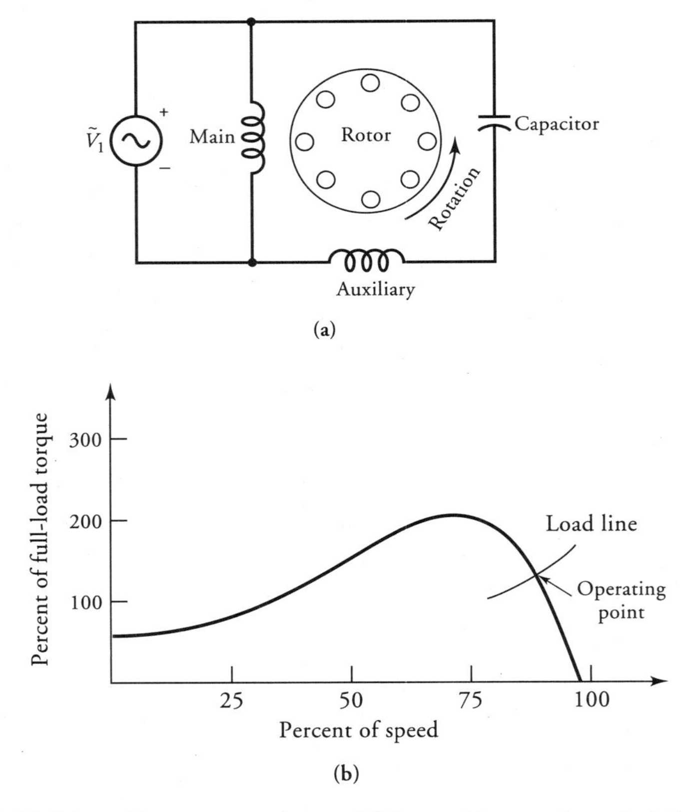

Figure 10.10 (a) Schematic representation and **(b)** speed-torque characteristic of a permanent-split capacitor motor.

motor is usually quieter and has a high efficiency at full load. The schematic representation of a PSC motor and its speed-torque characteristic are given in Figure 10.10.

10.5 Analysis of a Single-Phase Motor Using Both Windings

We already know how to determine the performance of a single-phase motor running on the main winding only. This analysis can be used to determine the performance at full load of a split-phase or a capacitor-start induction motor. However, the analysis of a single-phase motor is not complete without knowledge of its starting torque. The questions that still remain unanswered are the following:

(a) What is the starting torque developed by a split-phase or a capacitor-start motor?
(b) What is the voltage drop across the capacitor at starting? Is it within the maximum allowable limits?
(c) What is the current in-rush at the time of starting the motor? Does it cause severe fluctuations on the line?
(d) How can we determine the performance of a motor that uses both windings at all times?
(e) How does the switching of a capacitor from one value to another affect the performance of the motor prior to and immediately thereafter?

It must be clear by now that we cannot answer these questions based upon the information obtained from the single-winding analysis. Therefore, our analysis must include both windings.

Before proceeding further, some assumptions that are commonly accepted in this area are as follows:

1. The main and the auxiliary windings are in space quadrature with each other. This assumption implies that the flux produced by one winding does not induce an emf in the other winding. In other words, no transformer action exists between the two windings.
2. If we define the a-ratio as the ratio of the effective number of turns in the auxiliary winding to the effective number of turns in the main winding, then the leakage reactance, the magnetization reactance, and the rotor resistance for the auxiliary winding can be defined in terms of the main-winding parameters and the a-ratio.
3. When both main and the auxiliary windings are excited, they produce their own forward and backward revolving fields. Consequently, there are four revolving fields in a two-winding single-phase motor.
4. Each winding can be represented by an equivalent circuit with two parallel branches, one for the forward branch and the other for the backward branch.
5. A revolving field induces emf in both windings. It does not really matter

which winding sets up that revolving field. In other words, the forward and the backward revolving fields of the auxiliary winding induce emf in the main winding and vice versa. It is common to refer to these induced emfs as the **speed voltages**.

6. We assume that the main winding is displaced forward in space by 90° electrical with respect to the auxiliary winding. The forward field created by the auxiliary winding induces an emf in the main winding that lags by 90° the emf induced by the same field in the auxiliary winding. This is a very important concept and must be clearly understood in order to properly account for the induced emfs in one winding by the revolving fields of the other winding.

The above assumptions allow us to represent a two-winding single-phase motor by an equivalent circuit, as given in Figure 10.11, where

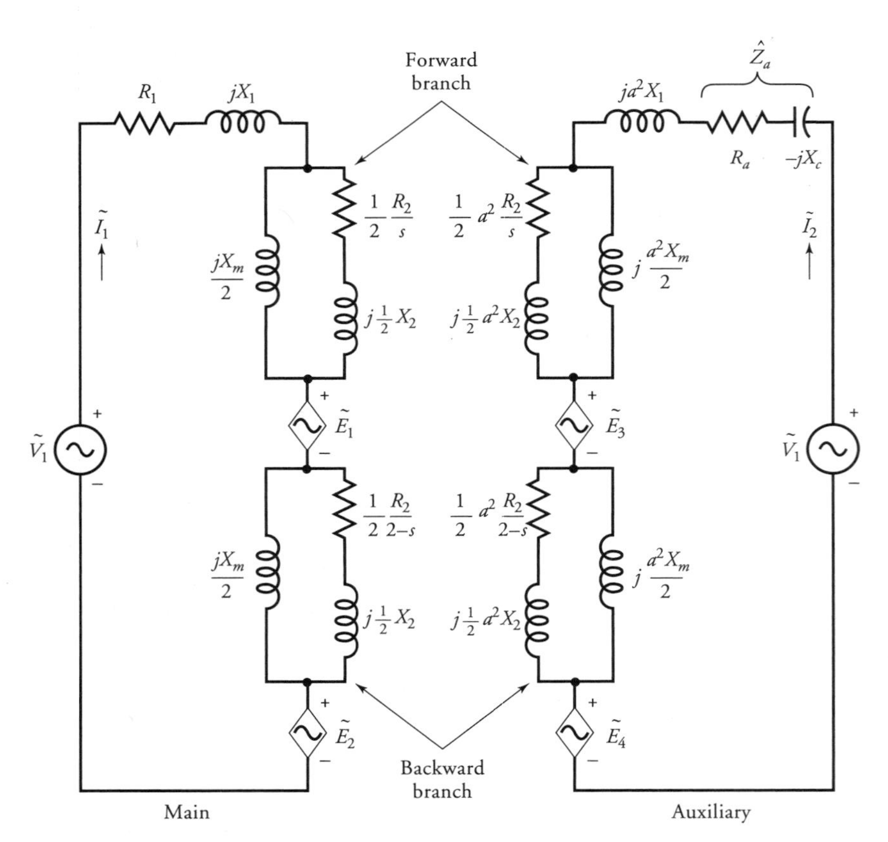

Figure 10.11 Equivalent circuit of a PSC motor.

R_1 = resistance of the main winding
X_1 = leakage reactance of the main winding
a = ratio of effective turns in the auxiliary winding to effective turns in the main winding
R_2 = rotor resistance as referred to the main winding at standstill
X_2 = rotor leakage reactance as referred to the main winding
X_m = the magnetization reactance of the motor as referred to the main winding
R_a = resistance of the auxiliary winding
$\tilde{E}_1$ = induced emf in the forward branch of the main winding by the forward revolving field of the auxiliary winding
$\tilde{E}_2$ = induced emf in the backward branch of the main winding by the backward revolving field of the auxiliary winding
$\tilde{E}_3$ = induced emf in the forward branch of the auxiliary winding by the forward revolving field of the main winding
$\tilde{E}_4$ = induced emf in the backward branch of the auxiliary winding by the backward revolving field of the main winding

The other parameters of the auxiliary winding have been defined in terms of the a-ratio. The equivalent circuit shown applies strictly to a PSC motor. We replace the capacitor impedance, $-jX_c$, by a short circuit if we want to analyze a split-phase motor. For a CSCR motor, $-jX_c$ has two different values, one for the start capacitor and the other for the run capacitor. In summary,

(a) For a split-phase motor

$$\hat{Z}_a = R_a \tag{10.22a}$$

and the auxiliary winding is in circuit for speeds below the operating speed of the centrifugal switch. Thereafter the motor operates on the main winding only.

(b) For a capacitor-start motor

$$\hat{Z}_a = R_a - jX_{cs} \tag{10.22b}$$

where X_{cs} is the reactance of the start capacitor. The auxiliary winding is included in the analysis as long as the speed is less than the operating speed of the centrifugal switch. After that the motor operates only on the main winding.

(c) For a CSCR motor

$$\hat{Z}_a = R_a - jX_{cs} \tag{10.22c}$$

as long as the motor speed is below the operating speed of the centrifugal switch. Thereafter,

$$\hat{Z}_a = R_a - jX_{cr} \tag{10.22d}$$

where X_{cr} is the reactance of the run capacitor.

(d) For a PSC motor

$$\hat{Z}_a = R_a - jX_c \tag{10.22e}$$

where X_c is the reactance of the capacitor in the auxiliary circuit. In this case, both windings are in circuit at all times.

The forward and backward impedances of the main winding are

$$\hat{Z}_f = R_f + jX_f = 0.5\,\frac{jX_m[R_2/s + jX_2]}{R_2/s + j(X_2 + X_m)} \tag{10.23}$$

$$\hat{Z}_b = R_b + jX_b = 0.5\,\frac{jX_m[R_2/(2 - s) + jX_2]}{R_2/(2 - s) + j(X_2 + X_m)} \tag{10.24}$$

A simplified equivalent circuit in terms of $\hat{Z}_f$ and $\hat{Z}_b$ of a two-winding single-phase induction motor is given in Figure 10.12.

The induced emfs in the main winding by its forward and backward revolving fields are

$$\tilde{E}_{fm} = \tilde{I}_1\hat{Z}_f \tag{10.25}$$

$$\tilde{E}_{bm} = \tilde{I}_1\hat{Z}_b \tag{10.26}$$

The induced emf in the auxiliary winding by its forward and backward revolving fields are

$$\tilde{E}_{fa} = \tilde{I}_2 a^2\hat{Z}_f \tag{10.27}$$

$$\tilde{E}_{ba} = \tilde{I}_2 a^2\hat{Z}_b \tag{10.28}$$

Since the main winding is displaced 90° electrical ahead of the auxiliary winding, the induced emf in the main winding by the forward revolving field of the auxiliary winding must lag by 90° the induced emf in the auxiliary. In addition, the induced emf in the main winding must be $1/a$ times the induced emf in the auxiliary. That is,

$$\tilde{E}_1 = -j\frac{1}{a}\tilde{E}_{fa} = -ja\tilde{I}_2\hat{Z}_f \tag{10.29}$$

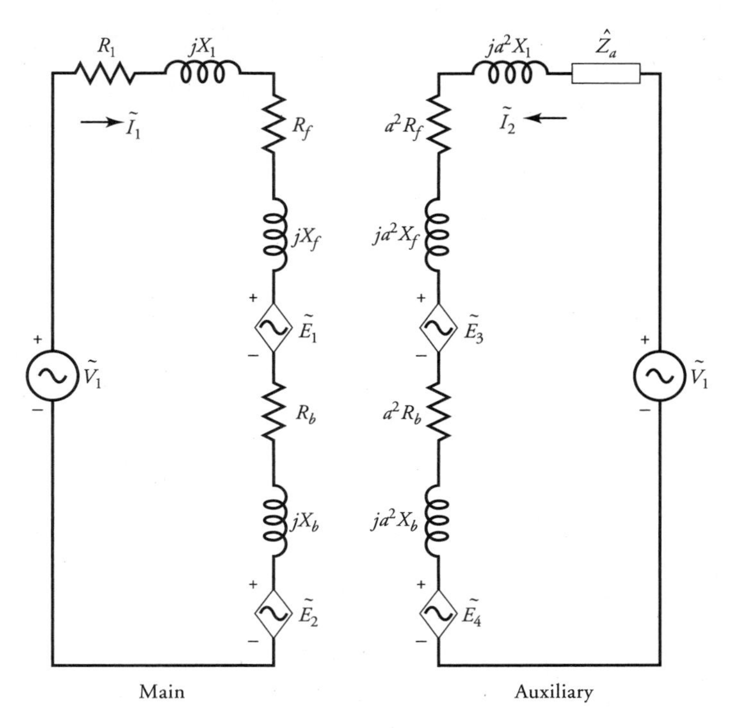

Figure 10.12 Simplified version of Figure 10.11.

By the same token, the induced emf in the main winding by the backward revolving field set up by the auxiliary winding must lead by 90° the emf it induces in the auxiliary winding. Thus,

$$\tilde{E}_2 = j\frac{1}{a}\tilde{E}_{ba} = ja\tilde{I}_2\hat{Z}_b \tag{10.30}$$

Similarly, the induced emfs in the forward and backward branches of the auxiliary winding by the forward and backward revolving fields of the main winding are

$$\tilde{E}_3 = ja\tilde{I}_1\hat{Z}_f \tag{10.31}$$

$$\tilde{E}_4 = -ja\tilde{I}_1\hat{Z}_b \tag{10.32}$$

Since all the induced emfs are now known, the application of Kirchhoff's voltage law to the coupled circuit yields

$$\tilde{I}_1(R_1 + jX_1) + \tilde{E}_{fm} + \tilde{E}_{bm} + \tilde{E}_1 + \tilde{E}_2 = \tilde{V}_1 \tag{10.33}$$

$$\tilde{I}_2(\hat{Z}_a + ja^2X_1) + \tilde{E}_{fa} + \tilde{E}_{ba} + \tilde{E}_3 + \tilde{E}_4 = \tilde{V}_1 \tag{10.34}$$

After substituting for the induced emfs, we can express the above equations in concise form as

$$\tilde{I}_1\hat{Z}_{11} + \tilde{I}_2\hat{Z}_{12} = \tilde{V}_1 \tag{10.35}$$

$$\tilde{I}_1\hat{Z}_{21} + \tilde{I}_2\hat{Z}_{22} = \tilde{V}_1 \tag{10.36}$$

where

$$\hat{Z}_{11} = R_1 + \hat{Z}_f + \hat{Z}_b + jX_1 \tag{10.37a}$$

$$\hat{Z}_{12} = -ja[\hat{Z}_f - \hat{Z}_b] \tag{10.37b}$$

$$\hat{Z}_{21} = ja[\hat{Z}_f - \hat{Z}_b] \tag{10.37c}$$

$$\hat{Z}_{22} = \hat{Z}_a + a^2[\hat{Z}_f + \hat{Z}_b + jX_1] \tag{10.37d}$$

The currents in the main and the auxiliary windings are

$$\tilde{I}_1 = \frac{\tilde{V}_1[\hat{Z}_{22} - \hat{Z}_{12}]}{\hat{Z}_{11}\hat{Z}_{22} - \hat{Z}_{12}\hat{Z}_{21}} \tag{10.38}$$

$$\tilde{I}_2 = \frac{\tilde{V}_1[\hat{Z}_{11} - \hat{Z}_{21}]}{\hat{Z}_{11}\hat{Z}_{22} - \hat{Z}_{12}\hat{Z}_{21}} \tag{10.39}$$

The line current is

$$\tilde{I}_L = \tilde{I}_1 + \tilde{I}_2 \tag{10.40}$$

The power supplied to the motor is

$$P_{in} = \text{Re}[\tilde{V}_1\tilde{I}_L^*] = V_1 I_L \cos\theta \tag{10.41}$$

where θ is the power-factor angle by which the line current lags the applied voltage.

The stator copper losses for both the windings are

$$P_{sc\ell} = I_1^2 R_1 + I_2^2 R_a \tag{10.42}$$

By subtracting the stator copper losses from the power supplied to the motor, we obtain the air-gap power. The air-gap power is distributed among the four revolving fields in the motor. We can also write an expression for the air-gap power just as we did for the motor operating on the main winding only. However, we have to take into account the presence of speed voltages and the power associated with them. On this basis, the air-gap power developed by the forward revolving field of the main winding is

$$\begin{aligned} P_{agfm} &= \text{Re}[(\tilde{E}_{fm} + \tilde{E}_1)\tilde{I}_1^*] \\ &= \text{Re}[(I_1^2 - ja\tilde{I}_1^*\tilde{I}_2)\hat{Z}_f] \end{aligned} \tag{10.43}$$

Similarly, the air-gap power produced by the forward revolving field of the auxiliary winding is

$$\begin{aligned} P_{agfa} &= \text{Re}[(\tilde{E}_{fa} + \tilde{E}_3)\tilde{I}_2^*] \\ &= \text{Re}[(I_2^2 a^2 + ja\tilde{I}_1\tilde{I}_2^*)\hat{Z}_f] \end{aligned} \tag{10.44}$$

The net air-gap power due to both forward revolving fields is

$$\begin{aligned} P_{agf} &= P_{agfm} + P_{agfa} \\ &= (I_1^2 + a^2 I_2^2)R_f + 2aI_1I_2R_f \sin\theta \end{aligned} \tag{10.45}$$

where

$$\tilde{I}_1 = I_1\angle\theta_1, \quad \tilde{I}_2 = I_2\angle\theta_2, \quad \text{and} \quad \theta = \theta_2 - \theta_1 \tag{10.46}$$

By the same token, the air-gap power developed by the backward revolving fields is

$$\begin{aligned} P_{agb} &= \text{Re}[(\tilde{E}_{bm} + \tilde{E}_2)\tilde{I}_1^* + (\tilde{E}_{ba} + \tilde{E}_4)I_2^*] \\ &= (I_1^2 + a^2 I_2^2)R_b - 2aI_1I_2R_b \sin\theta \end{aligned} \tag{10.47}$$

Hence, the net air-gap power developed by the motor is

$$\begin{aligned} P_{ag} &= P_{agf} - P_{agb} \\ &= (I_1^2 + a^2 I_2^2)(R_f - R_b) + 2a(R_f + R_b)I_1I_2 \sin\theta \end{aligned} \tag{10.48}$$

At standstill (i.e., the blocked-rotor condition, or at the time of starting) the per-unit slip of the motor is unity. The rotor impedance in the forward and the backward branches is the same. The net air-gap power developed by the motor, from the above equation when $R_f = R_b$, is

$$P_{ags} = 4aI_1I_2R_f \sin\theta \tag{10.49}$$

Note that the net power developed at the time of starting is proportional to the sine of the angle between the currents in the two windings. The power developed is maximum when the θ is 90°. For split-phase motors, θ may be as low as 30° and as high as 60°. For capacitor motors, θ is usually close to 90°. This is why a capacitor motor can develop more starting torque than a split-phase motor of the same size.

EXAMPLE 10.3

A 230-V, 50-Hz, 6-pole, single-phase PSC motor is rated at 940 rpm. The equivalent circuit parameters for the motor are $R_1 = 34.14\ \Omega$, $X_1 = 35.9\ \Omega$, $R_a = 149.78\ \Omega$, $X_2 = 29.32\ \Omega$, $X_m = 248.59\ \Omega$, $R_2 = 23.25\ \Omega$, $a = 1.73$, and $C = 4\ \mu\text{F}$, rated at 440 V. The core loss is 19.88 W, and the friction and windage loss is 1.9 W. Determine (a) the line current, (b) the power input, (c) the efficiency, (d) the shaft torque, (e) the voltage drop across the capacitor, and (f) the starting torque.

● SOLUTION

$$N_s = \frac{120 \times 50}{6} = 1000 \text{ rpm}$$

$$s = \frac{1000 - 940}{1000} = 0.06 \qquad \text{and} \qquad s_b = 2 - 0.06 = 1.94$$

$$\omega_m = \frac{2\pi \times 940}{60} = 98.437 \text{ rad/s}$$

$$X_c = \frac{10^6}{2\pi \times 50 \times 4} = 795.78\ \Omega$$

$$\hat{Z}_f = 0.5 \times \frac{j248.59[23.25/0.06 + j29.32]}{23.25/0.06 + j(29.32 + 248.59)} = 52.655 + j86.532\ \Omega$$

$$\hat{Z}_b = 0.5 \times \frac{j248.59[23.25/1.94 + j29.32]}{23.25/1.94 + j(29.32 + 248.59)} = 4.786 + j13.32\ \Omega$$

$$\hat{Z}_{11} = R_1 + jX_1 + \hat{Z}_f + \hat{Z}_b = 91.58 + j135.751\ \Omega$$

$$\hat{Z}_{12} = -ja(\hat{Z}_f - \hat{Z}_b) = 126.657 - j82.813\ \Omega$$

$$\hat{Z}_{21} = -\hat{Z}_{12} = -126.657 + j82.813\ \Omega$$

$$\hat{Z}_{22} = a^2[\hat{Z}_f + \hat{Z}_b + jX_1] + R_a - jX_c = 321.693 - j389.49\ \Omega$$

Substituting $\tilde{V}_1 = 230\underline{/0°}$ in Eqs. (10.38) and (10.39), we obtain

$$\tilde{I}_1 = 0.904\underline{/-49.47°} \text{ A}$$

$$\tilde{I}_2 = 0.559\underline{/21.71°} \text{ A}$$

(a) Thus, the line current: $\tilde{I}_L = \tilde{I}_1 + \tilde{I}_2 = 1.207\underline{/-23.48°}$ A

(b) The power input: $P_{in} = 230 \times 1.207 \times \cos(23.48°) = 254.561$ W

Let us compute the air-gap powers developed by the forward and backward fields as

$$P_{agf} = \text{Re}[(\tilde{I}_1\hat{Z}_f - ja\tilde{I}_2\hat{Z}_f)\tilde{I}_1^* + (\tilde{I}_2a^2\hat{Z}_f + ja\tilde{I}_1\hat{Z}_f)\tilde{I}_2^*] = 179.408 \text{ W}$$

$$P_{agb} = \text{Re}[(\tilde{I}_1\hat{Z}_b + ja\tilde{I}_2\hat{Z}_b)\tilde{I}_1^* + (\tilde{I}_2a^2\hat{Z}_b - ja\tilde{I}_1\hat{Z}_b)\tilde{I}_2^*] = 0.466 \text{ W}$$

$$P_{ag} = P_{agf} - P_{agb} = 178.942 \text{ W}$$

Power developed: $P_d = (1 - s)P_{ag} = (1 - 0.06)178.942 = 168.206$ W

Power output: $P_o = 168.206 - 19.88 - 1.9 = 146.426$ W

(c) Efficiency: $\eta = \dfrac{146.426}{254.561} = 0.575$ or 57.5%

(d) Shaft torque: $T_s = \dfrac{P_o}{\omega_m} = \dfrac{146.426}{98.437} = 1.488$ N·m

(e) The capacitor voltage: $\tilde{V}_c = -j\tilde{I}_2X_c = -j(0.559\underline{/21.71°})\,795.78$

$$= 444.666\underline{/-68.30°} \text{ V}$$

Since the voltage drop across the capacitor is higher than its normal rating of 440 V, for continuous operation either the capacitor with a higher voltage rating must be used or the windings must be redesigned.

(f) For the starting torque, $s = 1$ and $s_b = 1$. Using these values and following the above steps, we obtain

$$\hat{Z}_f = \hat{Z}_b = 9.237 + j13.886\ \Omega$$

$$\hat{Z}_{11} = 52.614 + j63.672\ \Omega$$

$$\hat{Z}_{12} = \hat{Z}_{21} = 0\ \Omega$$

$$\hat{Z}_{22} = 205.07 - j605.215\ \Omega$$

From Eqs. (10.38) and (10.39), the main- and the auxiliary-winding currents at standstill are

$$\tilde{I}_{1s} = \frac{\tilde{V}_1}{\hat{Z}_{11}} = 2.785\underline{/-50.43°} \text{ A}$$

$$\tilde{I}_{2s} = \frac{\tilde{V}_1}{\hat{Z}_{22}} = 0.36\underline{/71.28°} \text{ A}$$

$$\tilde{I}_{Ls} = \tilde{I}_1 + \tilde{I}_2 = 2.613\underline{/-43.70°} \text{ A}$$

$$P_{in} = 434.529 \text{ W} \qquad P_{agf} = 102.452 \text{ W} \qquad P_{agb} = 47.952 \text{ W}$$

$$P_{ag} = 54.497 \text{ W} \qquad \omega_s = \frac{2\pi \times 1000}{60} = 104.72 \text{ rad/s}$$

Hence, starting torque developed: $T_s = \dfrac{54.497}{104.72} = 0.52$ N·m

Note that the starting torque is only 1/3 of the full-load torque. For this reason, a PSC motor is usually employed in blower applications.

■

Exercises

10.5. A 120-V, 60-Hz, 4-pole, single-phase, capacitor-start motor has a main-winding impedance of $6 + j50\ \Omega$ and an auxiliary-winding impedance of $5 + j25\ \Omega$. The rotor impedance at standstill is $8 + j5\ \Omega$. The magnetization reactance is 150 Ω. If the starting capacitor is 100 μF, determine the starting torque developed by the motor.

10.6. What is the torque developed by the motor in Exercise 10.5 when it operates at a full-load speed of 1710 rpm? Note that the motor runs on the main winding only at its rated speed.

10.6 Testing Single-Phase Motors

The methods to determine the winding resistances of the main, R_1, and the auxiliary winding, R_a, are not discussed in this section because we can use any method that can accurately determine the resistances of these windings. The other equivalent circuit parameters of a single-phase induction motor can be determined by performing the blocked-rotor and the no-load tests.

Blocked-Rotor Test

The blocked-rotor test is performed with the rotor held at standstill by exciting one winding at a time while the other winding is left open. The test arrangement with the auxiliary winding open is shown in Figure 10.13. The test is performed by adjusting the applied voltage until the main winding carries the rated current. Since the slip at standstill in either direction is unity, the rotor circuit impedance is usually much smaller than the magnetization reactance. Therefore, for the blocked-rotor test we can use the approximate equivalent circuit of the main winding without the magnetization reactance, as depicted in Figure 10.14.

(a) **Auxiliary winding open**

Let V_{bm}, I_{bm}, and P_{bm} be the measured values of the applied voltage, the main-winding current, and the power supplied to the motor under blocked-rotor condition. The magnitude of the input impedance is

$$Z_{bm} = \frac{V_{bm}}{I_{bm}} \tag{10.50}$$

The total resistance in the circuit is

$$R_{bm} = \frac{P_{bm}}{I_{bm}^2} \tag{10.51}$$

Thus, the total reactance is

$$X_{bm} = \sqrt{Z_{bm}^2 - R_{bm}^2} \tag{10.52}$$

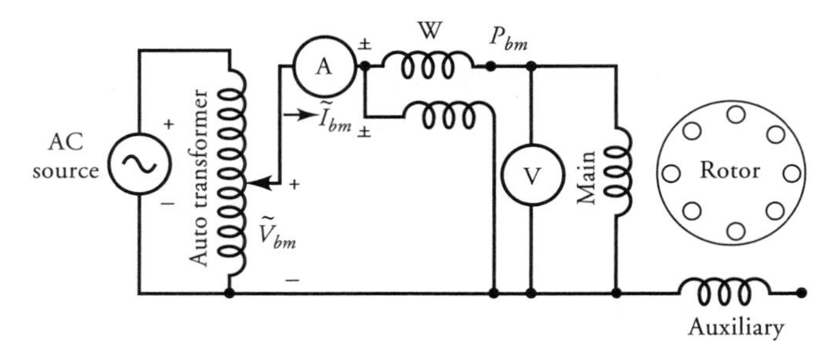

Figure 10.13 Experimental setup for blocked-rotor test with auxiliary winding open.

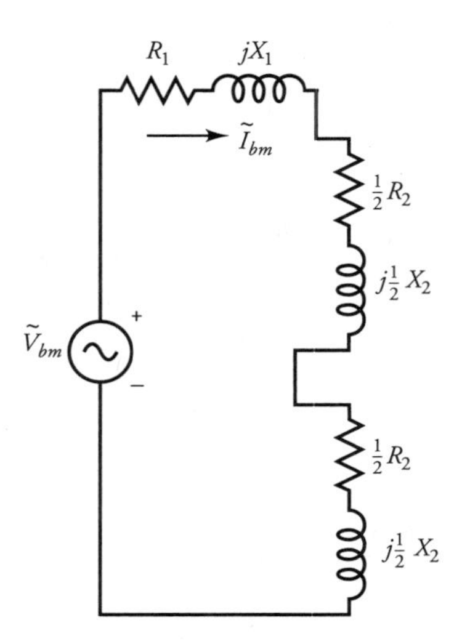

Figure 10.14 Approximate equivalent circuit as viewed from the main winding under blocked-rotor condition.

From the approximate equivalent circuit (Figure 10.14), we have

$$R_{bm} = R_1 + R_2 \tag{10.53}$$

and

$$X_{bm} = X_1 + X_2 \tag{10.54}$$

Since the main-winding resistance is already known, the rotor resistance at standstill, from Eq. (10.53), is

$$R_2 = R_{bm} - R_1 \tag{10.55}$$

To separate the leakage reactances of the main winding and the rotor, we once again make the same assumption that they are equal. That is,

$$X_1 = X_2 = 0.5X_{bm} \tag{10.56}$$

(b) **Main winding open**

We can also perform the blocked-rotor test by exciting the auxiliary winding with the main winding open. Let P_{ba}, V_{ba}, and I_{ba} be the power input, the applied voltage, and the current in the auxiliary winding when the rotor is at standstill. The total resistance of the auxiliary winding can now be computed as

$$R_{ba} = \frac{P_{ba}}{I_{ba}^2} \tag{10.57}$$

Since the rotor winding resistance is already known, we can compute the rotor resistance as referred to the auxiliary winding as

$$R_{2a} = R_{ba} - R_a \tag{10.58}$$

We can now determine the *a*-ratio, the ratio of effective turns in the auxiliary winding to the main winding, by the square root of the ratio of the rotor resistance as viewed from the auxiliary winding to the value of the rotor resistance as viewed from the main winding. Thus,

$$a = \sqrt{\frac{R_{2a}}{R_2}} \tag{10.59}$$

No-Load Test with Auxiliary Winding Open

In the three-phase induction motor operating under no load, we neglected the copper loss in the rotor circuit because it was assumed to be very small. In fact, we considered the rotor branch an open circuit because of very low slip at no load. In a single-phase motor running on main winding only, the no-load slip is considerably higher than that for a three-phase motor. If we still assume that the slip under no load is almost zero and replace the rotor circuit of the forward branch with an open circuit under no load, the error introduced in the calculation of the motor parameters based upon this test is somewhat greater than that for the three-phase motor. Making such an assumption, however, does simplify the equivalent circuit of the main winding under no load with auxiliary winding open. Such an equivalent circuit is given in Figure 10.15.

Let V_{nL}, I_{nL}, and P_{nL} be the measured values of the rated applied voltage, the current, and the power intake by the motor under no-load condition. Then the no-load impedance is

$$Z_{nL} = \frac{V_{nL}}{I_{nL}} \tag{10.60}$$

The equivalent resistance under no load is

$$R_{nL} = \frac{P_{nL}}{I_{nL}^2} \tag{10.61}$$

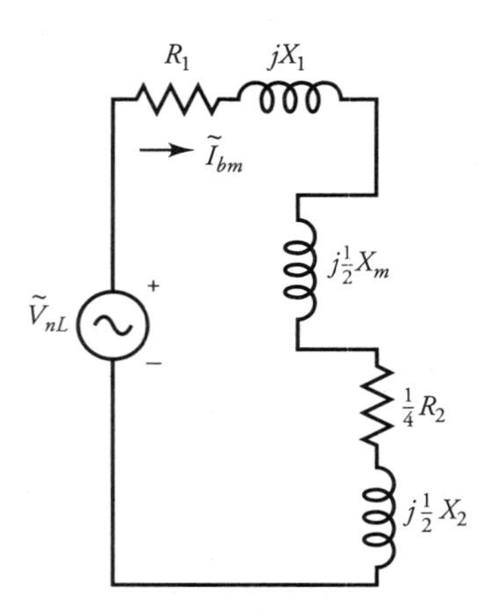

Figure 10.15 Approximate equivalent circuit as referred to the main winding under no-load condition ($s = 0$).

Hence, the no-load reactance is

$$X_{nL} = \sqrt{Z_{nL}^2 - R_{nL}^2} \tag{10.62}$$

However, from the equivalent circuit (Figure 10.15), we have

$$X_{nL} = X_1 + 0.5X_m + 0.5X_2$$

Since

$$X_1 = X_2 = 0.5X_{bm}$$

$$X_1 + 0.5X_2 = 0.75X_{bm}$$

Thus, from Eq. (10.62), the magnetization reactance is

$$X_m = 2X_{nL} - 1.5X_{bm} \tag{10.63}$$

Finally, the rotational loss is

$$P_{rot} = P_{nL} - I_{nL}^2(R_1 + 0.25R_2) \tag{10.64}$$

All the parameters of a single-phase induction motor are now known. Using these parameters we can now compute the torque developed, the power input, the power output, the winding currents, the line current, and the efficiency of the motor at any slip.

EXAMPLE 10.4

A 115-V, 60-Hz, single-phase, split-phase induction motor is tested to yield the following data.

	Voltage (V)	Current (A)	Power (W)
With auxiliary winding open			
No-load test	115	3.2	55.17
Blocked-rotor test	25	3.72	86.23
With main winding open			
Blocked-rotor test	121	1.21	145.35

The main-winding resistance is 2.5 Ω and the auxiliary-winding resistance is 100 Ω. Determine the equivalent circuit parameters of the motor.

● SOLUTION

From the blocked-rotor test on the main winding with the auxiliary winding open, we obtain

$$Z_{bm} = \frac{25}{3.72} = 6.72\ \Omega$$

$$R_{bm} = \frac{86.23}{3.72^2} = 6.23\ \Omega$$

$$X_{bm} = \sqrt{6.72^2 - 6.23^2} = 2.52\ \Omega$$

Thus,

$$X_1 = X_2 = 0.5 \times 2.52 = 1.26\ \Omega$$

and

$$R_2 = 6.23 - 2.5 = 3.73\ \Omega$$

From the no-load test data on the main winding with auxiliary winding open, we have

$$Z_{nL} = \frac{115}{3.2} = 35.94\ \Omega$$

$$R_{nL} = \frac{55.17}{3.2^2} = 5.39\ \Omega$$

$$X_{nL} = \sqrt{35.94^2 - 5.39^2} = 35.53\ \Omega$$

Hence,

$$X_m = 2 \times 35.53 - 0.75 \times 2.52 = 69.17\ \Omega$$

and

$$P_{rot} = 55.17 - 3.2^2(2.5 + 0.25 \times 3.73) \approx 20\ \text{W}$$

From the blocked-rotor test on the auxiliary winding with the main winding open, we get

$$Z_{ba} = \frac{121}{1.2} = 100.83\ \Omega$$

$$R_{ba} = \frac{145.35}{1.2^2} = 100.94\ \Omega$$

$$R_{2a} = 100.94 - 100 = 0.94\ \Omega$$

Finally, the a-ratio is

$$a = \sqrt{\frac{0.94}{3.73}} = 0.5$$

■

Exercises

10.7. A 230-V, 60-Hz, single-phase, CSCR motor is tested to yield the following data.

	Voltage (V)	Current (A)	Power (W)
With auxiliary winding open			
No-load test	230	2.2	63.1
Blocked-rotor test	120	8.22	743.1
With main winding open			
Blocked-rotor test	201	2.8	562.9

The main-winding resistance is 6 Ω and the auxiliary-winding resistance is 38 Ω. Determine the equivalent circuit parameters of the motor.

10.8. A 230-V, 60-Hz, reversible, garage door, PSC motor is tested as follows with the auxiliary winding open:
Blocked-rotor test: 230 V, 2.16 A, 406.84 W
No-load test: 230 V, 1.45 A, 138.52 W
Since the two windings are identical for a reversible motor, the a-ratio is unity and the resistance of each winding is 42.4 Ω. Determine the equivalent circuit parameters of the motor.

10.7 Shaded-Pole Motor

When the auxiliary winding of a single-phase induction motor is in the form of a copper ring, it is called the **shaded-pole motor**. Most often, a shaded-pole motor has a salient-pole construction similar to the stator of a dc machine. The pole is, however, always laminated to minimize the core loss. The pole is physically divided into two sections. A heavy, short-circuited copper ring, called the **shading coil**, is placed around the smaller section, as shown in Figure 10.16. This section usually covers one-third the pole arc and is called the **shaded part** of the pole. The larger section is referred to as the **unshaded part**. The main winding surrounds the entire pole, as shown in the figure. The rotor is die-cast, just like the rotor of any other single-phase induction motor.

As you may have already construed, a shaded-pole motor is very simple in construction and is thus the least expensive for fractional horsepower applications. Since it does not require a centrifugal switch, it is not only rugged but also very reliable in its operation. To keep the cost low, a shaded-pole motor usually operates in the magnetic saturation region. This is one of the reasons why the efficiency of a shaded-pole motor is low compared with that of other types of induction motors. These motors develop low starting torque and are suitable for blower applications. Shaded-pole motors are usually built to satisfy the load requirements up to 1/3 horsepower. For this reason, the low efficiency is of no interest.

Principle of Operation

Let us now examine how the shading coil helps a shaded-pole motor to set up a revolving field. To do so, we consider changes in the flux produced by the main winding at three time intervals:

(a) When the flux is increasing from zero to maximum
(b) When the flux is almost maximum
(c) When the flux is decreasing from maximum to zero

Any change in the flux in each pole of the motor is responsible for an induced emf in the shading coil in accordance with Faraday's law of induction. Since the shading coil forms a closed loop with very small resistance, a large current is

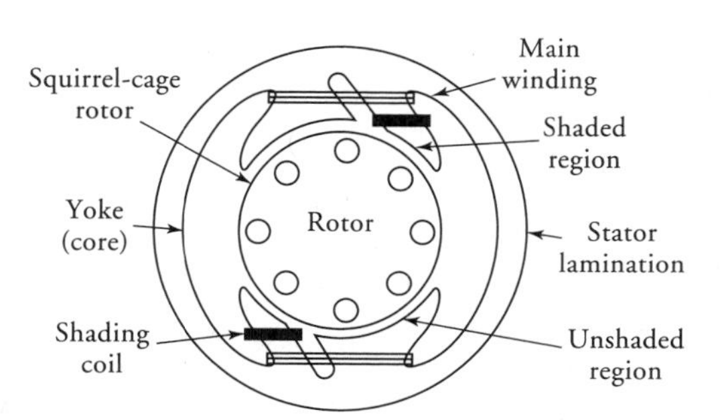

Figure 10.16 A shaded-pole motor.

induced in the shading coil. The direction of the current is such that it always creates a magnetic field that opposes the change in the flux in the shaded region of the pole. With this understanding, let us now analyze the effect of the shading coils during the time intervals mentioned above.

Interval a: During this time interval the flux in the pole is increasing and so is the current induced in the shading coil. The shading coil produces a flux that opposes the increase in the flux linking the coil. As a result, most of the flux flows through the unshaded part of the pole, as shown in Figure 10.17. The magnetic axis of the flux is then the center of the unshaded section of the pole.

Interval b: During this time interval the magnetic flux in the pole is near its maximum value. Therefore, the rate of change of flux is almost zero. Hence, the induced emf and the current in the shading coil are zero. As a result, the flux distributes itself uniformly through the entire pole. The magnetic axis, therefore, moves to the center of the pole. This shift in magnetic axis has the same effect as the physical motion (rotation) of the pole.

Interval c: During this time interval the magnetic flux produced by the main winding begins to decrease. The current induced in the shading coil reverses its direction in order to oppose the decrease in the flux. In other words, the shading coil produces the flux that tends to prevent a decrease in the flux produced by the main winding. As a result, most of the flux is confined in the shaded region of the pole. The magnetic axis of the flux has now moved to the center of the shaded region.

Note that without the shading coil, the center of the magnetic axis would always be at the center of the pole. The presence of the shading coil forces the flux to shift its magnetic axis from the unshaded region to the shaded region. The shift is gradual and has the effect of revolving magnetic poles. In other words, the magnetic field revolves from the unshaded part toward the shaded part of the

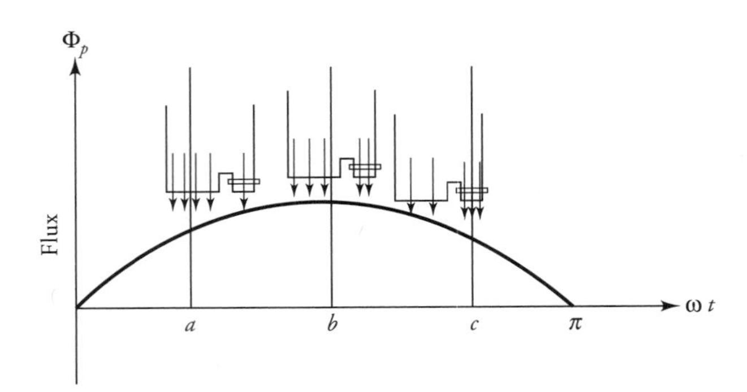

Figure 10.17 Shading-pole action during the positive half cycle of a flux waveform. a, $\omega t < \pi/2$: Almost all the flux passing through unshaded region; b, $\omega t = \pi/2$: No shading action, flux is uniformly distributed over the entire pole; c, $\omega t > \pi/2$: Most of the flux is passing through the shaded region.

motor. The revolving field, however, is neither continuous nor uniform. For this reason, the torque developed by the motor in not uniform but varies from instant to instant.

Since the rotor follows the revolving field, the direction of rotation of a shaded-pole motor cannot be reversed once the motor is built. To have a reversible motor, we must place two shading coils on both sides of the pole and selectively short one of them.

To increase the starting torque, the leading edge of the shaded-pole motor may have a wider air-gap than the rest of the pole. It has been found that if a part of the pole face has a wider gap than the remainder of the pole, the motor develops some starting torque without the auxiliary winding. Such a motor is called a **reluctance start motor**. Adding the reluctance feature to a shaded-pole motor increases its starting torque. This feature is commonly employed in the design of a shaded-pole motor.

A typical speed-torque curve of a shaded-pole motor is given in Figure 10.18. Take another look at this curve. It clearly shows one of the important drawbacks of a shaded-pole motor; the third harmonic dip. To cancel some of the third-harmonic effect, we can use a relatively high-resistance rotor. However, any increase in the rotor resistance is accompanied not only by a decrease in the operating speed of the motor but also by a drop of motor efficiency.

The analysis of a shaded-pole motor is quite involved and is not within the scope of this book. However, for the benefit of those readers who are interested in exploring further, we refer them to an IEEE publication, "Revolving-Field Analysis of a Shaded-pole Motor" by Bhag S. Guru, IEEE Trans., *Power Apparatus and Systems*, Vol. 102, No. 4, pp. 918–927, April 1983. This publication provides (a) a complete theoretical development including the harmonics analysis and non-quadrature placement of the shading coil with respect to the main winding, (b) all the necessary design equations, (c) a step-by-step procedure to develop a computer program for analyzing/designing a shaded-pole motor, and (d) a long list of major publications on the subject.

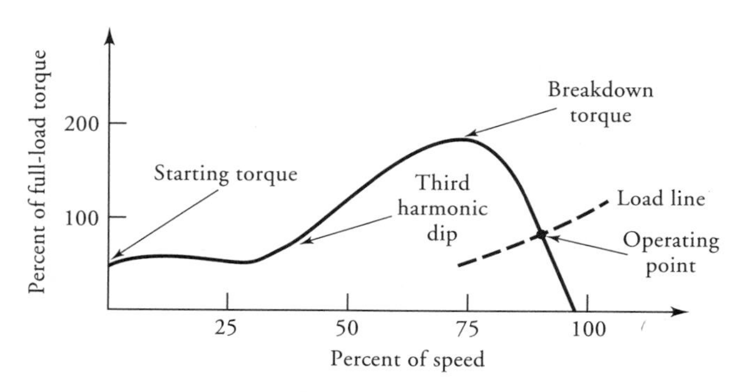

Figure 10.18 Speed-torque characteristic of a shaded-pole motor.

10.8 Universal Motor

A dc series motor specifically designed for ac operation is called a **universal motor**. A universal motor is wound and connected just like a dc series motor. That is, the field winding is connected in series with the armature winding. However, some modifications are necessary to transform a dc series motor into a universal motor. We discuss these modifications later. Let us first explore how a dc series motor can operate on an ac source.

Principle of Operation

When a series motor is operated from a dc source, the current is unidirectional in both the field and the armature windings. In other words, the flux produced by each pole and the direction of the current in the armature conductors under that pole remain in the same direction at all times. Hence, the torque developed by the motor is constant.

When a series motor is connected to an ac source, the current in the field and the armature windings reverses its direction every half cycle, as shown in Figure 10.19 for a two-pole series motor.

During the positive half cycle (Figure 10.19**a**), the flux produced by the field winding is from right to left. For the marked direction of the current in the armature conductors, the motor develops a torque in the counterclockwise direction. During the negative half cycle (Figure 10.19**b**), the applied voltage has reversed its polarity. Consequently, the current has reversed its direction. As a result, the

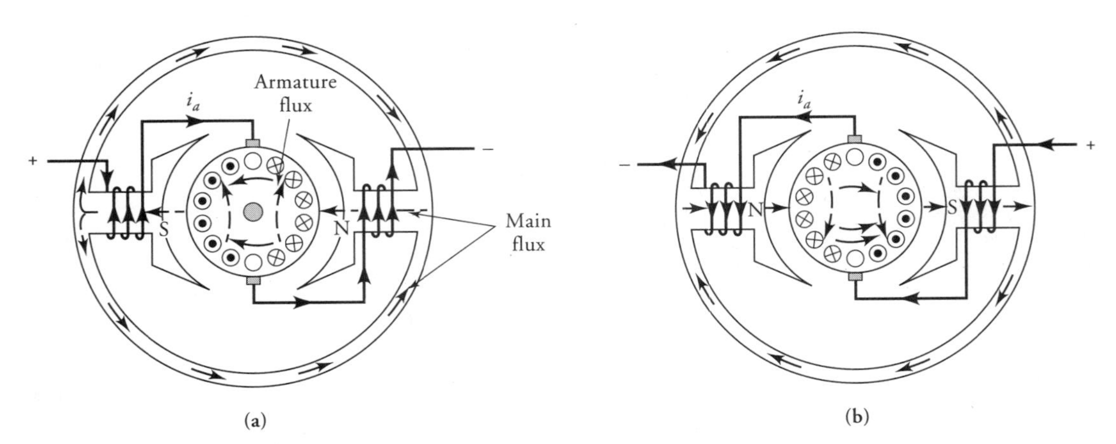

Figure 10.19 Current and flux directions in a universal motor during **(a)** the positive and **(b)** the negative half cycles.

flux produced by the poles is now directed from left to right. Since the reversal in the current in the armature conductors is also accompanied by reversal in the direction of flux in the motor, the direction of the torque developed by the motor remains unchanged. Hence, the motor continues its rotation in the counterclockwise direction.

If K_a is the machine constant, i_a is the current through the field and the armature windings at any instant, and Φ_p is the flux per pole at that instant, the instantaneous torque developed by the motor is $K_a i_a \Phi_p$. When the motor operates in the linear region (below the knee of the magnetization curve), the flux Φ_p must be proportional to the field current i_a. When the permeability of the magnetic core is relatively high, the presence of the air-gaps ensures that for all practical purposes the flux Φ_p is in phase with the current i_a. Thus, the instantaneous torque developed by the motor is proportional to the square of the armature current, as shown in Figure 10.20. In other words, the average value of the torque developed is proportional to the root-mean-square (rms) value of the current. It is obvious from Figure 10.20 that the torque developed by the universal motor varies with twice the frequency of the ac source. Such pulsations in torque cause vibrations and make the motor noisy.

Figure 10.21 shows the equivalent circuit, the phasor diagram, and the speed-torque characteristics of a universal motor. The back emf E_a, the winding current I_a, and the flux per pole Φ_p are in phase with each other as shown. R_s and X_s are the resistance and the reactance of the series field winding. R_a and X_a are the resistance and the reactance of the armature winding.

Design Considerations

1. When a series motor is to be designed as a universal motor, its poles and yoke must be laminated in order to minimize the core loss produced in them by the alternating flux. If a series motor with an unlaminated stator is connected to an ac supply, it quickly overheats owing to excessive core loss.

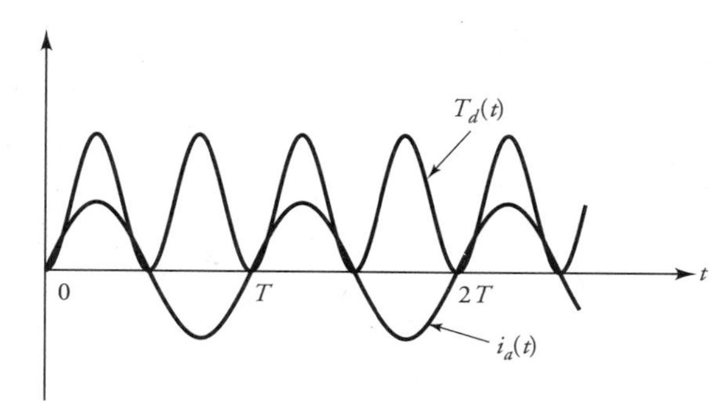

Figure 10.20 The current and the torque developed by a universal motor.

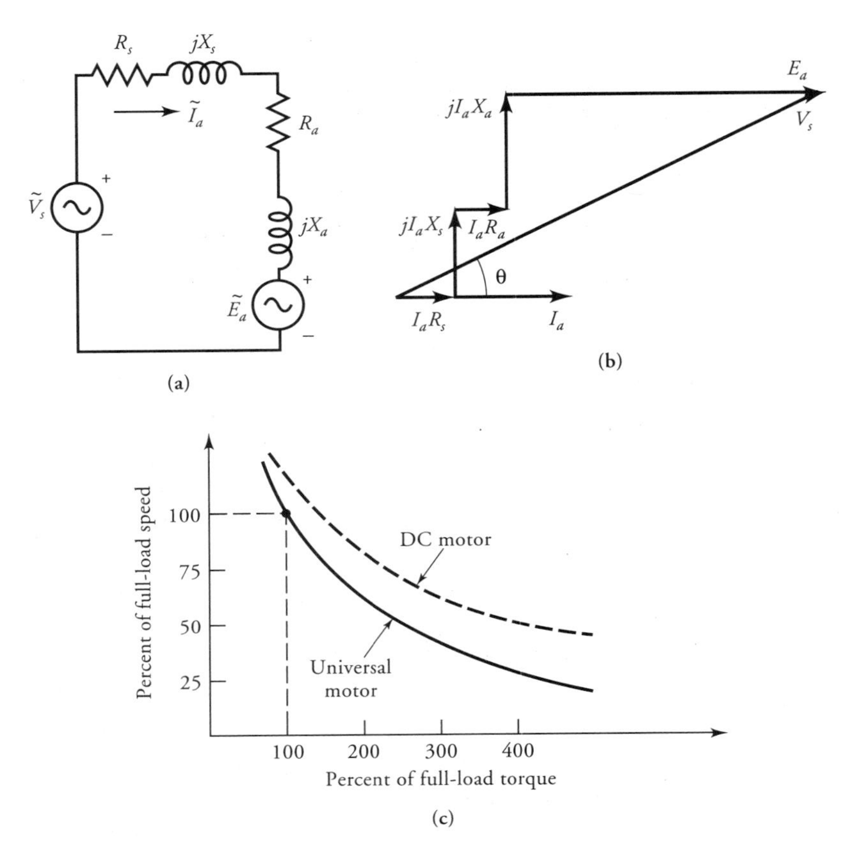

Figure 10.21 **(a)** Equivalent circuit of a universal motor, **(b)** phasor diagram, and **(c)** typical speed-torque characteristic.

2. Under steady-state operation of a dc series motor, the inductances of the series field and armature windings have little effect on its performance. However, the motor exhibits reactive voltage drops across these inductances when connected to an ac source. The high reactive voltage drops have a two-fold effect: (a) reducing the current in the circuit for the same applied voltage, and (b) lowering the power factor of the motor. The reactive voltage drop across the series field winding is made small by using fewer series field turns.
3. The decrease in the number of turns in the series field winding reduces the flux in the motor. This loss in flux is compensated by an increase in the number of armature conductors.
4. The increase in the armature conductors results in an increase in the armature reaction. The armature reaction can, however, be reduced by adding compen-

sating windings in the motor. A universal motor with a conductively coupled compensating winding is shown in Figure 10.22**a**. The corresponding phasor diagram (Figure 10.22**b**) shows the improvement in the power factor.

5. Under ac operation, an emf is induced by transformer action in the coils undergoing commutation. This induced emf (a) causes extra sparking at the brushes, (b) reduces brush life, and (c) results in more wear and tear of the commutator. To reduce these harmful effects, the number of commutator segments is increased and high-resistance brushes are used in universal motors.

One may ask a very logical question: With all these drawbacks, why do we use a universal motor? Some of the reasons are given below:

1. A universal motor is needed when it is required to operate with complete satisfaction on dc and ac supply.
2. The universal motor satisfies the requirements when we need a motor to operate on ac supply at a speed in excess of 3600 rpm (2-pole induction motor operating at 60 Hz). Since the power developed is proportional to the motor speed, a high-speed motor develops more power for the same size than a low-speed motor.
3. When we need a motor that automatically adjusts its speed under load, the universal motor is suitable for that purpose. Its speed is high when the load is light and low when the load is heavy.

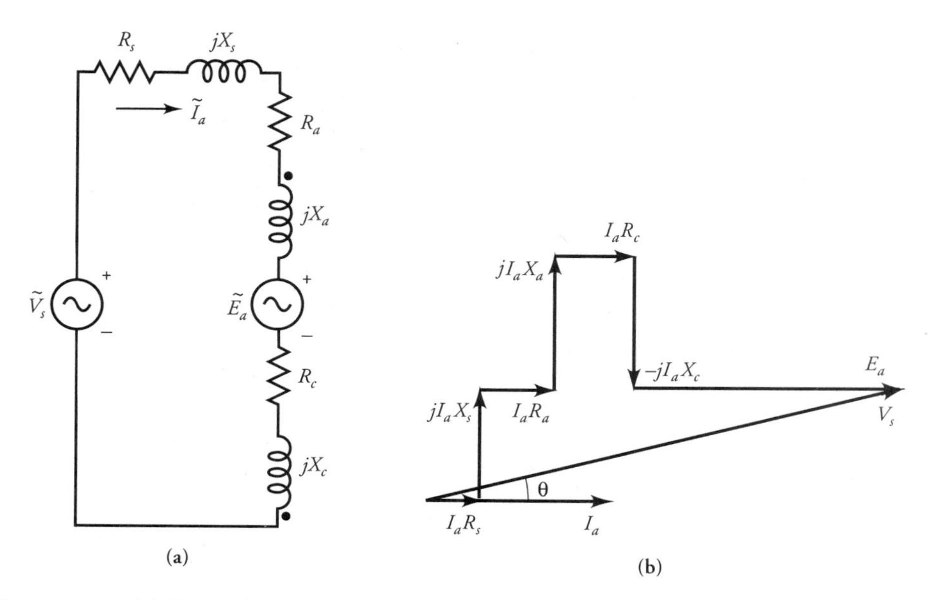

Figure 10.22 **(a)** Equivalent circuit and **(b)** phasor diagram for a conductively compensated universal motor.

In light of the above reasons, a universal motor is used quite extensively in the fractional horsepower range. Some applications that require variation in speed with load are saws and routers, sewing machines, portable machine tools, and vacuum cleaners.

EXAMPLE 10.5

A 120-V, 60-Hz, 2-pole, universal motor operates at a speed of 8000 rpm on full load and draws a current of 17.58 A at a lagging power factor of 0.912. The impedance of the series field winding is $0.65 + j1.2\ \Omega$. The impedance of the armature winding is $1.36 + j1.6\ \Omega$. Determine (a) the induced emf in the armature, (b) the power output, (c) the shaft torque, and (d) the efficiency if the rotational loss is 80 W.

● SOLUTION

(a) From the equivalent circuit of the motor (Figure 10.21**b**), we have

$$\tilde{E}_a = \tilde{V}_s - \tilde{I}_a(R_s + R_a + jX_s + jX_a)$$
$$= 120 - 17.58\underline{/-24.22^\circ}[0.65 + 1.36 + j(1.2 + 1.6)]$$
$$= 74.1\underline{/-24.22^\circ}\ \text{V}$$

As expected, the induced emf is in phase with the armature current.

(b) The power developed by the motor is

$$P_d = \text{Re}[\tilde{E}_a\tilde{I}_a^*] = 74.1 \times 17.58 = 1302.68\ \text{W}$$

The power output: $P_o = P_d - P_{rot} = 1302.68 - 80 = 1222.68$ W

(c) From the rated speed of the motor, we obtain

$$\omega_m = \frac{2\pi \times 8000}{60} = 837.76\ \text{rad/s}$$

Hence, the shaft torque: $T_s = 1222.68/837.76 = 1.46$ N·m

(d) The power input is

$$P_{in} = 120 \times 17.58 \times 0.912 = 1923.96\ \text{W}$$

The efficiency: $\eta = 1222.68/1923.96 = 0.636$ or 63.6%

■

Exercises

10.9. A 240-V, 50-Hz, 2-pole, universal motor operates at a speed of 12,000 rpm on full load and draws a current of 6.5 A at 0.94 pf lagging. The motor parameters are $R_a = 6.15\ \Omega$, $X_a = 9.4\ \Omega$, $R_s = 4.55\ \Omega$, and $X_s = 3.2\ \Omega$. Calculate (a) the induced emf of the motor, (b) the shaft torque, and (c) the efficiency if the rotational loss is 65 W.

10.10. If the motor of Exercise 10.9 draws a current of 12.81 A at 0.74 pf lagging when its load is increased, determine the operating speed of the motor. Assume that the motor is operating in the linear region.

SUMMARY

In this chapter we examined three different types of ac motors: the single-phase induction motor, the shaded-pole motor, and the universal motor.

We explained with the help of the double revolving-field theory that a single-phase induction motor continues rotating even when the flux is established by a single (main) winding. The split-phase motor and the capacitor-start motor operate on a single winding when the operating speed is usually above 75% of its synchronous speed. Based upon the double revolving-field theory, we were able to represent these motors by their equivalent circuits. The rotor circuit was resolved into two equivalent circuits: one for the forward revolving field and the other for the backward revolving field.

For a single-phase induction motor to develop starting torque, we need another winding called the auxiliary winding. This winding is usually placed in space quadrature with respect to the main winding.

In a split-phase motor, the phase difference between the currents in the two windings and thereby between the fluxes produced by them is obtained by making the auxiliary winding highly resistive while the main winding is highly inductive. A phase difference of as high as 60° electrical can be obtained for a well-designed split-phase motor.

In a capacitor motor, the phase difference is created by including a capacitor in series with the auxiliary winding. For a properly designed capacitor motor, the phase difference of nearly 90° electrical can be obtained at the time of starting. Thus, a capacitor motor comes close to a two-phase motor.

The power developed and the efficiency of a capacitor motor can be improved by retaining the auxiliary winding in the circuit at all times. Motors that fall into this category are permanent split-capacitor (PSC) motors and capacitor-start capacitor-run (CSCR) motors.

A CSCR motor uses two capacitors: One capacitor is optimized to develop high starting torque while the other is picked on the basis of high efficiency. The selection of the capacitor for a PSC motor is based upon a compromise between the starting torque and its full-load efficiency.

The auxiliary winding of a shaded-pole motor consists of a shorted copper ring. Most often, the two windings are not in space quadrature. The currents in the two windings set up an unbalanced revolving field, and the motor develops some starting torque. In order to increase the starting torque, most shaded-pole motors have a graded air-gap region.

The universal motor is a series dc motor properly designed for ac operation. The motor is capable of operating on both dc and ac supply. The performance, however, is better when the motor operates on the dc supply. Compensating winding may be used to improve the power factor.

The equivalent circuit parameters of an induction motor can be obtained by performing the blocked-rotor and no-load tests. When the blocked-rotor test is performed with the auxiliary winding open, we obtain the rotor resistance, the leakage reactance of the main winding, and the leakage reactance of the rotor. By performing the blocked-rotor test with the main winding open, we obtain the *a*-ratio. Performing the no-load test with the auxiliary winding open yields the magnetization reactance and the rotational loss.

Review Questions

10.1. Describe the construction of the following motors: split-phase motor, capacitor motor, shaded-pole motor, and universal motor.

10.2. Explain the principle of operation of a split-phase motor, a capacitor motor, a shaded-pole motor, and a universal motor.

10.3. Why do we refer to the split-phase motor, capacitor motor, and shaded-pole motor as induction motors?

10.4. Is a universal motor also an induction motor? Give reasons to justify your answer.

10.5. State some of the practical applications of universal motors other than those given in this text.

10.6. Is it always possible to replace a shaded-pole motor with other types of induction motors?

10.7. Can the direction of rotation of a shaded-pole motor be reversed?

10.8. How can the direction of rotation of a split-phase or a capacitor motor be reversed?

10.9. Why is it necessary to "cut off" the auxiliary winding from the circuit when a split-phase motor is operating at full load? What would happen if the centrifugal switch failed?

10.10. Why is a capacitor motor better than a split-phase motor?

10.11. Why is a capacitor-start capacitor-run motor better than a permanent split-capacitor motor?

10.12. What is the effect of armature reaction on the speed of a universal motor?

10.13. Why does a universal motor perform better on dc than ac supply?

10.14. What happens to the power factor of a universal motor when the load is increased?

10.15. What happens to the speed of a universal motor when the load is increased?

10.16. Is it possible for a universal motor to self-destruct under no load?

10.17. A 1/3-hp, 120-V, 60-Hz, shaded-pole motor takes 830 W at its full-load slip of 8%. Calculate (a) the speed and (b) the efficiency of the motor.

10.18. Determine the number of poles, the forward slip, and the backward slip for the following motors: (a) 1140 rpm, 60 Hz, (b) 1440 rpm, 50 Hz, and (c) 3200 rpm, 60 Hz.

10.19. How can a universal motor be reversed? Make a sketch using a double-pole double-throw (DPDT) switch.

10.20. Sketch a reversible shaded-pole motor.

10.21. When the operating speed of a motor under all conditions is less than 3600 rpm, is it still better to use a universal motor than an induction motor?

Problems

10.1. A 120-V, 60-Hz, 2-pole, single-phase induction motor operates at a slip of 4% and has a rotor resistance of 2.4 Ω at standstill. Determine (a) the speed of the motor, (b) the effective rotor resistance in the forward branch, and (c) the effective rotor resistance in the backward branch.

10.2. The effective rotor resistance of a 120-V, 50-Hz, 6-pole, single-phase induction motor in the forward branch at a slip of 5% is 120 Ω. What is the rotor resistance at standstill? What is the effective resistance in the backward branch? What is the operating speed of the motor?

10.3. A 230-V, 60-Hz, 6-pole, single-phase induction motor has a stator impedance of $1.5 + j3\ \Omega$ and a rotor impedance of $2 + j3\ \Omega$ at standstill. The magnetization reactance is 50 Ω. If the rotational loss is 150 W at a slip of 5%, determine the efficiency and the shaft torque of the motor.

10.4. A 208-V, 50-Hz, 4-pole, single-phase induction motor has $R_1 = 2.5\ \Omega$, $X_1 = 2.9\ \Omega$, $R_2 = 2.1\ \Omega$, $X_2 = 2.6\ \Omega$, and $X_m = 42\ \Omega$. If the friction, windage, and core loss is 50 W at a slip of 4%, determine the efficiency and the torque of the motor.

10.5. A 115-V, 60-Hz, 6-pole, single-phase induction motor operates at a speed of 1050 rpm and has $R_1 = 3.8\ \Omega$, $X_1 = 5.9\ \Omega$, $R_2 = 4.2\ \Omega$, $X_2 = 5.9\ \Omega$, and $X_m = 70.8\ \Omega$. Determine the torque and the efficiency of the motor if the rotational loss is 25 W.

10.6. Plot the speed-torque curve, current versus speed, efficiency versus speed, and power output versus speed characteristics of the single-phase induction motor of Problem 10.5. Assume that the rotational loss is proportional to the motor speed.

10.7. A 230-V, 60-Hz, 4-pole, two-value capacitor motor is rated at 1710 rpm. The motor parameters are $R_1 = 30\ \Omega$, $X_1 = 36\ \Omega$, $R_2 = 24\ \Omega$, $X_2 = 30\ \Omega$, $R_a = 120\ \Omega$, $X_m = 250\ \Omega$, $a = 1.75$, start capacitor $= 8\ \mu$F, and run capacitor $= 4\ \mu$F. The rotational loss at full load is 25 W. Determine the shaft-torque and the efficiency at full load. What is the starting torque developed by the motor?

10.8. Plot the speed-torque curve and speed-efficiency characteristic for the motor of Problem 10.7. Assume that (a) the switching action takes place at 75% of the synchronous speed and (b) the rotational loss is proportional to the speed.

10.9. A 115-V, 60-Hz, 6-pole, CSCR motor is rated at 1152 rpm. The motor parameters are $R_1 = 20\ \Omega$, $X_1 = 32\ \Omega$, $R_2 = 22\ \Omega$, $X_2 = 32\ \Omega$, $R_a = 55\ \Omega$, $X_m = 210\ \Omega$, $a = 1.8$, start capacitor $= 8\ \mu$F, and run capacitor $= 5\ \mu$F. The rotational loss is 10 W. Determine the shaft-torque and the efficiency at full load. What is the starting torque developed by the motor?

10.10. Plot the speed-torque curve and the speed-efficiency characteristic for the motor of Problem 10.9. Assume that (a) the switching action takes place at 75% of the synchronous speed and (b) the rotational loss is proportional to the speed.

10.11. A 115-V, 60-Hz, 6-pole, CSCR motor is rated at 1120 rpm. The motor parameters are $R_1 = 6\ \Omega$, $X_1 = 4.8\ \Omega$, $R_2 = 5\ \Omega$, $X_2 = 3.3\ \Omega$, $R_a = 38\ \Omega$, $X_m = 51\ \Omega$, $a = 2.6$, start capacitor $= 20\ \mu$F, and run capacitor $= 10\ \mu$F. The rotational loss is 28 W. Determine the shaft-torque and the efficiency at full load. What is the starting torque developed by the motor?

10.12. Plot the speed-torque curve and the speed-efficiency characteristic for the motor of Problem 10.11. Assume that (a) the switching action takes place at 75% of the synchronous speed and (b) the rotational loss is proportional to the speed.

10.13. A 115-V, 60-Hz, 2-pole, CSCR motor is rated at 3325 rpm. The motor parameters are $R_1 = 16\ \Omega$, $X_1 = 12\ \Omega$, $R_2 = 11\ \Omega$, $X_2 = 6.8\ \Omega$, $R_a = 26\ \Omega$, $X_m = 120\ \Omega$, $a = 1.2$, start capacitor $= 25\ \mu$F, and run capacitor $= 10\ \mu$F. The rotational loss is 10 W. Determine the shaft-torque and the efficiency at full load. What is the starting torque developed by the motor?

10.14. Plot the speed-torque curve and the speed-efficiency characteristic for the motor of Problem 10.13. Assume that (a) the switching action takes place at 75% of the synchronous speed and (b) the rotational loss is proportional to the speed.

10.15. A 220-V, 60-Hz, 4-pole, PSC motor is rated at 1710 rpm. The motor parameters are $R_1 = 20\ \Omega$, $X_1 = 30\ \Omega$, $R_2 = 24\ \Omega$, $X_2 = 30\ \Omega$, $R_a = 60\ \Omega$, $X_m = 200\ \Omega$, $a = 1.5$, and $C = 4\ \mu$F. The rotational loss at full load is 20 W.

Determine the shaft torque and the efficiency at full load. What is the starting torque developed by the motor?

10.16. Plot the speed-torque curve and the speed-efficiency characteristic for the motor of Problem 10.15. Assume that the rotational loss is proportional to the speed.

10.17. A 230-V, 60-Hz, 6-pole, PSC motor is rated at 1152 rpm. The motor parameters are $R_1 = 15\ \Omega$, $X_1 = 42\ \Omega$, $R_2 = 22\ \Omega$, $X_2 = 42\ \Omega$, $R_a = 25\ \Omega$, $X_m = 180\ \Omega$, $a = 1.4$, and $C = 7.5\ \mu F$. The rotational loss is 10 W. Determine the shaft torque and the efficiency at full load. What is the starting torque developed by the motor?

10.18. Plot the speed-torque curve and the speed-efficiency characteristic for the motor of Problem 10.17. Assume that the rotational loss is proportional to the speed.

10.19. A 120-V, 60-Hz, 6-pole, PSC motor is rated at 1120 rpm. The motor parameters are $R_1 = 5\ \Omega$, $X_1 = 5.8\ \Omega$, $R_2 = 7\ \Omega$, $X_2 = 5.8\ \Omega$, $R_a = 12\ \Omega$, $X_m = 80\ \Omega$, $a = 2$, and $C = 10\ \mu F$. The rotational loss is 20 W. Determine the shaft torque and the efficiency at full load. What is the starting torque developed by the motor?

10.20. Plot the speed-torque curve and the speed-efficiency characteristic for the motor of Problem 10.19. Assume that the rotational loss is proportional to the speed.

10.21. A 115-V, 60-Hz, 2-pole, reversible PSC motor is rated at 3325 rpm. The motor parameters are $R_1 = 4\ \Omega$, $X_1 = 8\ \Omega$, $R_2 = 15\ \Omega$, $X_2 = 8\ \Omega$, $R_a = 4\ \Omega$, $X_m = 120\ \Omega$, $a = 1$, and $C = 10\ \mu F$. Neglect the rotational loss. Determine the shaft torque and the efficiency at full load. What is the starting torque developed by the motor?

10.22. A 208-V, 50-Hz, 4-pole, split-phase induction motor has $R_1 = 12.5\ \Omega$, $X_1 = 25\ \Omega$, $R_2 = 25\ \Omega$, $X_2 = 25\ \Omega$, $R_a = 280\ \Omega$, $a = 0.5$, and $X_m = 150\ \Omega$. If the friction, windage, and core loss is 20 W at a speed of 1400 rpm, determine the efficiency and the torque of the motor. What is the starting torque developed by the motor? The switching speed is 75% of the synchronous speed.

10.23. Sketch the speed torque and speed-efficiency curves for the motor of Problem 10.22. Assume that the rotational loss is proportional to the speed.

10.24. A 120-V, 60-Hz, 4-pole, split-phase induction motor has $R_1 = 6.5\ \Omega$, $X_1 = 12.5\ \Omega$, $R_2 = 15\ \Omega$, $X_2 = 12.5\ \Omega$, $R_a = 120\ \Omega$, $a = 0.5$, and $X_m = 150\ \Omega$. If the friction, windage, and core loss is 25 W at a speed of 1650 rpm, determine the efficiency and the torque of the motor. What is the starting torque developed by the motor? The switching speed is 75% of the synchronous speed.

10.25. Sketch the speed-torque and speed-efficiency curves for the motor of Problem 10.24. Assume that the rotational loss is proportional to the speed.

10.26. A 120-V, 60-Hz, 4-pole, capacitor-start induction motor has $R_1 = 5\ \Omega$, $X_1 = 12\ \Omega$, $R_2 = 20\ \Omega$, $X_2 = 12\ \Omega$, $R_a = 20\ \Omega$, $a = 1.5$, and $X_m = 200\ \Omega$. The value of the start capacitor is 20 μF. If the friction, windage, and core loss is 20 W at a speed of 1650 rpm, determine the efficiency and the torque of the motor. What is the starting torque developed by the motor? The switching speed is 75% of the synchronous speed.

10.27. Sketch the speed-torque and the speed-efficiency curves for the motor of Problem 10.26. Assume that (a) the rotational loss is proportional to the speed and (b) the switching action takes place at 75% of the synchronous speed.

10.28. A 230-V, 60-Hz, 6-pole, capacitor-start motor is rated at 1152 rpm. The motor parameters are $R_1 = 10\ \Omega$, $X_1 = 22\ \Omega$, $R_2 = 20\ \Omega$, $X_2 = 22\ \Omega$, $R_a = 18\ \Omega$, $X_m = 220\ \Omega$, $a = 1.8$, and start capacitor = 8 μF. The rotational loss is 10 W. Determine the shaft torque and the efficiency at full load. What is the starting torque developed by the motor?

10.29. Plot the speed-torque curve and the speed-efficiency characteristic for the motor of Problem 10.28. Assume that (a) the switching action takes place at 75% of the synchronous speed and (b) the rotational loss is proportional to speed.

10.30. A 4-pole, 120-V, 60-Hz, capacitor-start motor is tested to yield the following data:
No-load test with auxiliary open: 120 V, 2.7 A, 56 W
Blocked-rotor test with auxiliary open: 120 V, 15 A, 1175 W
Blocked-rotor test with main open: 120 V, 5.2 A, 503.4 W
The main-winding resistance is 2.5 Ω and the auxiliary-winding resistance is 12.5 Ω. Determine the equivalent circuit parameters of the motor.

10.31. A 208-V, 4-pole, 50-Hz, reversible, PSC motor, when tested with auxiliary winding open, gave the following results:
No-load test: 208 V, 3.8 A, 96 W
Blocked-rotor test: 80 V, 8 A, 420 W
For a reversible motor to develop the same torque in either direction of rotation at full load, the main and the auxiliary windings are identical ($a = 1$). The resistance of each winding is 3.8 Ω. Determine the equivalent circuit parameters of the motor.

10.32. Calculate the torque developed and the efficiency of the motor of Problem 10.31 at a slip of 5%. What is the rated speed of the motor? What must be the voltage rating of the capacitor if the motor uses a 5-μF capacitor?

10.33. The following test results were obtained from a 120-V, 60-Hz, split-phase motor:
Main-winding resistance: 3.5 Ω
Auxiliary-winding resistance: 140 Ω
Blocked-rotor test with auxiliary open: 96.8 V, 10 A, 600 W

No-load test with auxiliary open: 120 V, 2.5 A, 45 W
Blocked-rotor test with main open: 120 V, 0.85 A, 102.2 W
Compute the equivalent circuit parameters of the motor.

10.34. A 120-V, 6-pole, 50-Hz, reversible, PSC motor, when tested with the auxiliary winding open, gave the following results:
No-load test: 120 V, 1.2 A, 60 W
Blocked-rotor test: 90 V, 5 A, 420 W
For a reversible motor to develop the same torque in either direction of rotation at full load, the main and the auxiliary windings are identical ($a = 1$). The resistance of each winding is 15 Ω. Determine the equivalent circuit parameters of the motor.

10.35. Calculate the torque developed and the efficiency of the motor of Problem 10.34 at a slip of 4%. What is the rated speed of the motor? What must be the voltage rating of the capacitor if the motor uses a 7.5-μF capacitor?

10.36. A 110-V, 60-Hz, 2-pole, universal motor operates at a speed of 12,500 rpm on full load and draws a current of 5 A at 0.74 pf lagging. The motor parameters are $R_s = 0.5\ \Omega$, $X_s = 2.235\ \Omega$, $R_a = 5.2\ \Omega$, and $X_a = 12.563\ \Omega$. Calculate (a) the induced emf in the armature, (b) the shaft torque, and (c) the efficiency of the motor.

10.37. If the motor of Problem 10.36 draws a current of 5.65 A at a power factor of 0.65 lagging when the load is increased, determine (a) the operating speed of the motor, (b) the power output, and (c) the shaft torque. Assume that the rotational loss is proportional to the motor speed.

CHAPTER 11

Dynamics of Electric Machines

Photograph of a motor for industrial instrumentation and precision applications (*Courtesy of Bodine Electric Company*)

11.1 Introduction

The discussion of electric machines presented in the preceding chapters was limited to the steady-state operating conditions. That is, the machine has either been running at a given condition for a long time or moving from one operating condition to another very slowly so that the energy imbalance between the electrical and mechanical elements would be insignificant during the transition period. However, when the change from one operating condition to another is sudden, changes in the stored magnetic energy and the stored energy in terms of the inertia of the rotating members do not occur instantaneously. As a result, a finite time, known as the transient (dynamic) period, is needed to restore the energy balance from the initial and to the final conditions. For example, if the applied voltage to a motor changes suddenly, the machine will undergo a transient period prior to attaining its new steady-state condition. In fact, such events occur quite often when machines are driven by electronic drives, as they generate periodic discontinuous voltage and current waveforms. Another condition that may cause a change in the dynamic equilibrium of a machine is a sudden change in its load.

From the above explanation, it is quite clear that the study of both electrical and mechanical transients is essential to better understand the operation of electric machines. Therefore, this chapter is devoted to the dynamics of direct-current (dc) and alternating-current (ac) machines.

11.2 DC Machine Dynamics

The saturation of the magnetic core makes the study of dynamics in dc machines quite difficult because the mathematical representation of the machine yields a set of nonlinear differential equations. In that case a closed-form solution would not be possible, and the need for a numerical method becomes inevitable. Hence, within the scope of our discussion, we consider a model that assumes the machine to be a piecewise linear device in which the magnetic flux is a linear function of the field current. Nevertheless, in practice, if a dc machine is operated in the unsaturated region of its magnetic characteristic, it can be approximated as a linear device.

Armature-Controlled DC Motors

The switch S connects a separately excited dc motor to a dc voltage supply, as shown in Figure 11.1. Owing to the sudden change in energy in the motor, until the energy readjusts itself, the motor experiences a transient state whose duration is essentially governed by the parameters of the motor. For a moment let us assume that there is no variation in the field current, but the armature current can

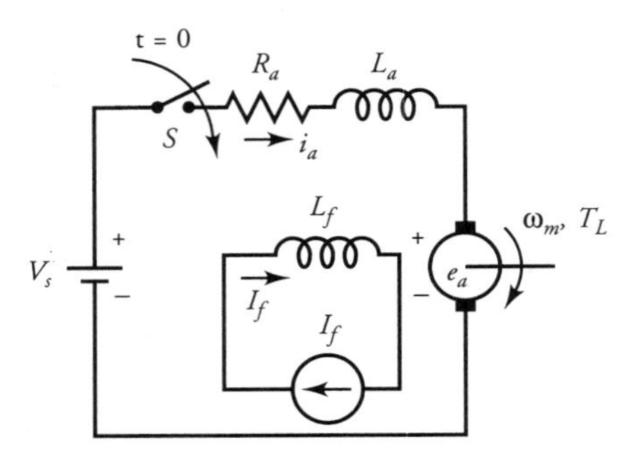

Figure 11.1 A separately excited dc motor with a constant field current.

vary depending on the applied voltage and the load. We can mathematically represent the variation of the armature current for such an operating condition by the following first-order differential equation:

$$R_a i_a(t) + L_a \frac{di_a(t)}{dt} + K\omega_m(t) = V_s \tag{11.1}$$

where R_a and L_a are the armature resistance and inductance, respectively. In this model $i_a(t)$ is the armature current while V_s is the applied voltage across the motor terminals. In Eq. (11.1) $K\omega_m(t)$ is the back electromotive force (emf) of the motor (e_a), where $K = K_a\Phi_p$ is a constant at a uniform field current I_f, and $\omega_m(t)$ is the angular velocity of the rotor.

A similar equation can be written to express the speed of the motor as follows:

$$T_L + D\omega_m(t) + J\frac{d\omega_m(t)}{dt} = Ki_a(t) \tag{11.2}$$

Here, $Ki_a(t)$ is the developed torque (T_d), and T_L is the load torque. Also, D and J are the viscous friction coefficient and the moment of inertia of the rotating members, respectively. Equations (11.1) and (11.2) can be rearranged in the matrix form as

$$\begin{bmatrix} \dfrac{d\omega_m(t)}{dt} \\ \\ \dfrac{di_a(t)}{dt} \end{bmatrix} = \begin{bmatrix} -\dfrac{D}{J} & \dfrac{K}{J} \\ \\ -\dfrac{K}{L_a} & -\dfrac{R_a}{L_a} \end{bmatrix} \begin{bmatrix} \omega_m(t) \\ \\ i_a(t) \end{bmatrix} + \begin{bmatrix} -\dfrac{1}{J} & 0 \\ \\ 0 & \dfrac{1}{L_a} \end{bmatrix} \begin{bmatrix} T_L \\ \\ V_s \end{bmatrix} \tag{11.3}$$

In the above equations the dynamics of a dc motor with a constant excitation is expressed in the state-space form, where $\omega_m(t)$ and $i_a(t)$ constitute the state variables and V_s and T_L are the input variables.

The solutions of the above equations provide us with the variations of the motor speed, ω_m, and the armature current, i_a, as a function of time. Since Eqs. (11.1) and (11.2) are a set of linear differential equations, they can be solved either by Laplace transformation technique or by a numerical method. In this chapter we use both techniques to determine $\omega_m(t)$ and $i_a(t)$.

Laplace Transformation Method

Laplace transformation allows us to transform linear differential equations into algebraic equations.

Taking the Laplace transform of Eqs. (11.1) and (11.2), we obtain

$$R_a I_a(s) + sL_a I_a(s) + K\Omega_m(s) = V_s(s) + L_a i_a(0) \tag{11.4}$$

and

$$T_L(s) + D\Omega_m(s) + sJ\Omega_m(s) = KI_a(s) + J\omega_m(0) \tag{11.5}$$

where $I_a(s)$, $\Omega_m(s)$, $V_s(s)$ and $T_L(s)$ are the Laplace transforms of $i_a(t)$, $\omega_m(t)$, V_s, and T_L, respectively.

When we solve Eqs. (11.4) and (11.5) simultaneously, we get

$$\Omega_m(s) = \frac{K[V_s(s) + L_a i_a(0)] + [J\omega_m(0) - T_L(s)](R_a + sL_a)}{(D + sJ)(R_a + sL_a) + K^2} \tag{11.6}$$

and

$$I_a(s) = \frac{-K[J\omega_m(0) - T_L(s)] + [L_a i_a(0) + V_s(s)](D + sJ)}{(D + sJ)(R_a + sL_a) + K^2} \tag{11.7}$$

If the motor is at standstill when it is energized at $t = 0$, $i_a(0)$ and $\omega_m(0)$ are zero. Consequently, Eqs. (11.6) and (11.7) become

$$\Omega_m(s) = \frac{KV_s(s) - T_L(s)(R_a + sL_a)}{(D + sJ)(R_a + sL_a) + K^2} \tag{11.8}$$

and

$$I_a(s) = \frac{KT_L(s) + V_s(s)(D + sJ)}{(D + sJ)(R_a + sL_a) + K^2} \tag{11.9}$$

The inverse Laplace transforms of Eqs. (11.6) and (11.7) or Eqs. (11.8) and (11.9) yield the motor speed and the armature current in time domain, as illustrated in Example 11.1.

EXAMPLE 11.1

An 80% efficient, separately excited dc motor is rated at 1150 rpm, 240 V, and 3 hp. The armature resistance and the inductance are 1.43 Ω and 10.4 mH, respectively. The moment of inertia of the motor is 0.068 kg-m² while K is 1.8, and the frictional losses are negligible. Determine the speed and the armature current at no load as a function of time after the motor is energized by applying the rated voltage at $t = 0$. Assume that the motor is at rest for $t \leq 0$.

● SOLUTION

Since the frictional losses are neglected, the friction coefficient D is zero. At $t = 0$, $\omega_m(0)$ and $i_a(0)$ are zero, and we can use Eqs. (11.8) and (11.9) to determine $\omega_m(t)$ and $i_a(t)$, respectively. Also, the load torque is zero because the motor operates at no load.

Substituting the given values in Eq. (11.8), we get

$$\Omega_m(s) = \frac{1.8\,\dfrac{240}{s}}{0.068s(1.43 + 10.4 \times 10^{-3}s) + 1.8^2}$$

or

$$\Omega_m(s) = \frac{610{,}859.72}{s(s^2 + 137.5s + 4581.45)}$$

In order to determine the inverse Laplace transform of $\Omega_m(s)$, we expand $\Omega_m(s)$ into partial fractions as

$$\Omega_m(s) = \frac{A}{s} + \frac{B}{s + 80.80} + \frac{C}{s + 56.70}$$

where A, B, and C can now be determined by root-substitution method. Thus,

$$A = s \left. \frac{610{,}859.72}{s(s+80.80)(s+56.70)} \right|_{s=0} = 133.33$$

$$B = (s+80.80) \left. \frac{610{,}859.72}{(s+80.80)s(s+56.70)} \right|_{s=-80.80} = 313.81$$

$$C = (s+56.70) \left. \frac{610{,}859.72}{(s+56.70)s(s+80.80)} \right|_{s=-56.70} = -447.14$$

Finally, we can take the inverse Laplace transform of

$$\Omega_m(s) = \frac{133.33}{s} + \frac{313.81}{s+80.80} - \frac{447.14}{s+56.70}$$

and get the angular velocity in rad/s as

$$\omega_m(t) = 133.33 + 313.81e^{-80.80t} - 447.14e^{-56.70t} \qquad \text{for } t \geq 0.$$

The graph of $\omega_m(t)$ is given in Figure 11.2.
From Eq. (11.9), the Laplace transform of the armature current is

$$I_a(s) = \frac{23{,}076.92}{s^2 + 137.50s + 4581.45}$$

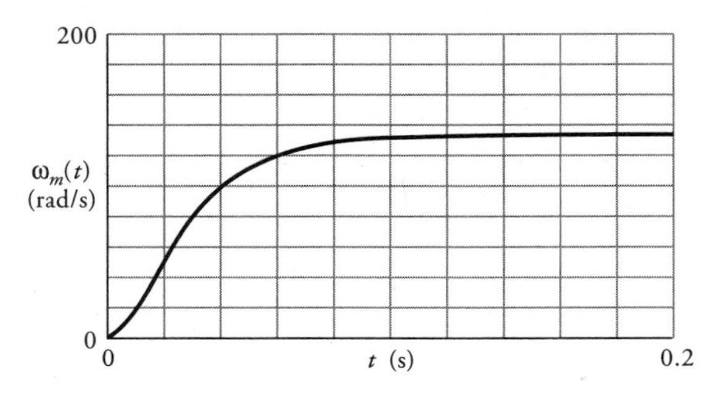

Figure 11.2 The motor speed as a function of time.

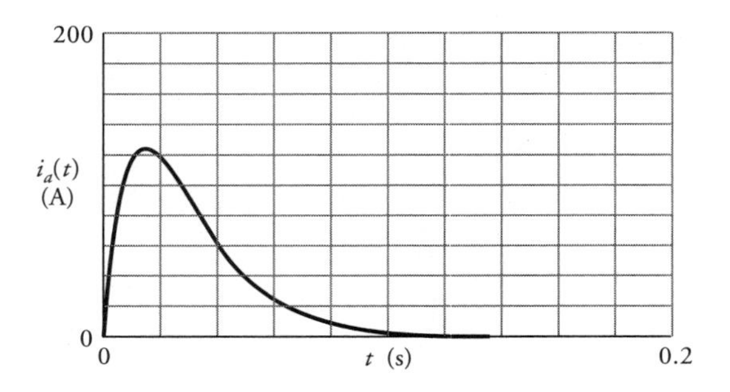

Figure 11.3 The armature current as a function of time.

In terms of its partial-fraction expansion, $I_a(s)$ can be written as

$$I_a(s) = \frac{957.85}{s + 56.70} - \frac{957.85}{s + 80.80}$$

Finally, we obtain the armature current as

$$i_a(t) = 957.85e^{-56.70t} - 957.85e^{-80.80t} \text{ A} \qquad \text{for } t \geq 0$$

which is shown graphically in Figure 11.3.

■

EXAMPLE 11.2

The motor given in Example 11.1 is coupled to a load having a torque of 18.58 N·m. Determine the variation of the motor speed as a function of time after the motor is suddenly energized at its rated voltage at $t = 0$.

● SOLUTION

Since the voltage is applied to the armature circuit at $t = 0$, the initial speed and the armature current are both zero.
From Eq. (11.8),

$$\Omega_m(s) = \frac{1.8\dfrac{240}{s} - \dfrac{18.58}{s}(1.43 + 10.43 \times 10^{-3}s)}{0.068s(1.43 + 10.4 \times 10^{-3}s) + 1.8^2}$$

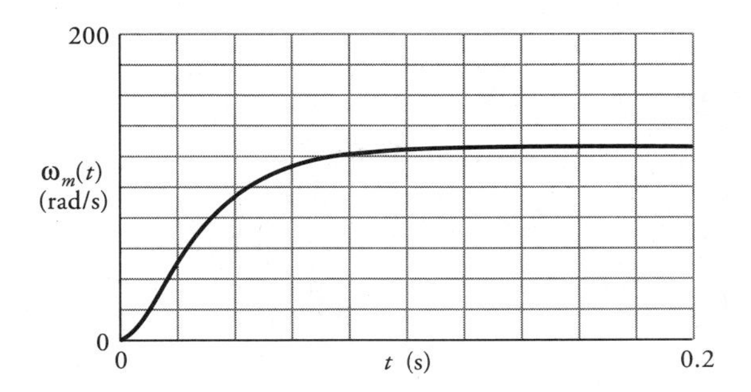

Figure 11.4 The motor speed as a function of time.

or

$$\Omega_m(s) = \frac{573{,}289.03 - 274.32s}{s(s^2 + 137.50s + 4581.45)}$$

$$= \frac{125.13}{s} + \frac{305.90}{s + 80.80} - \frac{431.03}{s + 56.70}$$

The inverse Laplace transform of $\Omega_m(s)$ yields

$$\omega_m(t) = 125.13 + 305.90e^{-80.80t} - 431.03e^{-56.70t} \text{ rad/s} \qquad \text{for } t \geq 0.$$

The variation in motor speed as a function of time is given in Figure 11.4.

For all practical purposes, the motor attains its steady-state operation after five time constants. Thus, this motor takes approximately 88.18 ms (based upon the largest time constant) to achieve its steady state.

■

EXAMPLE 11.3

The motor studied in Example 11.1 is suddenly energized with its rated voltage at $t = 0$ when it was at rest and coupled to a linear load of $T_L = 0.05\omega_m$.

(a) Determine the variation of the motor speed as a function of time for $t \geq 0$.

(b) Calculate the time needed to achieve the steady state.

● SOLUTION

(a) Since the motor was at rest before it was energized at $t = 0$, the initial speed and the initial armature current are zero. Thus, from Eq. (11.8) with $T_L(s) = 0.05\ \Omega_m(s)$, we get

$$\Omega_m(s) = \frac{1.8\dfrac{240}{s} - 0.05\ \Omega_m(s)(1.43 + 10.4 \times 10^{-3}s)}{0.068s(1.43 + 10.4 \times 10^{-3}s) + 1.8^2}$$

and after grouping the terms, we obtain

$$\Omega_m(s) = \frac{610{,}859.72}{s(s^2 + 138.29s + 4682.55)}$$

The roots of the polynomial in the denominator are

$$s_1 = 0,\ s_2 = -79.07,\ \text{and}\ s_3 = -59.22$$

Thus, in terms of partial-fraction expansion, $\Omega_m(s)$ can be written as

$$\Omega_m(s) = \frac{130.54}{s} + \frac{389.25}{s + 79.07} - \frac{519.71}{s + 59.22}$$

The inverse Laplace transform yields, for $t \geq 0$

$$\omega_m(t) = 130.54 + 389.25e^{-79.07t} - 519.71e^{-59.22t}\ \text{rad/s}$$

The variation of the speed with time is also given in graphic form in Figure 11.5.

(b) The largest time constant is

$$\tau = \frac{1}{59.22} = 1.69 \times 10^{-2}\ \text{s}$$

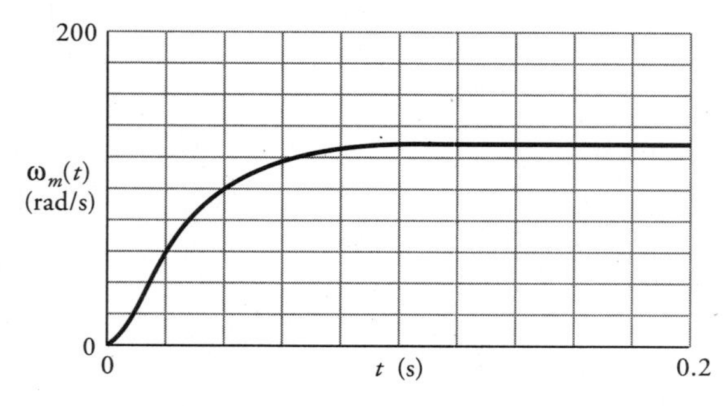

Figure 11.5 The motor speed as a function of time.

To achieve a steady state, the time taken should, at least, be 5τ. Hence,

$$t = 5 \times 1.69 \times 10^{-2} = 8.45 \times 10^{-2} s = 84.5 \text{ ms}$$

is the time needed to reach the steady state.

In Examples 11.1, 11.2 and 11.3, we deliberately used the same motor to give you the opportunity to observe the responses of the same motor under different operating conditions.

■

Exercises

11.1. The armature winding of the motor given in Example 11.1 can withstand twice its rated current for 6 ms. Determine the amount of resistance to be connected in series with the armature winding for a safe start-up at no load.

11.2. A 240-V, 2-hp, 850-rpm permanent-magnet (PM) dc motor operates at its rated conditions for a long time. Determine the variation of speed and armature current as a function of time if the load is suddenly removed from its shaft. The armature resistance and inductance are 2.6 Ω and 19 mH, respectively. The moment of inertia of the motor is 0.07 kg-m^2 while K is 2.5, and the frictional losses are negligible.

Field-Controlled DC Motors

In the armature-controlled dc motors we kept field current at a constant level and let the armature current vary with the changes in the load and/or the supply voltage. In a field-controlled dc motor, however, we assume a constant current in the armature circuit but have a varying field current in the field winding, which facilitates the control of the magnetic flux in the machine.

Let us consider the separately excited motor shown in Figure 11.6 driven by a constant current source in order to ensure constant armature current under any given load. In the linear region of the magnetization characteristic, the flux produced by the motor is proportional to the field current (i.e., $\Phi_p = K_f i_f$). The torque developed by the motor can be expressed as

$$T_d = K_a \Phi_p I_a = K_T \, i_f \tag{11.10}$$

where $K_T = K_a K_f I_a$ is a constant of proportionality. In the dynamic state, the torque developed by the motor can also be written as

$$K_T \, i_f(t) = T_L + D\omega_m(t) + J \frac{d\omega_m(t)}{dt} \tag{11.11}$$

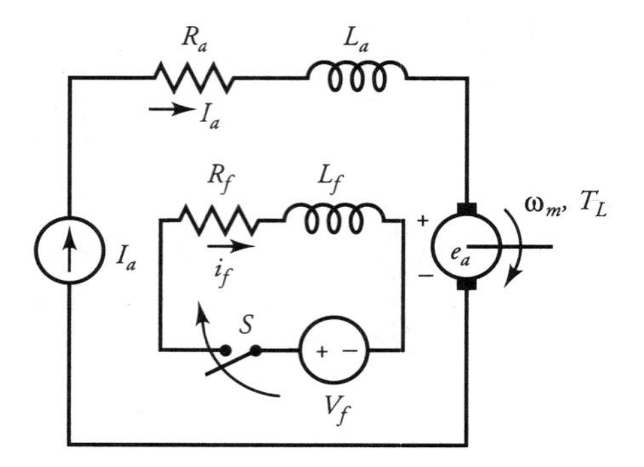

Figure 11.6 A separately excited dc motor with a constant armature current.

The voltage applied to the field circuit, from Figure 11.6, is

$$V_f = R_f i_f(t) + L_f \frac{di_f(t)}{dt} \tag{11.12}$$

where R_f and L_f are the effective resistance and inductance of the field circuit. These equations can be expressed in concise matrix form as

$$\begin{bmatrix} \dfrac{d\omega_m(t)}{dt} \\ \dfrac{di_f(t)}{dt} \end{bmatrix} = \begin{bmatrix} -\dfrac{D}{J} & \dfrac{K_T}{J} \\ 0 & -\dfrac{R_f}{L_f} \end{bmatrix} \begin{bmatrix} \omega_m(t) \\ i_f(t) \end{bmatrix} + \begin{bmatrix} -\dfrac{1}{J} & 0 \\ 0 & \dfrac{1}{L_f} \end{bmatrix} \begin{bmatrix} T_L(t) \\ V_f \end{bmatrix} \tag{11.13}$$

with $\omega_m(t)$ and $i_f(t)$ as the state variables, since they dictate the energy storage in terms of the moment of inertia of the rotor and the inductance of the field winding.

To obtain the dynamic response of the motor, Eqs. (11.11) and (11.12) can be transformed using the Laplace transformation as

$$K_T I_f(s) = (D + sJ)\Omega_m(s) + T_L(s) - J\omega_m(0) \tag{11.14}$$

and

$$V_f(s) = (R_f + sL_f)I_f(s) - L_f i_f(0) \tag{11.15}$$

From Eqs. (11.14) and (11.15), the Laplace-transformed angular velocity and the field current are

$$\Omega_m(s) = \frac{K_T[V_f(s) + L_f i_f(0)] + [J\omega_m(0) - T_L(s)](R_f + sL_f)}{(D + sJ)(R_f + sL_f)} \tag{11.16}$$

and

$$I_f(s) = \frac{V_f(s) + L_f i_f(0)}{R_f + sL_f} \tag{11.17}$$

EXAMPLE 11.4

A 240-V, separately excited dc motor with a field winding resistance and inductance of 120 Ω and 0.7 H, respectively, operates at no load in the linear region of its magnetization characteristic. The flux per pole in the machine is 0.2 times the field current, and the moment of inertia of the rotor is 0.08 kg-m^2. Determine the field current and the motor speed as a function of time when the field voltage is suddenly reduced from 240 V to 100 V. The motor constant $K_a = 5.5$, and the losses are negligible.

● SOLUTION

Since the motor has been operating at steady state before the field voltage is suddenly changed, we must first determine the initial conditions on i_f from the steady-state information.
The field current $i_f(0) = I_f = 240/120 = 2$ A
From Eq. (11.17) we can determine the Laplace transform of the field current as

$$I_f(s) = \frac{\frac{100}{s} + (0.7 \times 2)}{120 + 0.7s}$$

$$I_f(s) = \frac{2s + 142.85}{s(s + 171.43)}$$

or

$$I_f(s) = \frac{0.833}{s} + \frac{1.167}{s + 171.43}$$

in partial fraction expansion form. The inverse Laplace transform yields

$$i_f(t) = 0.833 + 1.167e^{-171.43t} \text{ A} \qquad \text{for } t \geq 0$$

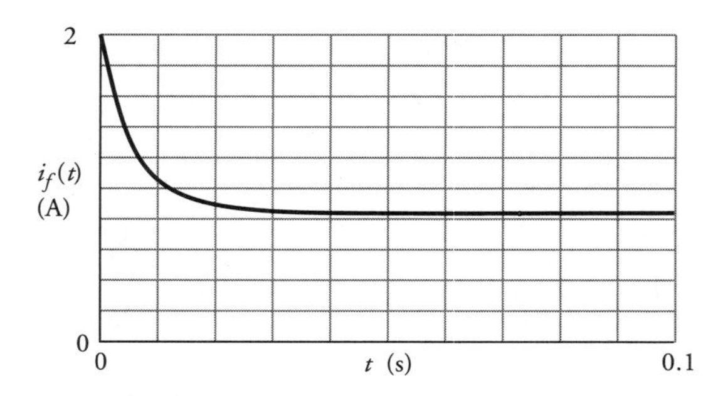

Figure 11.7 The field current as a function of time.

The field-current waveform is given in Figure 11.7.

Since the motor has been operating at no load and the losses are negligible, there is no current in the armature circuit, i.e., $I_a = 0$. In this case, the motor back emf is equal to the supply voltage as expressed below:

$$V_s = E_a = K_a K_f i_f(t) \omega_m(t)$$

From this equation we obtain the motor speed as

$$\omega_m(t) = \frac{V_s}{K_a K_f i_f(t)}$$

or

$$\omega_m(t) = \frac{V_s}{K_a K_f (0.833 + 1.167 e^{-171.43t})}$$

The graph in Figure 11.8 illustrates the changes in the motor speed. Note the increase in the speed of the motor from 115 rad/s at 240 V to 262 rad/s at 100 V. ■

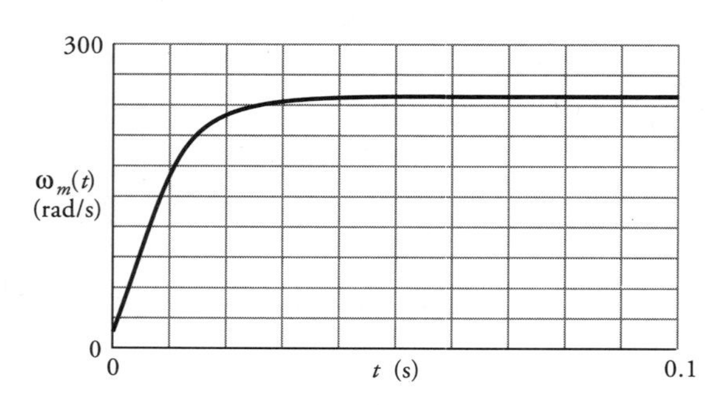

Figure 11.8 The motor speed as a function of time.

Exercises

11.3. A 24-V, 3000-rpm, 50-W, 70% efficient, separately excited dc motor operates at its rated conditions. If the speed is reduced to 1500 rpm at constant load and the time allowed to restore the steady state is 2 ms, determine the moment of inertia for this motor. The resistance of the armature winding is 5 Ω. The armature inductance and the effect of saturation are negligible. Assume a viscous friction coefficient of 0.01 N-m-s.

11.4. To run the motor given in Exercise 11.2 at 400 rpm at no load, calculate (a) the total resistance in the field circuit and (b) the time taken to reach the steady state. The rated field voltage is 24 V, and the field-winding resistance and inductance are 250 Ω and 100 mH, respectively. The flux per pole is given as $\Phi_p = 0.1 i_f$.

Dynamics of DC Generators

A separately excited dc generator delivering power to a static load is shown in Figure 11.9. Once again, we assume that the generator operates in the linear region. To determine the dynamic behavior of the generator with respect to the changes in the armature and field currents, a considerable simplification can be achieved by making an assumption that the shaft speed is practically constant.

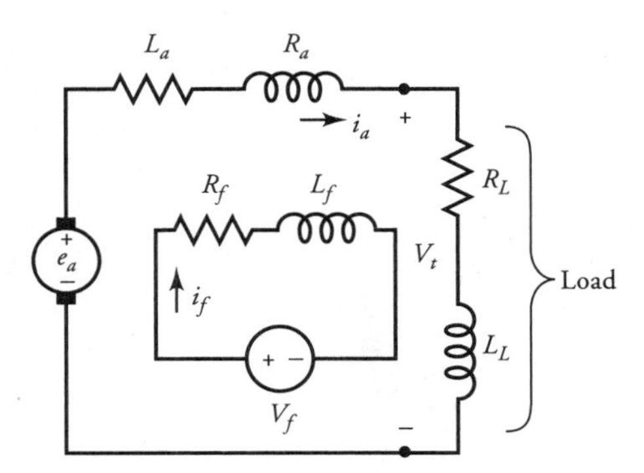

Figure 11.9 Equivalent circuit of a separately excited dc generator.

During the transient state, the field voltage is

$$V_f = R_f i_f(t) + L_f \frac{di_f(t)}{dt} \tag{11.18}$$

and the generated voltage is

$$e_a(t) = (R_a + R_L)i_a(t) + (L_a + L_L)\frac{di_a(t)}{dt} \tag{11.19}$$

However, the generated voltage can be expressed as

$$e_a(t) = K_e i_f(t) \tag{11.20}$$

where $K_e = K_a K_f \omega_m$ is a constant of proportionality. By substituting Eq. (11.20) into Eq. (11.19), we obtain

$$K_e i_f(t) = (R_a + R_L)i_a(t) + (L_a + L_L)\frac{di_a(t)}{dt} \tag{11.21}$$

The dynamic representation of the generator-load system can be written in the matrix form as

$$\begin{bmatrix} \dfrac{di_a(t)}{dt} \\ \dfrac{di_f(t)}{dt} \end{bmatrix} = \begin{bmatrix} -\dfrac{R_a + R_L}{L_a + L_L} & \dfrac{K_e}{L_a + L_L} \\ 0 & -\dfrac{R_f}{L_f} \end{bmatrix} \begin{bmatrix} i_a(t) \\ i_f(t) \end{bmatrix} + \begin{bmatrix} 0 \\ \dfrac{1}{L_f} \end{bmatrix} V_f \tag{11.22}$$

In the above set of equations, the armature and field currents are the state variables and V_f is the input variable.

With zero initial conditions, Eqs. (11.20) and (11.21) can be expressed in Laplace transform as

$$V_f(s) = (R_f + sL_f)I_f(s) \tag{11.23}$$

and

$$K_e I_f(s) = (R_a + R_L + sL_a + sL_L)I_a(s) \tag{11.24}$$

From these equations we obtain

$$I_f(s) = \frac{V_f(s)}{R_f + sL_f} \tag{11.25}$$

and

$$I_a(s) = \frac{K_e V_f(s)}{(R_f + sL_f)(R_a + R_L + sL_a + sL_f)} \tag{11.26}$$

EXAMPLE 11.5

A separately excited dc generator operating at 1500 rpm has the following parameters: $R_f = 3\ \Omega$, $L_f = 25$ mH, and $K_e = 30$ V/A. If a dc voltage of 120 V is suddenly applied to the field winding under no load, determine (a) the field current and the generated voltage as a function of time, (b) the approximate time to reach the steady-state condition, and (c) the steady-state values of the field current and induced voltage.

● SOLUTION

(a) From Eq. (11.25)

$$I_f(s) = \frac{\frac{120}{s}}{3 + 0.025s}$$

$$= \frac{40}{s} - \frac{40}{s + 120}$$

Therefore, the field current is

$$i_f(t) = 40(1 - e^{-120t})\ \text{A} \qquad \text{for } t \geq 0$$

and the generated voltage is

$$e_a(t) = K_e i_f(t) = 1200(1 - e^{-120t})\ \text{V} \qquad \text{for } t \geq 0.$$

The graphs of i_f and e_a are shown in Figures 11.10 and 11.11, respectively.

(b) For all practical purposes, the field current attains its steady-state value

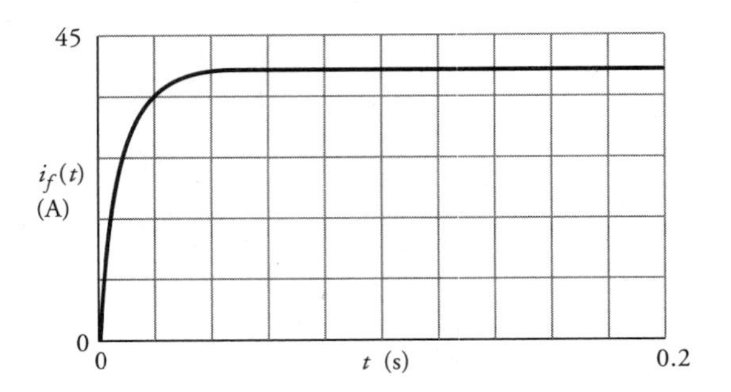

Figure 11.10 The field current as a function of time.

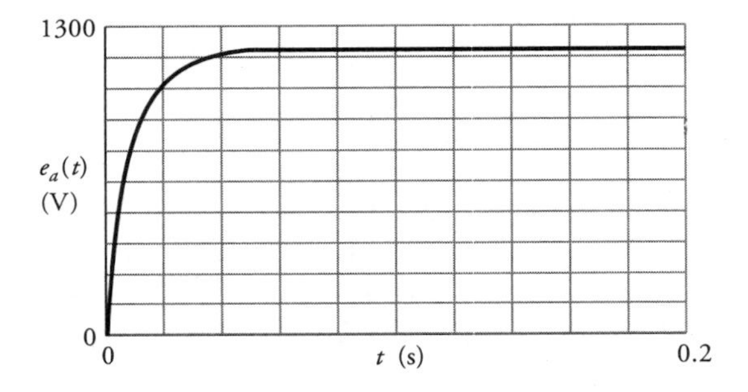

Figure 11.11 The induced voltage as a function of time.

after five time constants. Thus, the time required to reach the steady state is

$$T = \frac{5}{120} = 0.0417 \text{ s} \qquad \text{or} \qquad 41.7 \text{ ms}$$

(c) The final values of the field current and the induced (no-load) voltage are $I_f = 40$ A and $E_a = 1200$ V, respectively.

■

Exercise

11.5. Develop a mathematical model that represents the dynamic state of a dc generator driving a dc motor. Set up the differential equations in the matrix form. Neglect the saturation of the magnetic core, and assume that the motor field current and the generator speed are constant.

Numerical Analysis of DC Machine Dynamics

In the previous sections we obtained the closed-form solution for the dynamic response of dc machines using Laplace transforms. We resort to a numerical method to obtain the dynamic response when the machine is operating under nonlinear conditions. In this section, our aim is to investigate the dynamic response of dc machines by solving the differential equations numerically employing the fourth-order Runge-Kutta algorithm. Prior to its application, a brief description of the Runge-Kutta algorithm is given.

Fourth-Order Runge-Kutta Algorithm

Consider a set of first-order differential equations given in matrix form as

$$\frac{d\underline{x}(t)}{dt} = A\underline{x}(t) + B\underline{u}(t) \tag{11.27}$$

where $\underline{x}(t)$ is a state-variable vector (column matrix) and $\underline{u}(t)$ is the input-variable vector. A and B are constant coefficient matrices that include the parameters of the system. The solution of Eq. (11.27) can be given at discrete instants of time as follows:

$$\underline{x}_{n+1} = \underline{x}_n + \frac{\underline{K}_1 + 2\underline{K}_2 + 2\underline{K}_3 + \underline{K}_4}{6} \tag{11.28}$$

where $\underline{x}_n$ and $\underline{x}_{n+1}$ are the values of the state-variable vector at instants of n and $n + 1$, respectively. $\underline{K}_1$, $\underline{K}_2$, $\underline{K}_3$, and $\underline{K}_4$ are constant vectors computed at discrete instants of time as outlined next.

Let

$$f(\underline{x}_n) = A\underline{x}_n + B\underline{u}_n \tag{11.29}$$

then

$$\underline{K}_1 = hf(\underline{x}_n) \tag{11.30a}$$

$$\underline{K}_2 = hf(\underline{x}_n + 0.5\underline{K}_1) \tag{11.30b}$$

$$\underline{K}_3 = hf(\underline{x}_n + 0.5\underline{K}_2) \tag{11.30c}$$

$$\underline{K}_4 = hf(\underline{x}_n + \underline{K}_3) \tag{11.30d}$$

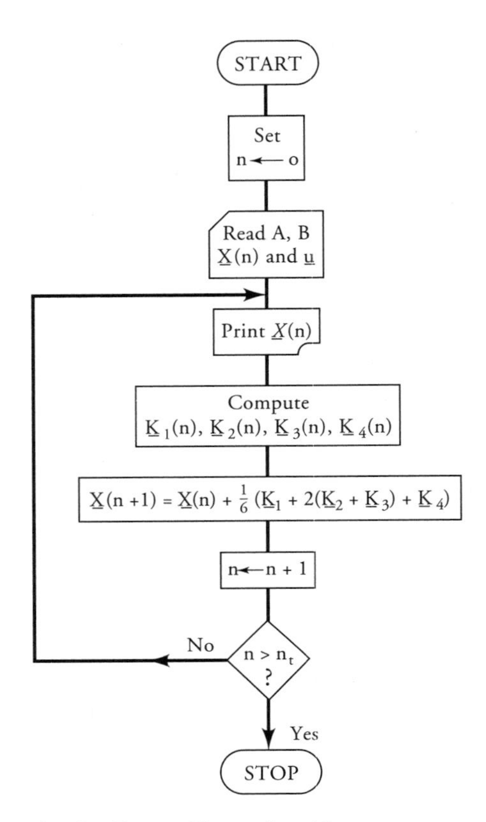

Figure 11.12 Flow chart for the Runge-Kutta algorithm.

where h is the step length, which is defined as the time between two discrete instants. The step length is highly dependent on the time constants of the system. In general, h is chosen less than the smallest time constant of the system in order to maintain the stability of the numerical method. Figure 11.12 shows the flow chart of the algorithm, and a sample Fortran program is given in Figure 11.13. There are, in all, five sets of inputs to the program: matrices A and B, the input variables, and the values for the step length and the total response time.

Since it is a numerical method, the Runge-Kutta algorithm can be used to solve a set of first-order linear and/or nonlinear differential equations. The following example illustrates the application of the Runge-Kutta method for a linear system.

```
      DIMENSION A(2,2) ,X(2) ,B(2,2) ,AX(2) ,BU(2) ,F(2) ,C1(2) ,C2(2) ,C3(2) ,
     *C4(2) ,G(2) ,XG(2) ,U(2)
      N=2
      READ(5,100) ((A(I ,J) ,J=1 ,N) , I=1 ,N)
      READ(5,100) (X(I) ,I=1 ,N)
      READ(5,100) ((B(I ,J) ,J=1 ,N) ,I=1 ,N)
      READ(5,100) (U(I) ,I=1 ,N)
  100 FORMAT(2F10.3)
      READ(5,110) T ,H
  110 FORMAT (2F15.9)
      M=T/H
      WRITE(6,115) M
  115 FORMAT(15)
      DO 5 K=1 ,M
      TS=H*(K-1)
      CALL PLOT(TS ,X(1))
      CALL MVMULT(A,X,AX,N)
      CALL MVMULT(B,U,BU,N)
      CALL VVADDT(AX,BU,F,N)
      DO 10 I=1 ,N
      C1(I)=H*F(I)
  10  G(I)=0.5*C1(I)
      CALL VVADDT(X,G,XG,N)
      CALL MVMULT(A,XG,AX,N)
      CALL VVADDT(AX,BU,F,N)
      DO 11 I=1 ,N
      C2(I)=H*F(I)
  11  G(I)=0.5*C2(I)
      CALL VVADDT(X,G,XG,N)
      CALL MVMULT(A,XG,AX,N)
      CALL VVADDT(AX,BU,F,N)
      DO 12 I=1 ,N
  12  C3(I)=H*F(I)
      CALL VVADDT(X,C3,XG,N)
      CALL MVMULT(A,XG,AX,N)
      CALL VVADDT(AX,BU,F,N)
      DO 13 I = 1,N
  13  C4(I) = H*F(I)
      DO 15 I =1,N
  15  X(I)=X(I)+(1./6.)*(C1(I)+2.*(C2(I)+C3(I))+C4(I))
  5   CONTINUE
      END
      SUBROUTINE MVMULT(A,B,C,N)
      DIMENSION A(N,N) ,B(N) ,C(N)
      DO 5 I=1 ,N
      C(I) = 0.0
      DO 5 J =1 ,N
  5   C(I)=C(I) + A(I ,J)*B(J)
      RETURN
      END
      SUBROUTINE VVADDT(A,B,C,N)
      DIMENSION A(N) ,B(N) ,C(N)
      DO 5 I =1 ,N
  5   C(I)=A(I) + B(I)
      RETURN
      END
```

Figure 11.13 Computer program to solve the state equations.

EXAMPLE 11.6

The parameters of a 240-V, PM motor are $R_a = 0.3\ \Omega$, $L_a = 2$ mH, $K = 0.8$, $J = 0.0678$ kg-m^2. Determine the motor speed and the armature current as a function of time when the motor is subjected to a torque of 100 N-m after 200 ms of starting at no load. Consider a step length of 0.01 s and observe the response for a period of 0.5 s. Neglect the frictional losses and assume that the motor operates in the linear region.

● SOLUTION

From Eq. (11.3), for $t < 200$ ms, we have

$$\underline{x}(t) = \begin{bmatrix} \omega_m(t) \\ i_a(t) \end{bmatrix}, \qquad \underline{u}(t) = \begin{bmatrix} 0 \\ 240 \end{bmatrix}$$

$$A = \begin{bmatrix} 0 & 11.8 \\ -400 & -150 \end{bmatrix} \quad \text{and} \quad B = \begin{bmatrix} -14.75 & 0 \\ 0 & 500 \end{bmatrix}$$

with initial values of $\omega_m(0) = 0$ and $i_a(t) = 0$.
For $t > 200$ ms,

$$\underline{u}(t) = \begin{bmatrix} 100 \\ 240 \end{bmatrix} \quad \text{and} \quad \underline{x}(0.2) = \begin{bmatrix} 299.93 \\ 0.25 \end{bmatrix}$$

The computed speed and armature current waveforms are shown in Figures 11.14 and 11.15, respectively. ■

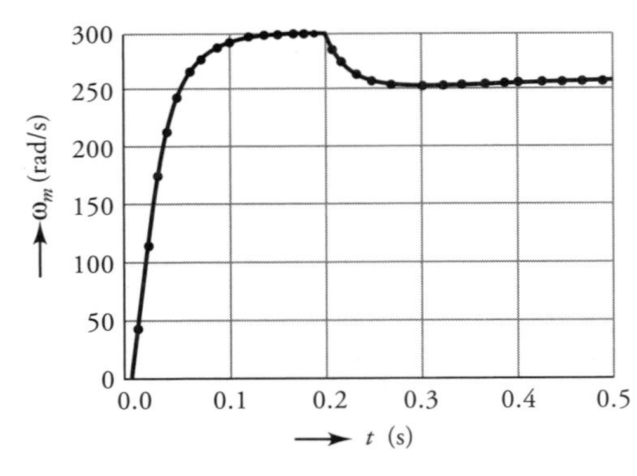

Figure 11.14 The motor speed as a function of time.

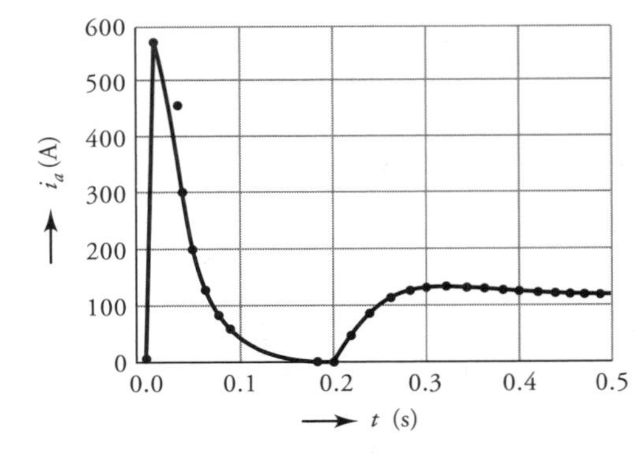

Figure 11.15 The armature current as a function of time.

Exercises

11.6. The dc machine given in Example 11.1 is coupled to a mechanical drive and runs at the rated speed to operate as a generator. The dc motor of Example 11.4 is connected to the armature of the generator with its field winding excited from a constant current source of 1 A. Write a computer program that predicts the variation of the motor speed after the field circuit of the generator is suddenly connected to a 240-V supply. The armature resistance and inductance of the motor are 7 Ω and 15 mH, respectively, while the field resistance and inductance of the generator are 90 Ω and 100 mH, respectively. The characteristics of the magnetic cores for both machines are the same.

11.7. Consider the motor given in Example 11.6. Write a computer program to determine the motor speed and the armature current if the motor drives a fan with a characteristic of $T_L = 0.5\ \omega_m^2$.

11.3 Synchronous Generator Transients

Whenever a sudden change occurs in the torque applied to the rotor or in the load current, a finite period of time is needed by a synchronous generator before returning to its steady-state operation. The operation of the generator during this finite period is known as the **transient operation**, and the transient may be mechanical or electrical in nature. In this section, each type of transient is considered separately.

Electrical Transients

The most severe transient that may occur in a synchronous generator is the development of a sudden short circuit across its three terminals. When such a short circuit takes place on a generator that has been operating at the synchronous speed under no load, the current in each phase, shown in Figure 11.16, is sustained by the generated voltage E_ϕ of the machine. It may be surprising from Figure 11.16 that each phase current has a decaying dc component even though only ac voltages are present in the armature circuit. The reason for this is that the machine tries to maintain constant flux linkages for each of the three phases. Another observation from Figure 11.16 is that the waveforms are not symmetric around the zero axis.

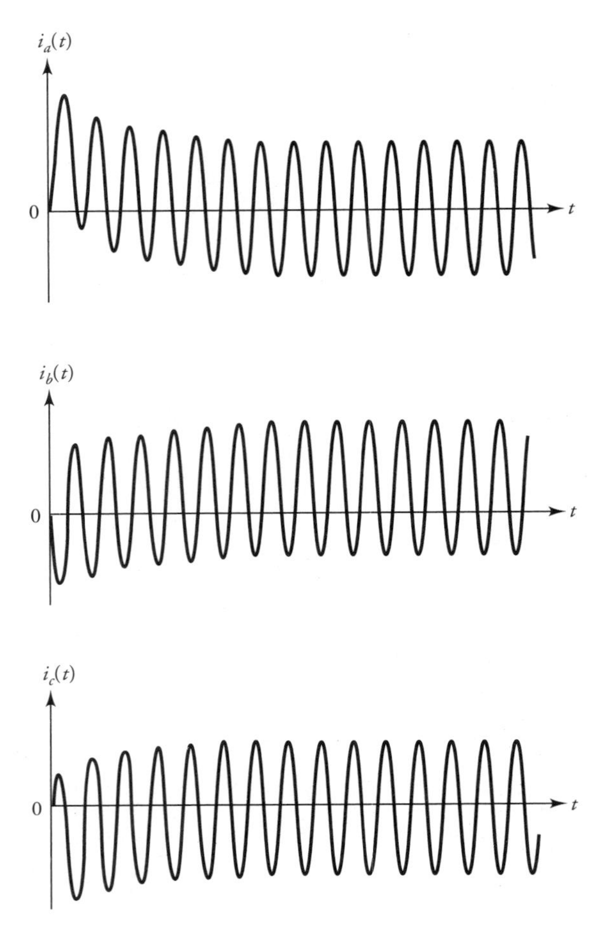

Figure 11.16 Phase currents in the armature winding after a sudden three-phase short circuit across the terminals of a synchronous generator operating at no load.

However, a symmetric waveform can be obtained by subtracting the dc component, as illustrated in Figure 11.17. This waveform can be divided into three periods—the **subtransient** period, the **transient** period, and the **steady-state** period. During the subtransient period, the current decrement is very rapid and it lasts only few cycles. The transient period, on the other hand, covers a longer time with a slower rate of decrease in the current. Finally, during the steady-state period the current is determined by the generated voltage and the synchronous

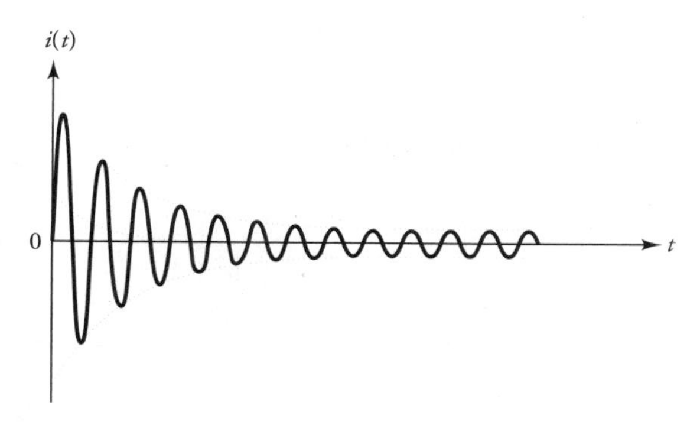

Figure 11.17 Symmetric short-circuit current.

reactance of the machine. All three periods of the exponentially decaying envelope of the symmetric short-circuit current are described in Figure 11.18 using a semi-log graph. We can now determine the equivalent reactances that control the current during the subtransient and transient periods.

In order to simplify the theoretical development, let us assume that the generator operates in the linear region of its magnetization characteristic and the winding resistances are negligible. Also, we assume that the machine does not carry any damper winding. Just prior to the short circuit, the total flux linkages of the field winding are

$$\lambda_f = N_f \Phi_f = L_f I_f \tag{11.31}$$

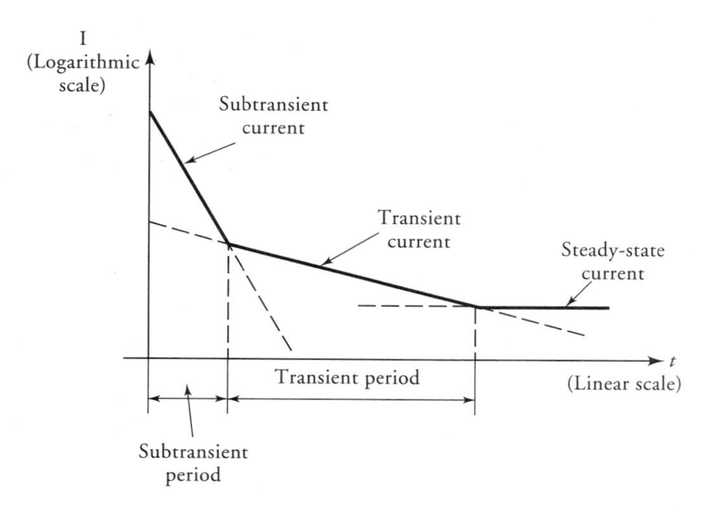

Figure 11.18 Semi-log graph of the envelope of the symmetric short-circuit current.

where I_f is the field current and L_f is the self-inductance of the field winding. Equation (11.31) can also be written as

$$\lambda_f = (L_{lf} + L_{af})I_f \tag{11.32}$$

after representing the self-inductance L_f as the sum of the leakage inductance L_{lf} of the field winding and the mutual inductance L_{af} between the field and the armature windings. Since the generator has been operating under no load before the short circuit, the flux linkage due to the armature winding is

$$\lambda_a = 0 \tag{11.33}$$

When a three-phase short circuit occurs across the armature terminals at $t = 0$, the magnetic axes of the field winding and armature winding, for example phase-a, are considered to be orthogonal, in accordance with Eq. (11.33) as indicated in Figure 11.19. After a short time the rotor attains a certain angular position α with respect to the magnetic axis of phase-a winding and causes currents of i_a and $I_f + i_f$ in phase-a and the field winding, respectively, in order to maintain the same total flux linkages. The flux linkages for phase-a can be expressed as

$$\lambda_a = i_a(L_{la} + L_{af}) + (I_f + i_f)L_{af} \sin(90° - \alpha) = 0 \tag{11.34}$$

where L_{la} is the leakage inductance of phase-a.

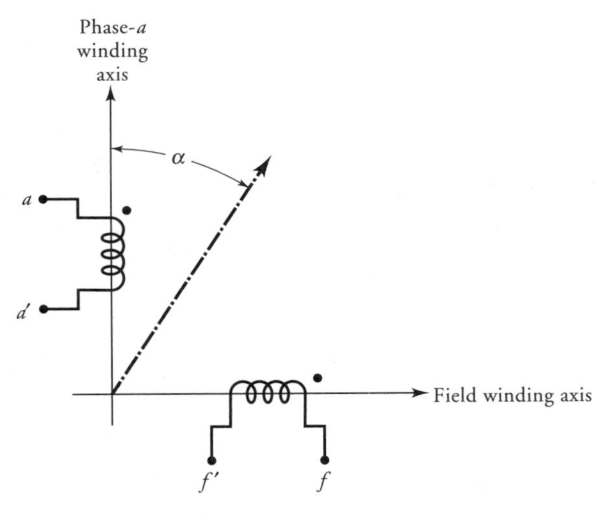

Figure 11.19 The relative positions of the phase and field windings at the inception ($t = 0$) of a short circuit.

The flux linkages associated with the field winding are

$$\lambda_f = (I_f + i_f)(L_{lf} + L_{af}) + L_{af} i_a \sin(90° - \alpha) \tag{11.35}$$

From Eqs. (11.32), (11.34), and (11.35), it can be shown that

$$i_a = \frac{L_{af} I_f (L_{lf} + L_{af}) \sin(90° - \alpha)}{L_{af}^2 \sin^2(90° - \alpha) - (L_{lf} + L_{af})(L_{la} + L_{af})} \tag{11.36}$$

and

$$i_f = -\frac{L_{af}^2 \sin^2(90° - \alpha) I_f}{L_{af}^2 \sin^2(90° - \alpha) - (L_{lf} + L_{af})(L_{la} + L_{af})} \tag{11.37}$$

The most severe transient condition takes place when the currents are maximum. From the preceding equations, the currents are maximum when $\alpha = 0$. Thus, corresponding to the most severe condition,

$$i_a = \frac{L_{af} I_f (L_{lf} + L_{af})}{L_{af}^2 - (L_{lf} + L_{af})(L_{la} + L_{af})} \tag{11.38}$$

and

$$i_f = -\frac{I_f L_{af}^2}{L_{af}^2 - (L_{lf} + L_{af})(L_{la} + L_{af})} \tag{11.39}$$

Equation (11.38) can be used to determine the **transient reactance** of the synchronous generator that governs its behavior during the transient period. Multiplying both the numerator and the denominator of Eq. (11.38) by ω^2, and after some simplifications, we obtain

$$i_a = -\frac{E_\phi (X_{lf} + X_{af})}{X_{lf} X_{la} + X_{lf} X_{af} + X_{af} X_{la}} \tag{11.40}$$

where $E_\phi = I_f \omega L_{af} = I_f X_{af}$ is the generated voltage prior to the short circuit under no-load condition. The transient reactance can be defined as the ratio of no-load voltage to the short-circuit current during the transient period, or

$$\frac{E_\phi}{-i_a} = X'_d = X_{la} + \frac{X_{af} X_{lf}}{X_{af} + X_{lf}} \tag{11.41}$$

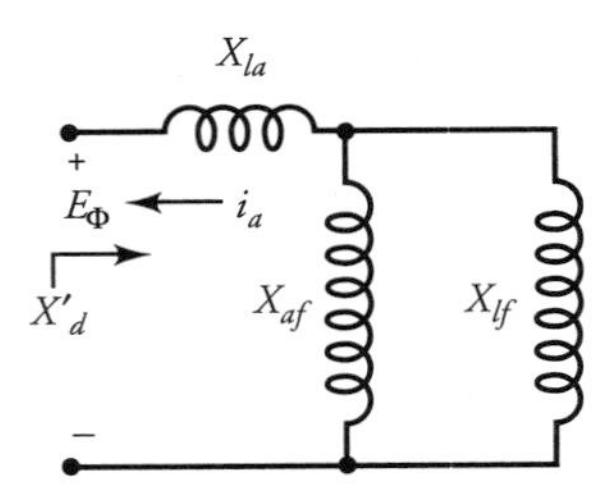

Figure 11.20 Equivalent circuit of the transient reactance.

and its equivalent-circuit representation is given in Figure 11.20. Note that E_ϕ and i_a are the peak values of the generated voltage and the armature current.

In the previous development we assumed no damper winding on the rotor. If we take the effect of the damper winding into consideration, with the similar approach we used to develop the transient reactance, we obtain

$$\frac{E_\phi}{-i_a} = X''_d = X_{la} + \frac{X_{af}X_{lf}X_{ld}}{X_{lf}X_{ld} + X_{af}X_{ld} + X_{af}X_{lf}} \tag{11.42}$$

which is referred to as the **subtransient reactance**. In this expression, i_a is the current that corresponds to the subtransient period, and X_{ld} is the leakage reactance of the damper winding and can be included in the equivalent circuit, as shown in Figure 11.21. It is evident from Figures 11.20 and 11.21 that X''_d is smaller than X'_d. Therefore, during the first few cycles after the short circuit has developed, the armature current is very large and we refer to this current as the **subtransient current**, and its root-mean-square (rms) value is given as

$$I''_d = \frac{E_\phi}{\sqrt{2}X''_d} \tag{11.43}$$

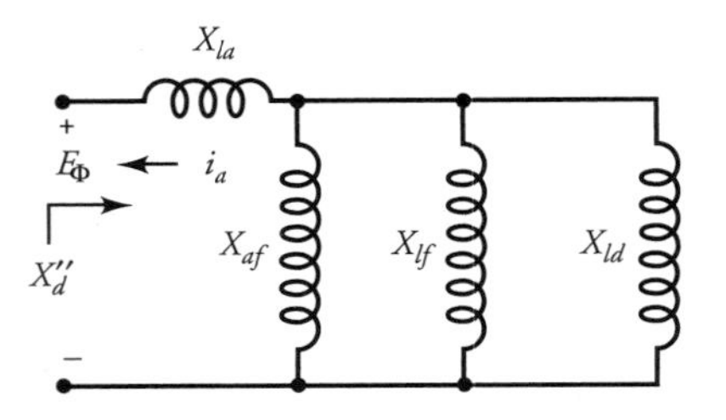

Figure 11.21 Equivalent circuit of the subtransient reactance.

The end of the subtransient period marks the beginning of the transient period, and the rms value of the **transient current** is

$$I'_d = \frac{E_\phi}{\sqrt{2}X'_d} \tag{11.44}$$

The transient period lasts typically for another seven to ten cycles. The transient current is followed by the steady-state current controlled by the synchronous reactance of the machine, and its rms value is

$$I_d = \frac{E_\phi}{\sqrt{2}X_s} \tag{11.45}$$

It must be borne in mind that our discussion of electrical transients assumes that all three phases of the synchronous generator developed the short circuit at the same time. If the short circuit occurs on only one or two phases, the nature of the response is more complex and is beyond the scope of this book.

EXAMPLE 11.7

The per-unit parameters of a 60-Hz, 71,500-kVA, 13,800-V, $\cos \Phi = 0.8$ synchronous generator are $X_{af} = 0.57$ pu, $X_{la} = 0.125$ pu, $X_{lf} = 0.239$ pu, and $X_{ld} = 0.172$ pu. Calculate the rms values of the symmetric subtransient and transient currents if a three-phase short circuit occurs across the armature terminals while the machine has been running at no load.

● SOLUTION

To calculate the subtransient short-circuit current we have to determine the subtransient reactance first from Eq. (11.42) as

$$X''_d = 0.125 + \frac{0.57 \times 0.239 \times 0.172}{(0.239 \times 0.172) + (0.57 \times 0.172) + (0.57 \times 0.239)} = 0.21 \text{ pu}$$

The generated voltage, $E_\phi = 1$ pu occurs at the rated rms voltage right before the short circuit. Thus,

$$I''_d = \frac{1}{0.21} = 4.76 \text{ pu}$$

Similarly, we can obtain the transient reactance from Eq. (11.41) as

$$X'_d = 0.125 + \frac{0.57 \times 0.239}{0.57 + 0.239} = 0.29 \text{ pu}$$

and the transient current as

$$I'_d = \frac{1}{0.29} = 3.45 \text{ pu}$$

The above currents are all normalized with respect to the rated values of the generator. Hence, if we calculate the rated current, we can determine the subtransient and transient currents in amperes as

$$I_{\text{rated}} = \frac{71{,}500 \times 10^3}{\sqrt{3} \times 13{,}800} = 2991.34 \text{ A}$$

$$I''_d = 4.76 \times 2991.34 = 14{,}238.79 \text{ A}$$

$$I'_d = 3.45 \times 2991.34 = 10{,}320.14 \text{ A}$$

■

Mechanical Transients

In this section we discuss the mechanical transients in synchronous generators followed by a major disturbance (possibly an electrical transient) occurring across the armature terminals. Although a mechanical transient is very slow, it is one of the most important transients because it may cause the machine to self-destruct.

Let us consider a synchronous generator coupled to a prime mover and receiving mechanical power input through its shaft. The armature terminals of this machine are connected to an infinite bus-bar* to deliver the power demanded by the network. If the mechanical power input and the electrical power output are equal to each other, assuming that the losses are negligibly small, the machine will run at its synchronous speed. But if the input and output powers are not equal to each other during the transient state, the power difference will be used in the machine to change the kinetic energy or the speed. In other words, the machine either slows down or speeds up to restore the synchronous speed. This speed adjustment also takes place if a sudden short circuit develops across the generator terminals.

* An infinite bus-bar is a bus that has a constant voltage at a constant frequency.

During the transient state, the torque in the machine can be expressed as

$$T_m = T_d(t) + D\omega_m(t) + J\frac{d\omega_m(t)}{dt} \tag{11.46}$$

where T_m is the mechanical torque applied to the generator and $T_d(t)$ is the torque developed in the synchronous generator at a given time.

In the absence of rotational losses, Eq. (11.46) can be rewritten as

$$J\frac{d\omega_m(t)}{dt} = T_m - T_d(t)$$

or

$$J\frac{d^2\theta_m(t)}{dt^2} = T_m - T_d(t) \tag{11.47}$$

where $\omega_m(t) = d\theta_m/dt$ and θ_m is the angular displacement of the rotor with respect to a stationary reference frame. $\theta_m(t)$ can be mathematically expressed as

$$\theta_m(t) = \omega_s t + \delta_m(t) \tag{11.48}$$

where $\delta_m(t)$ is the power angle at any time t.

Differentiating the above equation twice, we obtain

$$\frac{d^2\theta_m(t)}{dt^2} = \frac{d^2\delta_m(t)}{dt^2} \tag{11.49}$$

Equation (11.47) can now be written as

$$J\frac{d^2\delta_m(t)}{dt^2} = T_m - T_d(t) \tag{11.50}$$

Multiplying both sides of Eq. (11.50) by the rotor speed, $\omega_m(t)$, we get

$$J\omega_m(t)\frac{d^2\delta_m(t)}{dt^2} = \omega_m(t)T_m - \omega_m(t)T_d(t)$$

or

$$J\omega_m(t)\,\frac{d^2\delta_m(t)}{dt^2} = P_m(t) - P_{dm}\sin\delta_m(t) \tag{11.51}$$

where $$P_{dm} = \frac{3V_1E_\phi}{X_s}$$

is the maximum power developed by a round-rotor synchronous generator, as discussed in Chapter 7. Note that $J\omega_m(t)$ is the angular momentum, $M(t)$.

Equation (11.51) can be expressed in terms of per-unit quantities with respect to the rated power of the synchronous generator, S_n, as

$$\frac{J\omega_m(t)}{S_n}\,\frac{d^2\delta_m(t)}{dt^2} = \frac{P_m(t)}{S_n} - \frac{P_{dm}}{S_n}\sin\delta_m(t) \tag{11.52}$$

Let us define the inertia constant for the generator as

$$H = \frac{1}{2}\,\frac{J\omega_s^2}{S_n} = \frac{1}{2}\,\frac{J\omega_m^2}{S_n} \tag{11.53}$$

In Eq. (11.53), the synchronous speed has been replaced by the rotor speed without introducing any significant error. By doing so, we are assuming that the angular momentum is a constant. In terms of the inertia constant of the generator, Eq. (11.52) can be expressed as

$$\frac{2H}{\omega_s}\,\frac{d^2\delta_m(t)}{dt^2} = p_m - p_{dm}\sin\delta_m(t) \tag{11.54}$$

where p_m and p_{dm} are the per-unit powers. Equation (11.54) is a second-order nonlinear differential equation and is referred to as the **swing equation** in terms of the per-unit quantities. It helps us determine the stability of the synchronous generator during the transient state. The swing equation can be solved numerically using the fourth-order Runge-Kutta algorithm described earlier. For the sake of simplicity, the swing equation can be linearized by substituting $\delta_m(t)$ for $\sin\delta_m(t)$ as long as $\delta_m(t)$ is a small angle.

Equal-Area Criterion

The **equal-area criterion** is another method commonly employed to determine the stability of a synchronous generator during a transient state.

The power developed by a round-rotor synchronous generator is shown as a function of power angle in Figure 11.22. Let us assume that the generator operates

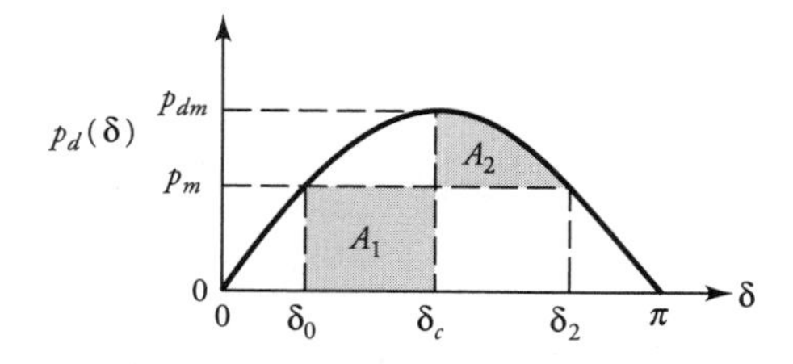

Figure 11.22 Power developed in a round-rotor synchronous generator as a function of power angle.

at a power p_0 corresponding to power angle δ_0. For a lossless machine, the mechanical power input must be the same as the power output or the power developed under steady state. If a sudden three-phase short circuit occurs, the power developed will become zero and the machine will be disconnected from the bus-bar. After the inception of the short circuit, Eq. (11.54) has to be modified as

$$\frac{d\omega_m(t)}{dt} = \frac{\omega_s}{2H} p_m \tag{11.55}$$

since the prime mover is still supplying mechanical power to the generator. The above equation yields

$$\omega_m(t) = \omega_s + \frac{\omega_s}{2H} p_m t \tag{11.56}$$

It is apparent that the shaft speed is higher than the synchronous speed and increases with time, suggesting that the machine can run away. If the short circuit is not cleared within a short period of time (approximately six cycles or so) and the generator is not reconnected to the bus-bar, the generator may self-destruct. The variation in power angle, from Eq. (11.56), is

$$\delta_m(t) = \delta_0 + \frac{\omega_s}{4H} p_m t^2 \tag{11.57}$$

where $\delta_0 = \omega_s t$.

If the fault is cleared at an instance t_c, the generator will start developing power at a higher level than the prefault state, since $\delta_m(t_c)$ becomes greater than δ_0, as is evident from Eq. (11.57). This causes the rotor to slow down.

From Eq. (11.48),

$$\frac{d\delta_m(t)}{dt} = \frac{d\theta_m(t)}{dt} - \omega_s = \omega_m(t) - \omega_s \tag{11.58}$$

and the swing equation can be rearranged as

$$\frac{2H}{\omega_s}\frac{d\omega_m(t)}{dt} = p_m - p_d \tag{11.59}$$

Multiplying both sides of Eq. (11.59) by $d\delta_m(t)/dt$ and then substituting Eq. (11.58) for $\delta_m(t)/dt$ to the left side yields

$$\frac{H}{\omega_s}\left[2\omega_m \frac{d\omega_m}{dt} - 2\omega_s \frac{d\omega_m}{dt}\right] = (p_m - p_d)\frac{d\delta_m}{dt} \tag{11.60}$$

By integrating Eq. (11.60), we obtain

$$\frac{H}{\omega_s}\left[\int_{\omega m1}^{\omega m2} 2\omega_m \, d\omega_m - \int_{\omega m1}^{\omega m2} 2\omega_s \, d\omega_m\right] = \int_{\delta 0}^{\delta 2} (p_m - p_d)\, d\delta_m$$

or

$$\frac{H}{\omega_s}[(\omega_{m2}^2 - \omega_{m1}^2) - 2\omega_s(\omega_{m2} - \omega_{m1})] = \int_{\delta 0}^{\delta 2} (p_m - p_d)\, d\delta_m \tag{11.61}$$

where ω_{m1} is the angular velocity prior to the inception of the short circuit, and ω_{m2} is the angular velocity corresponding to the maximum value of δ_2 after the short circuit has been cleared. Since there is no change in the power angle before the fault occurs, from Eq. (11.58) $\omega_{m1} = \omega_s$. On the other hand, at the maximum value of the power angle δ_2, $d\delta_m/dt = 0$. Thus from Eq. (11.58) $\omega_{m2} = \omega_s$.

Hence from Eq. (11.61),

$$\int_{\delta 0}^{\delta 2} (p_m - p_d)\, d\delta_m = 0 \tag{11.62}$$

We can rearrange Eq. (11.62) and write

$$\int_{\delta 0}^{\delta c} (p_m - p_d)\, d\delta_m = \int_{\delta c}^{\delta 2} (p_d - p_m)\, d\delta_m \tag{11.63}$$

where δ_c is the power angle at the instant, t_c, when the fault is cleared. Equation (11.63) clearly indicates that the area A_1 of Figure 11.22 corresponding to the left side of the equation is equal to the area A_2 of Figure 11.22 equivalent to the right side.

EXAMPLE 11.8

A 1000-kVA, 4.6-kV, 60-Hz, 4-pole synchronous generator delivers 0.9 pu of average power with power angle of 18° when a three-phase short circuit develops across its terminals. Calculate (a) the per-unit power generated by the generator when the fault is cleared four cycles after its inception and (b) the critical time to clear the fault in order not to lose the stability. The inertia constant is given as 10 J/VA.

● SOLUTION

The synchronous speed of the machine is $N_s = \dfrac{60 \times 120}{4} = 1800$ rpm

$$\omega_s = \frac{2\pi \times 1800}{60} = 188.5 \text{ rad/s}$$

$$p_{dm} = \frac{0.9}{\sin 18°} = 2.91 \text{ pu}$$

The fault-clearing time is four cycles; thus $t_c = 4(1/60) = 66.68$ ms. The initial rotor angle is $\delta_0 = 18° = 0.314$ rad. The rotor angle at the time the fault is cleared, from Eq. (11.57), is

$$\begin{aligned}\delta_m(66.68 \text{ ms}) &= 0.314 + \frac{188.5}{4 \times 10} \times 0.9 \times (66.68 \times 10^{-3})^2 \\ &= 0.333 \text{ rad} \quad \text{or} \quad 19.07°\end{aligned}$$

(a) The power generated is

$$p_d = 2.91 \sin 19.07° = 0.95 \text{ pu}$$

(b) After the fault occurrence, the developed power p_d for the left and right sides of the equal-area criterion given in Eq. (11.63) becomes $p_d = 0$ and $p_d = p_{dm} \sin \delta_m$, respectively.

Thus,
$$\int_{\delta_0}^{\delta_c} p_m \, d\delta_m = \int_{\delta_c}^{\delta_2} p_{dm} \sin \delta_m \, d\delta_m - \int_{\delta_c}^{\delta_2} P_m \, d\delta_m$$

or
$$\cos \delta_c = \frac{p_m}{p_{dm}} (\delta_2 - \delta_0) + \cos \delta_2$$

where $\delta_2 = \pi - \delta_0$ from Figure 11.22.

Hence, $$\cos \delta_c = \frac{0.9}{2.91}(2.83 - 0.314) + \cos(2.83) = -0.174$$

and $$\delta_c = 100° \text{ or } 1.745 \text{ rad}$$

By using Eq. (11.57), the critical fault-clearing time can be obtained as

$$t_c = \sqrt{(\delta_c - \delta_0)\frac{4H}{\omega_s p_m}} = \sqrt{(1.745 - 0.314)\frac{4 \times 10}{188.5 \times 0.9}}$$
$$= 0.581 \text{ s}$$

■

Exercises

11.8. Derive Eq. (11.42) when the damper winding is present on the rotor and its leakage reactance is given as X_{ld}.

11.9. A 30-MVA, 13-kV, 60-Hz hydrogenerator (synchronous generator used in hydroplants) has $X_s = 1.0$ pu, $X'_d = 0.35$ pu, and $X''_d = 0.25$ pu. The reactances are normalized based on the ratings of the generator. This generator delivers rated power to a load at the rated voltage with a lagging power factor of 0.85. (a) Calculate the current in the armature winding. (b) Calculate the subtransient and transient currents in the armature winding when a sudden three-phase short circuit occurs across the terminals of the generator.

11.10. A 60-Hz, 50-MVA , 12.5-kV, 4-pole synchronous generator with an inertia constant of $H = 10$ J/VA is delivering the rated power at a lagging power factor of 0.8. The load on the generator suddenly reduces to 20% of its rated value owing to a fault in the system. Calculate the accelerating torque after the fault. Neglect the losses and assume constant power input to the generator.

SUMMARY

When the operating condition of an electric machine changes suddenly, the machine cannot respond to the change instantaneously. Thus, it undergoes a transient (dynamic) state to readjust the energy balance from the time of the inception of the change to the time when the final steady state is achieved. The transient may be due to a sudden change in load and/or voltage.

The study of transients in electric machines is often complicated by their complex nature. However, some plausible simplifying assumptions reduce the complexity of the study significantly. Here, we assume that the machine is operating within the linear region of its magnetization characteristic. In that case, saturation of the magnetic core is not taken into consideration, which otherwise makes the study a nonlinear problem.

The transient response of a dc machine is exponential in nature and, for all practical purposes, disappears after five time constants. To determine the transient response in a dc machine analytically, the Laplace transform method appears to be a very useful technique. However, the method cannot be used if the operation with saturation is the main concern. Numerical methods, on the other hand, can solve both the linear and nonlinear problems. One of the numerical methods to study electric machine dynamics is based upon the fourth-order Runge-Kutta algorithm.

The most severe transient that may occur in a synchronous generator is the development of a sudden three-phase short circuit across its terminals. In the event of such a short circuit, the current in the armature winding indicates a damped oscillatory nature. The first few cycles of the short-circuit current waveform denote the **subtransient** period, and the corresponding current the subtransient current. The second period of the short-circuit current waveform is known as the **transient** period, and its current is called the transient current. Finally, the steady-state current occurs after the transient period is over. In synchronous generator models we can use subtransient, transient, and synchronous reactances for the three regimes of short-circuit current.

When the terminal conditions change suddenly on a synchronous generator, the rotor is not able to respond at the same time, and a mechanical transient on the machine occurs. Although mechanical transients on synchronous generators are the slowest, they are the most important because the machine may self-destruct. Following a mechanical transient, the power angle varies as a function of time and is modeled by a second-order differential equation often referred to as the **swing** equation. If the cause for the mechanical transient is cleared within a short period of time (several cycles), the machine does not lose its stability during the transient state, which otherwise may lead the machine to overspeed and to damage the rotor permanently. Equal-area criterion is a useful tool to determine the transient stability of a synchronous generator.

Review Questions

11.1. What is the reason for a transient state in an electric machine?
11.2. What can cause a transient in an electric machine?
11.3. How can we model a separately excited dc motor to analyze its transient (dynamic) response?
11.4. Why does the saturation make the analysis of dc machines more difficult?

11.5. What are the state variables and input variables in the linear dynamic representation of a separately excited dc motor?
11.6. Why may the phase currents be asymmetric in a synchronous generator during the initial stage of a three-phase short circuit?
11.7. What are the subtransient, transient, and steady-state periods after the occurrence of a three-phase short circuit in a synchronous generator?
11.8. How does the damper winding contribute to the short-circuit current?
11.9. The excitation of a synchronous generator is accomplished by placing permanent magnets on the rotor. In the presence of the damper winding on the rotor, comment on whether the steady state is achieved more rapidly or more slowly than in a synchronous generator with a conventional field winding when a three-phase short circuit occurs in the machine.
11.10. Would it be safe to operate a synchronous generator with a power angle of 90° at steady state?

Problems

11.1. A 250-V, 7.5-hp, separately excited dc motor is rated at 1500 rpm. The armature resistance and the inductance are 0.8 Ω and 9 mH, respectively. The constant $K = K_a\Phi_p$ is 1.6, while the moment of inertia of the rotating system is 0.15 kg-m^2. Determine the armature current and the motor speed as a function of time when the armature is subjected to the rated voltage under no load. Neglect the effects of saturation and the frictional losses.

11.2. The motor given in Problem 11.1 has been operating in steady-state condition at no load. Suddenly it is subjected to a torque of 5 N-m. Determine the decrease in speed of the motor as a function of time. What is the speed after the motor reaches the steady state?

11.3. A 380-V, 2-hp, 85% efficient, separately excited dc motor has the following parameters: $R_a = 12\ \Omega$, $L_a = 110$ mH, $R_f = 400\ \Omega$, $L_f = 1.5$ H, $K_a\Phi_p = 5$, and $J = 0.098$ kg-m^2. The characteristic of the mechanical load is given by $T_L = 1.05\omega_m$. Determine the torque developed by the motor as a function of time when the armature circuit is suddenly subjected to the rated voltage. Neglect the saturation and the frictional losses.

11.4. The motor-load system given in Problem 11.3 has been operating under a steady-state condition at its rated values. Determine the variations in the field current and the speed if the armature current is kept constant and the field voltage is suddenly reduced to 200 V. Neglect the frictional losses and consider $\Phi_p = 0.01I_f$.

11.5. A 440-V, 10-kW, separately excited dc generator has the following data: $R_a = 3\ \Omega$, $L_a = 80$ mH, $R_f = 400\ \Omega$, $L_f = 2$ H, $K_aK_f = 2.6$, and $J = 0.09$ kg-m^2. The armature connected to a load with a resistance of 75 Ω and an inductance of 20 mH was being driven at a speed of 1500 rpm when the rated voltage was suddenly applied to the field circuit. Determine the variations in the field and armature currents. Neglect the frictional losses and the effect of saturation.

11.6. A 120-V, 1-hp, 200-rpm, PM dc motor is suddenly subjected to its rated voltage and a load torque of 8 N-m. Develop the state equations for the state variables i_a and ω_m. The machine parameters are $R_a = 1\ \Omega$, $L_a = 8$ mH, $K = 3$, and $J = 0.3$ kg-m^2. Neglect the frictional losses and saturation.

11.7. Compute the instant at which the motor given in Problem 11.6 reaches its steady state by using the fourth-order Runge-Kutta algorithm.

11.8. Repeat Problems 11.6 and 11.7 if the motor is loaded with a linear load of $T_L = 1.2\,\omega_m$.

11.9. A 4.6-kV, 1000-MVA, round-rotor synchronous generator with subtransient and transient reactances of 80 Ω and 160 Ω, respectively, experiences a three-phase short circuit across its terminals while operating at no load. (a) Calculate the per-unit reactances. (b) Calculate the subtransient and transient short-circuit currents per unit and in amperes.

11.10. A 60-Hz, 30-MVA, 7.8-kV, synchronous generator has the following per-unit parameters: $X_{af} = 0.32$ pu, $X_{lf} = 0.18$ pu, and $X_{ld} = 0.13$ pu. The generator experiences a three-phase short circuit at its terminals while operating at no load. The subtransient short-circuit current is measured as 11,150 A. Calculate the leakage reactance of the armature.

11.11. A 1250-V, 100-kVA, synchronous generator operates a synchronous motor at its rated values of 25 kVA, 1000 V with a leading power factor of 0.9. The subtransient reactances of the generator and the motor are 0.2 pu and 0.15 pu, respectively. Calculate the fault currents if a three-phase short circuit occurs across the generator terminals.

11.12. A 5-MVA, 6.3-kV, 60-Hz, 4-pole synchronous generator experiences a three-phase short circuit at its terminals while delivering an average power of 0.8 pu with a power factor of 15°. The inertia constant of the synchronous generator is 8 J/VA. (a) Calculate the critical fault-clearing power angle. (b) Calculate the critical fault-clearing time.

CHAPTER 12

Special-Purpose Electric Machines

Sectional view of a brushless dc motor. (*Courtesy of Bodine Electric Company*)

12.1 Introduction

This chapter is devoted to special-purpose electric machines. Although all electric machines have the same basic principle of operation, special-purpose machines have some features that distinguish them from conventional machines.

It is not our intention to discuss all kinds of special-purpose machines in one chapter; rather, an attempt is made to introduce the basic operating principles of some special-purpose machines that are being used extensively in home, recreational, and industrial applications.

With the proliferation of power electronic circuits and digital control systems, precise speed and position control can be achieved in conjunction with special-purpose electric machines such as permanent-magnet (PM) motors, step motors, switched-reluctance motors, brushless direct-current (dc) motors, hysteresis motors, and linear motors. Some of these devices find applications in computer peripheral equipment or in process-control systems whereas others can be used in devices such as home appliances. For example, step motors are employed extensively in computers where precise positioning is required, as in the case of a magnetic head on a floppy disk drive. For applications that demand constant-speed drives, brushless dc motors offer excellent characteristics. Switched-reluctance motors, on the other hand, find applications where we traditionally use dc or induction motors.

In the following sections we discuss the construction, operating principles, and characteristics of each of the above-mentioned special-purpose electric machines.

12.2 Permanent-Magnet Motors

The development of new permanent-magnet materials has made PM motors a viable substitute for a shunt (dc) motor. In a PM motor the poles are made of permanent magnets, as shown in Figure 12.1. Although dc motors up to 75 hp have been designed with permanent magnets, the major application of permanent magnets is confined to fractional-horsepower motors for economic reasons. In a conventional dc motor with a wound-field circuit, flux per pole depends on the current through the field winding and can be controlled. However, flux in a PM motor is essentially constant and depends on the point of operation, as explained in Chapter 2.

For the same power output, a PM motor has higher efficiency and requires less material than a wound dc motor of the same ratings. However, the design of a PM motor should be such that the effect of demagnetization due to armature reaction, which is maximum at standstill, is as small as economically possible.

Since the flux in a PM motor is fixed, the speed- and current-torque

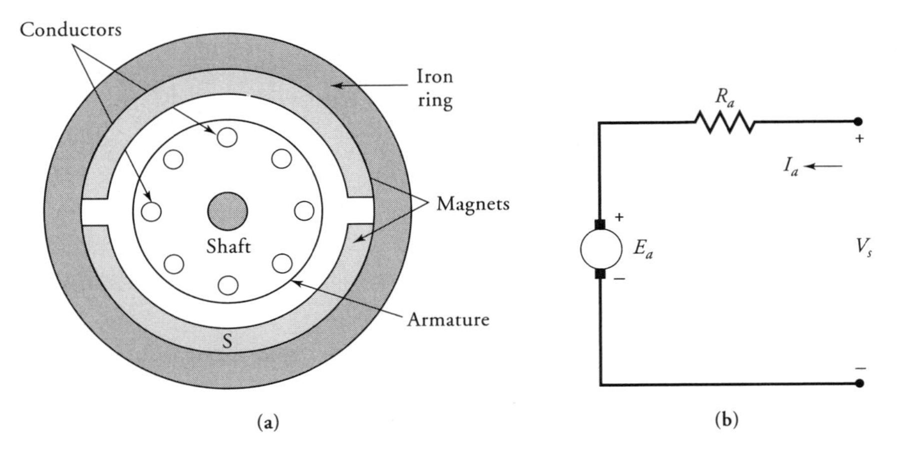

Figure 12.1 **(a)** Cross-sectional view of a PM motor; **(b)** equivalent circuit.

characteristics are basically straight lines, as shown in Figure 12.2. Mathematically, these relations can be expressed as

$$\omega_m = \frac{V_s}{K_a \Phi_p} - \frac{R_a}{(K_a \Phi_p)^2} T_d \tag{12.1}$$

and

$$I_a = \frac{1}{K_a \Phi_p} T_d \tag{12.2}$$

where K_a, V_s, Φ_p, R_a, and T_d are the machine constant, supply voltage, flux per pole, resistance of the armature winding, and developed torque.

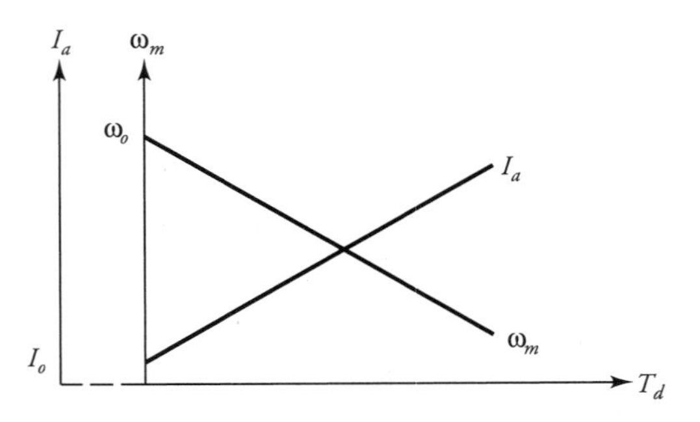

Figure 12.2 Speed- and current-torque characteristics of a PM motor.

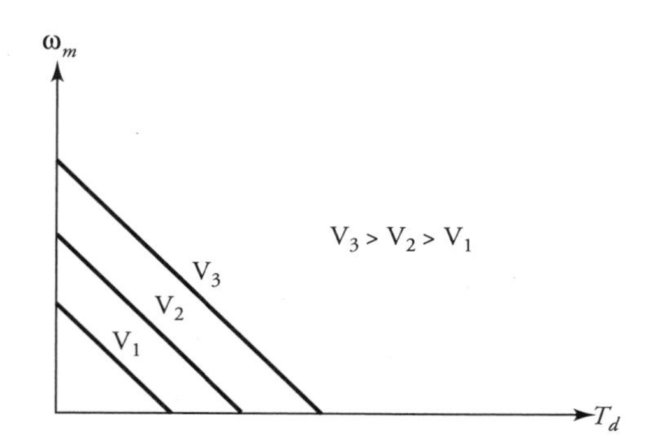

Figure 12.3 Operating characteristics for different supply voltages.

The speed-torque characteristic of a PM motor can be controlled by changing either the supply voltage or the effective resistance of the armature circuit. The change in the supply voltage varies the no-load speed of the motor without affecting the slope of the characteristic. Thus for different supply voltages, a set of parallel speed-torque characteristics can be obtained, as illustrated in Figure 12.3. On the other hand, with the change in the effective resistance of the armature circuit, the slope of the curve is controlled and the no-load speed of the motor remains the same, as indicated in Figure 12.4. Using magnets with different flux densities and the same cross-sectional areas, or vice versa, there are almost infinite possibilities for designing a PM motor for a given operating condition, as shown in Figure 12.5. From the same figure we can also conclude that an increase in blocked-rotor torque can be achieved only at the expense of a lower no-load speed.

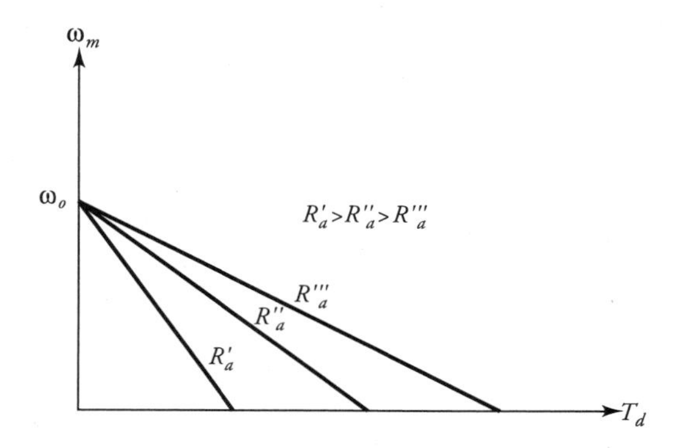

Figure 12.4 Operating characteristics for different resistances of armature circuit.

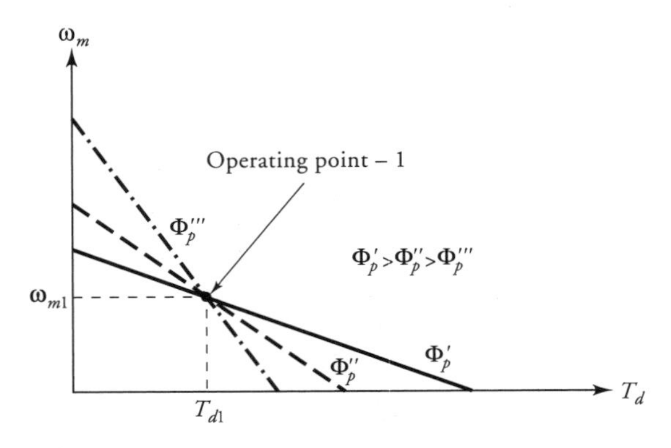

Figure 12.5 Operating characteristics for different fluxes in a PM motor.

EXAMPLE 12.1

A PM motor operates at a magnetic flux of 4 mWb. The armature resistance is 0.8 Ω and the applied voltage is 40 V. If the motor load is 1.2 N-m, determine (a) the speed of the motor and (b) the torque developed under a blocked-rotor condition. The motor constant K_a is 95.

● SOLUTION

(a) By using Eq. (12.1), we can obtain

$$\omega_m = \frac{40}{95 \times 0.004} - \frac{0.8 \times 1.2}{(95 \times 0.004)^2} = 98.62 \text{ rad/s}$$

or

$$N_m = \frac{98.62 \times 60}{2\pi} \approx 942 \text{ rpm}$$

(b) For the blocked-rotor condition $\omega_m = 0$. Hence,

$$0 = \frac{40}{95 \times 0.004} - \frac{0.8\, T_d}{(95 \times 0.004)^2}$$

or

$$T_d = 19 \text{ N-m}$$

■

EXAMPLE 12.2

Calculate the magnetic flux in a 200-W, 100-V PM motor operating at 1500 rpm. The motor constant is 85, the armature resistance is 2 Ω, and the rotational loss is 15 W.

SOLUTION

$$\omega_m = \frac{2\pi \times 1500}{60} = 157.08 \text{ rad/s}$$

Since the power developed is $P_d = 200 + 15 = 215$ W, the developed torque becomes

$$T_d = \frac{215}{157.08} = 1.37 \text{ N-m}$$

From Eq. (12.1) we can write

$$\Phi_p^2 - \frac{V_s}{K_a\omega_m}\Phi_p + \frac{R_aT_d}{K_a^2\omega_m} = 0$$

$$\Phi_p^2 - 0.0075\Phi_p + 2.41 \times 10^{-6} = 0$$

or

$$\Phi_p = 7.16 \text{ mWb}$$

■

As explained in Chapter 2, the operating point of a permanent magnet depends on the permeance of the magnetic circuit. The point of intersection of the operating line and the demagnetization curve determines the flux density in the magnetic circuit. The same situation takes place also in PM motors as long as the demagnetization effect of the armature reaction is neglected.

Let us assume that the point of operation for a PM motor is marked by X in Figure 12.6 when the effect of armature reaction is not considered. However, no matter what, the demagnetization effect of armature reaction should be included to determine the proper operating point of the magnet even though PM motors

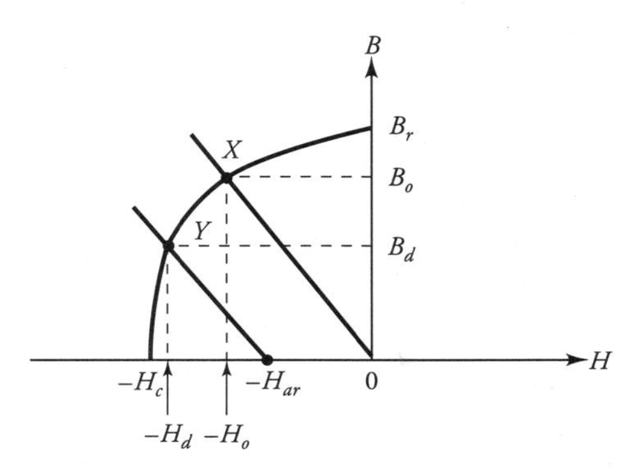

Figure 12.6 Demagnetization effect in a PM motor.

are designed with relatively large air-gaps to minimize the armature reaction. In this case the operating line moves to the left, as shown in Figure 12.6, where H_{ar} corresponds to the magnetic field intensity due to the armature. Thus, the actual operating point of the motor moves to point Y. From the figure, we can conclude that the useful magnetic flux density decreases with the increase in armature reaction.

The demagnetization effect in a PM motor due to the armature reaction is maximum under blocked-rotor condition. In order to study its effect, we consider the intrinsic demagnetization curve as given in Figure 12.7. This curve can be extracted from the normal demagnetization curve using $B_i = B_n + \mu_0 H_n$ where B_i is the intrinsic flux density. Here, B_n and H_n are the normal flux density and the corresponding field intensity, respectively. Figure 12.7 also highlights that if H_{ar} exceeds the intrinsic coercive force H_{ci}, the magnet in the PM motor is completely demagnetized.

Let us assume that the load line of a PM motor with no armature current intersects the demagnetization curve at point X, as illustrated in Figure 12.8. With an increase in armature current, the point of operation shifts to Y owing to armature reaction. We would expect the operating point to move back to X again as soon as the armature current is switched off. This, in fact, is not so, and the new operating point will be Z on the original operating line. The line from Y to Z is known as the recoil line. The recoil line is approximately parallel to the slope of the demagnetization curve at point B_r. The overall influence of the armature reaction is a reduction in the operating flux density in the motor. However, if ceramic magnets are used, the reduction is insignificant, as the demagnetization curve is essentially a straight line.

The effect of temperature should also be taken into consideration when designing a PM motor. Figure 12.9 illustrates the changes in the demagnetization characteristic at two different temperatures. As the temperature increases, the residual flux density in the magnet decreases and the intrinsic coercive force in-

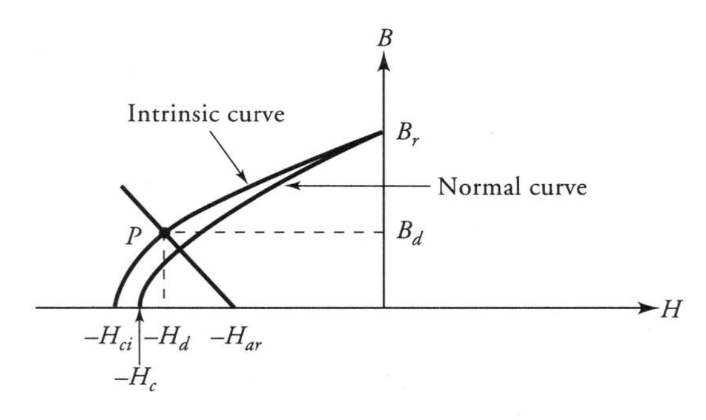

Figure 12.7 Intrinsic operating point of the trailing tip of the magnet during start or stall.

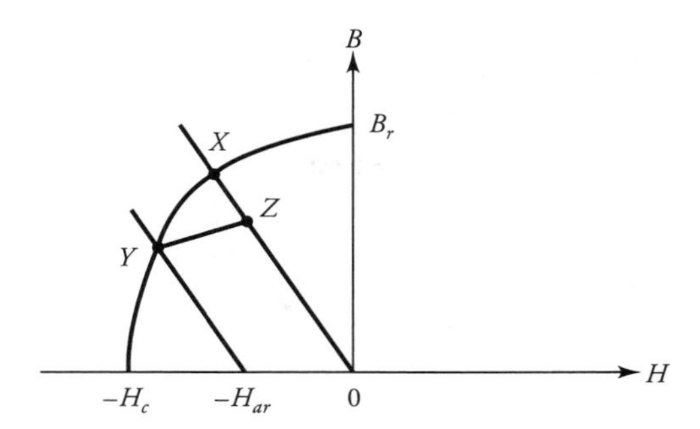

Figure 12.8 Recoil effect on the operation of a PM motor.

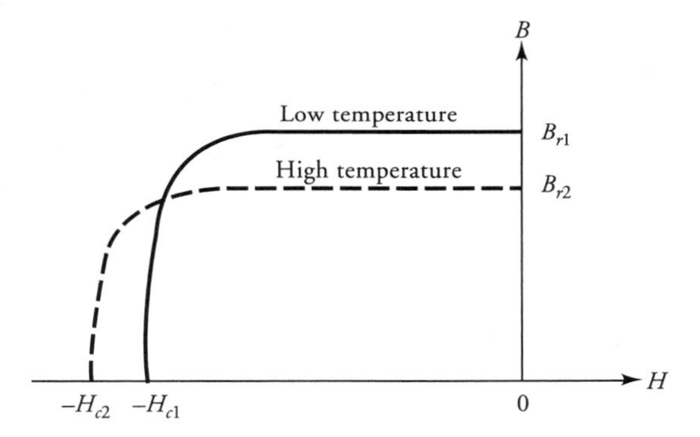

Figure 12.9 Temperature effect on the intrinsic curve.

creases. On the other hand, the lower the temperature, the more pronounced the demagnetization effect of the armature reaction.

Exercises

12.1. A 0.5-hp, 120-V, 2-pole, 70% efficient PM motor utilizes samarium-based permanent magnets. Its ideal no-load speed is 1000 rpm, and armature resistance is 1.5 Ω. The pole length and the average radius of the motor are 55 mm and 45 mm, respectively. Determine the operating line of the motor if all the losses except the copper loss are negligible. Consider the motor constant to be 80.

12.2. A 100-V, 2-pole, PM dc motor with Alnico magnets drives a load of 0.25 hp at an efficiency of 72%. When the motor operates at no load, its speed is 1000 rpm. The pole length and the average radius of the magnets are 65 mm and 55 mm, respectively, and the resistance of the armature circuit is 1.2 Ω. Determine the operating line of the motor. The demagnetization characteristic of the Alnico magnet is given in Figure 2.28. K_a is 65.

12.3 Step Motors

Step motors, also known as stepping or stepper motors, are essentially incremental motion devices. A step motor receives a rectangular pulse train and responds by rotating its shaft a certain number of degrees as dictated by the number of pulses in the pulse train. Usually the pulse train is controlled by means of a microcomputer or an electronic circuit. As a result, a step motor is very much compatible with digital electronic circuits and may form an interface between a microcomputer and a mechanical system. Since the motion in a step motor is generally governed by counting the number of pulses, no feedback loops and sensors are needed for controlling a step motor. Therefore, step motors are excellent devices for position control in an open loop system. They are relatively inexpensive and simple in construction and can be made to step in equal increments in either direction. Step motors are excellent candidates for such applications as printers, *XY* plotters, electric typewriters, control of floppy disk drives, robots, and numerical control of machine tools. However, step motors do not offer the flexibility of adjusting the angle of advance, and their step response is oscillatory in nature with a considerable overshoot. Step motors can be classified into three categories—variable-reluctance, permanent-magnet, and hybrid.

Variable-Reluctance Step Motors

Variable-reluctance step motors operate on the same principle as a reluctance motor, which was explained in Chapter 3. The principle involves the minimization of the reluctance along the path of the applied magnetic field, as shown in Figure 12.10.

The stator of a variable-reluctance step motor comprises a magnetic core construction made of a single stack of steel laminations with phase windings wound on each stator tooth, as illustrated in Figure 12.11. The rotor, which is also made of a stack of steel laminations, does not carry any winding. In order to make one set of stator and rotor teeth align, the number of teeth in the rotor is made different from that of the stator. The step motor in Figure 12.11 has six stator teeth and four rotor teeth. The stator windings are excited at different times, leading to a multiphase stator winding. The stator of the step motor shown in Figure 12.11 has three phases—*A*, *B*, and *C*, with teeth 1 and 4, 2 and 5, and 3 and 6, respectively.

In Figure 12.11**a**, rotor teeth 1 and 2 are aligned with stator teeth 1 and 4 when phase-*A* winding is excited by a constant current. As long as phase-*A* is energized

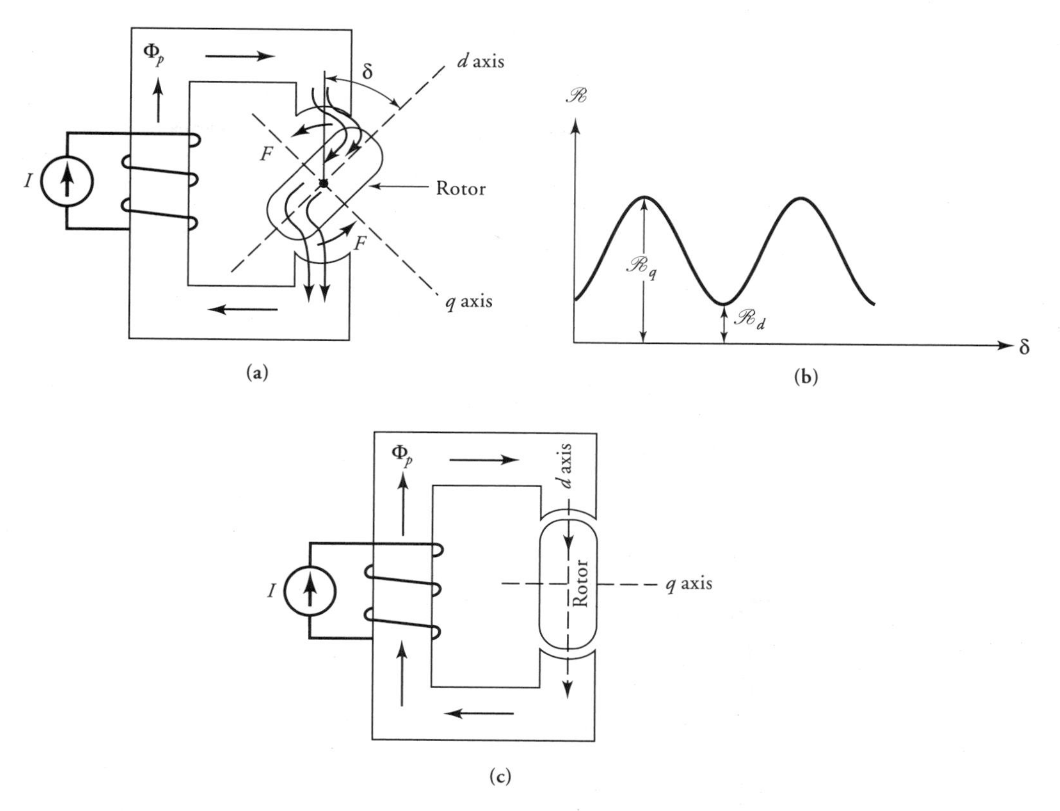

Figure 12.10 **(a)** Magnetic circuit with a freely rotating member, **(b)** reluctance as a function of position, and **(c)** minimum reluctance, equilibrium, or no rotation position.

while all the other phases are not, the rotor is stationary and counteracts the torque caused by the mechanical load on the shaft. Since the angle between the magnetic axis of phase-*B* or -*C* and the axis of rotor teeth 3 and 4 is 30°, if phase-*A* is switched off and phase-*B* winding is excited, this time rotor teeth 4 and 3 align under stator teeth 3 and 6, leading to 30° of displacement of the rotor. Finally, if we excite phase-*C* winding after de-energizing phase-*B*, the rotor rotates another 30° and aligns with phase-*C*, as shown in Figure 12.11**c**. The rotor can be made to rotate continuously in the clockwise direction by following the switching sequence described above. To achieve a counterclockwise rotation, however, phases should be sequentially switched in the order of *A*, *C*, *B*. Figure 12.12 illustrates the phase voltages applied to the variable-reluctance step motor discussed, and Table 12.1 shows the proper switching sequence for a clockwise rotation. For this particular motor, applied voltage must have at least five cycles for one revolution.

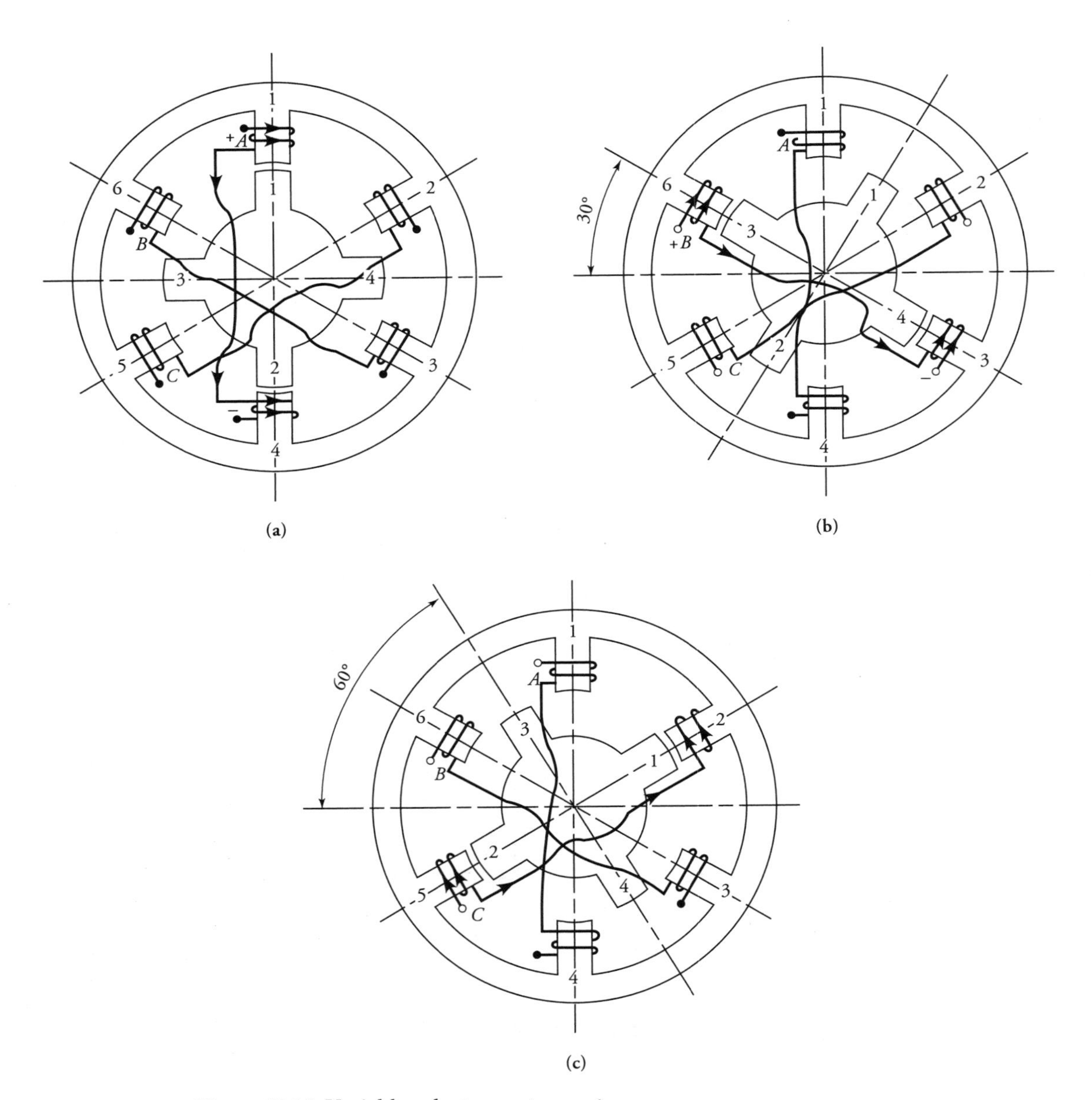

Figure 12.11 Variable-reluctance step motor.

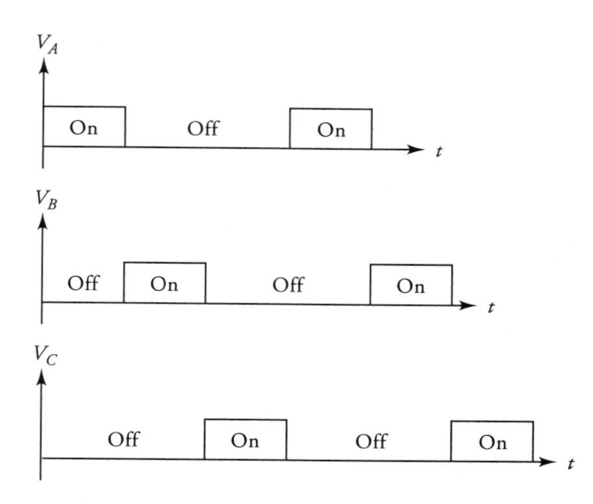

Figure 12.12 Phase-voltage waveforms for a variable-reluctance motor.

Table 12.1: Switching Sequence for a Variable-Reluctance Motor. "1" and "0" correspond to positive and zero current in a phase winding, respectively.

Cycle	Phase A	Phase B	Phase C	Position δ°
1	1	0	0	0
	0	1	0	30
	0	0	1	60
2	1	0	0	90
	0	1	0	120
	0	0	1	150
3	1	0	0	180
	0	1	0	210
	0	0	1	240
4	1	0	0	270
	0	1	0	300
	0	0	1	330
5	1	0	0	360

The step angle, δ, for a variable-reluctance step motor is determined by

$$\delta = \frac{2\pi}{np} \tag{12.3}$$

where n and p are the number of phases and the number of rotor poles, respectively.

Permanent-Magnet Step Motors

A PM step motor differs from its variable-reluctance counterpart in that its rotor is made of permanent magnets. The stator construction of a PM step motor is the same as that of a variable-reluctance step motor. A two-phase, 2-pole rotor PM step motor is illustrated in Figure 12.13. In this motor the rotor is radially magnetized so that the rotor poles align with the appropriate stator teeth.

When phase-A winding is excited by a constant current, as shown in the figure, tooth 1 acts as a south pole. This makes the north pole of the PM rotor align with the south pole of the stator. Later in time, phase-A is de-energized while the phase-B winding is activated, causing a 90° displacement in the counterclockwise direction to align the rotor's north pole with stator tooth 2. If we reverse the polarity of the applied current and start exciting phase-A again, the rotor will further rotate 90° along the counterclockwise direction, this time to align the rotor's north pole with the stator tooth 3. So far the motor has completed one-half revolution, and with the continuation of the appropriate switching, the rotor continues to rotate and completes its revolution. Figure 12.14 illustrates the input waveforms to phase-A and phase-B of a two-phase step motor, while Table 12.2 describes the switching sequence for one full revolution of the motor.

Table 12.2: Switching Sequence for a Two-Phase PM Step Motor. "1", "−1", and "0" correspond to positive, negative, and zero current in a phase winding, respectively.

	Phase		
Cycle	A	B	Position δ°
+	1	0	0
	0	1	90
−	−1	0	180
	0	−1	270
+	1	0	360

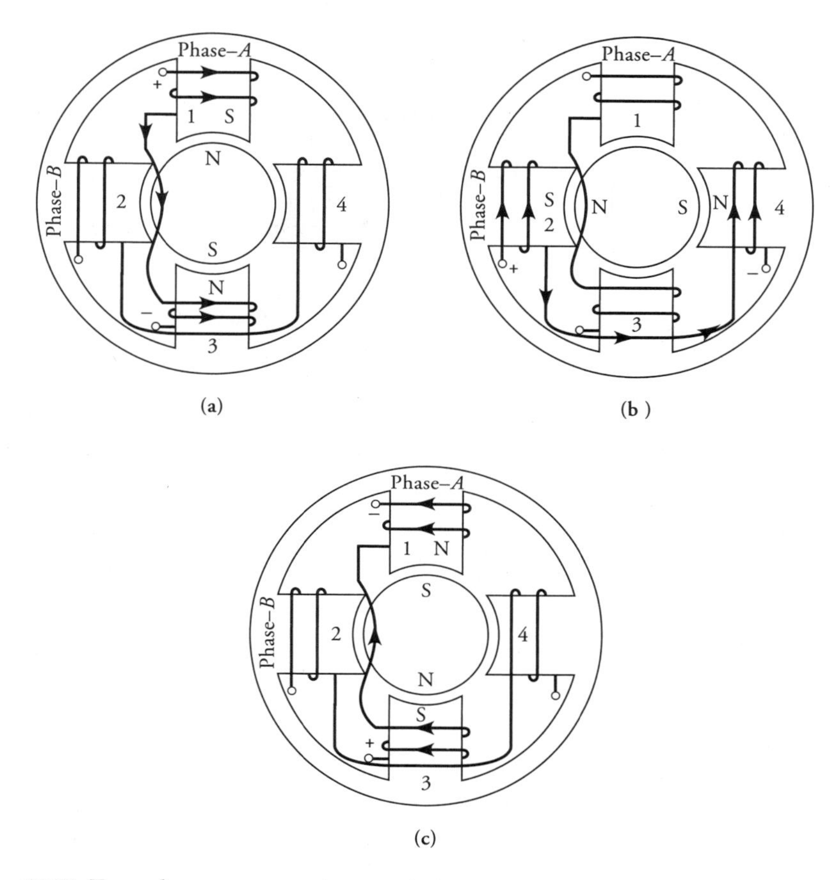

Figure 12.13 Two-phase permanent magnet step motor.

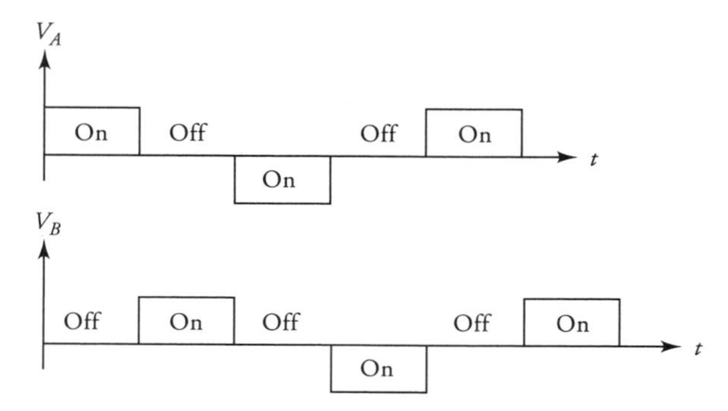

Figure 12.14 Applied voltage waveforms for a two-phase PM step motor.

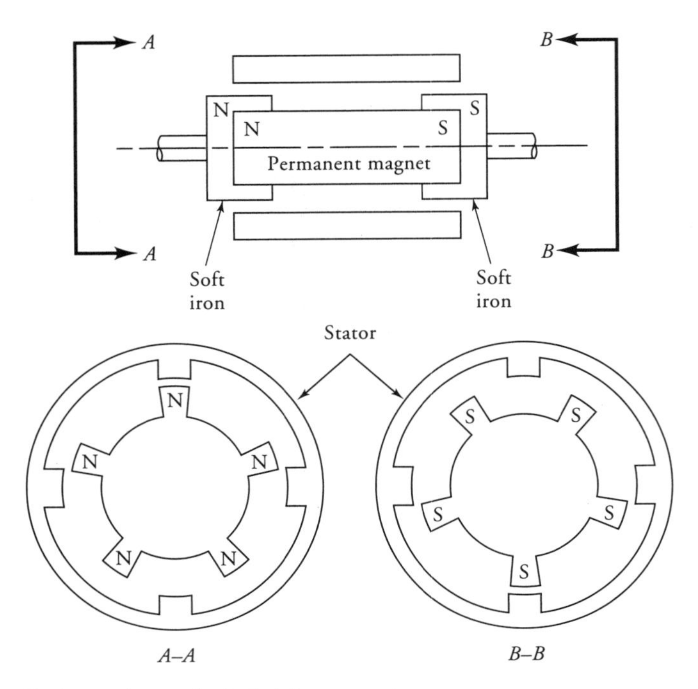

Figure 12.15 Various views of a hybrid step motor.

Hybrid Step Motors

The stator construction of a hybrid step motor is no different from that of a variable-reluctance or PM step motor. However, the rotor construction integrates the design of the rotors of a variable-reluctance and a PM step motor. The rotor of a hybrid step motor consists of two identical stacks of soft iron as well as an axially magnetized round permanent magnet. Soft iron stacks are attached to the north and south poles of the permanent magnet, as shown in Figure 12.15. The rotor teeth are machined on the soft iron stacks. Thus the rotor teeth on one end become the north pole, while those at the other end become the south pole. The rotor teeth at both north and south poles are displaced in angle for the proper alignment of the rotor pole with that of the stator, as shown in Figure 12.15. The operating mode of the hybrid step motor is very similar to that of a PM step motor.

Torque-Speed Characteristic

Step motors are generally used in a range from 1 W to about 3 hp, and their step sizes vary from approximately 0.72° to 90°. However, the most common step sizes are 1.8°, 7.5°, and 15°.

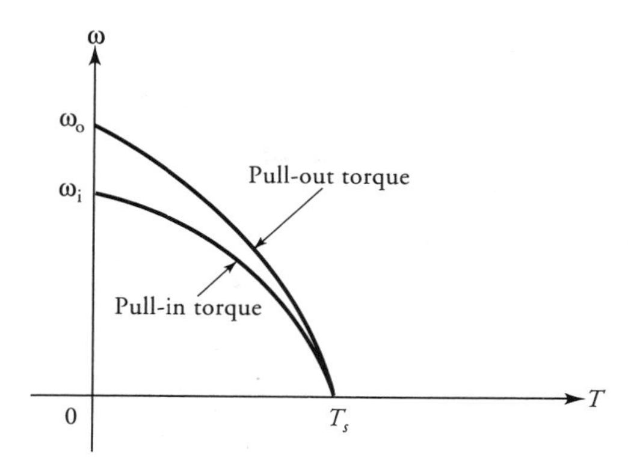

Figure 12.16 Speed-torque characteristic of a step motor.

Since a step motor rotates when a series of pulses is applied to its phase windings, the duration of each pulse should be sufficiently long to accurately rotate the motor at the desired speed. If the pulse duration is too short, the rotor will miss the steps and be unable to follow the applied pulses accurately. Thus either the motor will not rotate or the required speed will not be achieved. To avoid such an operation, usually the pulse duration is selected so that it is greater than the inertial time-constant of the combination of the rotor and the mechanical load. Therefore, it is anticipated that a large motor with a high moment of inertia requires a slower pulse rate for accurate operation.

The pull-in torque characteristic shown in Figure 12.16 illustrates the permissible range of rate of steps for a given load and a motor in order not to miss a step. When the motor achieves its steady-state operation, the speed is uniform and no starting and stopping take place at each step. We can load the motor up to a limit, which is defined by the pull-out torque characteristic shown in Figure 12.16. Above that torque level the motor starts missing the steps, thereby losing speed.

12.4 Switched-Reluctance Motors

In principle, a switched-reluctance motor operates similarly to the variable-reluctance step motor discussed in the previous section. However, the operation differs mainly in the complicated control mechanism of the motor. In order to develop torque in the motor, the rotor position should be determined by sensors so that the excitation timing of the phase windings is precise. Although its construction is one of the simplest possible among electric machines, because of the complexities involved in the control and electric drive circuitry, switched-reluctance motors have not been able to find widespread applications for a long time. However, with the introduction of new power electronic and microelectronic switching circuits, the control and drive circuitry of a switched reluctance motor have become economically justifiable for many applications where traditionally dc or induction motors have been used.

A switched-reluctance motor has a wound stator but interestingly it has no windings on its rotor, which is made of soft magnetic material. Figure 12.17 is a schematic diagram of a switched-reluctance motor in its simplest form. The change in reluctance around the periphery of the stator forces the rotor poles to align with those of the stator. Consequently, torque develops in the motor and rotation takes place.

The total flux linkages of phase-A in Figure 12.17 is $\lambda_a = L_a(\theta)i_a$ and of phase-B is $\lambda_b = L_b(\theta)i_b$, with the assumption that the magnetic materials are infinitely permeable. Since the magnetic axes of both windings are orthogonal, no mutual flux linkages are expected.

The co-energy in the motor is

$$W = \frac{1}{2} L_a(\theta)i_a^2 + \frac{1}{2} L_b(\theta)i_b^2 \tag{12.4}$$

and the developed torque is

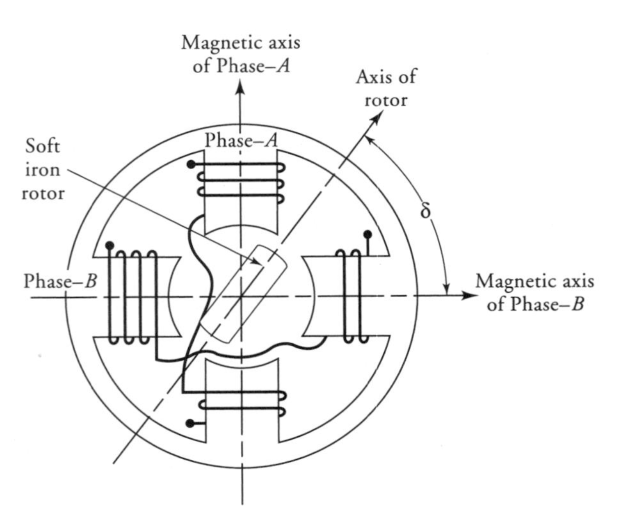

Figure 12.17 Schematic diagram of a switched-reluctance motor in its simplest form.

$$T = \frac{dW}{d\theta}$$

$$T = \frac{1}{2} i_a^2 \frac{dL_a}{d\theta} + \frac{1}{2} i_b^2 \frac{dL_b}{d\theta} \tag{12.5}$$

From Eq. (12.5) we can conclude that the developed torque in the motor is independent of the direction of the supply current because it is proportional to the square of the phase currents. However, rotor position has a significant impact on the developed torque in a switched-reluctance motor. Thus, a reliable rotor position sensor and a control circuit are necessary to energize the motor at the proper instant for a uniform developed torque.

12.5 Brushless DC Motors

Owing to their inherent characteristics, as discussed in previous chapters, dc motors find considerable applications where controlling a system is a primary objective. However, electric arcs produced by the mechanical commutator-brush arrangement are a major disadvantage and limit the operating speed and voltage. A motor that retains the characteristics of a dc motor but eliminates the commutator and the brushes is called a **brushless dc motor**.

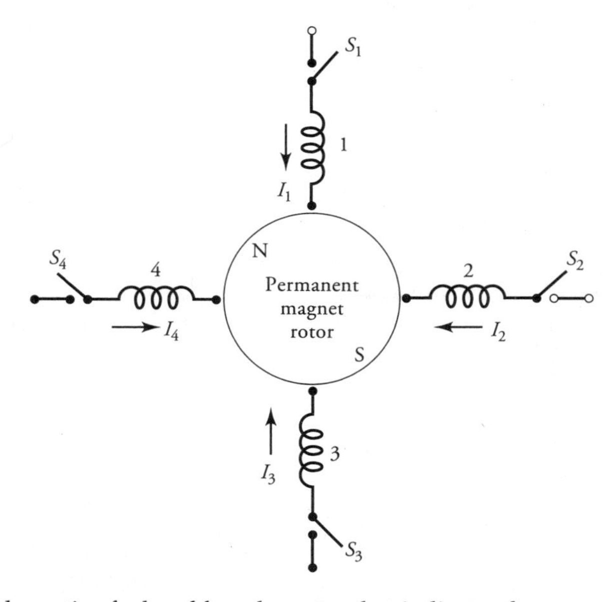

Figure 12.18 Schematic of a brushless dc motor that indicates the operating principle.

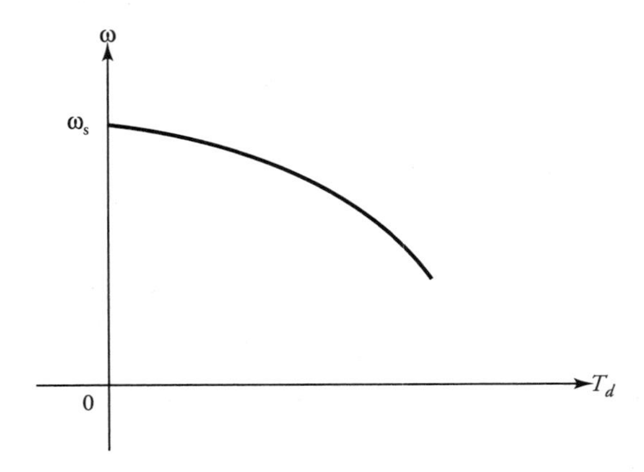

Figure 12.19 Speed-torque characteristic of a brushless dc motor.

A brushless dc motor consists of a multiphase winding wound on a nonsalient stator and a radially magnetized PM rotor. Figure 12.18 is a schematic diagram of a brushless dc motor. The multiphase winding may be a single coil or distributed over the pole span. Direct voltage or alternating voltage is applied to individual phase windings through a sequential switching operation to achieve the necessary commutation for the rotation of the motor. The switching is done electronically using power transistors or thyristors. For example, if winding 1 is energized, the PM rotor aligns with the magnetic field produced by winding 1. When winding 1 is switched off while winding 2 is turned on, the rotor is made to rotate to line up with the magnetic field of winding 2. As can be seen, the operation of a brushless dc motor is very similar to that of a PM step motor. The major difference is the timing of the switching operation, which is determined by the rotor position to provide the synchronism between the magnetic field of the permanent magnet and the magnetic field produced by the phase windings. The rotor position can be detected by using either Hall-effect or photoelectric devices. The signal generated by the rotor position sensor is sent to a logic circuit to make the decision for the switching, and then an appropriate signal triggers the power circuit to excite the respective phase winding. The control of the magnitude and the rate of switching of the phase currents essentially determine the speed-torque characteristic of a brushless dc motor, which is shown in Figure 12.19.

12.6 Hysteresis Motors

Hysteresis motors utilize the property of hysteresis of magnetic materials to develop torque. The stator may have a uniformly distributed three-phase or single-phase winding. In a single-phase hysteresis motor, the stator winding is connected as a permanent split capacitor (PSC) motor. The capacitor is selected in such a way that a balanced two-phase condition can be approximately achieved, so that almost a uniform rotating field can be established in the motor. The rotor is a solid hard magnetic material with no teeth or windings. Figure 12.20 is a schematic diagram of a single-phase hysteresis motor.

When the stator winding is excited, a rotating field is set up in the air-gap of the motor that revolves at the synchronous speed. The rotating field magnetizes the rotor and induces as many poles on its periphery as there are in the stator. Owing to the large hysteresis loss in the rotor, the magnetic flux developed in the rotor lags the stator magnetomotive force (mmf). Thus a rotor angle, δ, takes place between the rotor and stator magnetic axes. Figure 12.20**b** illustrates the relative positions of the rotor and the stator magnetic axes for a 2-pole hysteresis motor. The greater the loss due to hysteresis, the larger the angle between the magnetic axes of the rotor and the stator. Owing to the tendency of the magnetic poles of the rotor to align themselves with those of the stator, a finite torque, called **hysteresis torque**, is produced. This torque is proportional to the product of the rotor flux and the stator mmf and the sine of the rotor angle δ. Thus, it should be noted that a rotor with a large hysteresis loop results in a higher hysteresis torque.

Since the rotor is a solid magnetic material, eddy currents are induced in the rotor by the magnetic field of the stator as long as there is a relative motion between the stator magnetic field and the rotor. These eddy currents produce their own magnetic fields and thereby their own torque, which further enhances the total torque developed by the motor. The torque due to the eddy currents is proportional to the slip of the motor, and it is maximum at standstill and zero when the synchronous speed is reached.

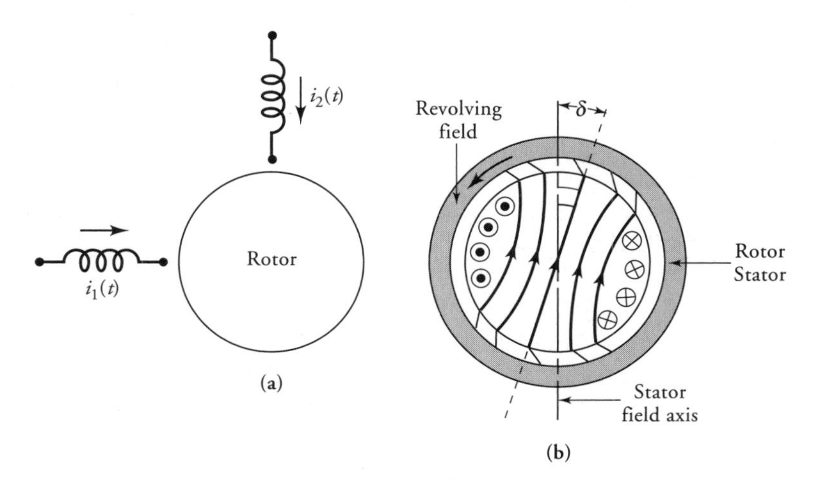

Figure 12.20 **(a)** Schematic of a two-phase hysteresis motor; **(b)** Flux distribution showing hysteresis effect.

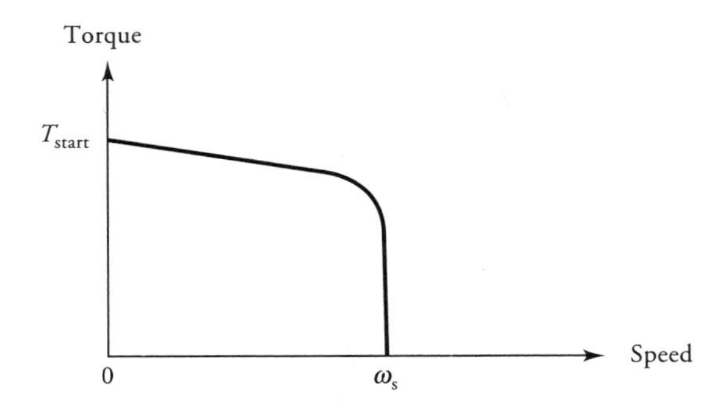

Figure 12.21 Speed-torque characteristic of a hysteresis motor.

When the motor is excited at a certain voltage while it has been at standstill, a constant torque is developed by the motor, and if it is greater than the torque required by the load, the motor will start rotating. With the fixed applied voltage, the hysteresis torque remains almost constant over the entire speed range of the motor up to the synchronous speed because the rotor angle, δ, essentially depends on the rotor material. However, with the influence of the eddy-current torque, a slight reduction in the total developed torque is observed as the motor speed increases. After synchronous speed is achieved, the motor adjusts its rotor angle, δ, so that the required torque can be developed by the motor. A typical torque-speed characteristic of a hysteresis motor is shown in Figure 12.21. From this characteristic it can be observed that the torque developed is higher at any speed other than the synchronous speed and is maximum at standstill. Thus the starting torque is never a problem in hysteresis motors. Furthermore, because the developed torque is almost uniform from standstill to synchronous speed, a hysteresis motor can accelerate a high-inertia load.

12.7 Linear Induction Motors

So far we have examined the fundamental operating principles of electric machines that produce circular motion. During the last few decades, extensive research in the area of propulsion has led to the development of linear motors. Theoretically each type of rotating machine may find a linear counterpart. However, it is the linear induction motor that is being used in a broad spectrum of such industrial applications as high-speed ground transportation, sliding door systems, curtain pullers, and conveyors.

If an induction motor is cut and laid flat, a linear induction motor is obtained. The stator and rotor of the rotating motor correspond to the primary and secondary sides, respectively, of the linear induction motor. The primary side consists of a magnetic core with a three-phase winding, and the secondary side may be just a metal sheet or a three-phase winding wound around a magnetic core. The basic difference between a linear induction motor and its rotating counterpart is that the latter exhibits endless air-gap and magnetic structure, whereas the former is open-ended owing to the finite lengths of the primary and secondary sides. Also, the angular velocity becomes linear velocity, and the torque becomes the thrust in a linear induction motor. In order to maintain a constant thrust (force) over a considerable distance, one side is kept shorter than the other. For example, in high-speed ground transportation, a short primary and a long secondary are being used. In such a system, the primary is an integral part of the vehicle, whereas the track manifests as the secondary.

A linear induction motor may be single-sided or double-sided, as shown in Figure 12.22. In order to reduce the total reluctance of the magnetic path in a single-sided linear induction motor with a metal sheet as the secondary winding, the metal sheet is backed by a ferromagnetic material such as iron.

When the supply voltage is applied to the primary winding of a three-phase linear induction motor, the magnetic field produced in the air-gap region travels at the synchronous speed. The interaction of the magnetic field with the induced currents in the secondary exerts a thrust on the secondary to move in the same direction if the primary is held stationary. On the other hand, if the secondary side is stationary and the primary is free to move, the primary will move in the direction opposite to that of the magnetic field.

Let us consider the simplified schematic diagram for a linear induction motor shown in Figure 12.23. In this figure, only one phase winding, say phase-*a*, of the three-phase primary winding is shown. The *N*-turn phase winding experiences an mmf of *NI*, as indicated in Figure 12.23**b**. If we focus our attention only on the fundamental of the mmf waveform, we get

$$\mathcal{F}_a = k_w \frac{2}{n\pi} N i_a \cos \frac{2\pi}{\lambda} z \tag{12.6}$$

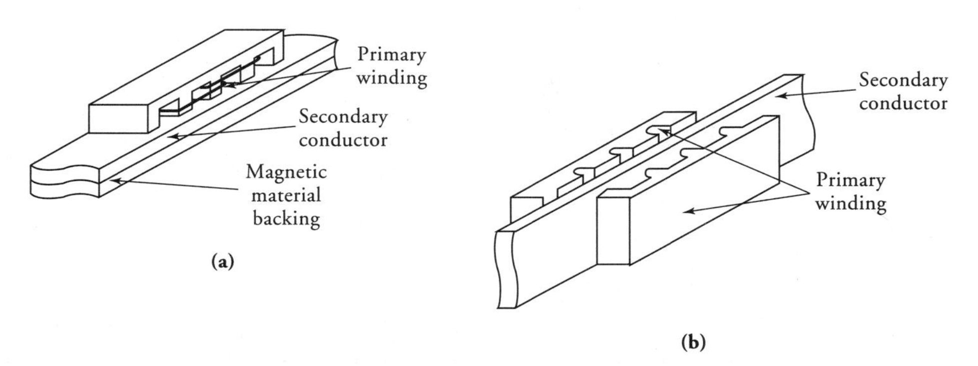

Figure 12.22 **(a)** Single-sided and **(b)** double-sided linear induction motors.

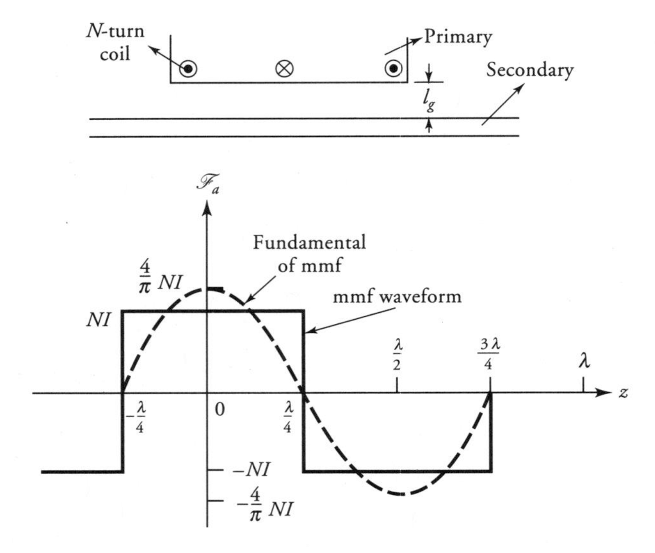

Figure 12.23 Schematic diagram of a linear induction motor and its mmf waveform.

where k_w is the winding factor, i_a is the instantaneous value of the fundamental current in phase-*a*, λ is the wavelength of the field (in essence it is the winding pitch), n is the number of periods over the length of the motor, and z is an arbitrary location in the linear motor. Each phase winding is displaced from the others by a distance of $\lambda/3$ and excited by a balanced three-phase supply of angular frequency ω. Thus the net mmf in the motor consists of only a forward-traveling wave component as given by

$$\mathscr{F}(z, t) = \frac{3}{2} F_m \cos\left(\omega t - \frac{2\pi}{\lambda} z\right) \tag{12.7}$$

where

$$F_m = \frac{2}{n\pi} k_w N I_m \tag{12.8}$$

The synchronous velocity of the traveling mmf can be determined by setting the argument of the cosine term of Eq. (12.7) to some constant K

$$\omega t - \frac{2\pi}{\lambda} z = K \tag{12.9}$$

and then differentiating Eq. (12.9) with respect to t to obtain v_s as

$$v_s = \frac{dz}{dt} = \frac{\omega\lambda}{2\pi} \tag{12.10}$$

or

$$v_s = \lambda f \tag{12.11}$$

where f is the operating frequency of the supply. Equation (12.11) can also be expressed in terms of the pole pitch τ as

$$v_s = 2\tau f \tag{12.12}$$

Both Eqs. (12.11) and (12.12) suggest that the synchronous velocity is independent of the number of poles in the primary winding. Moreover, the number of poles need not be an even number.

Similar to the rotating induction motors, the slip in a linear induction motor is defined as

$$s = \frac{v_s - v_m}{v_s} \tag{12.13}$$

where v_m is the velocity of the motor.

The power and thrust in a linear induction motor can be calculated by using the same equivalent circuit employed for its rotating counterpart. Thus, the air-gap power, P_g, is

$$P_g = 3I_2^2 \frac{r_2}{s} \tag{12.14}$$

the developed power, P_d, is

$$P_d = (1 - s)P_g \tag{12.15}$$

and the developed thrust, F_d, is

$$F_d = \frac{P_d}{v_m}$$

$$= \frac{P_g}{v_s} = 3I_2^2 \frac{r_2}{sv_s} \tag{12.16}$$

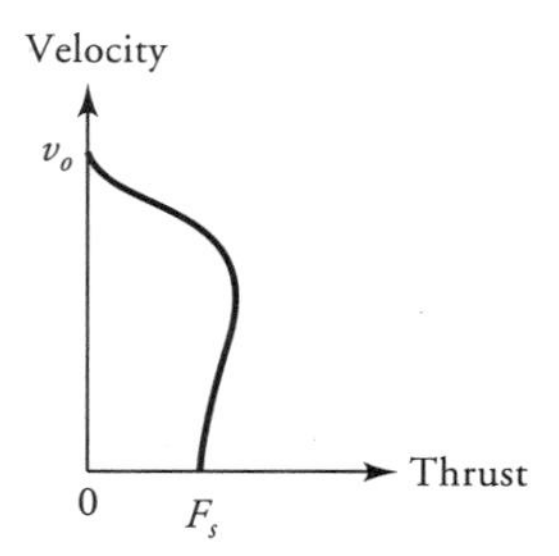

Figure 12.24 Typical velocity versus thrust characteristic of a linear induction motor.

The speed-torque characteristic of a conventional induction motor is equivalent to the velocity-thrust characteristic of a linear induction motor shown in Figure 12.24. The velocity in a linear induction motor decreases rapidly with the increasing thrust, as can be observed from Figure 12.24. For this reason these motors often operate with low slip, leading to a relatively low efficiency.

Owing to its open-ended structure, a linear induction motor displays a phenomenon known as end effects. The end effects can be classified as static and dynamic. Static end effect occurs solely because of the asymmetric geometry of the primary. In this case, the mutual inductances of the phase windings are not equal to one another. This results in asymmetric flux distribution in the air-gap region and gives rise to unequal induced voltages in the phase windings. The dynamic end effect occurs as a result of the relative motion of the primary side with respect to the secondary. As the primary moves over the secondary, at every instant a new secondary conductor is brought under the leading edge of the primary, while another secondary conductor is leaving the trailing edge of the primary. The conductor coming under the leading edge opposes the magnetic flux in the air-gap, while the conductor leaving the trailing edge tries to sustain the flux. Therefore, the flux distribution is distorted. It's weaker in the leading edge region as compared to the trailing edge region. Furthermore, the conductor leaving the trailing edge, although still carrying the current and contributing to the losses, does not contribute to the trust. Therefore, the increased losses in the secondary reduce the efficiency of the motor.

EXAMPLE 12.3

The pole pitch of a linear induction motor is 0.5 m, and the frequency of the applied three-phase voltage is 60 Hz. The speed of the primary side of the motor is 200 km/h, and the developed thrust is 100 kN. Calculate the developed power by the motor and the copper loss in the secondary side.

● SOLUTION

The speed of the motor

$$v_m = \frac{200 \times 10^3}{3600} = 55.55 \text{ m/s}$$

The developed power

$$P_d = F_d v_m = 100 \times 10^3 \times 55.55 = 5555 \times 10^3 \text{ W}$$

$$P_d = 5555 \text{ kW}$$

Slip for this operating condition is

$$s = \frac{v_s - v_m}{v_s} = \frac{60 - 55.55}{60} = 0.074$$

Copper loss in the secondary side is

$$P_{cus} = 3I_2^2 r_2 = F_d s v_s = 100 \times 10^3 \times 0.074 \times 60 = 444{,}000 \text{ W}$$

or $$P_{cus} = 444 \text{ kW}$$

■

Exercises

12.3. A linear induction motor with a pole pitch of 50 cm is used on a trolley that travels a distance of 10 km. The resistance and the current of the secondary side as referred to the primary are determined to be 4 Ω and 500 A, respectively. Determine the developed thrust by the motor when the slip is 25% while operating at 60 Hz.

12.4. A 660-V (line), 50-Hz linear induction motor drives a vehicle a slip of 20%. The motor has 5 poles and the pole pitch is 30 cm. The parameters of its equivalent circuit are:
Primary side: $r_1 = 0.15\ \Omega$, $x_1 = 0.5\ \Omega$
Secondary side: $r_2 = 0.3\ \Omega$, $x_2 = 0.3\ \Omega$
Magnetizing reactance: $X_m = 3\ \Omega$
Determine (a) the synchronous speed, (b) the power delivered to the load, (c) the thrust, (d) the input current, and (e) the power factor. Neglect the core losses.

SUMMARY

Recent developments in the area of power electronics technology and digital control systems have facilitated the use of special-purpose motors for the accurate control of speed and/or position. The special-purpose motors studied in this chapter are permanent-magnet (PM) motors, step motors, switched-reluctance motors, brushless dc motors, and linear induction motors. These motors find a wide variety of applications ranging from computer peripheral equipment to high-speed ground transportation and process control.

PM motors operate as separately excited dc motors except that they do not have a field winding. Instead, a PM motor has permanent magnets to set up the necessary magnetic field for the electromechanical energy conversion. However, special care should be taken in order not to demagnetize the magnets by exceeding their coercive force due to the armature reaction.

A step motor is an incremental motion device and is widely used in computer peripherals. Its rotation is dictated by the number of pulses applied to the stator winding. Thus, no feedback loops or sensors are required for the operation of a step motor. However, the step response of a step motor is oscillatory, with a considerable overshoot. The three kinds of step motors are variable-reluctance, permanent-magnet, and hybrid.

A switched-reluctance motor operates similarly to the variable-reluctance step motor. However, a sensor accurately detects the position of the rotor to maintain precise timing of the phase windings for a given operating condition. Switched-reluctance motors have found widespread applications where ac or dc motors have traditionally been employed.

Brushless dc motors do not exhibit the disadvantages of conventional dc motors because they do not have the commutator-brush arrangement. Yet their performance characteristics are very similar to those of a conventional dc shunt motor. Thus they find a variety of system-control applications where a dc shunt motor could be used. Construction of a brushless dc motor is similar to the construction of a PM step motor. However, in a brushless dc motor, the rotor position is accurately sensed by a sensor in order to properly time the switching of the stator windings. Hall-effect or optoelectronic devices are used to sense the rotor position in a brushless dc motor.

A hysteresis motor is a synchronous motor that uses the hysteresis property of magnetic materials to develop the torque. The torque developed by a hysteresis motor is inherently higher at any speed other than the synchronous speed. Consequently, the starting torque is never a problem for a hysteresis motor.

Even though each type of rotating machine can find a linear counterpart, the linear induction motor is the one that covers a wide spectrum of applications such as high-speed ground transportation, conveyors, and sliding doors. A linear induction motor has a primary and a secondary side. The primary consists of a three-phase winding and a magnetic core, and the secondary is either a metal sheet or a three-phase winding wound around a magnetic core. For high-speed ground transportation, a short primary (the vehicle) and a long secondary (the track) are used. The synchronous velocity v_s can be calculated from $v_s = 2\tau f$, where τ is the pole pitch (m) and f is the frequency (Hz) of the applied three-phase voltage. The developed thrust, on the other hand, is $F_d = P_d/v_m$, where P_d(W) is the developed power and v_m (m/s) is the velocity of the motor.

Review Questions

12.1. Why is the efficiency of a PM motor higher than that of a wound dc motor?
12.2. Why does the useful magnetic flux density decrease in a PM motor while it is operating under a load?
12.3. Under what conditions can a PM motor be completely demagnetized?
12.4. What is the recoil line in a PM motor?
12.5. How does the temperature affect the operation of a PM motor?
12.6. How does a step motor operate?
12.7. What are the different kinds of step motors?
12.8. What is the expression for the step angle of a variable-reluctance step motor?
12.9. What is the basic difference between a PM and a variable-reluctance step motor?
12.10. What is the difference between a variable-reluctance step motor and a switched-reluctance motor?
12.11. In what kind of applications can one use a switched-reluctance motor?
12.12. What is the difference between the PM step motor and a brushless dc motor?
12.13. How does the speed vary in a brushless dc motor with changing torque?
12.14. Explain the operating principle of a hysteresis motor.
12.15. What is the hysteresis torque?
12.16. Is a hysteresis motor a synchronous motor?
12.17. How does a linear induction motor operate?
12.18. How is the synchronous velocity expressed in a linear induction motor?
12.19. How does the slip vary as a function of thrust in a linear induction motor?
12.20. Can a backward-traveling field be supported in a three-phase linear induction motor? Give reasons.

Problems

12.1. A 20-V, PM dc motor develops a torque of 1 N-m at the rated voltage. The magnetic flux in the motor is 2 mWb. The armature resistance is 0.93 Ω, and the motor constant is 95. Calculate the operating speed of the motor. Neglect the rotational losses.

12.2. What is the armature current of the motor given in Problem 12.1 under the blocked-rotor condition?

12.3. Determine the magnitude of the applied voltage when the motor given in Problem 12.1 develops a torque of 10 N-m under blocked-rotor condition.

12.4. A 100-V, PM dc motor operates at 1200 rpm and the rated voltage. The flux per pole due to the magnets is 1.5 mWb, and the armature resistance is 0.7 Ω. Determine the developed torque if the motor constant is 82. Neglect the rotational losses.

12.5. A 12-V, 2-pole, PM dc motor manufactured with ceramic magnets drives a load of 0.134 hp with an efficiency of 54%. The ideal no-load speed of the motor is 800 rpm, and the armature resistance is 2 Ω. Determine the operating line of the motor if the pole length and the average radius are 35 mm and 25 mm, respectively. The motor constant is 75, and all the losses are assumed to be negligible except the copper loss. The demagnetization curve is given in Figure 2.28.

12.6. Determine the performance of the motor given in Problem 12.5 if samarium-based rare-earth magnets are substituted for the ceramic magnets without changing the dimensions of the motor.

12.7. Determine the magnetic flux in a 120-V, 1-hp, PM motor operating at a speed of 1500 rpm. The motor constant is 85, the armature resistance is 0.7 Ω, and the rotational losses are 50 W.

12.8. A 120-V, PM dc motor operates at a speed of 400 rad/s at no load. If the resistance of the armature circuit is 1.3 Ω, determine the speed of the motor when the load demands 5 N-m at 50 V. Draw the speed-torque characteristics for both 50-V and 100-V operations. Assume that the motor maintains constant magnetic flux with no rotational losses.

12.9. A three-phase linear induction motor has a pole pitch of 1 m. Determine the velocity of the resultant traveling mmf wave if the motor is excited by a three-phase supply having a frequency of 50 Hz.

12.10. The synchronous speed, the operating frequency, and the peak current of a linear induction motor are 10 m/s, 60 Hz, and 100 A, respectively. Determine the net mmf traveling wave if the number of turns per phase is 300 with a winding factor of 0.9. Assume that the number of periods over the length of the motor is 2.

12.11. A linear induction motor drives a conveyor belt at a speed of 20 km/h with a slip of 20% at 60 Hz and develops a thrust of 200 N. (a) Determine the pole pitch of the motor. (b) Calculate the power developed by the motor. (c) Calculate the amount of copper loss in the secondary side.

0.8 pf lagging while operating with a slip of 30% and an efficiency of 72%. The pole pitch of the motor is 60 cm. If the magnetizing current is 15% of the applied current and lags the applied voltage by an angle of 88°, determine the developed thrust by the motor, and the winding resistance of the secondary as referred to the primary. Neglect the core, friction, and windage losses.

APPENDIX A

System of Units

Although the use of the International System of Units (SI) has become more and more prevalent in all areas of electrical engineering, English units are still widely used throughout the profession in dealing with electric machines. For example, it is easy to say that the speed of the motor is 1500 rpm (revolutions per minute) rather than saying 157 rad/s (radians per second). Similarly, a 0.25-hp (horsepower) motor is very rarely referred to as a 187-W (watt) motor. However, most equations in this book are given in SI units, and it sometimes becomes necessary to convert from one unit to another. In the International System of Units, the units of length, mass, and time are meter (m), kilogram (kg), and second (s), respectively. The basic unit of the charge is expressed in coulombs (C). The current is the time rate of change of charge and is expressed in amperes (A). Thus, 1 A = 1 C/s. Among other fundamental units are the unit of temperature, kelvins (K), and the luminous intensity, candelas (cd). In the English system of units, force is expressed in pounds, length in inches or feet, torque in foot-pounds, and time in seconds. The conversion from one unit to the other is given in Table A.1.

The units for other quantities that we refer to in this book are given in Table A.2. These are known as the derived units, as they can be expressed in terms of the basic units. For example, the fundamental unit of power is the watt (W). It is the rate at which work is done or energy is expended. Thus, the watt is defined as 1 J/s. On the other hand, joule (J) is the fundamental unit of work. A joule is the work done by a constant force of 1 newton (N) applied through a distance of 1 meter (m). Hence, 1 joule is equivalent to 1 newton-meter (N-m). Above all, 1 newton is the force required to accelerate a mass of 1 kilogram by 1 meter per second per second. That is, 1 N = 1 kg-m/s^2. Note that the newton is expressed in terms of the basic units. Therefore, we can now express joules and watts in terms of the basic units.

Table A.1 Unit Conversion Factors

From	Multiply by	To obtain
gilbert	0.79577	ampere turns (A-t)
ampere turns/cm	2.54	ampere turns/inch
ampere turns/in	39.37	ampere turns/meter
oersted	79.577	ampere turns/meter
lines (maxwells)	10^{-8}	webers (Wb)
lines/cm^2 (gauss)	6.4516	lines/in^2
lines/in^2	0.155×10^{-4}	webers/m^2 (T)
webers/m^2 (tesla)	6.4516×10^4	lines/in^2
webers/m^2	10^4	lines/cm^2
inch	2.54	centimeter (cm)
feet	30.48	cm
meter	100	cm
square inch	6.4516	square cm
square inch	1.27324	circular mils
ounce	28.35	gram
pound	0.4536	kilogram
pound-force	4.4482	newton
ounce-force	0.27801	newton
newton	3.597	ounce-force
newton-meter	141.62	ounce-inch
newton-meter	0.73757	pound-feet
revolutions/minute	$2\pi/60$	radians/second
horsepower	746	watt
watts/pound	2.205	watts/kilogram

Table A.2 Derived Units for Some Quantities

Symbol	Quantity	Unit	Abbreviation
Y	admittance	siemens	S
ω	angular frequency	radian/second	rad/s
C	capacitance	farad	F
Q	charge	coulomb	C
ρ	charge density	coulomb/meter^3	C/m^3
G	conductance	siemens	S
σ	conductivity	siemens/meter	S/m
J	current density	ampere/meter^2	A/m^2
ϵ_r	dielectric constant	dimensionless	—
E	electric field intensity	volt/meter	V/m
D	electric flux density	coulomb/meter^2	C/m^2
V	electric potential	volt	V
E	electromotive force	volt	V
W	energy	joule	J
F	force	newton	N
f	frequency	hertz	Hz
Z, z	impedance	ohm	Ω
L	inductance	henry	H
Φ	magnetic flux	weber	Wb
B	magnetic flux density	weber/meter^2	Wb/m^2
		tesla	T
H	magnetic field intensity	ampere-turn/meter	A-t/m
$\mathcal{F}$	magnetomotive force	ampere-turn	A-t
μ	permeability	henry/meter	H/m
ϵ	permittivity	farad/meter	F/m
P	power	watt	W
X, x	reactance	ohm	Ω
$\mathcal{R}$	reluctance	henry^{-1}	H^{-1}
R, r	resistance	ohm	Ω
B	susceptance	siemens	S
T	torque	newton-meter	N-m
v	velocity	meter/second	m/s
V, v	voltage	volt	V
W	work (energy)	joule	J

APPENDIX B

The Laplace Transform

The Laplace transform of a single-valued function $f(t)$ is defined as

$$\mathcal{L}[f(t)] = F(s) = \int_{0^-}^{\infty} f(t)e^{-st}\, dt \tag{B.1}$$

where $s = \sigma + j\omega$ is a complex number. In order to include the switching action or a unit impulse function at $t = 0$, the lower limit is usually taken as 0^-.

To obtain the function $f(t)$ from $F(s)$, we may use the inverse Laplace transform defined as

$$\mathcal{L}^{-1}[F(s)] = f(t) = \frac{1}{2\pi j}\int_{c-j\infty}^{c+j\infty} F(s)e^{st}\, ds \tag{B.2}$$

where $\mathcal{L}^{-1}[F(s)]$ is read as the "inverse Laplace transform of $F(s)$."

The integration in Eq. (B.2) is not easy to perform because it requires knowledge of functions of complex variables and is avoided. Since there exists a one-to-one correlation between $f(t)$ and $F(s)$, we use a "look-up" table for the Laplace transforms. For this reason, the Laplace transforms for some of the well-known functions are given in Table B.1.

Table B.1 Laplace Transforms of Some Functions

$f(t)$	$F(s)$	$f(t)$	$F(s)$
$\delta(t)$	1	$u(t)$	$\dfrac{1}{s}$
$\delta'(t)$	s	$u(t-a)$	$\dfrac{e^{-as}}{s}$
$\delta^{(n)}(t)$	s^n	$tu(t)$	$\dfrac{1}{s^2}$
$\delta(t-a)$	e^{-as}	$t^n u(t)$	$\dfrac{n!}{s^{n+1}}$
$e^{-at}u(t)$	$\dfrac{1}{s+a}$	$(t-a)u(t-a)$	$\dfrac{e^{-as}}{s^2}$
$te^{-at}u(t)$	$\left(\dfrac{1}{s+a}\right)^2$	$tu(t-a)$	$e^{-as}\left(\dfrac{a}{s}+\dfrac{1}{s^2}\right)$
$e^{-at}t^n u(t)$	$\dfrac{n!}{(s+a)^{n+1}}$	$\sin \omega t$	$\dfrac{\omega}{s^2+\omega^2}$
$\cosh \beta t$	$\dfrac{s}{s^2-\beta^2}$	$\cos \omega t$	$\dfrac{s}{s^2+\omega^2}$
$\sinh \beta t$	$\dfrac{\beta}{s^2-\beta^2}$	$e^{-at}\sin \omega t$	$\dfrac{\omega}{(s+a)^2+\omega^2}$
$e^{-at}f(t)$	$F(s+a)$	$e^{-at}\cos \omega t$	$\dfrac{s+a}{(s+a)^2+\omega^2}$
$f(t-a)u(t-a)$	$e^{-as}F(s)$	$tf(t)$	$-\dfrac{dF(s)}{ds}$
$f(at)$	$\dfrac{1}{a}F\left(\dfrac{s}{a}\right)$	$t^n f(t)$	$(-1)^n \dfrac{d^nF(s)}{ds^n}$
$\dfrac{df(t)}{dt}$	$sF(s)-f(0^-)$	$\dfrac{d^2f(t)}{dt^2}$	$s^2F(s)-sf(0^-)-\dfrac{df(0^-)}{dt}$
$\int_0^t f(t)\,dt$	$\dfrac{F(s)}{s}$	$\dfrac{f(t)}{t}$	$\int_s^\infty F(s)\,ds$

Bibliography

Brown, D. & Hamilton, E. P., III. *Electromechanical Energy Conversion.* New York: Macmillan, 1984.

Chapman, S. J. *Electric Machinery.* New York: McGraw-Hill, 1985.

Del Toro, V. *Electric Machines and Power Systems.* Englewood Cliffs, NJ: Prentice-Hall, Inc., 1985.

El-Hawary, M. E. *Electric Machines with Power Electronic Applications.* Englewood Cliffs, NJ: Prentice-Hall, Inc., 1986.

Fitzgerald, A. E., Kingsley, C., & Umans, S. D. *Electric Machinery.* New York: McGraw-Hill, 1983.

Lindsay, J. F. & Rashid, M. H. *Electromechanics and Electric Machinery.* Englewood Cliffs, NJ: Prentice-Hall, Inc., 1986.

Matsch, L. W. & Morgan, J. D. *Electromagnetic and Electromechanical Machines.* New York: Harper & Row, 1986.

McPherson, G. *An Introduction to Electrical Machines and Transformers.* New York: John Wiley & Sons, Inc., 1981.

Nasar, S. A. & Unnewehr, L. E. *Electromechanics and Electric Machines.* New York: John Wiley & Sons, Inc., 1983.

Say, M. G. *Alternating Current Machines.* New York: Pitmann, 1976.

Slemon, G. R. & Straughen, A. *Electric Machines.* Reading, MA: Addison Wesley, 1981.

Index